# HAZARDOUS MATERIALS AND WASTEWATER:

## *TREATMENT, REMOVAL AND ANALYSIS*

# HAZARDOUS MATERIALS AND WASTEWATER:

## *TREATMENT, REMOVAL AND ANALYSIS*

ALLISON A. LEWINSKY
EDITOR

**Nova Science Publishers, Inc.**
*New York*

For permission to use material from this book please contact us:
Telephone 631-231-7269; Fax 631-231-8175
Web Site: http://www.novapublishers.com



**LIBRARY OF CONGRESS CATALOGING-IN-PUBLICATION DATA**

Hazardous materials and wastewater : treatment, removal and analysis / Allison A. Lewinsky (editor).
p. cm.
Includes index.
ISBN 13 978-1-60021-257-4
ISBN 10 1-60021-257-3
1. Hazardous wastes. I. Lewinsky, Allison A.
TD1030.H382 2006
628.4'2--dc22 2006016758

*Published by Nova Science Publishers, Inc. ✢ New York*

# CONTENTS

# PREFACE

Hazardous waste is a waste with properties that make it dangerous or potentially harmful to human health or the environment. Hazardous waste generally exhibits one or more of these characteristics: ignitability, corrosivity, reactivity or toxicity. The universe of hazardous wastes is large and diverse. Hazardous wastes can be liquids, solids, contained gases, or sludges. They can be the by-products of manufacturing processes or simply discarded commercial products, like cleaning fluids or pesticides. One major type is radioactive waste. This new book brings together the latest research in this diverse field.

In Chapter 1, chromium-containing aqueous solutions were prepared and tested in a bench-scale electrochemical process to acquire design, scale-up and operation parameters. The process employs DC current to perform electrochemical reduction, electrochemical coagulation and electroflotation in one unit. When a DC electric field is applied between a set of iron anodes and cathodes, iron is dissolved from the anodes to produce ferrous ions. At the same time, water molecules are electrolyzed to hydrogen gas and hydroxyl ions. When the solution is treated in the process, hexavalent chromium is reduced to trivalent chromium by ferrous ions. The trivalent chromium is then precipitated out as hydroxide. The sludge, ferric and chromium hydroxides, is then floated to the surface by the hydrogen bubbles generated on the cathode. The process is capable of treating the hexavalent chromium down to less than 1 mg/L.

A pilot scale electrochemical system, an electrocoaguator followed by an electroflotator was designed and constructed to incorporate the complex mechanism of the process. The system reduced chromium concentration from 60 mg/L to 10 mg/L in two minutes in a batch operation. The concentration continue to drop to less than 1 mg/L in a 6-minute treatment. The electrocoagulator was confirmed in a continuous operation. It took 15s minute to stabilize the system, i.e. an outlet concentration of less than 1 mg/L. With a throughput of 3.3 liter per minute, a current of 60 amps and a voltage of 3 volts, the operation cost is estimated to be $425 for 1 million gallons of chromium contaminated wastewater.

Recently, there has been a great interest for applying natural sorbent materials in the wastewater treatment. As an example, orange bagasse (waste of orange juice industries, containing pulp, skin and seeds) is proposed for effluent remediation, focusing on the heavy metal ions species. The physical and chemical characterization of the material pointed out the presence of functional groups as well as morphological aspects, which make sorption process possible. Therefore, in Chapter 2, the sorption capacity of the biomass was measured and a number of parameters were studied to find the best conditions for the heavy metal ions sorption. Batch experiments were conducted at room temperature to investigate the sorption of Cu(II), Ni(II), Pb(II) and Zn(II) ions onto dried orange bagasse as a function of initial pH, ionic strength, initial metal ion concentration and sorbent amount. The metal ions sorption increased with the pH, and the sorption efficiencies were greatest between pH 4.0 and pH 7.0. Cu(II), Ni(II) and Zn(II) sorption were strongly influenced by the ionic strength. The removal of these metal ions decreased when the ionic strength have been changed from 0.1 mol $L^{-1}$ to 1.0 mol $L^{-1}$. Probably, this phenomenon is due to a competition process among the heavy metal ions and the electrolyte. However, the ionic strength had only a very small effect on the Pb(II) sorption process. The sorption kinetics was very fast, specially for Pb(II) and Zn(II) ions, and the equilibrium was achieved within 30 min. The sorption process followed a pseudo second-order kinetics, thus indicating a chemical sorption mechanism. This study revealed that the diffusion of the ions inside the biomass is less important for the sorption rate than the external mass transfer. The experimental data were well fitted by the Langmuir equation ($R^2 > 0.99$). So, the sorption process of metal ions onto orange bagasse follows the Langmuir isotherm model, where the ions are independently taken up on a single type of binding site, and the uptake of the first metal ion does not affect the sorption of another ions. The sorption capacity – 66 mg Pb(II) $g^{-1}$, 18 mg Ni(II) $g^{-1}$, 17 mg Cu(II) $g^{-1}$ and 13 mg Zn(II) $g^{-1}$ – suggest the handling of high volumes of wastewater containing 1 mg $L^{-1}$ of these metal ions. Finally, the negative Gibbs free energy values for the metal ions sorption onto orange bagasse confirm the feasibility and the spontaneous nature of the sorption process for these species.

Hazardous metal cations enter water through the natural geochemical route or the industrial wastes. These non-biodegradable and persistent metals accumulate in living tissues getting biomagnified in the food chain. The present work in Chapter 3 investigates the interaction of Cd(II), Cu(II), Ni(II) and Pb(II), with clay minerals, kaolinite, montmorillonite and their acid-activated forms. As an important constituent of soil, clay minerals play the important role of a scavenger of toxic materials from the soil. The clay-heavy metal interactions were studied under different conditions of pH, concentration of the metal ions, amount of clay, interaction time and temperature. The interactions were dependent on pH and the uptake was controlled by amount of clay and initial concentration of the metal ions. 2 g of clay (kaolinite, acid-activated kaolinite, montmorillonite, acid-activated montmorillonite) per liter of solution could remove 15.9 to 92.1 % Cd(II), 15.7 to 87.7 % Cu(II), 20.7 to 69.2 % Ni, and 20.9 to 90.8 % Pb(II) from a solution of initial concentration 50 mg/L at equilibrium at 303 K. Adsorption was in the order of kaolinite < acid-activated kaolinite < montmorillonite < acid-activated montmorillonite. A detailed study was undertaken with respect to the kinetics, Freundlich and Langmuir isotherms, and the thermodynamics of the processes. The interactions were adsorptive in nature and were relatively fast. Equilibrium was approached within 180, 360, 240 and 180 min for Pb(II), Cu(II), Cd(II) and Ni(II) respectively. The rate processes did not match the pseudo first order kinetics of Lagergren and therefore, they were further investigated with the simple second order kinetics, Elovich equation, liquid film diffusion and intra-particle diffusion mechanisms, with the second order kinetic model giving the best results. The adsorption data fitted both Langmuir and Freundlich isotherms indicating that strong adsorptive forces were involved in the interactions and that the interaction sites were energetically non-uniform. Langmuir monolayer capacity was 11.5 to 31.4, 4.3 to 28.0, 6.8 to 30.7 and 7.1 to 21.3 mg/g respectively for the four clays. The thermodynamics of the rate processes showed that the adsorption of Pb(II) and Ni(II) was exothermic accompanied by decreases in entropy and Gibbs energy, but the adsorption of

Cd(II) and Cu(II) was endothermic with an increase in entropy and an appreciable decrease in Gibbs energy. The results have established good potentiality for kaolinite, montmorillonite and their acid activated forms to be used as adsorbents for Cd(II), Cu(II), Ni(II) and Pb(II) ions from aqueous medium.

The rapid development and changing technologies, industrial products and practices of the present day result in the generation of high quantum of hazardous heavy metal laden wastewater materials, and sludges, which need to be appropriately dealt with to prevent any adverse effects on public health and the natural environment system. Although source reduction and Waste minimization are the long term solutions for reducing or eliminating the release of toxic heavy metal ions to the environment, treatment technologies will continue to play a crucial role in controlling heavy metal discharges from current operations and in cleaning up the contaminated sites from previous operations.

The objective of Chapter 4 is to overview the processes currently available for wastewaters contaminated with heavy metal ions and seeks to highlight the many of the significant contributions to the field of treatment of hazardous heavy metals in the environment.

This chapter reviews the applicable separation technologies relating to removal of heavy metals from industrial wastewater in order to present the state of the art in the field.

A number of specialized processes have been developed for the removal of heavy metals from waste discharges. Technologies for removal of hazardous heavy metal waste management can be illustrated into two major classes – chemical process technologies and bioprocess technologies. Chemical process technologies are of basically of two segments - conventional treatment technologies like chemical precipitation (including hydroxide, carbonate, or sulfide reagents) coagulation/flocculation, sedimentation, chelating complexation and filtration and alternative technologies, for end of the pipe treatment of hazardous heavy metal waste like adsorption, solvent extraction, ion exchange, electrochemical operations, membrane filtration and evaporation. Each technology is briefly described and typical operating conditions and technology performance are presented.

An attempt has been made to introduce engineers and scientist to the spectrum of practiced and emerging process technologies for removal of heavy metals from wastewaters.

In this chapter, the status of technologies adopted by the various industries and further requirement to meet the objective of treatment of hazardous heavy metal waste management have been assessed.

The manufacturing and the application of organic dyes involve the production and the handling of many organic compounds hazardous to human health. Many of these substances are considered toxic, even carcinogenic. Over the past couple of decades, manufacturers and users of dyes have faced increasingly stringent legal regulations promulgated to safeguard human health and the environment. So, there is a clear need to treat dye wastewater prior to discharge into the primary effluent. The limitation of traditional wastewater treatment technologies (biological and physical), such as low rate, disability to degrade many of recalcitrant organic dyes and the production of secondary waste which demands further treatment, can be overcome by the utilization of advanced oxidation processes (AOPs). These wastewater treatment methods are considered as low- or even non-waste generation technologies. AOPs are based on the production of very reactive species, such as hydroxyl radicals, able to decolorize and to reduce recalcitrant colored wastewater loads due to the high oxidation power and the lack of selectivity of OH radicals towards a broad range of organic pollutants present in wastewater. Generally, AOPs can be broadly classified concerning the way of OH radicals generation into chemical, photochemical, photocatalytic, mechanical and electrical technologies. This research in Chapter 5 deals with the application of chemical and photochemical AOPs for the minimization of recalcitrant colored pollutants present in wastewaters. The comparative study of several processes, $Fe^0/H_2O_2$, $UV/Fe^0/H_2O_2$, $UV/O_3$ and $UV/O_3/H_2O_2$, was performed through an experimental research of decolorization and mineralization of a reactive azo dye C.I. Reactive Blue 137, as a model wastewater pollutant. Applied processes were optimized according to their process parameters, $Fe^0$ dosages, initial $H_2O_2$ dosages and initial pH values of treated solutions. The influence of initial organic dye concentration, as well as the addition of solid particles, synthetic zeolites, on the process effectiveness was also investigated. Studied AOPs were evaluated on the basis of their eco-effectiveness, by the means of color ($A_{610}$), aromaticity ($UV_{280}$), TOC and AOX value decrease, and their cost-effectiveness as well.

Wastewater from various industries and domestic households has been a big environmental problem for human society. Wastewater discharged from industries and homes includes a large quantity of various toxic substances such as inorganic heavy metals and organic compounds. Treatment of wastewater involves several processes and several technologies have been developed such as filtration, chemical precipitation, advanced oxidation, reverse osmosis or adsorption. Among these processes, the adsorption with the selection of a suitable adsorbent can be a simple and effective technique for the removal of inorganic and organic pollutants from wastewater. Activated carbon is usually used as an effective adsorbent. However, due to the high cost and regeneration problem, other low-cost and efficient adsorbents are required. Fly ash as a byproduct generated from thermal power plants is solid waste but available in large quantity. Utilisation of the fly ash will help reduce the environmental burden and enhance economic benefit. In Chapter 6, the authors report a comprehensive investigation employing various treatment techniques to convert fly ash to various effective adsorbents for removal of organic dyes and heavy metals from wastewater. Fly ash is usually composed of inorganic minerals and unburned carbon. In their investigation, the mineral section of fly ash has been treated by physical and chemical methods to convert it to zeolite or inorganic polymer. Unburned carbon in the fly ash was also used as precursor for production of activated carbon for wastewater treatment. The adsorption kinetics and isotherms of these adsorbents (Fly ash, zeolite, inorganic polymer and unburned carbon) in aqueous solution for uptake of dyes and heavy metals were investigated. The results indicate that modified fly ashes exhibit comparable adsorption capacity as activated carbon.

The objective of this study in Chapter 7 was to prepare organically modified adsorbent (organoclay) from a low-cost natural clay mineral, which is namely sepiolite by treated with a surfactant for enhancement of the removal of anionic textile dyes due to the fact that these dyes generate a considerable amount of hazardous colored wastewater, which cause for environmental pollution. Reactive Blue 19 (RB19) was selected as adsorbate for evaluating the adsorption performance of sepiolite samples, which was affected significantly by a surfactant [dodecyltrimethyl ammonium (DTMA)-bromide] modification. The surface modification of DTMA-sepiolite was examined using the elemental analysis and FTIR technique. The effects of temperature, pH, contact time and concentration were tested in the experiments. The maximum adsorption capacities, which are obtained from Langmuir isotherm equation, of RB19 onto natural sepiolite and DTMA-sepiolite at pH 1.5 and 20 $^{o}C$

were $2.30x10^{-4}$ and $3.11x10^{-4}$ mol $g^{-1}$, respectively. These results indicate that DTMA-sepiolite has 1.35 times higher adsorption capacity than natural sepiolite. The Langmuir, Freundlich and Dubinin-Radushkevich (D-R) adsorption models were applied to describe the equilibrium isotherms and the isotherm constants were also deduced from experimental data. The results indicated that the Langmuir isotherm model appears to fit better than other adsorption isotherm models for the adsorption of RB19 onto natural sepiolite and DTMA-sepiolite. The first-order, pseudo-second-order kinetic and the intraparticle diffusion models were used to describe the kinetic data and the rate constants were evaluated as well. The experimental data fitted very well by the pseudo-second-order kinetic model with the high correlation coefficients of 0.999 for natural sepiolite and DTMA-sepiolite and also followed by the intraparticle diffusion model up to 60 min, whereas diffusion is not only the rate-controlling step. The rate constants calculated from the pseudo-second-order kinetic model were used to derive the activation energy ($E_a$) of the adsorption. The numerical values of activation energies are 13.58 kJ $mol^{-1}$ for natural sepiolite and 12.49 kJ $mol^{-1}$ for DTMA-sepiolite and these results show that the adsorption process follows a physisorption mechanism. The equilibrium constants were used to calculate thermodynamic parameters including the changes of free energy, enthalpy and entropy. The numerical values of the changes of free energy for natural sepiolite and DTMA-sepiolite were 2.593 and 1.936 kJ $mol^{-1}$ at 20 °C, respectively. The changes of enthalpy values for natural sepiolite and DTMA-sepiolite at a temperature range of 20 – 40 °C were –3.723 and –6.835 kJ $mol^{-1}$, respectively. These results indicated that the adsorption of RB19 onto natural sepiolite and DTMA-sepiolite are not spontaneous according to the change of free energy and are exothermic according to the change of enthalpy. All over these results indicate that effective adsorbents for the removal of RB19 could be prepared by appropriate modification of sepiolite, although natural sepiolite itself can be considered as a potential adsorbent in wastewater treatment systems.

Aromatic amines, such as aniline and its substituted derivatives, are widely used as raw materials in many industrial processes, and very harmful chemical species constituting the focus of many environmental concerns. Although it is well known that they can easily seep through soil and contaminate groundwater due to their high solubility in water, unfortunately, these compounds are still being released to the environment directly through some industrial processes and/or indirectly through decomposition of herbicides and pesticides. Therefore, removal of such types of compounds from industrial wastewaters has become an of prime importance environmental concern in many research areas. In the past decades, adsorption on activated carbon had been a primary way to overcome mentioned environmental troubles; however, nowadays this is not the case due to the irreversible adsorptive characteristics of activated carbon, and accordingly high regeneration costs coming out. Nowadays, attempts have, therefore, intensified towards developing alternative adsorbents, ligand-exchange based adsorbents, which can be easily regenerated even at relatively low temperatures. Nowadays, the ligand-exchange processes have been increasingly employed to overcome inherent drawbacks of some traditional processes. Ligand-exchange behaviors of aromatic amines have been investigated using ligand-exchange sporopollenin in several column studies, which deal with chromatographic separation, adsorption and kinetics. Adsorption and kinetics studies are essentially based on the adsorption data obtained from breakthrough curves, and the adsorption behaviors of ligands have been analyzed by Langmuir and Freundlich isotherm models. The characteristics of particular adsorption process have also been investigated using

Scatchard plot analysis. In Chapter 8, adsorption of aniline and its some substituted derivatives on a modified sporopollenin exhibiting ligand exchanger properties is discussed in detail. The scope of the chapter is constructed on the basis of experimental results obtained through some previously conducted systematic column studies, and the results of these studies are evaluated in a collective manner. Thus, some proposed modification methods of sporopollenin, the theory of ligand-exchange, breakthrough technique, related adsorption isotherms and Scatchard plot analysis are described, and the adsorption phenomena of the studied resin, including column operation and rate-controlling step of processes are also discussed in detail.

Undesirable ions such as Cr(III), and Cr(VI) are removed from water or wastewater by adsorption. The method is technologically simple and enables efficient removal of even traces of impurities from aqueous solutions. Mechanisms including ion exchange, complexation, and adsorption to the surface are possible in the sorption process. Examples of removal of chromium ions by adsorption are presented in Chapter 9. Suitability of synthetic ion exchangers such as Chelex-100 (chelating resin), Lewatit TP 207 (chelating resin), Lewatit S 100 (cation exchanger resin), Lewatit MP 62 (anion exchanger resin), Lewatit M 610 (anion exchanger resin), a novel chelating resin b-DAEG-sporopollenin (bis-diaminoethyl glyoxim-sporopollenin, CEP-sporopollenin (carboxylated epichlorohydrine-sporopollenin), and natural adsorbents (brown coals, montmorillonite) as highly efficient adsorbent for the removal of chromium was studied and compared with each other. Batch sorption experiments were carried out with chromium to quantify the sorption kinetics. Sorption was found as a function of initial concentration, pH, temperature, shaking time, and adsorbent dosage. The Langmuir and Freundlich adsorption isotherms were used to describe the observed sorption phenomena. Sorption equilibria, calculated from adsorption data, explained favorable performance of adsorption system.

Natural and controlled processes of biosorption and bioaccumulation may be efficiently used in the assessment of environmental pollution as well as in pollution prevention and cleaning the polluted environment. The property that is used in such pollution control processes is the ability of all types of biomasses to bind toxic metals, in different extent though, depending on morphology and physiology of an organism. If the processes are performed at controlled conditions, the efficiency can be greatly improved. If we would like natural processes to work for us, it would be indispensable to understand their mechanism, rules that govern them, we should know how to model their kinetics and equilibrium. Basically, we should be able to predict the course of the processes at given process parameters in order to design a complete technique and to make a practical use of it.

Biosorption and bioaccumulation differ when considering the mechanism and for this reason the potential applications will also be diverse. While in the process of biosorption, pollutants are bound to the surface of cellular wall of the biomass, in bioaccumulation contaminants are also transferred into cellular interiors. The latter can affect the metabolic functions of an organism. Biosorption can be performed by non-living biomass (materials of biological origin), but the essential condition for bioaccumulation to occur is that an organism should be metabolically-active.

Chapter 10 discusses two main environmental applications of biosorption and bioaccumulation processes. The first is environmental biomonitoring performed by living organisms. The property that is used here is that the level of a given contaminant in biological tissues is related to the concentration of this substance in the environment. This is expressed by the assessment of bioaccumulation factors. The chapter reports examples of the application of human hair and consumable tissues of animals, and various parts of plants (green parts as well as wood) in environmental pollution monitoring.

Biosorption as well as bioaccumulation can find an application in removal of contaminants from aqueous solutions. Biosorption by different materials of plant (microalgal biomass, aquatic plants, plant leaves, straw, grass) and animal origin (eggshells, bones) can be applied in industrial wastewater treatment processes. Also, bioaccumulation by aquatic organisms: microalgae, macroalgae and aquatic plants in metal ions removal from effluents can be used to bind metal ions simultaneously with nutrients.

An attempt to make the processes predictable is discussed here, including screening for good biosorbents/bioaccumulators, determining biosorption and bioaccumulation capacity as well as bioavailability as the function of the concentration of metals in the environment in which an organism is present, together with growth requirements, mathematical modelling, carrying out processes at different process conditions. This should constitute the basis for the assessment of standardized procedures, in the design of biosorption or bioaccumulation treatment plants and for environmental biomonitoring.

In: Hazardous Materials and Wastewater
Editor: Allison A. Lewinsky, pp. 1-49

ISBN 1-60021-257-3


*Chapter 1*

# ELECTROCHEMICAL TREATMENT OF CR-CONTAINING AQUEOUS SOLUTIONS

***Mariam Ismail***
Department of Chemistry and Biochemistry, University of Massachusetts – Dartmouth, Massachusetts, USA

***Chen-Lu Yang****
Advanced Technology and Manufacturing Center, University of Massachusetts – Dartmouth, Massachusetts, USA

## ABSTRACT

Chromium-containing aqueous solutions were prepared and tested in a bench-scale electrochemical process to acquire design, scale-up and operation parameters. The process employs DC current to perform electrochemical reduction, electrochemical coagulation and electroflotation in one unit. When a DC electric field is applied between a set of iron anodes and cathodes, iron is dissolved from the anodes to produce ferrous ions. At the same time, water molecules are electrolyzed to hydrogen gas and hydroxyl ions. When the solution is treated in the process, hexavalent chromium is reduced to trivalent chromium by ferrous ions. The trivalent chromium is then precipitated out as hydroxide. The sludge, ferric and chromium hydroxides, is then floated to the surface by the hydrogen bubbles generated on the cathode. The process is capable of treating the hexavalent chromium down to less than 1 mg/L.

A pilot scale electrochemical system, an electrocoaguator followed by an electroflotator was designed and constructed to incorporate the complex mechanism of the process. The system reduced chromium concentration from 60 mg/L to 10 mg/L in two minutes in a batch operation. The concentration continue to drop to less than 1 mg/L in a 6-minute treatment. The electrocoagulator was confirmed in a continuous operation. It took 15 minute to stabilize the system, i.e. an outlet concentration of less than 1 mg/L.

* Corresponding author: Tel.: (508) 910-9857, E-mail address: cyang@umassd.edu

With a throughput of 3.3 liter per minute, a current of 60 amps and a voltage of 3 volts, the operation cost is estimated to be $425 for 1 million gallons of chromium contaminated wastewater.

## 1. INTRODUCTION

The treatment of metal-containing wastewater has been an issue for the metal finishing industry for decades. Metal finishers that discharge to receiving streams are regulated under the National Pollutant Discharge Elimination System (NPDES), which is a Federal EPA Program normally administrated by the state regulatory agency. Therefore, the dischargers need both federal and state permits. The most comprehensive and expensive pollution control law affecting metal finishers is the Resource Conservation and Recovery Act (RCRA). This is the federal law to control hazardous wastes from cradle to grave. The original act was passed in 1976 and greatly expanded in 1984 to regulate small-quantity generators. Most metal finishers discharge to Public Owned Treatment Works (POTW) and therefore must meet local pretreatment limits. Based on the RCRA, the EPA published electroplating pretreatment standards in 1981. The discharge limitations are shown in Table 1. In many cases, these federal pretreatment standards are not adequate to meet the water quality limitations placed upon the POTW discharge. This has resulted in local municipal authorities adapting much more stringent discharge limitations. Before installing wastewater treatment equipment, the metal finishers need to meet with local authorities and obtain discharge limits. Table 2 shows the more stringent metal finishing regulations for electroplating shops.

There are three categories of waste to be considered in electroplating shops: drag-out to rinse water, bath dumps and floor spills. The periodic bath dumps are infrequent and low volume, and can be hauled over to a waste service provider. Floor spills are nearly impossible to manage due to their unpredictable and intermittent nature and to the fact that they are so heterogeneous. Currently, most of the effort on pretreatment technologies has focused on the rinse water, since they are the majority of the flow leaving an operation.

**Table 1. Electroplating effluent guidelines (mg/L) – pretreatment standards existing sources (PSES) greater than 10,000 gpd**

| Pollutant | Maximum for any 1 day (mg/L) | Average of daily values for four consecutive monitoring days should not exceed (mg/L) |
|---|---|---|
| Cyanide, total | 1.9 | 1.0 |
| Copper | 4.5 | 2.7 |
| Nickel | 4.1 | 2.6 |
| Chrome, total | 7.0 | 4.0 |
| Zinc | 4.2 | 2.6 |
| Lead | 0.6 | 0.4 |
| Cadmium | 1.2 | 0.7 |
| Total metals | 10.5 | 6.8 |
| TTO | 2.13 | --- |

**Table 2. Captive manufacturers pretreatment standards (mg/L) discharge to POTWs**

| Parameter | Existing source (PSES) | | New source (PSNS) | |
|---|---|---|---|---|
| | 1 day max | 30 day max | 1 day max | 30 day max |
| Cadmium | 0.69 | 0.26 | 0.11 | 0.07 |
| Chrome | 2.77 | 1.71 | 2.77 | 1.71 |
| Copper | 3.38 | 2.07 | 3.38 | 2.07 |
| Lead | 0.69 | 0.43 | 0.69 | 0.43 |
| Nickel | 3.98 | 2.38 | 3.98 | 2.38 |
| Silver | 0.43 | 0.24 | 0.43 | 0.24 |
| Zinc | 2.61 | 1.48 | 2.61 | 1.48 |
| Cyanide (T) | 1.20 | 0.65 | 1.20 | 0.65 |
| Cyanide (A) | 0.86 | 0.32 | 0.86 | 0.32 |
| TTO | 2.13 | --- | 2.13 | --- |

Several methods have been applied for the treatment of Cr-containing rinse water. Among them, evaporation, reverse osmosis, electrodialysis and ion exchange are worth mentioning. However, the most commonly used procedure involves adding chemicals to reduce the hexavalent chromium, Cr(VI), to trivalent chromium, Cr(III), and then adding alkali to create a pH greater than eight, at which chromium hydroxide, $Cr(OH)_3$, slurry forms. Due to the large amount of chemicals used, these methods usually leave the waste streams with several milligrams per liter (mg/L) of chromium. The regulations have been complied with but the total pollutants have increased.

The main objective of this research is to study the feasibility of treating Cr-containing wastewater with an electric field and to obtain design, scale-up and operation parameters. An integrated electrochemical process has been developed to reduce the chromium content in the synthetic wastewater. The process utilizes consumable electrodes to produce a reducing agent for Cr(VI) reduction, a coagulant for Cr(III) precipitation, and hydrogen and oxygen bubbles for sludge separation.

## 2. Literature Review

Hexavalent chromium, which exists in several forms, is widely used in industry. Industrial application of chromates include pigment production, electroplating, dyes, paints, leather tanning, etc. Improper disposal of chromium can cause groundwater contamination [1, 2]. Chromium is known to be present in natural environments in two stable oxidation states, Cr(VI) and Cr(III) [3]. The toxicity of Cr(III) and Cr(VI) differs significantly. Cr(III) is considered to be non malignant, while Cr(VI) is known to be a human carcinogen. The federal regulatory maximum contaminant level (MLC) for total Cr, i.e. Cr(VI) and Cr(III), is 100μg/L, while MLC in California is 50μg/L [4]. Since Cr(III) can be reoxidized to Cr(VI) by chlorine and chloramines disinfectants [2], the removal of total Cr is crucial.

The conventional processes used to remove hexavalent chromium is either its precipitation as chromium hydroxide or ion exchange. Precipitation makes the reuse in production impossible [5], and causes secondary waste products which are due to various reagents used in a series of treatments such as reduction of hexavalent chromium, and they cause instability in the ion exchange membrane due to the powerful oxidation of Cr(VI). Several technologies used for Cr(VI) treatment, such as electrochemical treatment [6], reduction/coagulation/filtration (RCF) unit [2], and ion exchanged systems [4,5,7], will be discussed in this section.

Electrochemical treatment of metal wastewater offers several advantages for remediation. Further electrochemical forms of remediation offer an end process for recovery of metals from contaminated wastewaters and effluents. Some of the materials used as electrodes for the removal of heavy metals were, carbon felt, scrap iron, activated charcoal cloth, and reticulated vitreous carbon (RVC) [8]. In a study done at the University of Delhi, India, an electrochemical process using carbon aerogel electrodes was developed to treat chromium contaminated waters. These electrodes were used for the direct current electrolytic treatment of effluent containing chromium species. The carbon aerogels were used because of their thermal, mechanical, and electrical properties. These properties are directly related to their unusual nanostructure. The nanostructure is composed of interconnected particles with microscopic interstitial pores. Its low electrical resistivity, high specific surface area, and controllable pore size distribution, make carbon aerogel an excellent electrode. The viability of this process depends on the life of the carbon aerogels [8, 9].

Adsorption is another method for Cr(VI) removal. It uses condenses-tannin gels derived from a natural polymer with many polyhydrophenyl groups. The adsorption mechanism consists of four reaction steps: the esterification of chromate with tannin molecules, the reduction of Cr(VI) to Cr(III), the formation of carboxyl group by the oxidation of tannin molecules, and the ion exchange of the reduced Cr(III) with the carboxyl and hydroxyl groups. A large amount of proton was found to be consumed, along with the reduction of Cr(VI) so that the acidic solution containing Cr(VI) was transferred automatically to neutral one by choosing the appropriate initial pH. This adsorption capacity, 287 mg Cr/g, was five to ten times higher than that obtained by the ion exchange between ordinary Cr(III) and tannin molecules for the tannin gels prepared under similar conditions. This method not only demonstrated the mechanisms involved in the removal of hexavalent chromium by condensed-tannin gel, but it also developed a recovery process of hexavalent chromium based on a zero-emission-oriented concept [10].

Ion exchange is one method for recovery of hexavalent chromium. This method involves the process of high and moderate concentrated electroplating wastewater after pretreatment by quantitative electrochemical oxidation of trivalent chromium to chromates. This leads to the recovery of chromium, copper, and water and prevents the formation of chromium containing sludge [5]. Ion exchange technologies, either in solid or liquid phases, provide an alternative method for precipitation, since they allow potential recovery of the metal, higher selectivity, and lower volumes of the sludge produced [7]. Several studies have dealt with the application of ion exchange technology in the removal of Cr(VI). One drinking water study showed that Cr(VI) can be removed by strong-base anion (SBA) exchange. Also, initial tests in Germany and China indicated that weak-base anion (WBA) exchange resin may be effective in the removal of Cr(VI). Several pilot units, which received spiked untreated water at a specified flow rate, were analyzed by the different exchange technologies.

The first pilot unit consisted of two treated paths for SBA and WBA exchange. The SBA resin tested was a type 2 chloride-based resin (SBA1) which was contained within two vessels in lead-lag configuration. This test leads to Cr removal of less than 5μg/L. It requires frequent regeneration with sodium chloride brine. It was observed that with higher sodium chloride brine concentration, the more effective Cr(VI) removal from the resin was. The WBA exchange resin was also tested in a lead-log configuration of two vessels. The WBA resin required a pH of less than 6 for Cr(VI) removal, which was obtained by using a preceding cation exchange column that exchanged influent calcium and magnesium ions for hydrogen ions. The WBA resin removal of Cr(VI) was strongly influenced by the pH.

The second pilot unit included four SBA resin and two granular activated carbon (GAC) columns. The magnetized ion exchange process consisted of continuously stirred mixed tanks for anion exchange. The SBA6 resin contained a magnetic component in its structure which allowed rapid adherence and settling of the resin from the treated water [4]. The reason behind this is due to the resin dose, resin contact time and regeneration rate. Optimal Cr(VI) removal occurred with a resin dose of 50-60 ml of SBA6/liter of water, contact time of 30 minutes, and 5% regeneration rate [4]. Another method of Cr(VI) removal was accomplished by ferrous sulfate, coagulation, and filtration method.

For concentrations in milligram per liter range, Cr(VI) removal is established by reduction to Cr(III) with ferrous ion (Fe(II)), then by precipitation, or coagulation. Since Cr(III) can be oxidized back to Cr(VI) by chlorine and chloramines disinfectants, the removal of total Cr, both Cr(VI)and Cr(III), is a key to successful Cr treatment. A reduction/coagulation/filtration (RCF) pilot was constructed to evaluate the treatment of Cr contaminated water. The system consisted of peristaltic chemical feed pumps for ferrous sulfate, sulfuric acid, sodium hydroxide, and polymer filter aid addition, a reduction tank, four aeration column in series, two parallel granular dual-media filter columns, and a back wash module [2]. The system received spiked Cr(VI) water at a given flow rate.

Industrial-grade $FeSO_4$ was used as a reducing agent, and analytic-grade $H_2SO_4$ to adjust the pH in certain tests. The oxidation of Fe(II) to Fe(III) occurred in the reduction tank and aeration columns, which was due to the presence of dissolved oxygen (DO) in the influent water. The purpose of the aeration columns was to advance the oxidation of excess Fe(II) to Fe(III) with DO and to assist the coagulation of Cr(III) with Fe(III). Cr(VI) samples were analyzed using a Dionex DX-320 ion chromatograph with an AD25 post column UV-visible detector. Cr(VI) reduction coupled with Fe(II) oxidation requires a 3:1 molar ratio of Cr(VI) to Fe(II) at pH greater than 6.5, which yields a solid made up of this molar ratio that can possibly contain Cr(III) either sorbed to or co-precipitated with Fe(III) hydroxide. High concentration of Fe(II) result in near completion Cr(VI) reduction in the presence of DO [2]. The effect of the pH in Cr(VI) removal varied significantly. Higher values of pH were found to increase the Fe(II) oxidation rate by DO. For pH values less than 7.5, total Cr removal was greater than 98%, while for filtration at pH 7.5, the total Cr removal was decreased to less than 50% [2]. The removal of total Cr is based on many variables including, mass ratio of Fe(II) to Cr(VI), pH of reduction, pH of coagulation and filtration, and filter loading rates [4]. Usually, RCF systems generate around 95% of total Cr removal.

Ferragels, supported zero-valent iron nanoparticles, rapidly separate and immobilize Cr(VI) as well as Pb(II) from aqueous solutions by various mechanisms including adsorption and reductive precipitation [11, 12]. Borohydride reduction of an aqueous iron salt in the presence of a support material, which could be a polymeric resin, silica gel, or sand, give rise to supported zero-valent iron nanoparticles that are 10-30 nm in diameter. The Ferragels also reduce Cr(VI) to Cr(III) and Pb to Pb(0) while oxidizing the Fe to goethite ($\alpha$-FeOOH) [12]. The typical result from this redox reaction is a chromite, with some of the iron substituted for the chromium. Powder X-ray diffraction confirmed that the material was in fact goethite, with no evident impurities [3]. X-ray diffraction analysis also show that Cr(VI) was in fact reduced to Cr(III). Remediation rates of Cr(VI) for Ferragels were compared to that of commercial iron filings and iron powder. Over a two month period, reduction of Cr(VI) was nearly 4.8 times greater for Ferragels than for an equal sample weight of iron filings. Zero-valent iron removes aqueous contaminations by reduction to an insoluble form, in the case of aqueous metal ions.

## 3. Process Chemistry

An electrochemical reaction is a heterogeneous reaction that involves a number of physical-chemical processes such as simple electron transfer, metal deposition, gas evolution, surface film transformation, anodic dissolution, oxide formation, gas reduction, and electron transfer with coupled chemical reactions. When wastewater is treated with direct current, the electron transfer causes oxidation and reduction of impurities in wastewater, dissolution of metallic anodes, discharge and coagulation of colloidal particles, electrophoresis, electroflotation of particles, precipitation of metallic ions on the cathodes, regeneration and concentration of acids and alkalis, and water desalination.

The concept of electrochemical coagulation to treat a hexavalent Cr-containing wastewater is to precipitate the dissolved chromium as an oxide or hydroxide species by anodically generated ferrous ions, Fe(II), and cathodically generated hydroxyl ions ($OH^-$). The simplified scheme is:

Anode: $$Fe \rightarrow Fe^{+2} + 2e^- \quad (1)$$

Cathode: $$2H_2O + 2e^- \rightarrow H_2 + 2OH^- \quad (2)$$

Electrolyte: $$3Fe^{+2} + CrO_4^{-2} + 4H_2O \rightarrow 3Fe^{+3} + Cr^{+3} + 8OH^- \quad (3)$$

Overall: $$Cr_2O_7^{-2} + 6Fe(OH)_2 + 7H_2O \rightarrow 2Cr(OH)_3 + 6Fe(OH)_3 + 2OH^- \quad (4)$$

After the treatment, the hydroxide precipitate or sludge is removed from the solution by filtration, centrifugation, sedimentation or flotation. Electroflotation is an electrochemical version of flotation. It differs from flotation mainly in the mechanism of bubble generation. The bubbles generated by electrolysis are very uniform. This makes the separation of suspended particles very effective.

Anode: $$2H_2O \rightarrow O_2 + 4H^+ + 4e^- \quad (5)$$

Cathode: $$2H_2O + 2e^- \rightarrow H_2 + 2OH^- \quad (2)$$

Overall: $$2H_2O \rightarrow 2H_2 + O_2 \quad (6)$$

When sodium chloride is added to the solution or wastewater to increase conductivity, chloride ions involve in the reaction and produces chlorine gas ($Cl_2$) at the anode. The alkaline solution produced at the cathode during water reduction can be allowed to mix with the $Cl_2$ to produce hypochlorous acid (HClO).

Anode: $2Cl^- \rightarrow Cl_2 + 2e^-$ (7)

Cathode: $2H_2O + 2e^- \rightarrow H_2 + 2OH^-$ (2)

Electrolyte: $Cl_2 + OH^- \rightarrow HClO + Cl^-$ (8)

It is essential to keep the pH of the solution at neutral or above to prevent the formation and evolution of chlorine gas.

# 4. Analysis

When hexavalent chromium is dissolved in water, it produces an intense yellow color that can be determined directly at 450 nanometer (nm). The intensity is used to quantify the concentration of Cr(VI) in the solution. A Perkin Elmer 280 Atomic Absorption Spectrophotometer was used to confirm the analytical results from the photometric method. An Imhoff settling cone was used to measure the sludge formation in each run. Due to the low sensitivity of the settling cone, photometric methods were developed to assist the sludge quantification.

## 4.1. Chromium Analysis

When dissolved in water, Cr(VI) gives an intense brown-yellow color. A UV-visible range spectrophotometer (DR/4000U, HACH Company) was used to quantify the concentration of Cr(VI) in the solution. Figure 1a shows the spectrum of Cr(VI) in an aqueous solution. The spectrum was constructed by subtracting the water spectrum from that of the Cr(VI) aqueous solution. The standard solution was prepared by dissolving reagent grade chromates ($CrO_3$) in deionized water. Figure 1b is the calibration curve, concentration vs. absorbance, of the solution at a wavelength of 450 nm. Usually a local maximum of a spectrum is used for analytical method development because a local maximum leads to a more sensitive signal and a lower detection limit for the analytical method. However, the Cr(VI) solution was found to have an intense yellow color at very low concentrations. Therefore, to avoid errors from dilution and over-scale of absorbance readings, a local minimum at 450 nm was used to construct the calibration curve. The maximum absorbance of the DR/4000U spectrophotometer is 2 absorbance unit (AU). In order to keep a good linearity in the calibration curve, the instrument was always operated at a low AU range, below 0.25 AU.

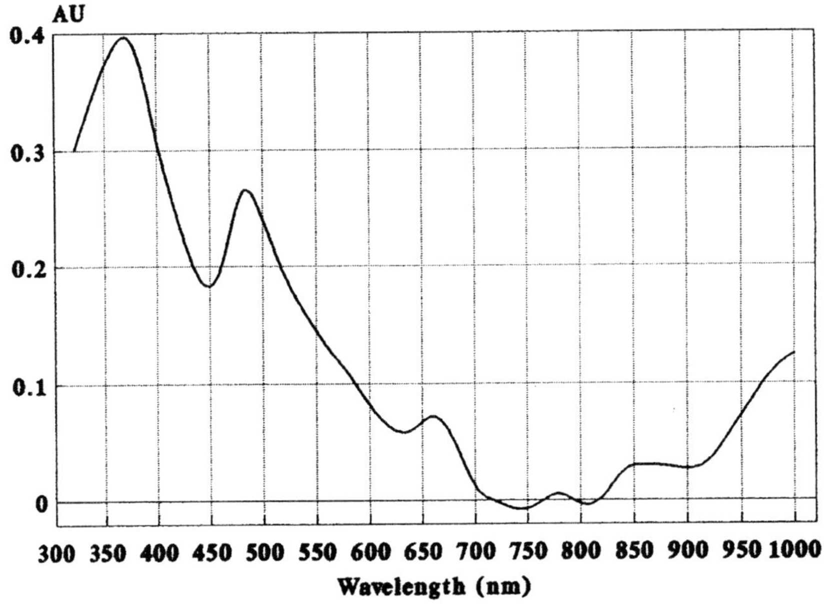


Figure 1a. Spectrum of hexavalent chromium in aqueous solution

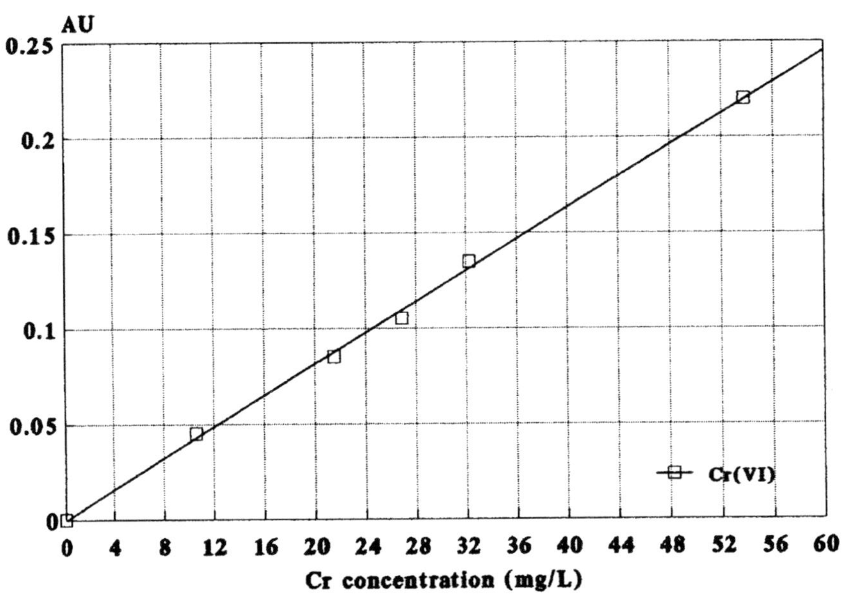


Figure 1b. Calibration curve (concentration vs. absorbance) of Cr(VI) aqueous solution

## 4.2. Sludge Quantification

Both black and brown sludge were produced in the electrochemical process. In the beginning, coagulant or suspended particles are the only concern with regard to effluent to be reused or discharged. But since black sludge is produced in the system as well, turbidity becomes another concern with regard to the treated effluent. Brown sludge was first quantified in an Imhoff settling cone in terms of milliliter per liter (mL/L). Because of the relatively low sensitivity of the settling cone, the low sludge content was quantified with the help of the HACH DR/4000U spectrophotometer. The sludge content in terms of mL/L was correlated to the transmittance at 720 nanometer (nm). Light at a wavelength of 720 nm shows no response to the existence of the Cr(VI) and Cr(III) that are expected to be found in the solution.

Figure 2 shows the relationship of sludge content and transmittance at 720 nm of a known sludge content. The plot is expected to be very useful for the electroflotation study, since flotation is expected to bring the sludge content down to a very low level, well beyond the measurement limit of an Imhoff settling cone. The figure covers a sludge content range of 0 to 40 mL/L with the lowest transmittance of about 70% at 720 nm.

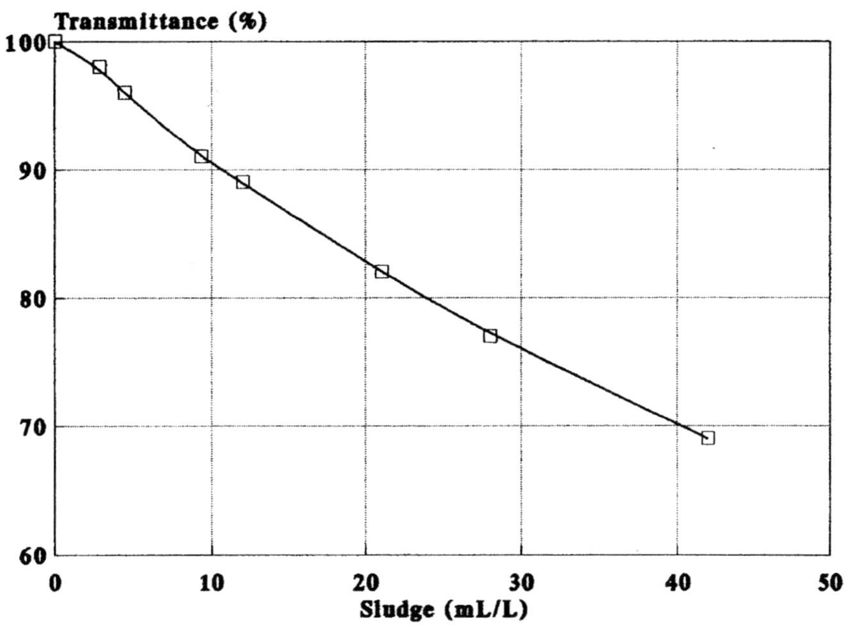


Figure 2. Relationship of brown sludge content and transmittance at 720 nm

Figure 3 shows the relationship of sludge content and transmittance at 720 nm of a black sludge produced in the process. As can be seen in the figure, a 20 mL/L black sludge can have a transmittance at 720 nm of as low as 3%. The index of effluent quality, clarity, was therefore defined as a percentage of the transmittance at 720 nm. Three types of sludge were produced in the electrochemical treatment of Cr-containing aqueous solutions: black sludge, brown sludge and a colloid system. Five effluent solutions were collected at current density of 0.75, 1.5, 2.25, 3.0 and 3.75 mA/cm$^2$ to study the characteristics of the sludge produced in each process condition. The colors of the sludge can be found in Table 3. The sludge was passed through a series of filter paper and membrane filters. Figure 4 shows the clarity of each solution after passing through each type of filter. The filters have pore sizes of 10 microns, 5 microns, and 0.2 microns. As can be seen in the figure, the effluent from high current density has smaller particles in the sludge. Therefore, after passing through the 10-micron filter, the transmittance increase from 60% to 80%, which is considered not very effective. Even after passing through the 5-micron filter, the transmittance at 720 nm is still only 90%. Finally, a 0.2-micron syringe filter was able to remove almost all of the particle in the solution. All samples for Cr(VI) measurement were filtered with 0.2-micron filter in order to avoid the interference of suspended particles. With a 5-mocron syringe membrane filter, the sludge in the effluent under both 0.75 and 1.5 mA/cm$^2$ were completely removed. The sludge produced under both 2.25 and 3.0 mA/cm$^2$ were between the two extreme cases. A bar chart for these experiments is presented in Figure 5. Clarity in terms of transmittance at 720 nm was improved as the samples went through the sequence of the filters.

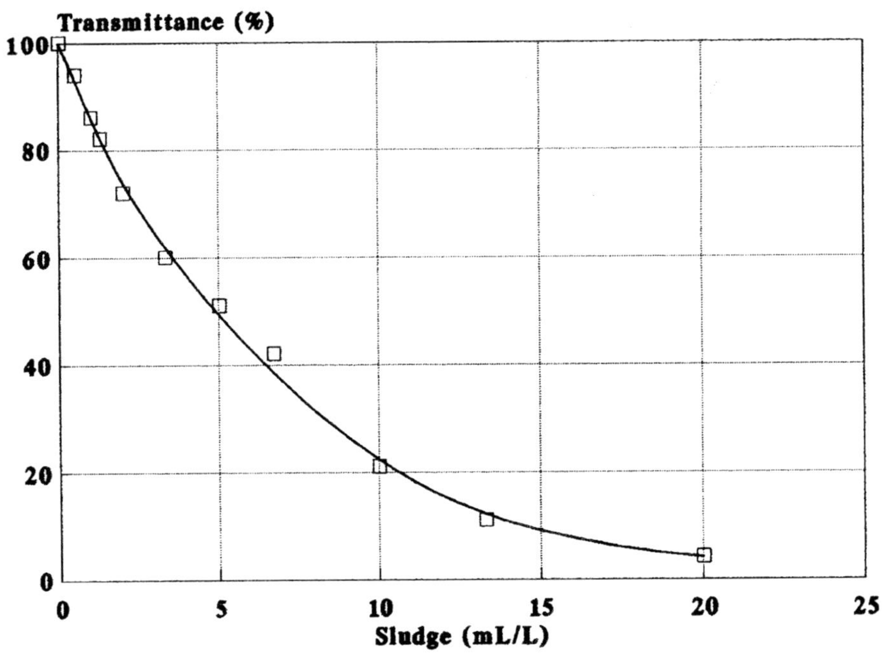


Figure 3. Relationship of black sludge content and transmittance at 720 nm

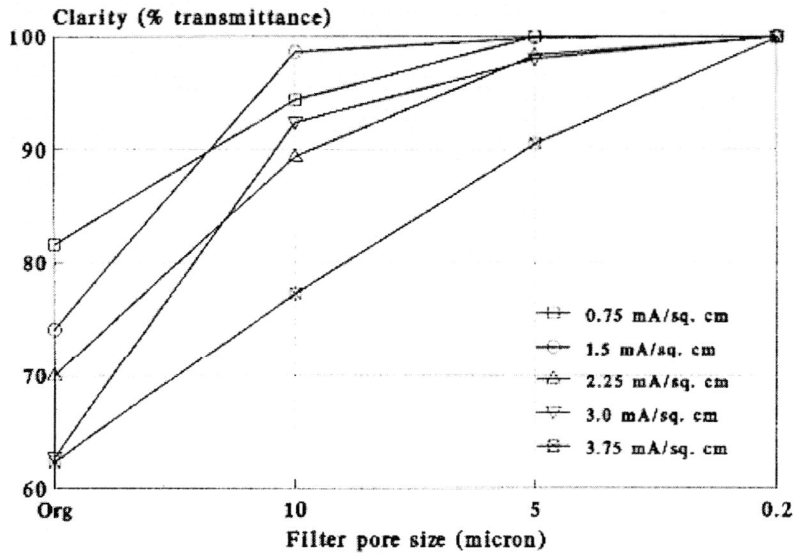


Figure 4. Clarity of sludge-containing solutions after passing through filters

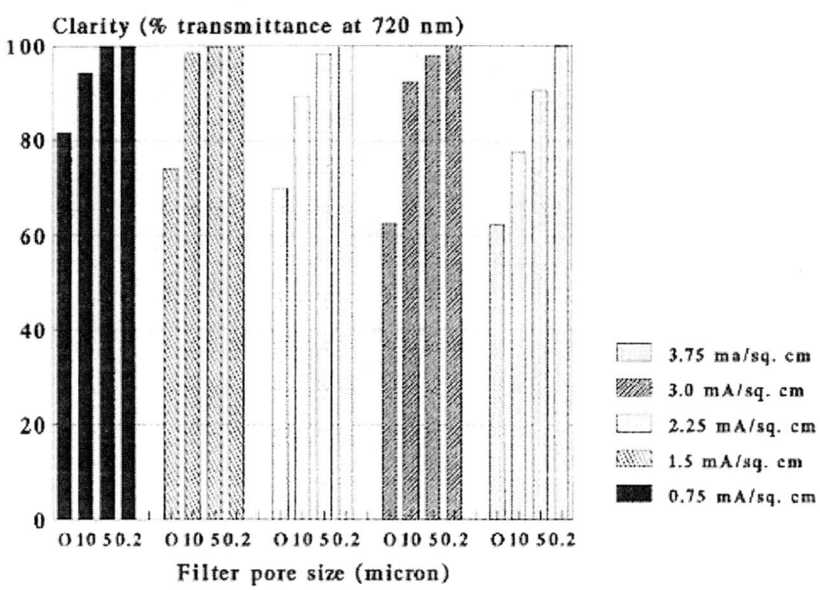


Figure 5. Improvement of transmittance at 720 nm with filtration

## 5. Electrochemical Coagulation

Electrochemical coagulation refers to a process that electrically dissolves anodes of steel or aluminum to form ferric or aluminum hydroxides. These metal hydroxides are widely used coagulants for water and wastewater treatment. As a result of coagulation contaminants, especially heavy metal ions, are removed together with the hydroxides as sludge. Because of the special morphology of these hydroxides, the sludge or precipitates are easily separated by flotation or sedimentation.

Figure 6 shows the parallel-plate flow cell test system. The electrocoagulator is made of acrylic plastic with dimensions of 160 millimeter (mm) by 67 mm by 92 mm. However, due to the space taken by the equalizer at the bottom and the siphon system at the top, the real volume of the cell is 100 mm by 67 mm by 92 mm, which is equal to 616 $cm^3$. With eleven cast iron electrodes in the chamber, the cell has an active volume of 560 $cm^3$. The total electrode surface area is 1,204 $cm^2$ for both cathodes and anodes. The electrode sets consist of five pieces of cast iron for anode and six pieces for cathode, situated approximately 4 mm apart from each other and submerged in the solution. Under normal operation all the electrodes are physically connected to either the positive or negative outlet of the DC power supply.

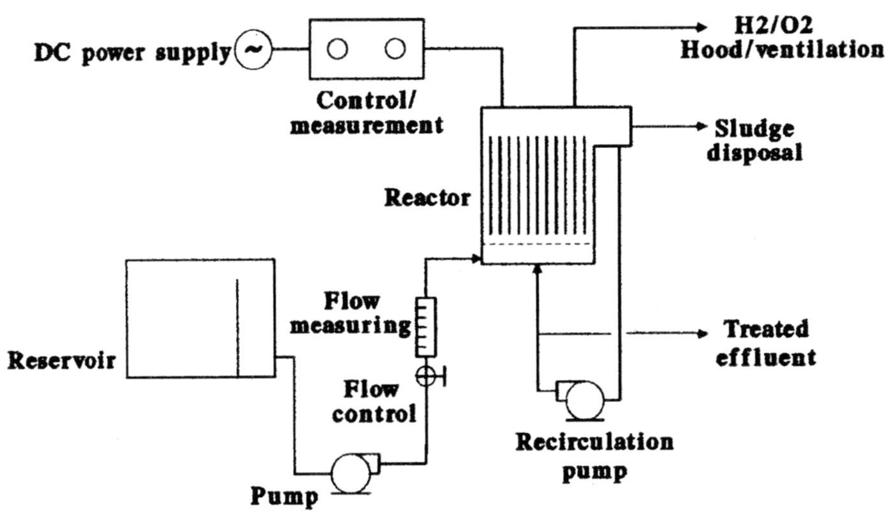


Figure 6. Schematic of electrochemical coagulation test system

## 5.1. Electrical and Chemical Characteristics

Energy consumption is always the number one criterion for a process development project, especially for pollution control processes. Since the feasibility of treating Cr-containing solutions with a DC current was proved promising in the literature, the first concern is the cost. A set of experiments was performed to acquire electrical characteristics for cost estimation. The experiments were conducted under the following conditions:

- Electrode: Cast iron for both cathode and anode
- Cathode: 602 $cm^2$
- Anode: 602 $cm^2$
- Electrode gaps: 4 mm between each anode and cathode
- Voltages applied: 0 to 30 volts DC

- Reactor volume: 560 $cm^3$
- Solution pH: 2.9 initial
- Cr(VI): 54 mg/L

Figure 7 shows the relationship of current and applied voltages in the cell. The system needs a voltage of 5.6 to draw a current of 1.5 amperes initially. The voltage kept increasing during the operation. At the end of the treatment, a voltage of 25 volts DC is needed to draw a current of 1.5 amperes in the cell. The chemical reaction rate is proportional to the current in the reactor. However, there are more than one pathways in the electrochemical system. Therefore, a higher current does not imply a shorter treatment time for certain removal efficiency. Even with a current of 1.5 amperes in the cell, the current density, 2.5 $mA/cm^2$ is still reasonable. Normally, an electrochemical reactor is operated at a current density between 2 and 10 $mA/cm^2$.

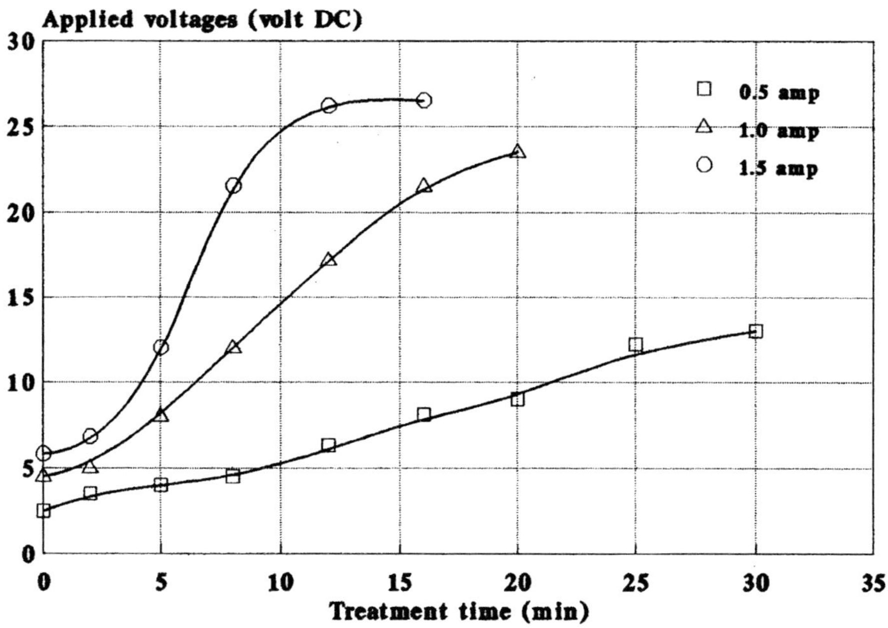


Figure 7. Relationship of current and applied voltage in the electrochemical coagulation cell

### *Change of Solution pH*

On the base of the mechanism proposed in a previous section, $OH^-$ is produced as a byproduct in the electrolysis process. The $OH^-$ later reacts with Cr(III) and Fe(III) and precipitates out as sludge. Before precipitation, the $OH^-$ has to build up to a certain concentration. The threshold concentration can be calculated from $K_{sp}$, the solubility products of $Cr(OH)_3$, and $Fe(OH)_3$. To study the effectiveness of the electrochemical process for Cr(VI) removal, the production of $OH^-$ in the solution is critical. Figure 8 shows the pH change while the Cr-containing aqueous solution was subjected to electrochemical treatment. As was expected, the greater the current, the faster the chemical reaction, therefore the faster the pH change in the solution. The change to a higher pH indicates the production of $OH^-$.

With a 1.5 amps DC current pass through the cell, the pH increased from 2.9 to 7.0 in 10 minutes. The initial pH at 2.9 was due to the formation of chromic acid while $CrO_3$ was dissolved in water to prepare the stock solution. It is expected that when the solution becomes alkaline, Cr(III) will precipitated out and removed from the solution.

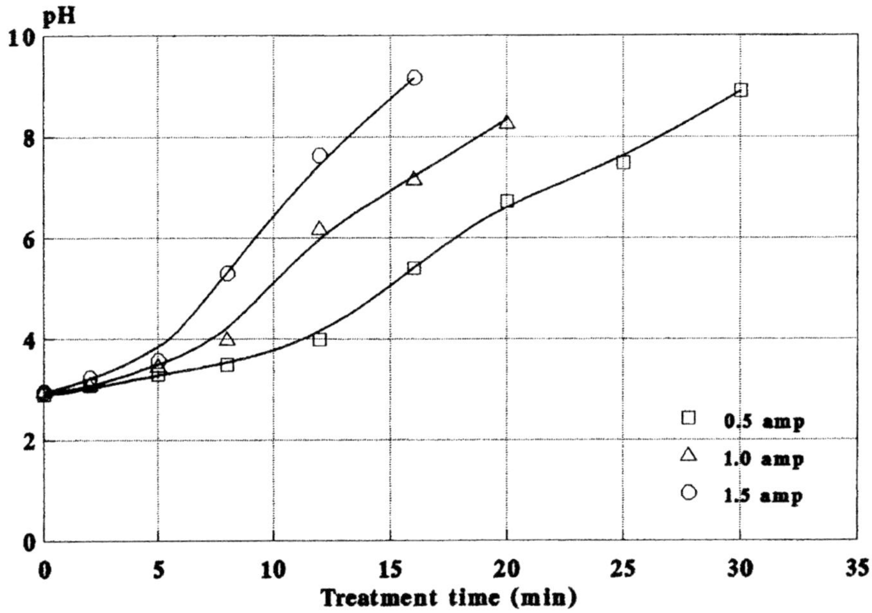


Figure 8. Change of pH in the electrochemical treatment of Cr-containing solutions

## 5.2. Coagulant Generation

The same procedure as the one described in the previous section was repeated to study the production of coagulant for metal ion removal. The current was controlled at 0.5, 1.0, 1.5 amperes. Figure 9 shows the formation of coagulant in the solution. Brown coagulant was observed within a few minutes after the DC current was applied to the reactor. As expected, the higher the current, the faster the coagulant was produced. However, beyond a certain point, 8 minutes in this case, the coagulant production at 0.5 amps surpassed the one at 1 amp. With a treatment time of 16 minutes, the coagulant production at 0.5 amps even surpassed the one at 1.5 amps. The same result was observed from time to time. This is because when a Cr-containing solution was treated in the cell, the voltages were increased to keep a constant current. When the applied voltage reached a threshold, oxygen evolved from the anode. The coagulant content leveled off due to the inhibition of the Fe(II) and Fe(III) dissolution on the anode by oxygen generation. With a low current, such as 0.5 amps, the voltage did not pass the threshold for oxygen formation and therefore iron dissolution occurred during the entire course of treatment. As a result, the coagulant content kept increasing and surpassed that of 1 ampere and then even surpassed that of 1.5 amps.

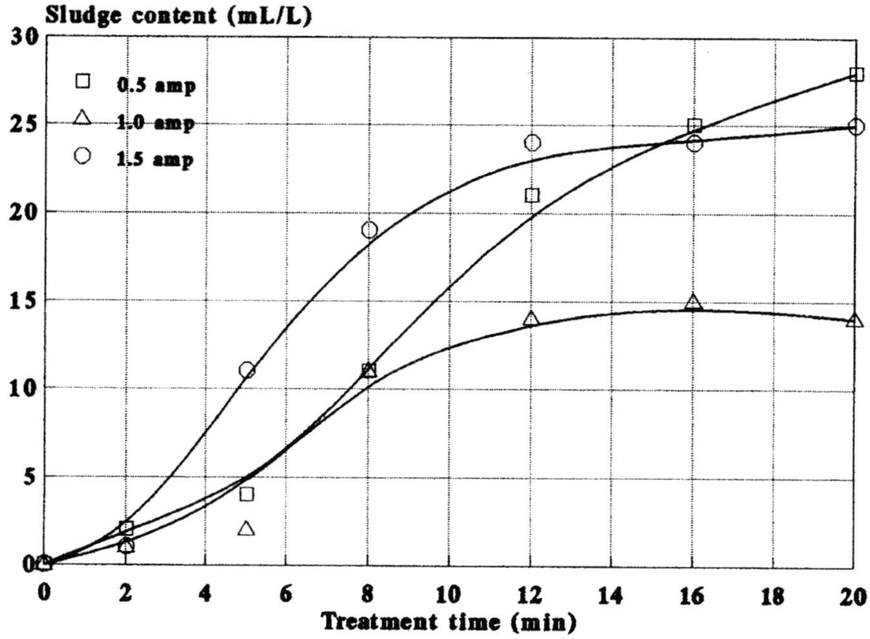


Figure 9. Formation of coagulant in the electrochemical treatment process

The samples were tested in the UV/Vis spectrophotometer, Figure 10 shows the absorbance of these samples at a wavelength of 720 nm. The photometer was set at zero at 720 nm with DI water. The curves of absorbance in Figure 10 are similar to those of the coagulant content in Figure 9, because the coagulant did not change color.

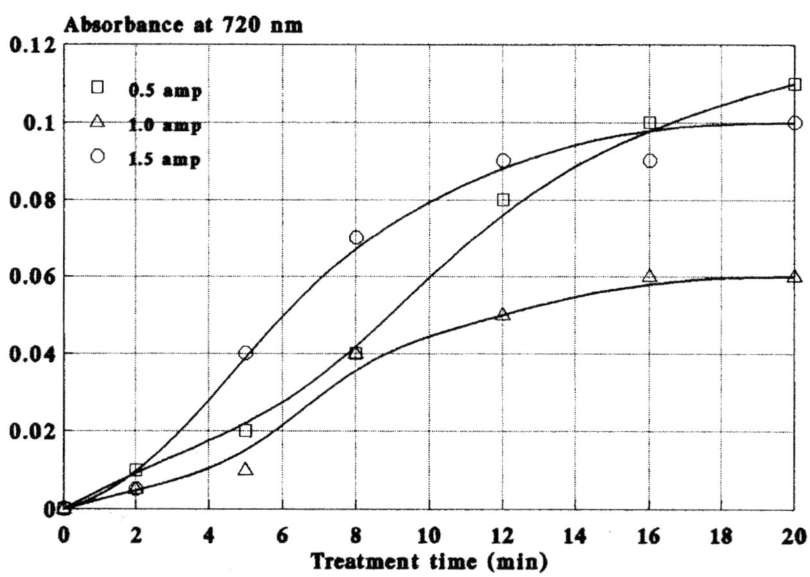


Figure 10. Absorbance of brown sludge at 720 nm

### *NaOH Assisted Coagulant Generation*

The experiments were repeated with a Cr-containing solution pretreated with sodium hydroxide (NaOH). The experiment was designed to study how NaOH promotes the formation of coagulant. Two milliliters of 5% NaOH were added to a 600 $cm^3$ solution to raise the pH of the solution to 9.0. The solution was then treated in the electrochemical cell with 0.5, 1.0, and 1.5 amps of DC current. The treatment increased the pH to 10.8 in 20 minutes. Figure 11 shows the absorbance of the NaOH pretreated solution in the course of the electrochemical treatment. After the NaOH was added, the solution became turbid. The turbidity was quantified by the spectrophotometer. Before the treatment, the solution had absorbance at 720 nm of about 0.01 AU. After the treatment, the absorbance increased from 0.01 AU to 0.05 AU. The production of coagulant was not significant. The 0.05 AU amount is equivalent to less than 10 mL/L of Fe-coagulant.

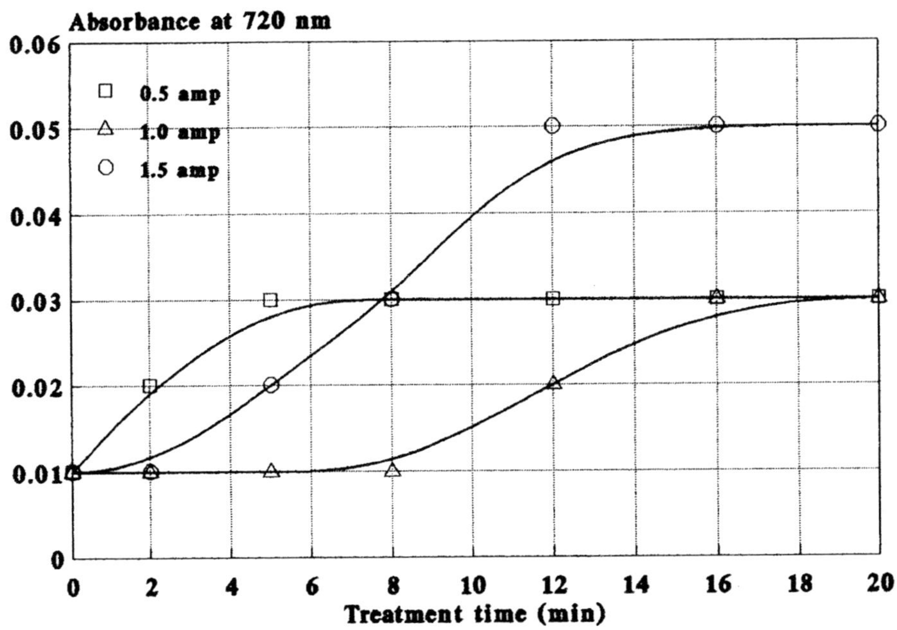


Figure 11. Absorbance of NaOH pretreated solution in the course of electrochemical treatment

### *NaCl Assisted Coagulant Generation*

Two milliliters of 5% NaCl were added to a 600 $cm^3$ solution of 54 mg/L Cr(VI) to study the effect of NaCl on coagulant generation. The solution was then subjected to the same treatment described in the previous section. The NaCl did not change the initial pH of the solution. Figure 12 shows the results of the solution treated with a current of 0.5 amps, 1 amp and 1.5 amps. As expected, NaCl lowered the voltages applied to the cell to keep a certain current flow throughout the cell. Within 10 minutes, the formation of coagulant was the same as in the case without NaCl in the solution. When treated with 1.5 amps, the coagulant became black in 12 minutes of treatment. It takes 16 minutes for the coagulant to become black when treated with 1 amp DC current. However, when treated with 0.5 amps, the coagulant was always brown. Even though the absorbance increased dramatically when the coagulant turned black. The actual coagulant content did not increase. The first observation with regard to these black coagulant was that they are magnetic. This property makes it very easy to settle under a magnetic field. Further study is needed to compare the advantages and disadvantages of black coagulant.

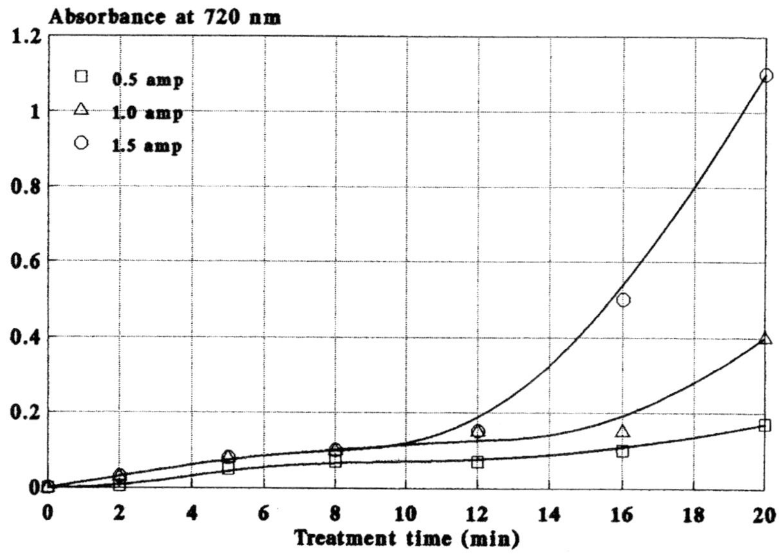


Figure 12. Effect of current density on coagulant generation

A set of experiments was conducted with a circulation system to replace the magnetic stirring system. Figure 13 shows that the production of coagulant was very stable and vigorous. It is believed that the parallel-plate flow cell is controlled by a mass transfer regime. Therefore, recirculation has a strong and positive impact on coagulant generation.

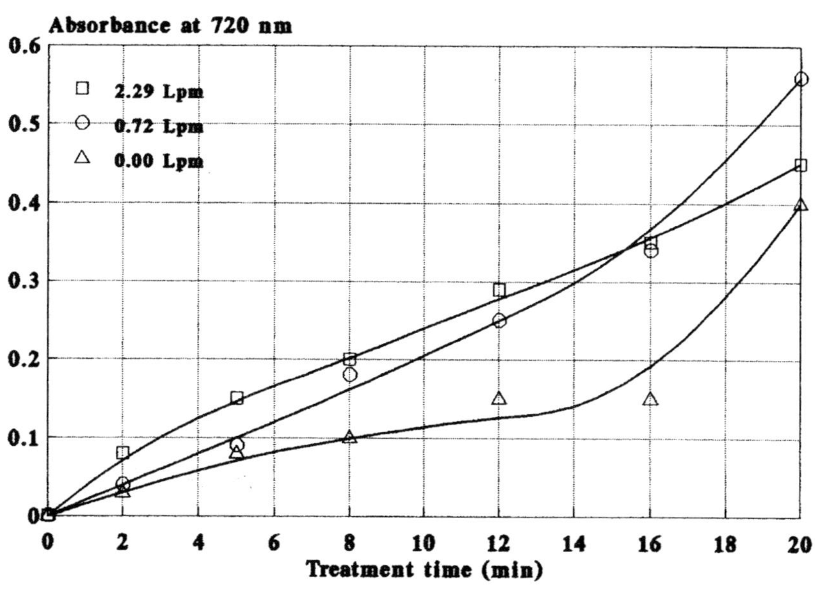


Figure 13. Effect of recirculation on coagulant generation

## 5.3. Chromium Removal

The critical step in this process is that the electrochemical cell has to be able to produce Fe(II) ions effectively because Fe(II) is needed to reduce Cr(VI) to Cr(III). The Cr(III) is removed by precipitation as $Cr(OH)_3$. At the same time, $H_2$ gas is produced to perform the electroflotation for suspended solid separation.

***Chemical Profile***

In accordance with those equations presented in a previous section, Cr(VI) and Fe(II) ions, $H_2$ gas, $OH^-$, as well as $Cr(OH)_3$, $Fe(OH)_2$, and $Fe(OH)_3$ precipitates would be observed in the operation. A set of experiments were conducted to confirm the mechanism. The concentration of OH- anions was quantified in terms of pH. $Cr(OH)_3$, $Fe(OH)_2$ and $Fe(OH)_3$ were measured together as water insoluble sludge by Imhoff settling cone or photometric method. Figure 14 and Figure 15 show the profile of each parameter along with treatment time. The concentration of Cr(VI) dropped from 60 mg/L to less than 1 mg/L in a treatment time of 13 minutes. Right before the disappearance of Cr(VI), Fe(II) was detected in the solution and its concentration kept increasing until it stabilized at 20 mg/L. This can be explained by the mechanism described in the previous section. The Fe(II) produced in the solution was consumed by the Cr(VI). Because the reaction is fast, before Cr(VI) was depleted, there was no Fe(II) in the solution. Once the Cr(VI) was used up, Fe(II) began to accumulate. Because of the existence of $OH^-$ anions, Fe(II) precipitated out. The formation of sludge is strongly affected by pH, because all the metal ions, Cr(III), Fe(III), and Fe(II), formed low solubility hydroxides. The system can be optimized at a treatment time of 9 minutes where the Cr(VI) content and pH are acceptable for reuse and discharge and Fe(II) and sludge are minimal.

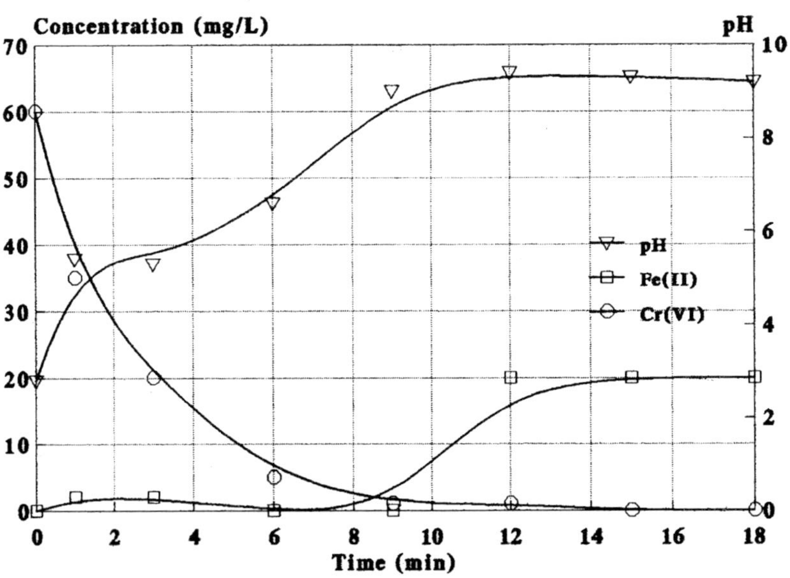


Figure 14. Changes of pH, and Fe(II) and Cr(VI) concentrations in the course of electrochemical treatment

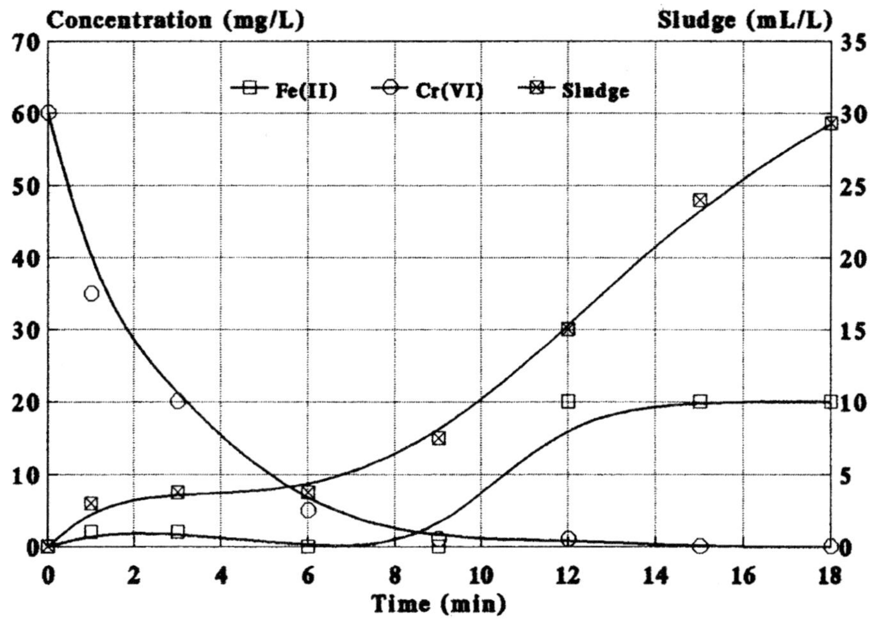


Figure 15. Sludge generation and Fe(II) formation in electrochemical treatment

***Bipolar Electrochemical Cell***

When a cell contains more than two electrodes, there are two ways of making the electrical connection. The cell may be monopolar or bipolar. There is an external electrical contact to each electrode in the monopolar cell and the cell voltage is applied between each cathode and anode. Bipolar connection requires only two external electrical contacts to the two end electrodes. Ferrous ions and $OH^-$ anions are crucial for the electrochemical method to be effective for Cr(VI) removal. In view of the performance of a bipolar electrochemical cell for Fe(II) and OH- production, it is believed that the cell will be effective for Cr(VI) removal. A Cr(VI) containing aqueous solution was prepared with $CrO_3$ crystals from Aldrich Chemicals. When dissolved in water, $CrO_3$ formed chromic acid. Therefore, the initial pH of the solution was 2.5. Figure 16 shows the effectiveness of a bipolar electrochemical cell for Cr(VI) removal. During the operation, chromic acid is gradually neutralized by $OH^-$ produced in the cell. At the same time, Cr(VI) is reduced to Cr(III) by Fe(II). When there is enough $OH^-$ to neutralize all hydrogen ions produced by the chromic acid, $Cr(OH)_3$ precipitates out along with $Fe(OH)_2$ and $Fe(OH)_3$ as sludge. Under the operation conditions, Cr(VI) was reduced from 60 mg/L to less than 1 mg/L in 55 minutes.

Another test was carried out to determine the enhancement of chemical pretreatment for Cr(VI) removal. NaOH was added to a 60 mg/L Cr-containing aqueous solution to have a pH of 8.6. The solution was then introduced into the cell foe electrochemical treatment. In 30 minutes, the Cr(VI) concentration dropped to less than 1 mg/L. At the same time, the pH increase from 8.6 to 11. Figure 17 shows the concentration of Cr(VI) and pH as a function of treatment time. It is to be noted that in both experiments the current in the cell was 400 milliamps.

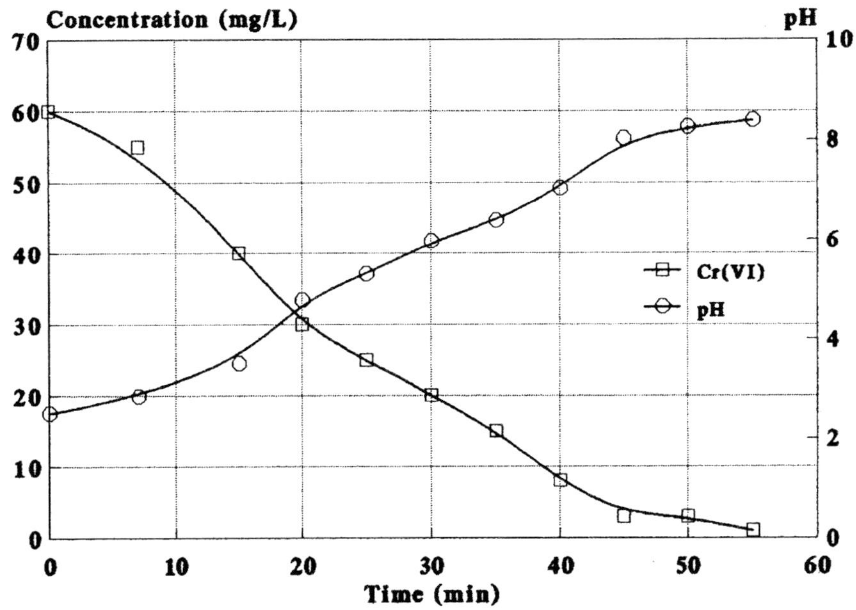


Figure 16. Effectiveness of bipolar electrochemical cell for Cr(VI) removal

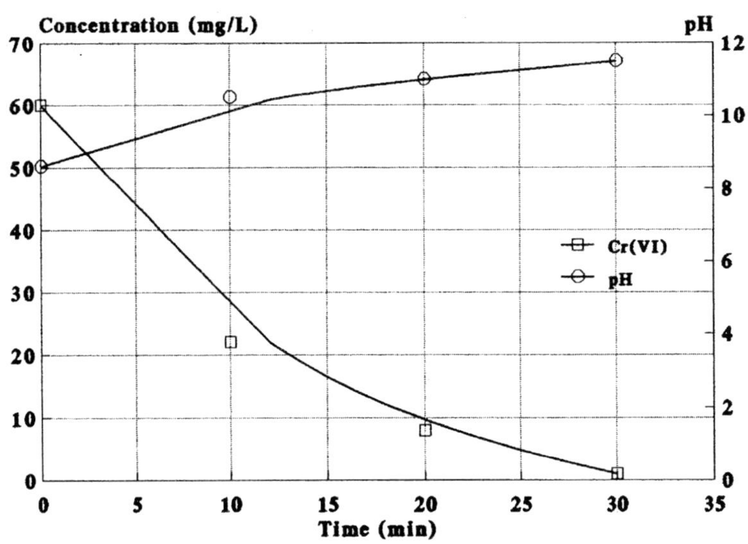


Figure 17. Effect of NaOH pretreatment on Cr(VI) removal in a bipolar electrochemical cell

***Monopolar Electrochemical Cell***

One of the main advantages of electrochemical coagulation is the much more compact apparatus, the electrochemical cell. In order to cut down the treatment time, the system was tested in a monopolar state. All ten electrodes were divided into two sets, anode and cathode. Each electrode was physically connected to the positive or negative outlet of the DC power supply. Figure 18 shows the changes of Cr(VI) and pH in the solution during the course of electrochemical treatment. The monopolar electrochemical cell draws a larger current than a bipolar cell under the same voltage. Therefore, a better result in terms of treatment time can be expected. It takes 16 minutes to reduce 54 mg/L Cr(VI) down to less than 1 mg/L and the pH to stabilize at 9.6. The treatment time is reduced to almost half of a bipolar cell. The pH of the treated waste solution is suitable for reuse or discharge, even without pH adjustment.

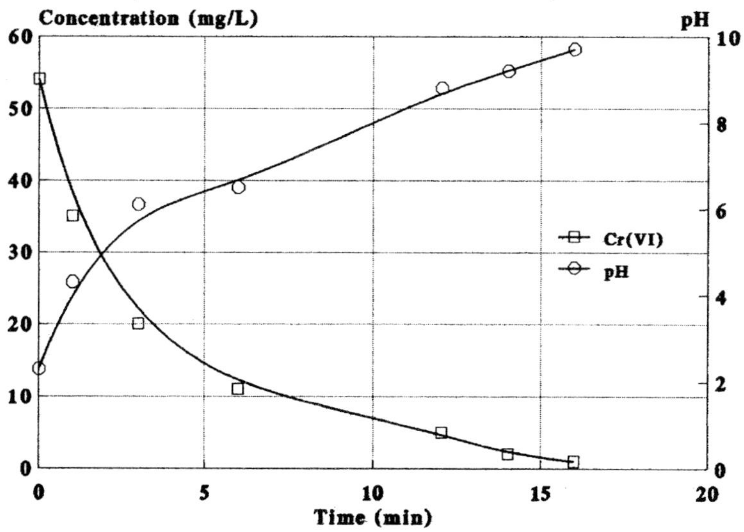


Figure 18. Effectiveness of monopolar electrochemical cell for Cr(VI) removal

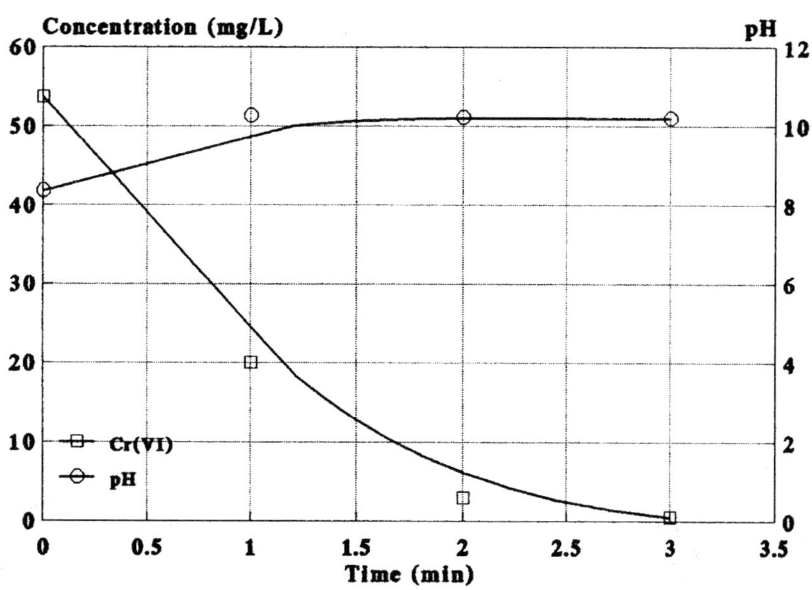


Figure 19. Effect of NaOH pretreatment on Cr(VI) removal in a monopolar electrochemical cell

The experiment was repeated with NaOH in the solution to confirm the enhancement of chemical pretreatment for Cr(VI) removal. NaOH was added to adjust the Cr-containing solution to have a pH of 8.4. Figure 19 shows the effectiveness of chemical pretreatment on Cr(VI) removal. In a treatment time of 3 minutes, the Cr(VI) concentration was reduced from 54 mg/L to less than 1 mg/L. A monopolar cell is 20 times more effective than a bipolar cell in terms of treatment time. The experimental results of both a monopolar and a bipolar cell with and without NaOH pretreatment are depicted together in Figure 20. It is very clear that the monopolar electrochemical cell is very effective in treating a Cr-containing alkaline aqueous solution. To avoid the drawback of cost and extra apparatus of chemical pretreatment, the monopolar cell for Cr-containing acid solutions was chosen for further study.

# 6. Electroflotation

One of the widely used electrochemical methods for water and wastewater treatment is electroflotation. Electroflotation may not be able to compete with flotation in large scale ore dressing processes. However, it is very attractive in this application based on the experience on all the different electrochemical methods that have been developed in this project. Electroflotation shares all the accessory units in the process with electrochemical treatment units.

## 6.1. Electroflotator

An electroflotator was constructed to perform the separation of sludge produced in the electrocoagulator. The cell was designed to have evenly distributed bubbles evolve from the bottom of the cell so that the bubbles have a maximum contact time with the sludge or fine particles suspended in the solution. In electroflotation, the gas bobbles are generated electrically in the cell. A pair of closely spaced, 2 mm apart, horizontal gauze and graphite electrodes were placed towards the base of the cell to cover the whole area. The volume of the cell is 250 milliliter (mL) so that each batch of the solution treated in the electrocoagulator, 560 mL in volume, can be tested in the flotation cell twice.

A sampling port was installed in the middle of the cell. During the operation, samples were taken by a syringe to measure the sludge content or clarity. As described in a previous section, the Imhoff settling cone was not sensitive enough to measure the sludge content after the flotation. A clarity index was applied to evaluate the effectiveness of the process. Clarity was defined as the percent transmittance at a wavelength of 720 nm. The 720 nm wavelength was selected to avoid the interference of Cr(VI), Cr(III), Fe(II), and Fe(III), which were expected to be in the process solution. Deionized water was used to set the clarity at 100%. For a solution of more than 95% clarity, its appearance was like clear water.

## 6.2. Sludge Characteristics

The treatment of Cr(VI) in an aqueous solution was studied thoroughly in batch, circulation batch and continuous systems. Tables 3, 4, and 5 summarize the results and present observations on the operation conditions under which all kinds of sludge were produced. High current density, long treatment time and low circulation rate favored the production of black sludge. In a continuous system, a low feed rate favors black sludge. Black sludge tends to be in particle form rather than in coagulant form. However, black sludge was found to have magnetic property.

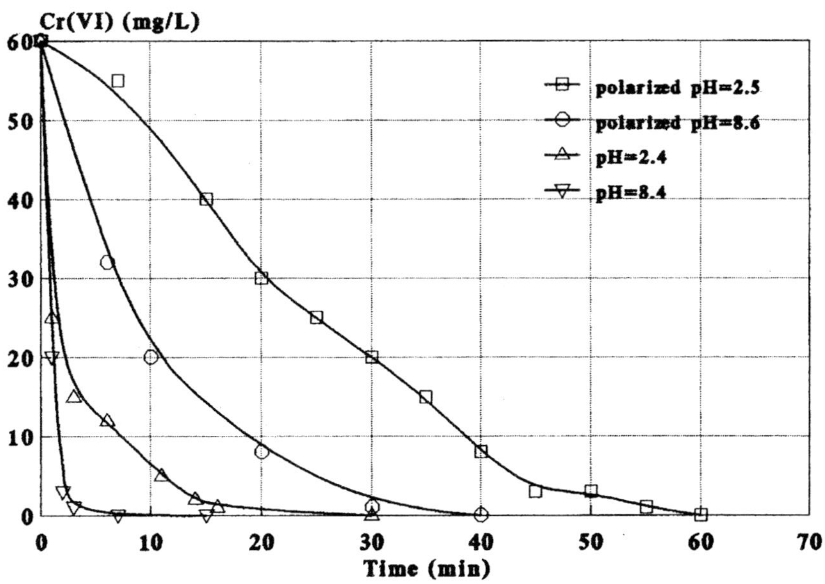


Figure 20. Effect of chemical pretreatment and electrical connections on treatment times for Cr(VI) removal

The electrochemical process produces both brown and black sludge. In the beginning, sludge or suspended particles are the only concern with regard to an effluent to be reused or discharged. Since black sludge is produced in the system as well, turbidity becomes another concern with regard to treated effluent. Brown sludge was first quantified in an Imhoff settling cone. A HACH DR/4000U spectrophotometer was used to determine the sludge content at low range when the solution was treated in the electroflotator. The details of the analytical method development are described in a previous section. Figure 21 was constructed by plotting the Cr(VI) concentrations in Table 3 as a function of current density. As expected, a low feed rate favors the production of black sludge. In each experiment, at least three samples were taken from the effluent after the system was operated for more than 1.2 retention time. Black sludge was observed at a low feed rate and high current density. Figure 22 shows the corresponding pH changes in the experiments referred to in Figure 21 and Table 3. The electrochemical treatment brings the pH from 2.8, the initial pH in the feed, to 10 in the effluent. However, the treatment with short retention time needs a high current density to bring the pH to 10 and the Cr(VI) concentration to an undetectable level.

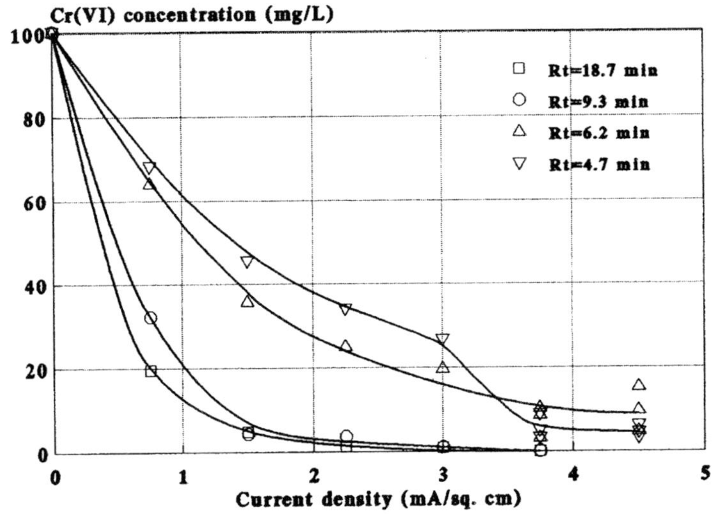


Figure 21. Effect of retention time on Cr(VI) removal in electrochemical treatment

**Table 3. Effect of throughput on sludge characteristics and Cr(VI) removal**

| Flowrate (mL/min) | Sludge color/Cr(VI) concentration (mg/L) | | | | | |
|---|---|---|---|---|---|---|
| | Current Density | | | | | |
| | 0.75 | 1.5 | 2.25 | 3.0 | 3.75 | 4.5 |
| 30 | Bn/19.5 | Bn/4.8 | Bk/Bn/1.3 | Bn/ND | Bk/ND | --- |
| 60 | Bn/32.2 | Bn/4.2 | Bn/3.7 | Bk/Bn/1.0 | Bk/Bn/ND | --- |
| 90 | Bn/64.2 | Bn/35.8 | Bn/25.1 | Bn/19.8 | Bn/8.7 | Bn/9.8 |
| 120 | Bn/68.2 | Bn/45.5 | Bn/34.0 | Bn/26.6 | Bn/5.0 | Bn/4.5 |

Bn, brown sludge; Bk, black sludge; ND, Undetectable

**Table 4. Effect of current density on sludge characteristics and Cr(VI) removal**

| Current (amp) | Sludge color/Cr(VI) concentration (mg/L) | | | | | |
|---|---|---|---|---|---|---|
| | Treatment time (min) | | | | | |
| | 2 | 5 | 8 | 12 | 16 | 20 |
| 0.5 | Bn/54.0 | Bn/54.0 | Bn/38.6 | Bn/16.3 | Bn/8.4 | Bn/4.8 |
| 1.0 | Gn/54.0 | Bn/34.2 | Bn/23.9 | Bk/Bn/5.3 | Bk/1.1 | Bk/ND |
| 1.5 | Bn/54.0 | Bn/27.1 | Bk/Bn/5.3 | Bk/1.4 | Bk/ND | Bk/ND |
| 2.0 | Bn/42.3 | Bn/1.6 | Bk/Bn/ND | Bk/ND | Bk/ND | Bk/ND |
| 2.5 | Bn/28.3 | Bn/13.4 | Bk/Bn/ND | Bk/ND | Bk/1.5 | Bk/1.4 |

Bn, brown sludge; Bk, black sludge; Gn, green sludge; ND, Undetectable

**Table 5. Effect of recirculation rate on sludge characteristics and Cr(VI) removal**

| Circulation rate (Lpm) | Sludge color/Cr(VI) concentration (mg/L) | | | | | |
|---|---|---|---|---|---|---|
| | Treatment time (min) | | | | | |
| | 2 | 5 | 8 | 12 | 16 | 20 |
| 0.72 | Bn/67.3 | Bn/44.2 | Bn/42.8 | Bn/10.6 | Bk/Bn/1.1 | Bk/1.7 |
| 1.54 | Gn/66.6 | Bn/32.8 | Bn/10.4 | Bn/1.4 | Bk/Bn/ND | Bk/Bn/ND |
| 2.28 | Bn/32.1 | Bn/26.2 | Bn/15.0 | Bn/1.5 | Bn/ND | Bn/ND |

Bn, brown sludge; Bk, black sludge; Gn, green sludge; ND, Undetectable

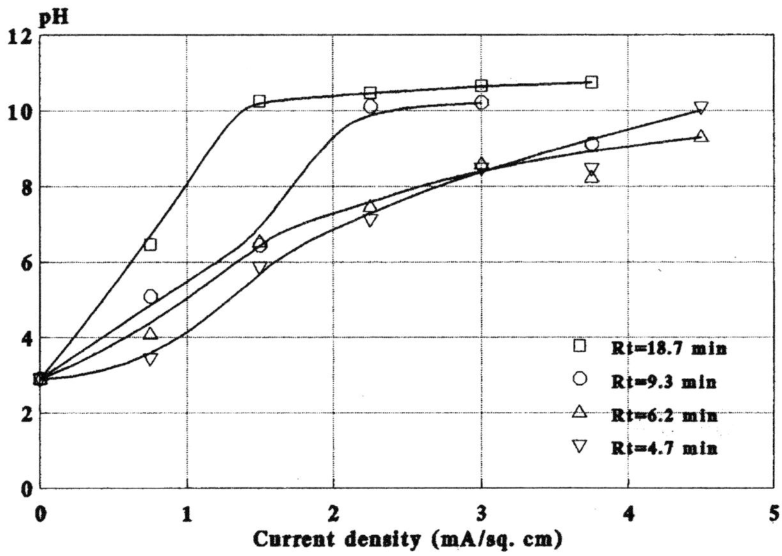


Figure 22. Effect of retention time on pH in electrochemical treatment of Cr-containing solutions

## 6.3. Floatability and Sedimentation

After the treatment of Cr-containing solution in the electrocoagulator, a significant amount of insoluble sludge was produced in the solution. The sludge indicates the effectiveness of the cell to produce coagulants and remove Cr(VI) ions. However, it poses a serious problem for the treated solution to be reused. Sedimentation is a classic yet effective method for sludge separation. Fresh brown sludge was successfully separated by sedimentation in a few minutes, but not black sludge. Black sludge form colloidal system with water. Sedimentation fails to provide separation in a reasonable treatment time. A comparison of electroflotation and sedimentation was studied in fresh brown sludge produced in the electrochemical treatment of a Cr(VI) solution. Two liters of a sludge-containing solution were collected and split into two portions to be tested in the settling cone and the flotation cell simultaneously. Samples were taken periodically to determine the transmittance at 720 nm. Figure 23 shows that it takes four minutes for electroflotation to increase transmittance from zero percent to more than 90%, while it takes 25 minutes for sedimentation to reach the same level. The transmittance was set at zero percent with a brown sludge content of 35 mL/L. After the electroflotation, sludge from both the surface of the liquid and the liquid itself was filtered and dried to determine the absolute sludge content. The sludge content is 0.14 g/L in the liquid and 10.77 g/L in the surface sludge.

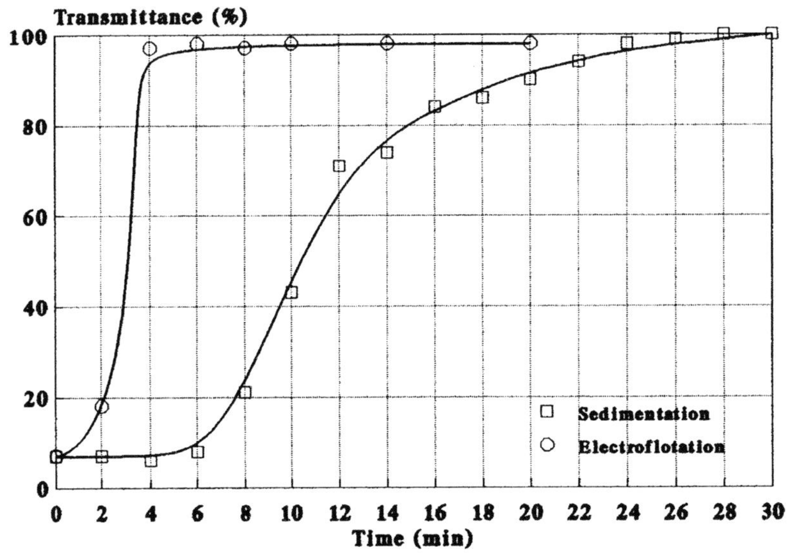


Figure 23. Electroflotation and sedimentation for brown sludge separation

## ***Floatability of Brown Sludge***

Fresh brown sludge can easily be separated from a solution by either sedimentation or electroflotation. When stored for 24 hours, the sludge is transformed into fine particles. It is still takes about 25 minutes for the sludge to settle in an Imhoff settling cone, It takes much more treatment time for electroflotation to separate the sludge effectively. Two critical mechanisms for flotation to be effective are attachment to gas bubbles and the stability of the bubbles. The particle type sludge was found to move up to the surface with bubbles but did not stay on the surface of the liquid. Therefore, in some cases, there is no efficiency for sludge separation at all. Organic chemicals, used in flotation for ore dressing, were studied to enhance the separation efficiency of electroflotation for brown sludge. Two mL, 5% of lauric, mystic, palmitic and oleic acids were added to a 250 mL solution for each test. Figure 24 shows the results of chemical-assisted electroflotation for brown sludge separation. Only lauric acid is effective in enhancing the sludge separation. Transmittance at 720 nm increases from 40% to 80% . The other organic acids show insignificant or no effect in enhancing the separation process.

The experiments were repeated with a group of commercially available surfactants, such as Dermassage, Ajax and Dawn. The enhancement of organic acids on electroflotation is by means of ionic reaction. Surfactants have both positive and negative legends, and hydrophobic and hydrophilic groups on each molecule. They are believed to be effective and enhancing flotation and detaining the sludge on the surface of the liquid. Figure 25 shows the results of surfactant-assisted electroflotation for brown sludge separation. The efficiency is expressed in terms of percent of transmittance at 720 nm. For those treated solutions with transmittance greater than 95%, their appearances are like clear water. The measuring instrument is calibrated to deionized water as 100% transmittance at 720 nm. Figure 25 shows that both Dermassage and Ajax were effective in enhancing electroflotation for brown sludge separation. The treatment time of Dermassage and Ajax are 12 minutes and 14 minutes, respectively.

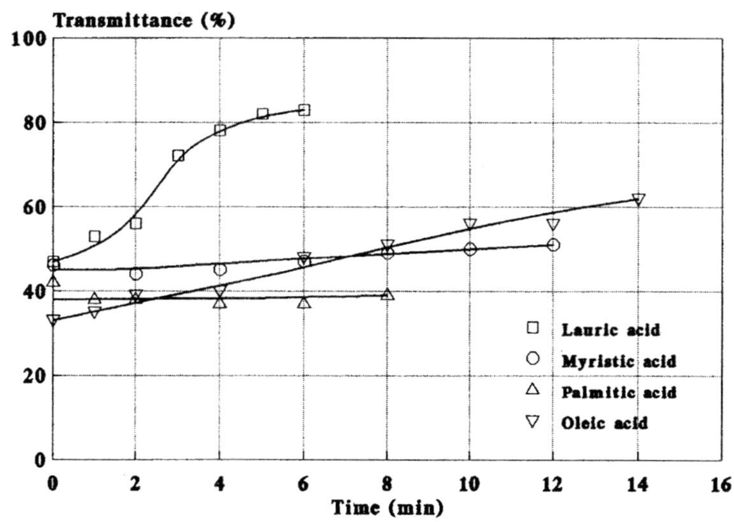


Figure 24. Effect of organic acids on brown sludge floatability

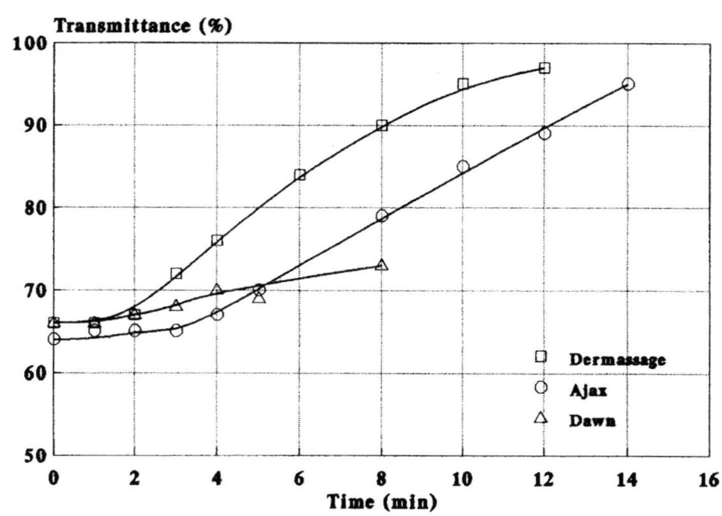


Figure 25. Effect of commercial surfactants on brown sludge floatability

## *Floatability of Black Sludge*

Black sludge formed a stable black colloid system in an aqueous solution. With only 20 mL/L of sludge, the transmittance of the solution is about zero percent. Compared to brown sludge, turbidity is a big concern in reusing or discharging water with the black sludge. Both surfactants and organic acids were tested to enhanced the separation efficiency of electroflotation. However, only Dermassage has some positive effect but the results were not consistent. Sodium chloride (NaCl) was added to the solution before electrochemical treatment. NaCl is a strong electrolyte. It is known that NaCl can depassivate the system and produce sludge that will float more easily. Two milliliters of 5% NaCl solution were added to a solution of 560 milliliters before electrical treatment. Figure 26 shows the positive results with the addition of NaCl. Although, the process still produces dark black sludge, the sludge is easier to float without surfactants or organic acids. The transmittance at 720 nm was increased from zero to more than 90% in 8 minutes. The effluent from the electroflotation cell is ready for reuse.

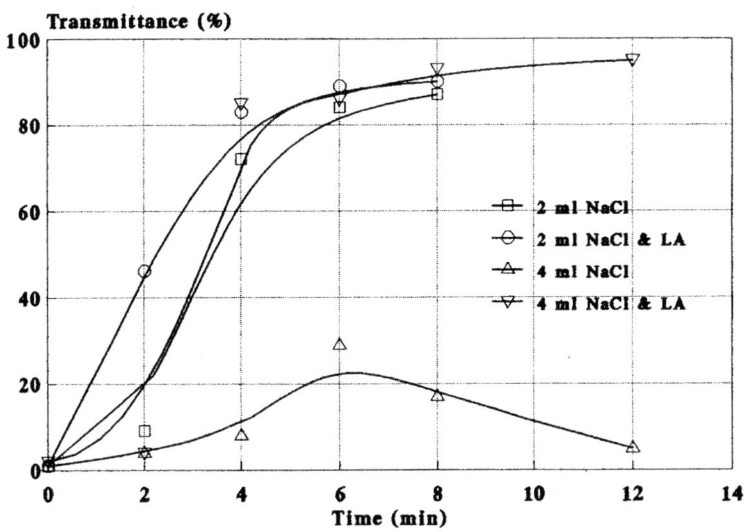


Figure 26. Effect of NaCl on sludge floatability in electroflotation

## 6.4. Brown Sludge Separation

Nine industrial surfactants were collected and tested. The surfactants are Citranox from Allconox, Inc., All-purpose Cleaner from Surpass, Protasorb L-20-NF from Protameen Chemicals, Legsolve from Walter G. Legge Co., Alconox from Alconox, Inc., Zymit™ Low-foam Enzyme Cleaner from International Product, Oil/grease Emulsifier from Walter G. Legge Co., Procol LA-4 from Protameen Chemicals Inc., and Micro-90 from International Product. Along with these tests was the test with Dermassage, which was found most effective in the commercial detergent category. The Dermassage test was performed for a cross comparison of a commercial detergent and industrial surfactants. In this series, the typical test conditions were:

- Solution: 250 mL solution with brown sludge
- Chemical: 1 mL 5% surfactant, 3 mL 5% sodium chloride
- Initial pH: 6.5 to 7.5

- Current: 400 mA
- Applied voltage: 10 to 13 volt DC
- Treatment time: 12 minutes

Figure 27 shows the efficiency of surfactant-assisted electroflotation for brown sludge separation. The separation efficiency is presented as a percentage of transmittance at 720 nm. The initial sludge content was 36% transmittance at 720 nm. While adding the surfactant, the transmittance increased from 36% to about 55%. It is believed that coagulation or flocculation occurred even before the current was applied. Figure 27 also shows that only Legsolve is more effective than Dermassage in assisting sludge separation in the electroflotation process. It took only two minutes to complete this operation. The treated solution has a transmittance at 720 nm of more than 99%. This group of industrial surfactants is effective in assisting electroflotation up to 12 minutes. All of these surfactants are able to accumulate sludge on the surface of the water body and leave the treated solution with more than 95% transmittance at 720 nm.

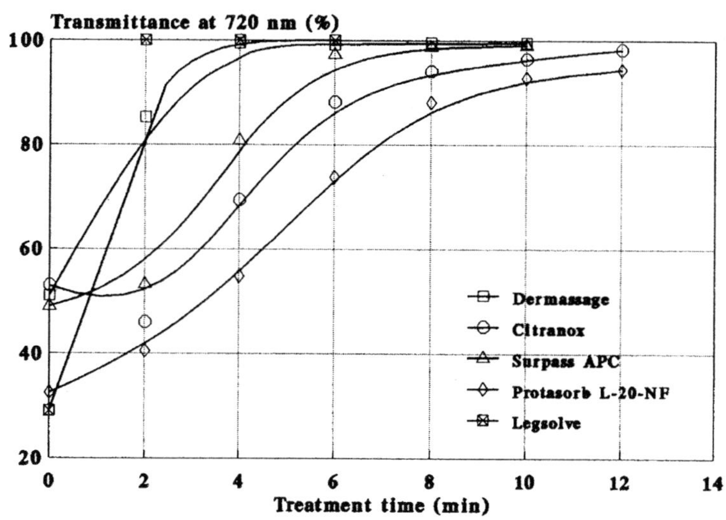


Figure 27. Surfactant assisted electroflotation for brown sludge separation

Figure 28 shows the results of another group of surfactants in sludge separation. Only the Oil/Grease Emulsifier from Walter G. Legge Co. is effective in collecting sludge from the solution. Alconox and Procol LA-4 are not efficient at all. For comparison, sedimentation was performed on the brown sludge. Figure 29 shows that in up to 4 minutes, transmittance increases from 35% to 80%. After 4 minutes, the settling process slow down and the transmittance approaches a maximum of about 92%. It took more than 24 hours to have a transmittance of about 97%. With 1 mL of 5% Legsolve, the sedimentation process was accelerated. In 2 minutes, the solution had a transmittance approaches 80%. This is due to the coagulation of the sludge by Legsolve. Transmittance approaches 96% in a sedimentation time of 8 minutes. The Legsolve is very efficient in assisting electroflotation and sedimentation for brown sludge separation. Figure 30 shows the effectiveness of detergents for assisting brown sludge separation in a electrofloataion cell. The efficiency is presented as a percentage of transmittance at 720 nm. The initial sludge content was set at 36% transmittance at 720 nm. While adding Dermassage, the transmittance increased from 36% to 50%. It is believed that coagulation or flocculation occurred even before the current was applied. The transmittance increased from 50% to 99% in four minutes in the electroflotation assisted by a trace amount of Dermassage. Coagulation or flocculation did not occur when industrial detergents such as O/G Emulsifier and Legsolve were added to the brown sludge. However, the industrial detergent effectively enhanced the sludge separation in the electroflotation process. With trace amounts of Legsolve, electroflotation increased the transmittance of the sludge-containing solutions from 36% to 99% in two minutes. It took O/G Emulsifier six minutes to accomplish the same degree of separation.

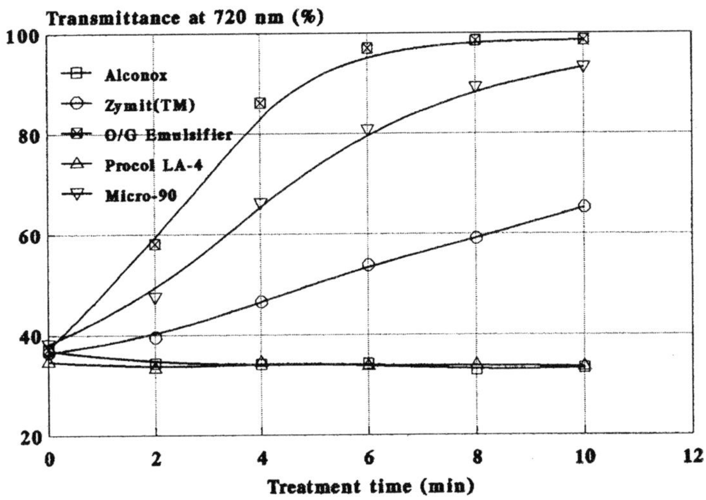


Figure 28. Effect of industrial detergents on brown sludge separation

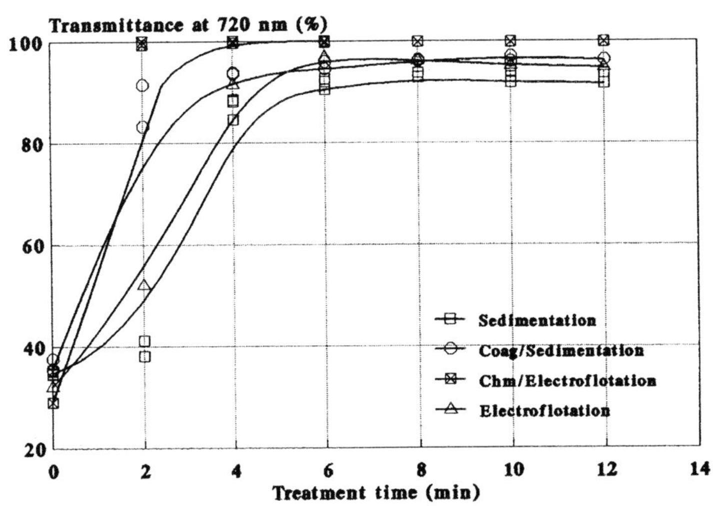


Figure 29. Sedimentation, coagulation/sedimentation, electroflotation, and chemical assisted electroflotation for brown sludge separation

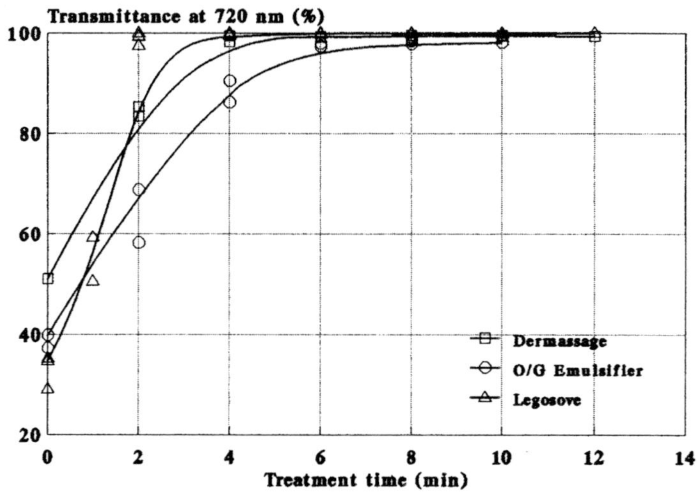


Figure 30. Effectiveness of Dermassage, O/G Emulsifier and Legsolve on sludge flotation

## 6.5. Black Sludge Separation

For electroflotation, the major challenge is the separation of black sludge. As stated in a previous section, the black sludge can be removed to more than 99% transmittance only if a 0.2-micron membrane filter is used. A series of experiments were designed to evaluate the efficiency of electroflotation on black sludge separation. Figure 31 shows that electroflotation is not effective in black sludge separation. The transmittance of the solution was increased from 2% to 38% in a 20-minute flotation operation. Sedimentation showed some efficiency in black sludge separation. However, the efficiency leveled off at a transmittance of about 78%. A solution with transmittance of 78% looks black, which is not suitable for discharge. Again, Legsolve showed its effectiveness in enhancing the separation process in electroflotation. The solution was cleared up to more than 92% in two minutes and ended up with a transmittance of more than 95%.

Legsolve was found very effective in enhancing electroflotation for sludge separation. The cost of the product is acceptable in the environmental business. However, for scientific research, it is desirable to pin down the active ingredient in this product. Legsolve is a mixture of EDTA, triethanol amine, tail oil fatty acid, sodium metasilicate and diethylene glycol monoethyl ether. All of these chemicals were purchased and tested for their effectiveness in enhancing electroflotation for black sludge separation. One milliliter of 5% of each of the ingredient in Legsolve was added to a solution of 250 mL for electroflotation. Figure 32 shows the effectiveness of the components in Legsolve for black sludge separation. To be effective in enhancing electroflotation, the chemicals have to have the characteristics of a collector and frother. A collector is a substance that will form an attracting force to bond sludge particles to up-rising bubbles. A frother is a substance that is able to retain sludge particles on the surface of the solution. Figure 32 shows that tail oil fatty acid is a good collector, because the solution was cleared up from transmittance 3.5% to 90% in less than 4 minutes. However, after a 4-minute operation, transmittance began to decrease. This indicates that tail oil fatty acid was not able to retain sludge froth on the surface of the solution. EDTA is less effective than tail oil fatty acid both as a collector and as a frother. Sodium metasilicate shows another property of surfactants: emulsification. After adding 1 mL 5% sodium metasilicate in the solution, the colloidal system became very stable. The solution remained at about 4% transmittance after 12 minutes of electroflotation. On the other hand, triethanol amine shows the properties of both a collector and a frother. With 1 mL 5% triethanol amine, the solution was cleared up to more than 80% transmittance in two minutes of electroflotation. The operation further cleared up the solution to more than 98% transmittance and retained the sludge froth on the surface of the solution to the end of the test. Triethanol amine was identified as the active ingredient in Legsolve to enhance the efficiency of electroflotation for black sludge separation.

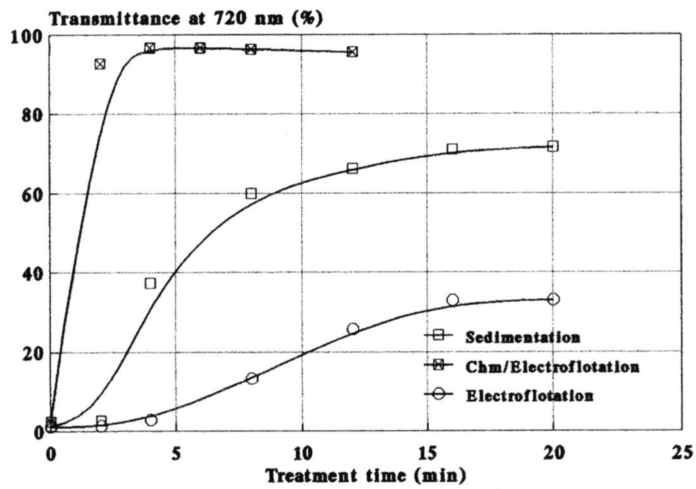


Figure 31. Sedimentation, electroflotation, and chemical assisted electroflotation for black sludge separation

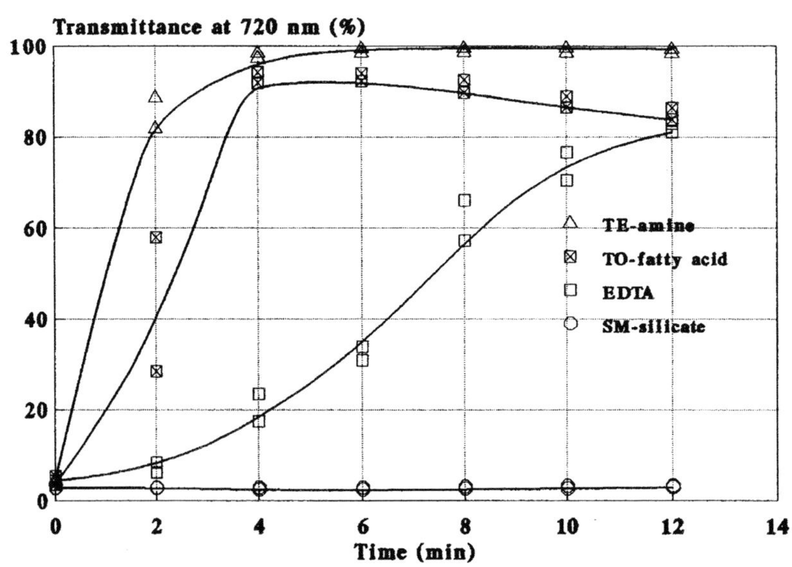


Figure 32. Effectiveness of components in Legsolve for black sludge flotation

Since black sludge is a metal oxide, more precisely ferric/ferrous oxides, it is reasonable to believe that collectors used in ore dressing in the flotation process can be good candidates to enhance black sludge separation. Another approach to finding a suitable collector and forther is to test the collectors recommended in literature, such as thiourea, potassium ethyl xanthate, terpineol and n-dodecylamine. Based on its ionic characteristics, Thiourea (cationic), potassium ethyl xanthate (anionic), terpinol (nonionic) and n-dodecylamine (cationic) were selected for the test. Figure 33 shows the results of these tests. The n-dodecylamine was found effective as a collector and frother. Thiourea is a good collector but average as a frother. Terpineol is a better frother than thiourea but thieurea is a better collector then terpienol. Potassium ethyl xanthate is acceptable as a collector and frother. Overall, the group is effective in enhancing the electroflotation for black sludge separation.

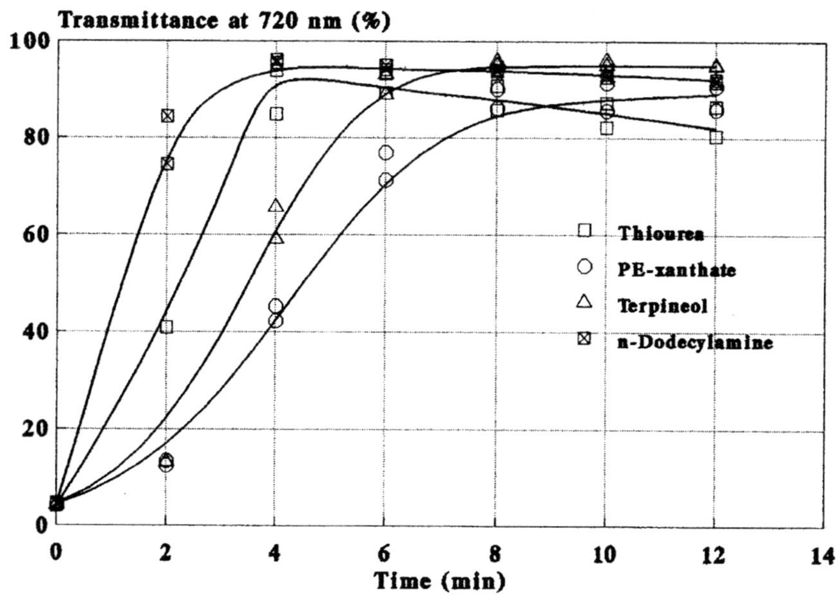


Figure 33. Effect of industrial collectors on black sludge flotation

Both amines, triethanol amine and n-dodecylamine, were effective in enhancing electroflotation. Another amine, 1-octadecylamine, was tested under the same conditions to confirm the results. Figure 34 shows that 1-octadecylamine is effective as well. However, triethanol amine has the best efficiency. Triethanol amine is a tertiary amine with six carbons in the molecule. 1-octadecylamine and n-dodecylamine are long chain amines with 18 and 12 carbons in the backbone chains.

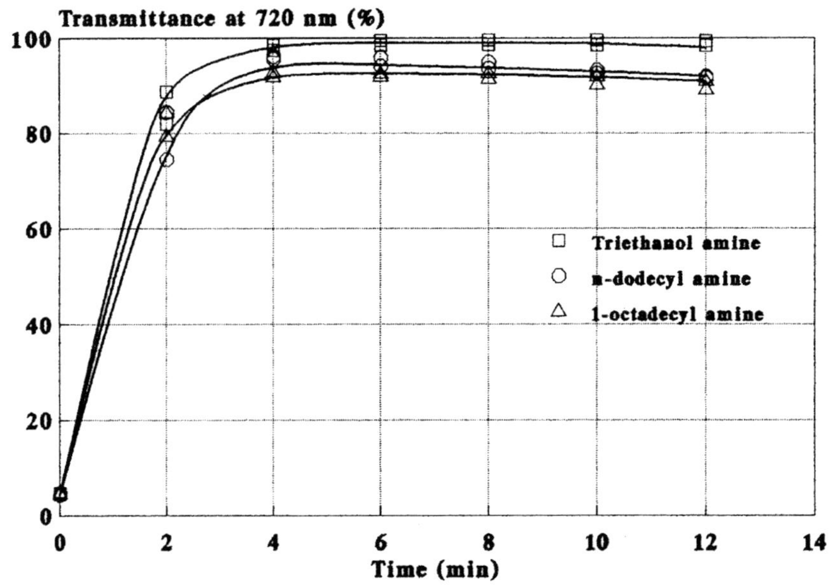


Figure 34. Effectiveness of amine grope for black sludge flotation

# 7. Pilot Unit Study

A pilot-scale electrochemical system was designed and fabricated to study the scale-up of electrochemical coagulation and electroflotation for Cr(VI) removal. The electrocoagulator is a plate-in-tank cell where three banks of consumable electrodes are installed and submerged in the solution. The Cr-containing solution is pumped into the electrocoagulator from the bottom and flow out of the cell from the top. The Cr(VI) ions are converted to hydroxide precipitates. The whole stream, solution and suspended sludge, is pumped to the electroflotator where the sludge will be carried to the surface of the solution by electrically generated bubbles. A converyer was designed to remove the sludge from the top of the electroflotator. A mixing tank was installed up stream of the electrocoagulator. A small quantity of NaCl aqueous solution was added to improve the conductivity of the solution and to prevent the passivation of electrodes. Another mixing tank was installed between the electrocoagulator and the electroflotator. A small amount of surfactant was added to increase the energy efficiency of electroflotation. The system was designed based on results from the bench-scale experiments.

## 7.1. Electrochemical Cells

The electrochemical coagulator is designed to treat a waste stream of 1 to 10 liters per minute with hexavalent chromium as its major contaminant. It is designed so that it will only be necessary to connect the inlet and outlet water lines, run the interconnecting piping and conduit and bring the 115 volt, single phase, 60 cycle AC electricity to the power supply. The major components and specifications of the unit ware shown in Table 6. Figures 35 to 37 show the details of the electrocoagulator. A hood is built right on top of the cell where some hydrogen gas might be produced during the operation. A pare of feed pump and circulation pump are installed to provide the proper flow rate and mixing. The waste stream is pumped into the reservoir at the bottom of the cell through a distributor, the solution push up through the cell, between the electrodes where the Cr(VI) is converted to Cr(III) and then $Cr(OH)_3$. The effluent flows out of the cell from the top of the electrodes sets. During the first trial, the Cr(VI) concentration was clearly lower as bright yellow color is diminished to brown sludge, which is believed to be the ferric oxide.

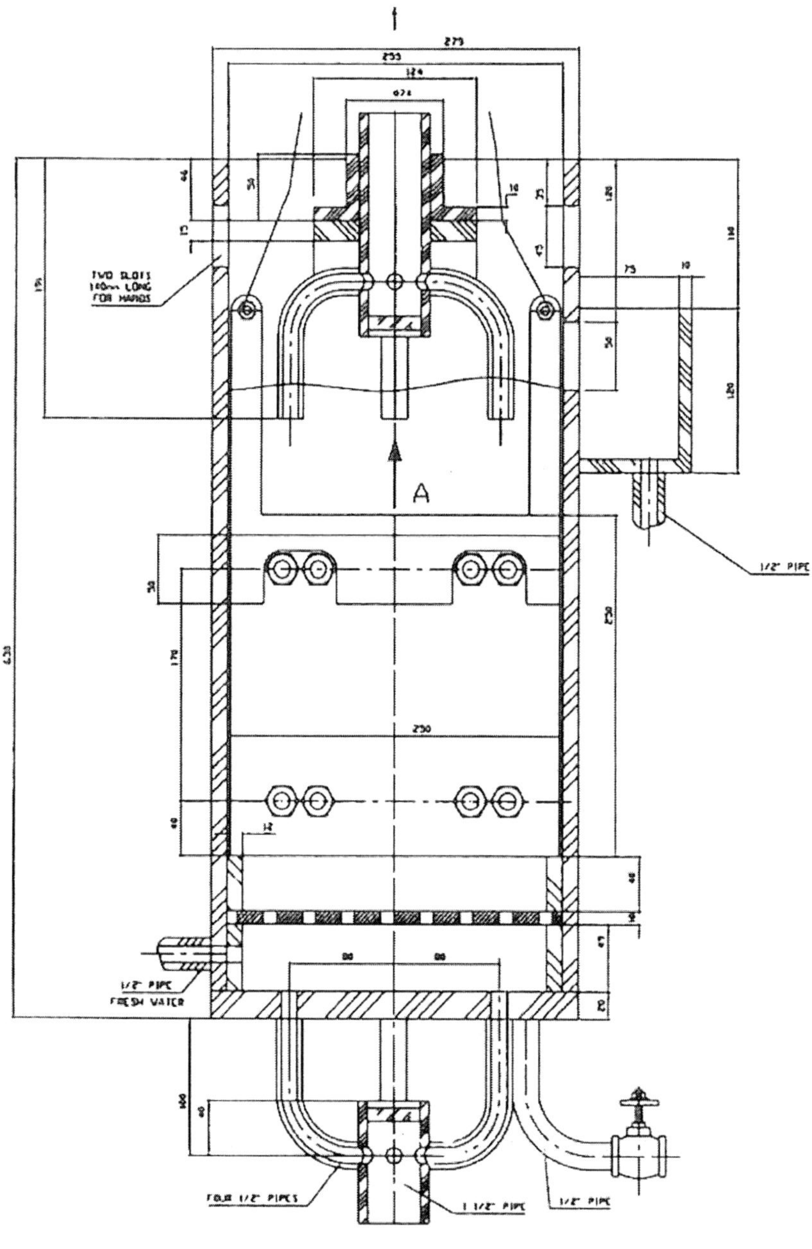


Figure 35. Design of the pilot-scale electrocoagulator

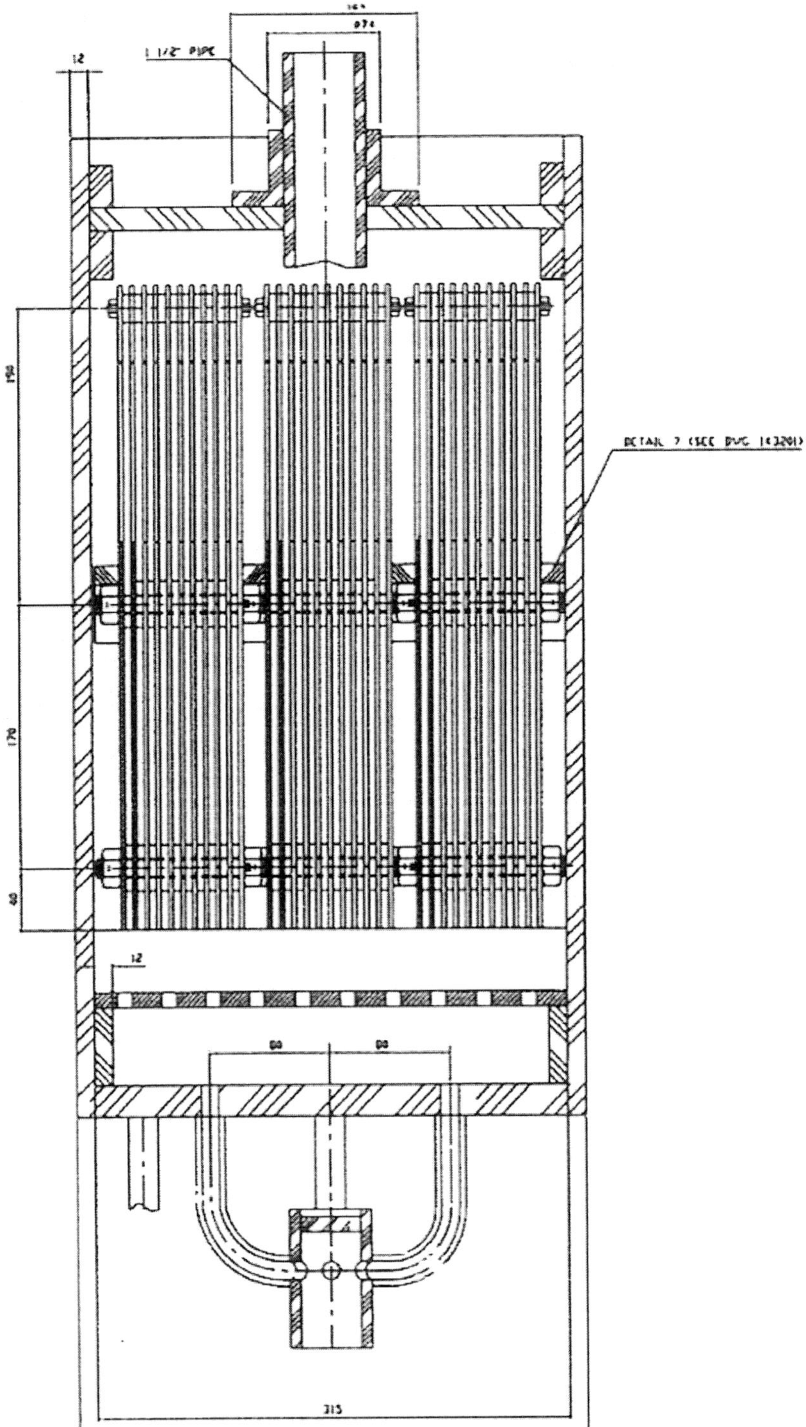


Figure 36. Fabrication details of the pilot-scale electrocoagulator

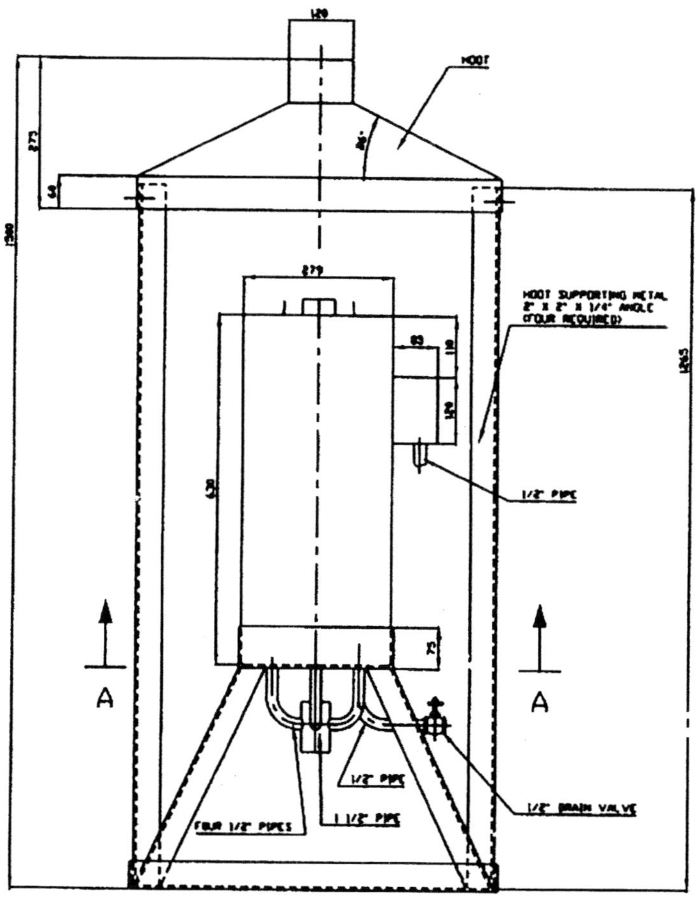


Figure 37. Installation details of the pilot-scale electrocoagulator

**Table 6. Specifications of the pilot-scale electrocoagulator**

| Item | Major components | Specifications | Quantity |
|---|---|---|---|
| 1 | Parallel plate electrochemical cell | With recirculation and distribution devices, 1'x1'x2' | 1 |
| 2 | Anode set | 17 carbon steel sheet electrodes, 1'x1' | 1 |
| 3 | Cathode set | 16 carbon steel sheet electrodes, 1'x1' | 1 |
| 4 | Power supply | 12 volt DC, 60 amps | 2 |
| 5 | Power control and measurement | 30 volt DC, 100 amps | 2 |
| 6 | Ventilation hood | 190 cubic feet per minute | 1 |
| 7 | Blower (with ductwork) | ¼ HP | 1 |
| 8 | Reservoir (bottom of the cell) | 1'x1'x0.5' | 1 |
| 9 | Feed pump and flow control | 0 – 6 liters per minute | 1 |
| 10 | Circulation pump and flow control | 0 – 12 liters per minutes | 1 |

Figure 38 to 40 shows the details of the electroflotator and its ventilation system. A hood is mounted on the top of the unit and a blower with a placement of 190 cubic feet per minute (cfm). The design of the distributor is such that the flow will evenly distributed into the cell. The specifications of the pilot-scale electroflotator are presented in Table 7.

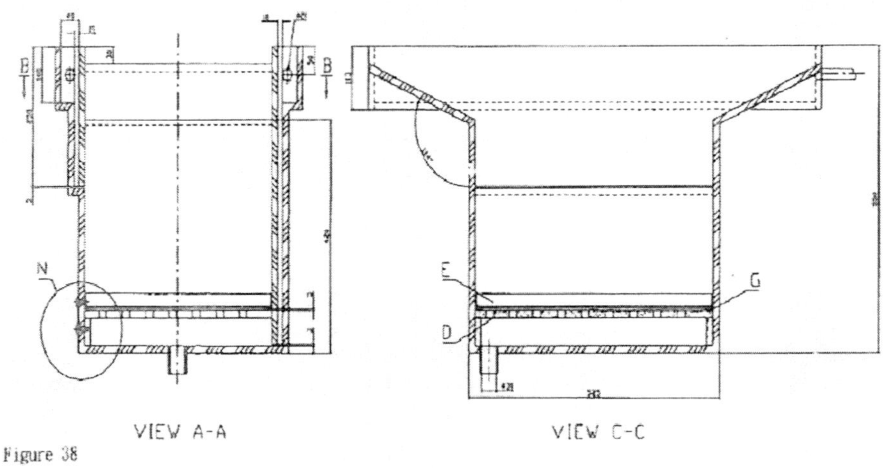


Figure 38. Design of the pilot-scale electroflotator

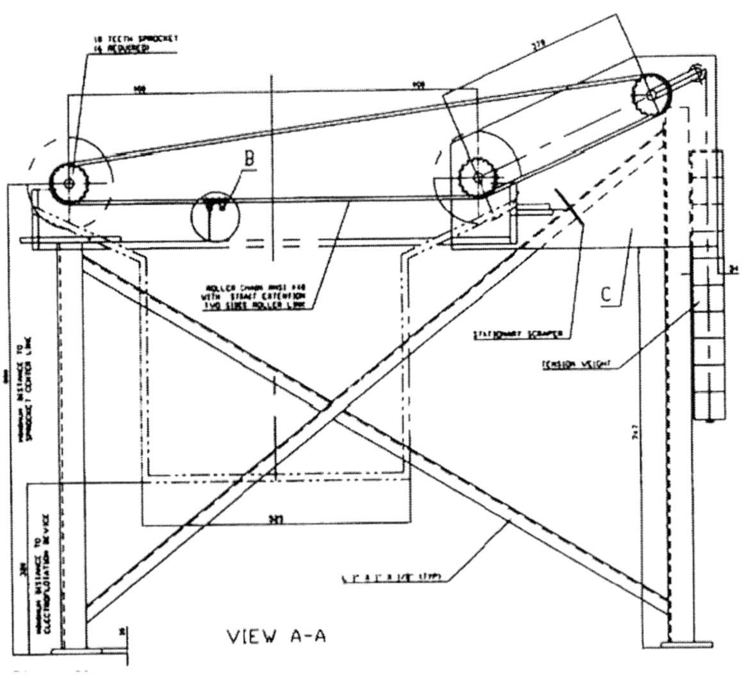


Figure 39. Fabrication details of the pilot-scale electroflotator

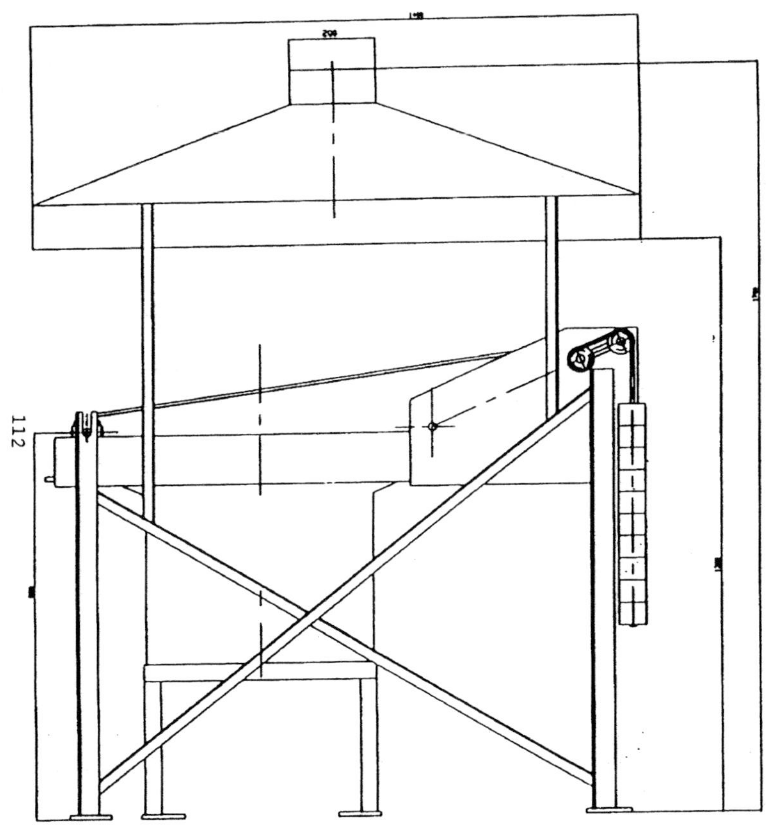

Figure 40. Installation details of the pilot-scale electroflotator

**Table 7. Specifications of the pilot-scale electroflotator**

| Item | Major components | Specifications | Quantity |
|---|---|---|---|
| 1 | Horizontal mesh-plate electrochemical cell | With recirculation and distribution devices, 15.5"x10.5"x12" | 1 |
| 2 | Anode set | 10.5"x15.5"x1" graphite | 1 |
| 3 | Cathode set | 10.5"x15.5" stainless steel woven mesh | 1 |
| 4 | Power supply | 25 volt DC, 20 amps | 2 |
| 5 | Power control and measurement | 30 volt DC, 50 amps | 2 |
| 6 | Ventilation hood | 190 cubic feet per minute | 1 |
| 7 | Blower (with ductwork) | ¼ HP | 1 |

## 7.2. Electrochemical Coagulation for Cr(VI) Removal

In the previous section, the specifications of the pilot-scale electrochemical system were presented along with some quantitative test results. A series of tests were performed on the unit to evaluate the effectiveness of the process for Cr(VI) treatment.

***Batch Operation***

An aqueous solution with 60 mg/L Cr(VI) was prepared for the tests. The unit was first tested in batch mode. Sixteen liters of the solution were pumped into the cell to cover all the electrodes. With 500 mg/L NaCl in the solution, the system was able to provide a constant current of 60 amps at a voltage of 3.2 Volts DC. Samples were withdrawn from the top, between the electrodes, in two-minute periods. Seven samples were collected in each 12-minute test. As usual, the samples were filtered with a 5-micron membrane filter before the measurement of the Cr(VI) concentration. The sample taken at 6 minutes was colloidal in nature and the colloid system was stable for more than 24 hours. Figure 41 shows the measurement of Cr(VI) in the spectrophotometer. An unreasonable reading appeared on the colloid sample. It is believed that the colloid system is formed by very fine, less than 5 microns, particles, so that even when passing through a 5-micron membrane filter, the particles penetrate the membrane and appear on the reading. A 0.2-micron membrane was then used to confirm the hypothesis. A reasonable treatment curve is shown in the Figure 41 with the Cr(VI) concentration represented by circles.

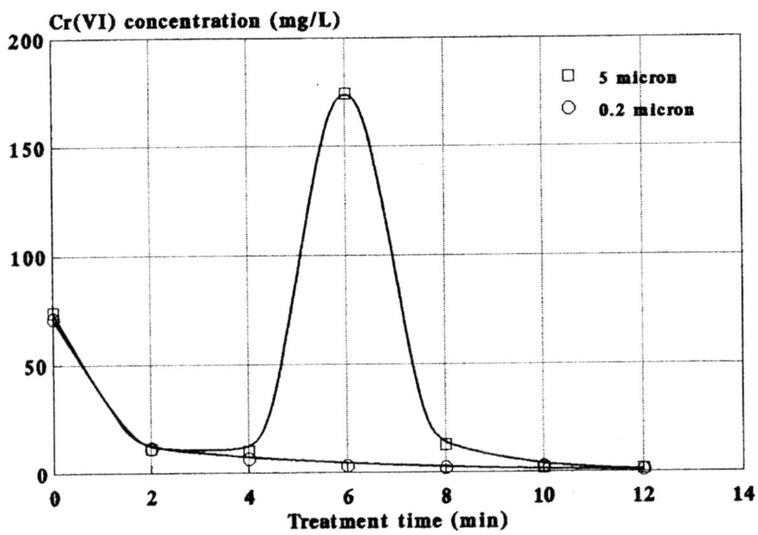


Figure 41. Formation of colloid system in the batch operation

A new concern arises upon using 0.2-micron membrane filters. Will the sample treatment affect the Cr(VI) concentration? A set of standard solutions was prepared from reagent grade $CrO_3$ crystals. A portion of these solutions was passed through a 5-micron membrane filter before the Cr(VI) measurement. The results are presented as squares in Figure 42. Small deviations were observed in the high Cr(VI) concentrations. At low concentrations, the results were virtually the same, whether or not the solutions were passed through the 5-micron membrane filter. The experiments were repeated with 0.2-micron membrane filters. More significant, but acceptable, errors appeared on these readings. The results are presented in Figure 42 with circles.

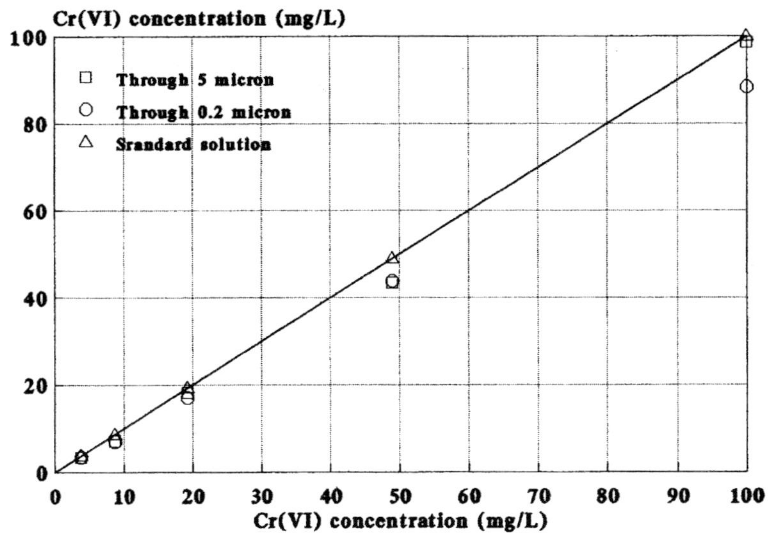


Figure 42. Concentration deviations in sample treatment procedure

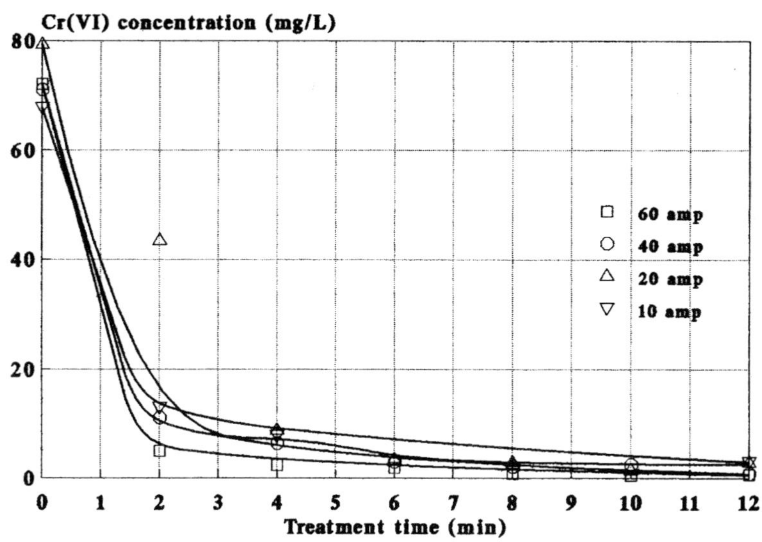


Figure 43. Effectiveness of the pilot-scale electrocoagulator for Cr(VI) removal

Figure 43 shows the treatability of the solution in the pilot-scale electrochemical cell. With 60 amps or 40 amps in the cell, the concentration of Cr(VI) dropped from 60 mg/L to less than 10 mg/L in two minutes. The concentration continued dropping to less than 1 mg/L in a 6-minute treatment. However, when the current was reduced to 20 amps, it took four minutes to reduce the Cr(VI) concentration less than 10 mg/L and at the end of the 12-minute treatment the Cr(VI) concentration was 2.5 mg/L. Figure 44 shows that it took 8 minutes to reduce the Cr(VI) concentration from 260 mg/L to less than 10 mg/L when the cell was operated at 60 amps. Electrochemical coagulation is effective in treating high concentration

Cr(VI) as well. However, the treatment time has to be adjusted to compensate the increase of inlet Cr(VI) concentration.

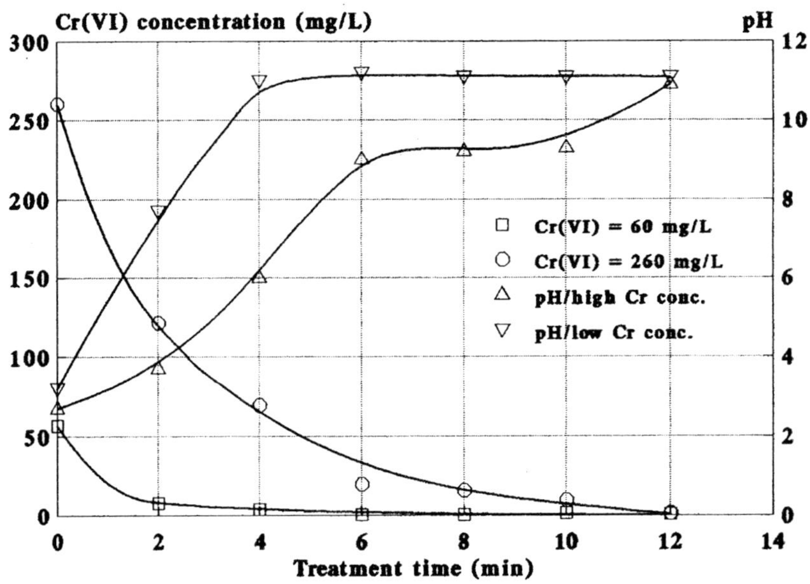


Figure 44. Applicability of the electrocoagulator for treatment of high concentration Cr(VI) solutions

### *Continuous Mode*

In the previous section, the experimental results from the pilot-scale electrochemical coagulator were presented in details. All of the results were acquired when the cell was operated in batch mode, which was the direct scale-up fro the bench-scale experiments. However, when this unit is installed in a production line, in full flow or on a by-pass, the unit will be operated in continuous mode. A set of experiment was conducted in continuous mode to acquire data on Cr(VI) removal and gain experience in the operation of the system. The important parameters are time needed to stabilize the system and the steady state outlet concentration. Most of the these experiments were conducted under the following conditions:

- Cell volume: 20 liters (the active zone, 10 liters; equalizer below the active zone, 4 liters; and stabilization zone above the active zone, 6 liters)
- Feed rate: 1.1, 2.2 and 3.3 liters per minute
- Circulation: 0 or 12 liter per minute
- Concentration: 100 mg/L, Cr(VI)
- NaCl: 500 mg/L
- Electrode: anode/cathode: 15 pieces, 10" by 10" carbon steel
- Applied voltage: 0 to 20 volts DC
- Current: 10 to 80 amps

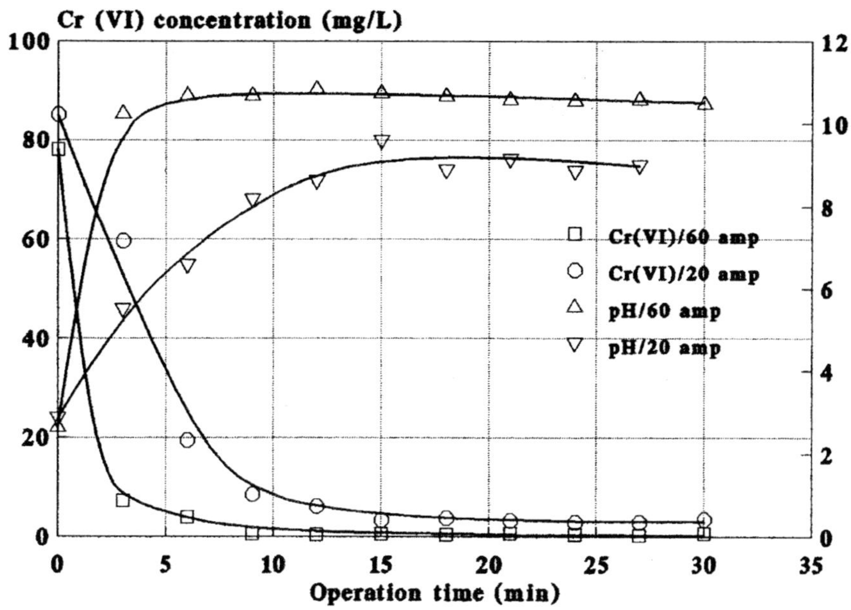


Figure 45. Effect of current on Cr(VI) concentration in effluents of the electrocoagulator

Figure 45 shows the concentration change of Cr(VI) in the solution when the cell was operated at a feed of 1.1 liters per minute and a current of 60 and 20 amps, respectively. When a solution of 100 mg/L Cr(VI) was treated with 60 amps, the concentration dropped to 5 mg/L in the first three minutes. The Cr(VI) concentration dropped to less than 1 mg/L in 15 minutes. It is to be remembered that the cell was operated continuously. Therefore, 15 minutes was the time needed to stabilize the system. The outlet, steady state, concentration was less than 1 mg/L. When the cell was operated at 20 amps, it took 15 minutes, the same as with 60 amps, to stabilize the system. However, the outlet concentration was about 3.5 mg/L. The figure also shows the pH change after the treatment. A higher current produced a higher pH in the effluent. This is also favorable when there are other metal ions, such as Cd(II), Ni(II)and Pb(II), that can precipitate out.

Another set of experiments was arranged to investigate the effect of recirculation on the system. Figure 46 shows that under 20 amps, the cell was able to treat a Cr(VI) concentration from 80 mg/L to 10 mg/L. With recirculation the outlet concentration was actually higher than that without recirculation. The system stabilized in a 15-minute operation with a outlet concentration of about 20 mg/L. When there is no recirculation, the cell can be considered as a plug flow reactor, and with a 12—liter-per-minute recirculation, the cell is a continuous stir tank reactor (CSTR). This is why the cell without recirculation performed better than the cell with recirculation. Figure 47 shows the results of the experiments designed to determine the effect of retention time on Cr(VI) removal efficiency. As expected, the lower the feed, the longer the retention time, the better the removal efficiency. With a throughput of 3.3 liters in the cell, a current of 60 amps and a voltage of 3 volts DC, the operating cost can be calculated as follows:

$$W = IV = 60(amp) \times 3(V) = 180(watt) \tag{9}$$

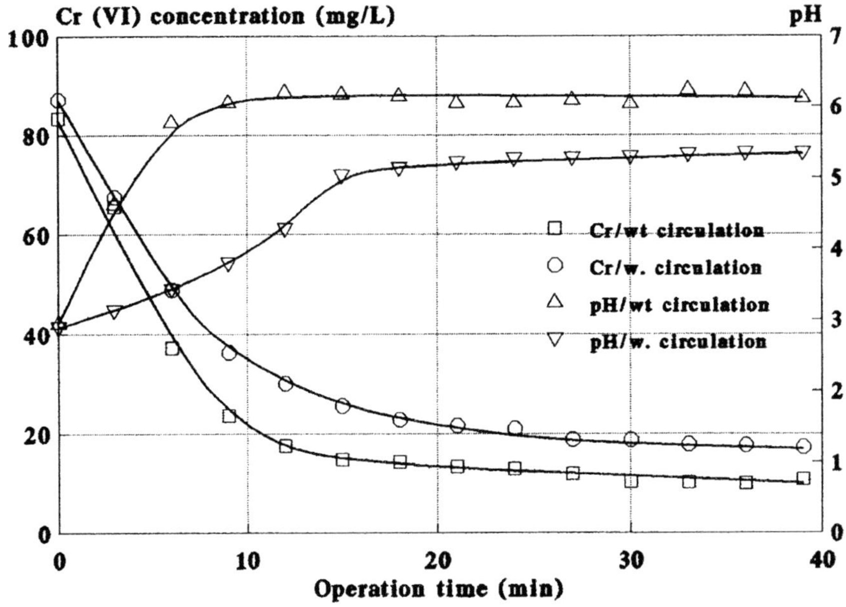


Figure 46. Effect of recirculation on Cr(VI) concentration in coagulation effluents

Converting from 120 V, 60 Hz AC current at 80% efficiency,

$$W = \frac{180(watt)}{0.8} = 225(watt) \tag{10}$$

$$1,000,000(gal) \times \frac{3.75(liter)}{1(gal)} \times \frac{225(watt)}{3.3\frac{(liter)}{(min)}} \times \frac{1(kW)}{1000(W)} \times \frac{1(hr)}{60(min)} \times \frac{\$0.1}{1(kW-hr)} = \$425 \tag{11}$$

The electrocoagulator requires \$425 in electricity at \$0.1 per kilowatt-hr to treat one million gallons of Cr(VI) contaminated wastewater. The cost is competitive with the existing technologies.

## 7.3. Electroflotation for Fe-sludge Separation

As stated in the previous section, the pilot-scale electrochemical coagulator is effective for Cr(VI) removal. The effluent from the coagulator contains a very low concentration of Cr(VI) ions. There was a significant amount of Fe-coagulant in the treated effluent. Most of the sludge are brown in color and can be separated from the solution by sedimentation. However, it was desired to develop a more efficient separator for the Fe-sludge separation. The electroflotation was confirmed to be effective in this task in a bench-scale unit. the pilot-scale electroflotator was designed and fabricated to conduct the scale-up study. In order to prepare Fe-sludge for flotation study, a Cr-containing solution was prepared and was run through the electrocoagulator. The effluent from the treatment has a transmittance of 40%, a Cr(VI) concentration of 0.8 mg/L with a pH of 7.5. A trace amount of surfactant was added to the solution to enhance the sludge separation. Experiments were conducted under the following conditions:

| | |
|---|---|
| Electroflotator: | 30 liters |
| Electrode: | anode, graphite; cathode, stainless steel mesh |
| Solution: | sludge containing effluent from electrocoagulator |
| Surfactant: | 20 mL Legsolve in 120 liters of solution |
| Initial pH: | 7.6 |
| Feed rate: | 2.2 liters per minute |
| Applied voltage: | 17 volts DC |
| Current: | 30 amps |

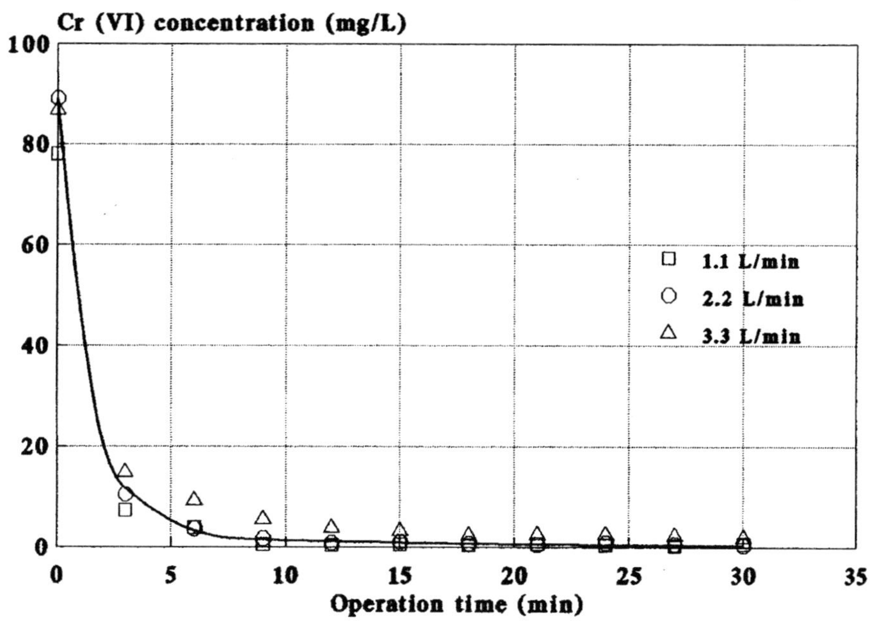


Figure 47. Effect of throughput on effluent Cr(VI) concentration

Figure 48 shows that the system stabilizes, in terms of effluent clarity and pH, in 20 minutes, A effluent of more than 95% transmittance is achievable. With a current of 30 amps and 17 volts DC in the electroflotator, the operating cost can be calculated as follows:

$$W = 30(amp) \times 17(volt) = 510(watt) \tag{12}$$

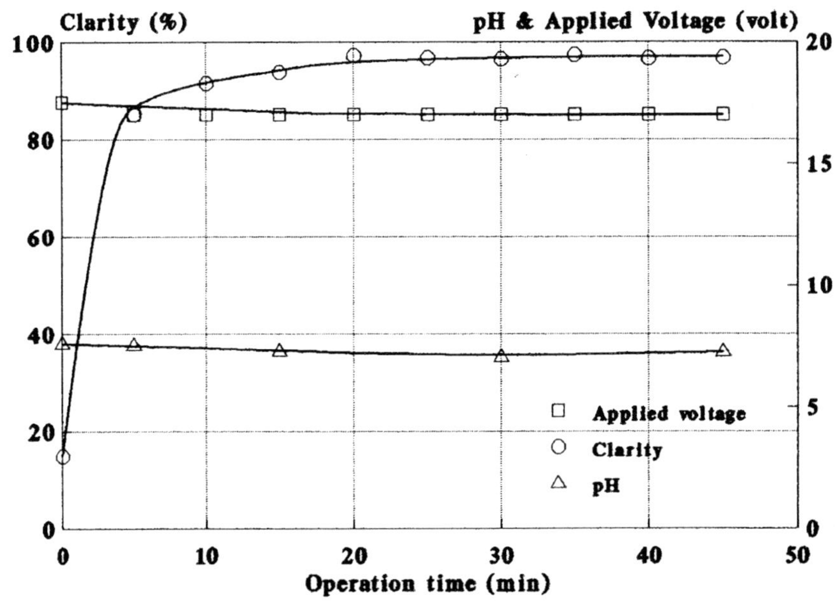


Figure 48. Stabilization of the pilot-scale electroflotation system

Converting from 120 volt, 60 Hz AC at 80% efficiency,

$$W = \frac{510(watt)}{0.8} = 637.5(watt) \tag{13}$$

$$1,000,000(gal) \times \frac{3.75(liter)}{1(gal)} \times \frac{637.5(watt)}{2.2\frac{(liter)}{(min)}} \times \frac{1(kW)}{1000(W)} \times \frac{1(hr)}{60(min)} \times \frac{\$0.1}{1(kW-hr)} = \$1,811 \tag{14}$$

The electroflotator consumes 510 watts to treat a feed of 2.2 liter. The treatment cost is estimated at $1,811 per million gallons provided the cost of electricity is 10 cents per kilowatt-hour. Unless the space is limited, the electroflotation is not able to compete with dissolved air flotation for sludge separation. Another possibility would be add an electrolyte such as NaCl to increase conductivity thus decrease the energy consumption.

An attempt to lower power consumption has been explored by reducing the power input in the system. Figure 49 shows the quality of effluent after a treatment of 10, 20 and 30 amps in the electroflotator. The operations at 10 and 20 amps represent a power input of 80 and 270 watts respectively. The lower power input can produce an effluent with only 80% clarity.

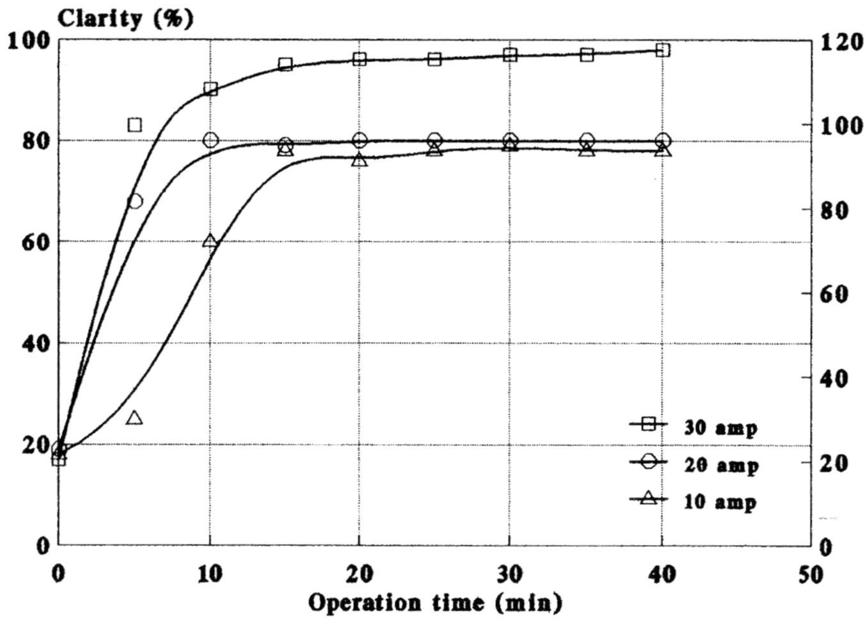


Figure 49. Effluent quality after a treatment of 10, 20, and 30 amps in the pilot-scale electroflotator

## 8. CONCLUSIONS

In an electrochemical cell, chemical reaction rates are proportional to current density. However, there is more than one reaction pathway in the system. Therefore, a higher current does not imply a shorter treatment time for a certain percentage of removal efficiency. To selectively remove metal ions from wastewater, it is favorable to operate the cell at low current density. With an electrode surface area of 602 $cm^2$, a current of 2 amps in the bench-scale cell corresponds to a current density of 3.3 $mA/cm^2$. The current density is well within the maximum operating current that is limited by the excessive oxygen evolution at the anode at the expense of iron dissolution. This usually occurs at a current density above 10 $mA/cm^2$ of electrode surface.

When operated under a constant current, power input to the cell is directly proportional to the applied voltage across the electrodes. The applied voltage in the cell depends on the desirable current, the conductivity of the solution, and the extent of anodic fouling. Generally, a voltage of 25 volt is needed to draw a current of 1 amp in the cell with a gap of 4 mm between electrodes provided the operation solution is tap water. A trace amount of NaOH in the solution reduces the applied voltage to 5 volts. A trace amount of NaCl in the tap water can further reduce the applied voltage to 3 volts to have the same result. In some instances, the applied voltage increases dramatically during the operation. This is the symptom of anodic fouling. The operation has to be stopped and corrective measures have to be taken to reduce the applied voltage to normal levels. One way of avoiding anodic fouling is by adding the NaCl into the solution before treatment. Chloride ions in the solution effectively penetrate the build-up on the anode and continuously renew the electrode surface, thus preventing the fouling.

The concept of electrochemical coagulation to treat a hexavalent Cr-containing wastewater is to precipitate the dissolved chrome as an oxide or hydroxide species by anodically generated Fe(II) and cathidically generated hydroxyl ions. The mechanism is confirmed experimentally in the project. The concentration of Cr(VI) dropped from 60 mg/L to less than 1 mg/L in a treatment time of 13 minutes. Right before the disappearance of Cr(VI), Fe(II) was detected in the solution and its concentration kept increasing until it stabilized at 20 mg/L. The Fe(II) produced in the solution in consumed by the Cr(VI). Because the reaction is fast, before Cr(VI) is depleted, there is no Fe(II) detected in the solution. Once the Cr(VI) is used up, Fe(II) begins to accumulate. Because of the existence of hydroxyl ions, Fe(II) precipitates out. The formation of sludge is strongly affected by the pH, because all the metal ions, Cr(III), Fe(III), and Fe(II), form low solubility hydroxides. The system can be optimized at a treatment time of 9 minutes where the Cr(VI) content and pH are acceptable for reuse and discharge, and Fe(II) and sludge are minimal.

During the electrochemical treatment of Cr-containing solution, the process generates significant amount of coagulant. After the treatment, the coagulant has to be removed from the solution as sludge. Most of the sludge can be easily removed by sedimentation while others form colloid system that can only be separated by more advanced separation techniques. While treating the samples, most of suspended particles can be removed by 10-micron filters. The small percentage of suspended particles that penetrate 5-micron filters can be caught only by 0.2-micron membrane filters. All of the sludge generated in the electrochemical process can be separated from the solution by electroflotation or chemical-assisted electroflotation. The amine family was found to be the most effective collector and flothor for Fe-sludge separation. A transmittance of more than 96% is achievable on all the treated effluents provided the right chemical is used to assist the electroflotation.

Based on the bench-scale experiments, an electrocoagulator was designed to treat a throughput of about 2 liters per minute or 40 times the capacity of the bench-scale unit. The electrode surface area is scaled-up from 602 to 24,080 $cm^2$, that is in direct proportion to the flowrate. Since the bench-scale experiments indicated that it is the product of current density, surface area and retention time which determines the efficiency of the unit, the pilot unit performed as predicted. Recirculation turned out to have a negative effect on the performance of the unit. The same procedure was applied to scale up the electroflotator. In addition to using the right chemicals, the design of the solution inlet significantly influences the performance of the sludge separation.

The operating cost of the chemical oxidation for industrial wastewater treatment is from \$200 to \$10,000 per million gallons. The cost for precipitation is from \$50 to \$2,000 per million gallons and that of flotation is between \$20 and \$100 per million gallons. Based on the results of pilot-scale test, the cost of the electricity consumed in the electrochemical coagulation for Cr(VI) removal is about \$425 per million gallons. Considering the applicability, simplicity and both the capital and operating costs, electrochemical coagulation shows a great potential to be put into practice, whereas the electroflotation may not be able to compete with dissolved air or other flotation processes for sludge separation.

## REFERENCES

[1] Seaman, C. J., Bertsch, P. M., and Schwallie, L., (1999). In situ Cr(VI) reduction within coarse-textured, oxide-coated soil and aquifer systems using Fe(II) solutions, *Environ. Sci. and Tech.*, 33, 938-944.

[2] Qin, G., McGuire, M. J., Blute, N. K., Seidel, C., and Fong, L., (2005). Hexavalent chromium removal by reduction with ferrous sulfate, coagulation, and filtration: A Pilot-Scale Study. *Environ. Sci. Tech.*, 39, 6321-6327.

[3] Deng, Y., Stjernstrom, M., and Banwart, S., (1996). Accumulation and remobilization of aqueous chromium (VI) at iron oxide surfaces: Application of a thin-film continuous flow-through reactor. *J. Contaminant Hydrology*. 21, 141-151.

[4] McGuire, M., Blute, N. K., Seidel, C., Qin, G., and Fong, L., (2006). Pilot-scale studies of Hexavalent chromium removal from drinking water. *J. American Water Works Association*. 98, 134-143.

[5] Urbanaki, T.S., Chmielewski, A.G., Kazmierczak, W., Ciesielski, T., Kierat, Z., Flak, M., and Szmander, H., (1993). An Electrochemical oxidation and ion exchange method for recovery of chromium. *Metal Finishing*. 55-59.

[6] Yang, C.-L., and Kravets, G., (2000). Removal of chromium from abrasive blast media by leaching and electrochemical precipitation, *J. of Air and Waste Manage. Assoc*, 50, 536-542.

[7] Rivero, M. J., Primo, O., and Ortiz, M. I., (2004). Modeling of Cr(VI) removal from polluted ground waters by ion exchange. *J. Chemical Technol Biotechnol.*79, 822-829.

[8] Rana-Madaria, P., Nagarajan, M., Rajagopal, C., and Garg, B. S., (2005). Removal of chromium from aqueous solutions by treatment with carbon aerogel electrodes using response surface methodology. *Ind. Eng. Chem. Res*. 44, 6549-6559.

[9] Farmer, J. C., Fix, D. V., Mack, G.V., Pekala, R. W. and Poco, J. F., (1996). Capacitive deionization of NaCl and $NaNO_3$ solutions with carbon aerogel electrodes. *J. Electrochem. Soc*., 143-1, 159-169.

[10] Nakano, Y., Takeshita, K., and Tsutsumi, T., (2001). Adsorption mechanism of Hexavalent Chromium by redox within condensed-tannin gel. *Wat. Res*. Vol. 35, No. 2, 496-500.

[11] Wilkin, R. T., Su, C., Ford, R. G., and Paul, C. J., (2005). Chromium-Removal processes during groundwater remediation by a zero-valent iron permeable reactive barrier. *Environ. Sci. Tech*., 39, 4599-4605.

[12] Ponder, S. M., Darab, J. G., and Mallouk, T. E., (2000). Remediation of Cr(VI) and Pb(II) aqueous solutions using supported, nanoscale zero-valent iron. *Environ. Sci. Tech.*, 34, 2564-2569.

In: Hazardous Materials and Wastewater
Editor: Allison A. Lewinsky, pp. 51-78

ISBN 1-60021-257-3


***Chapter 2***

# UTILIZATION OF AGRO-RESIDUES IN THE METAL IONS REMOVAL FROM AQUEOUS SOLUTIONS

***Silvana Fernandes Montanher, Elisângela A. de Oliveira and Maria do Carmo Rollemberg***
Department of Chemistry, State University of Maringá – UEM, Maringá, PR – Brazil

## ABSTRACT

Recently, there has been a great interest for applying natural sorbent materials in the wastewater treatment. As an example, orange bagasse (waste of orange juice industries, containing pulp, skin and seeds) is proposed for effluent remediation, focusing on the heavy metal ions species. The physical and chemical characterization of the material pointed out the presence of functional groups as well as morphological aspects, which make sorption process possible. Therefore, the sorption capacity of the biomass was measured and a number of parameters were studied to find the best conditions for the heavy metal ions sorption. Batch experiments were conducted at room temperature to investigate the sorption of Cu(II), Ni(II), Pb(II) and Zn(II) ions onto dried orange bagasse as a function of initial pH, ionic strength, initial metal ion concentration and sorbent amount. The metal ions sorption increased with the pH, and the sorption efficiencies were greatest between pH 4.0 and pH 7.0. Cu(II), Ni(II) and Zn(II) sorption were strongly influenced by the ionic strength. The removal of these metal ions decreased when the ionic strength have been changed from 0.1 mol $L^{-1}$ to 1.0 mol $L^{-1}$. Probably, this phenomenon is due to a competition process among the heavy metal ions and the electrolyte. However, the ionic strength had only a very small effect on the Pb(II) sorption process. The sorption kinetics was very fast, specially for Pb(II) and Zn(II) ions, and the equilibrium was achieved within 30 min. The sorption process followed a pseudo second-order kinetics, thus indicating a chemical sorption mechanism. This study revealed that the diffusion of the ions inside the biomass is less important for the sorption rate than the external mass transfer. The experimental data were well fitted by the

Langmuir equation ($R^2$ > 0.99). So, the sorption process of metal ions onto orange bagasse follows the Langmuir isotherm model, where the ions are independently taken up on a single type of binding site, and the uptake of the first metal ion does not affect the sorption of another ions. The sorption capacity – 66 mg Pb(II) $g^{-1}$, 18 mg Ni(II) $g^{-1}$, 17 mg Cu(II) $g^{-1}$ and 13 mg Zn(II) $g^{-1}$ – suggest the handling of high volumes of wastewater containing 1 mg $L^{-1}$ of these metal ions. Finally, the negative Gibbs free energy values for the metal ions sorption onto orange bagasse confirm the feasibility and the spontaneous nature of the sorption process for these species.

## 1. INTRODUCTION

Controlling heavy metal discharges and removing toxic heavy metals from aqueous solutions has become a challenge for the 21st century. The modern industrial activity is, to a large degree, responsible for environmental contamination, and among several different pollutants and/or toxic species reaching hazardous levels are the heavy metals, which discharge into watercourses becomes a serious pollution problem to water supply and quality. The presence of increasing concentrations of heavy metals in the water constitute a severe health hazard due to their non-degradability, toxicity, accumulation and magnification throughout the food chain, in which the humans are at the top. As a result of this contamination, some inland water bodies have been closed for fishing, ecological disasters in marine environments have been frequently reported, and fresh waters have been found contaminated due to ruthless or illegal activities [1].

Metal contamination exists in aqueous waste streams of many industries, such as metal plating, mining, refining and tanneries. Such industries can generate a low concentration of metals for which standard techniques are not well suited to reduce the heavy metal levels to acceptable regulatory standards. Therefore, the problem of coping with the presence of metals has become a top priority in water treatment, and more stringent control over the allowable limits in drinking water has been imposed. However, with the increasing population, the need for controlling heavy metal emissions into the environment is more pronounced; this is best done right at the source of such emission, before metals enter the ecosystem. However, the only answer to the environmental disasters of yesteryears is remediation and, therefore, the removal of metal ions from effluents is of great importance to many countries both environmentally and for water re-use [2].

Although several technologies – such as chemical precipitation, filtration, reverse osmosis, oxidation-reduction, ion exchange, electrochemical methods based on electrolysis or membrane separation – have proved to be adequate to remove the heavy metal contamination [3-6], they have been found to be limited, since they often involve high capital and operational costs and may also be associated with the generation of secondary wastes, which present treatment problems. The selection of an effluent treatment method has been largely based on the concentration of waste and the cost of whole treatment, and those technologies cited above have been often inadequate economically and/or technologically. It is not within the framework of this text to discuss these conventional techniques, but it would suffice to say that as the emission standards tighten, they are becoming progressively more inadequate. On the other hand, better technologies are more costly and often not feasible [7]. Therefore, in the last decades it becomes clear that cost-effective alternative treatment processes are desirable, thus resulting in the search for new techniques capable of inexpensive treatment of wastewaters polluted with metals. Besides, with the increase in environmental awareness and governmental policies there has been an emphasis on the development of new environmentally friendly ways to decontaminate waters using low-cost methods and materials.

One emerging technology that is receiving more and more attention is the use of sorption process involving low-cost natural sorbents. The use of sorption process has been proposed as an interesting alternative, not only for its ability to remove metals from aqueous solutions, but also for the possibility of recovering metals. The accelerating depletion of natural mineral sources led to the need for recycling metals – and this is also interesting from an economic point of view. Also, sorption process has been considered an economical and efficient method for metal ions removal from wastewater [8]. The most generally used solid sorbent is activated carbon, which is a very efficient sorbent in many different applications [4]. However, activated carbon is expensive, and for effluents containing metal ions activated carbon requires chelating agents to enhance its performance, thus increasing treatment cost.

In this context, natural materials that are available in large amounts, or certain waste products from industrial and/or agricultural activities, may have potential as interesting and inexpensive sorbents. A *low-cost* sorbent requires little processing, is abundant in nature, or is a by-product or a waste material from other industries [9].

In the 1990s, a new scientific area appeared to help in the heavy metal recovery and water decontamination: *biosorption*. The first reports on biosorption described how abundant materials could be used to remove, at very low cost, even small amounts of heavy metals from industrial effluents. Once the metals can be removed from solutions only when they are appropriately immobilized, the procedure of metal removal form aqueous solutions often leads to effectively concentrating the metal, and this aspect of biosorption makes the eventual metal recovery easier and economical [10].

This contribution reports an outline of sorption theory and mechanisms associated with sorption process, and points out some parameters that should be taken into account when wastewater remediation is carried out – such as physical and chemical characterization, and sorption isotherms.

As an alternative natural sorbent, the potential of orange bagasse for heavy metal removal from aqueous solutions is demonstrated. This biomass can be useful in the effluent treatment due to its physical and chemical characteristics, negligible cost and abundance. In order to know the morphological characteristics and chemical groups able of heavy metal removal, spectroscopic analyses were carried out using different techniques such as $^{13}C$ NMR (nuclear magnetic resonance), FT-IR (Fourier transform infrared) and SEM (scanning electron microscopy). The parameters of the sorption process, such as pH of the sample, contact time, and amount of biomass were also evaluated to get the best conditions for building sorption isotherms. The maximum sorption capacity was obtained from the Langmuir isotherm model fitting.

## 2. Natural Sorbents for Wastewater Treatment

Several works have already been done with a wide variety of sorbents aiming the metal ions removal. In these works, the sorbents used include tannin-rich materials, lignin-rich materials, chiti/chitosan, dead biomass, algae/seaweed, zeolites, clays, moss, wool, cotton and a number of other natural materials or by-products of the agricultural industry. Bailey [9] review concluded that these materials could be an alternative for the most costly process, which involves the use of activated carbon and ion exchangers.

*Biosorption* is the use of *biosorbents* – that is, sorbents of biological origin – to remove metals from wastewaters. *Biosorption* can be mainly used for treatment of wastewaters with low or medium initial heavy metal concentrations. Another advantages of biosorption are that it avoids the generation of toxic sludge and can be used under a broad range of operating conditions (pH, temperature, metal concentration, presence of other ions in the solution, etc…). *Biosorbents* are a very special kind of natural sorbents because, unlike ion exchange resins, which contain single functionalities, the biosorbents may contain several binding organic groups. Thus, biosorbents may have higher sorptive capacity for metals than commercial synthetic sorbents. Besides, biosorbents are less subject to interferences from alkali and/or alkali-earth metals that are resins due to the diversity of binding sites [11].

Compared with conventional techniques for metals removal, the biosorption offers advantages. Volesky [12] pointed out some advantages of biosorbents:

- Cost effectiveness: since they are often made from abundant or waste material, the cost for biosorbents is low – around $10/kg. This is the main attraction of metal sorption, and it should be kept that way. Not only should biosorbent be used directly, but sorbents derived from it in a simple process should be most low-priced for economical metal removal applications.
- Selectivity: the sorption efficiency of different types of biosorbents can be more or less selective on different metals; this dependence is due to a number of factors, such as type of biomass, type of biomass preparation, physico-chemical environment, contact with the solution
- Regeneration: biosorbents can be reused after the metal is recycled
- Without secondary wastes (as is the case with other techniques, *e.g.*, precipitation)
- Metal recovery
- Competitiveness: biosorbents present a comparable efficiency to the most similar technique – ion exchange treatment, which is rather costly.

Biosorbents are mainly constituted by macromolecules such as humic and fulvic substances, lignin, cellulose, hemicellulose and/or proteins, which present sorptive groups like carbonyl, carboxyl, amine and hydroxyl. Therefore, metal sorption by complexation or ion exchange mechanisms is possible [13].

The use of sorbents of biological origin for removing heavy metals from dilute aqueous solutions has long been recognized [14], and several biomaterials have shown sufficient heavy metal binding capacity to be considered for the use in full-scale biosorption process. However, a successful biosorption technology not only depends on the metal-sorbing potential of the biomass, but also on continuous supply of the biomass for the process.

Therefore, for this purpose, the biomass should either be an industrial waste or available in plenty in nature [15].

Alginate has been one of the most extensively investigated biomass for binding heavy metals from aqueous solutions [16,17]. Also chitosan has been widely studied. Chitosan is prepared from chitin – a natural polymer extracted from crustacean shells – by deacetylating its acetamide groups with a strong alkali solution. Among many new materials, the use agricultural and forest waste by-products is very prominent, as these biomaterials (vegetal materials) could be assumed as low-cost since they require little processing and are abundant in nature. Therefore, a wide variety of agricultural waste-based sorbents have been evaluated, and each one has its drawbacks and advantages [18-23].

Several agricultural wastes (lignocellulosic materials) have been considered a significant waste disposable problem by agro-industry. These lignocellulosic materials include a wide variety of agricultural wastes such as peat, wood, barks, sawdust, wool, fruit residues (banana pith, fruit seeds, fruit peels), husks (soybean, peanut, rice), shells (peanuts, hazelnuts, almonds), hulls (soybean, cottonseed, rice), bran (rice, soybean, maize, wheat), seeds (cottonseed), and others [1, 24-30]. Due to its chemical composition, mainly lignin, cellulose and hemicellulose, metal sorption studies with lignocellulosic materials have shown that both complexation and ion exchange mechanisms are involved in the metal sorption process. These materials can be used in their natural form or chemically modified to improve the sorption performance, and different chemical modifiers have been studied, such as phosphoric and citric acids and alkaline solutions [31-35]

It is unquestionable that the wastewater treatment with an agricultural-based biosorbent is an attractive option with a double benefit: the reduction of residues amounts whose disposal becomes a costly problem, and the conversion of the wastes into useful inexpensive sorbents for water cleaning. The choice for an sorbent for metal ions removal from water must be guided by at least two criteria: sorption must be thermodynamically possible and kinetics limitations must not prevent it even with a very low contact time. However, the complexity of the natural sorbents has made them unpredictable regarding the impact of existing conditions on binding mechanisms, capacity and affinity. Therefore, the efficient application of such materials for heavy metal remediation is greatly dependent on the understanding of the basic chemistry involved in the metal binding process [33].

Whereas the use of synthetic ion exchangers can be considered a mature technology, biosorption is in its developmental stages, and further improvement in both performance and costs can be expected. Figure 1 schematically summarizes process pathways to produce biosorbent materials that are effective and durable in repeated long-term applications aimed at removing metals from large volumes of effluents. Obviously, the biosorbent preparation is very important, but it is based on trial-and-error routines, but challenging because of the many different raw materials [36].

Numerous chemical groups have been suggested to contribute to metal biosorption. These groups comprise hydroxyl, carbonyl, carboxyl, sulfhydryl, thioether, sulfonate, amine, imine, amide, imidazole, phosphonate, and phosphodiester groups. The importance of any given group for meal biosorption by a certain biomass depends on factors such as the number of binding sites in the sorbent material, the accessibility of the sites, the availability of the sites (that is, its chemical state) and the affinity between site and metal (that is, the binding strength). For covalent metal binding, even an already occupied site is theoretically available; the extent to which the site can be used by a metal depends on its binding strength and concentration as compared to the metal already occupying the site. For electrostatic metal binding, a site is only available if the metal is ionised [1].

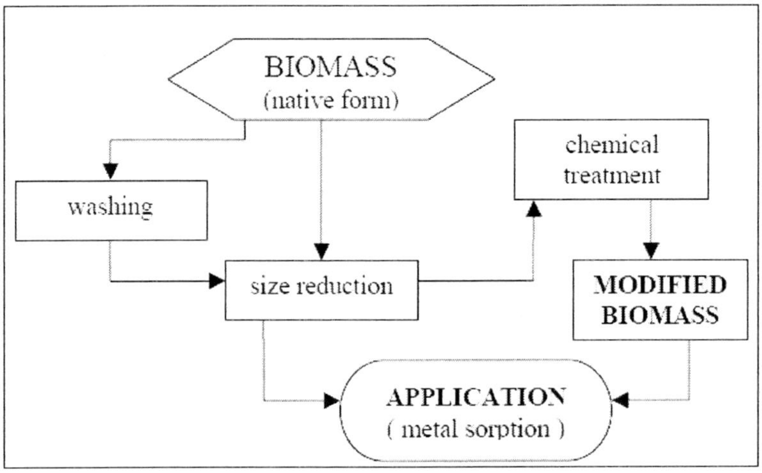


Figure 1: Schedule for the biosorbent preparation

## 3. Sorption Process

Sorption process comprises the selective transfer of solutes from the bulk solution onto the surface of the sorbent. The sorbing species is called *sorbate*, and the solid media to which the sorbate is attracted is known as the *sorbent*.

The underlying principle behind this attraction results from some form of bonding between the sorbate and the sorption sites on the sorbent. The extension of the sorption depends on particular characteristics of the sorbate and the sorbent.

Sorption reactions generally occur over a short period of time. However, if the sorbed species begin to be incorporated into the sorbent structure, a slow reaction, known as *absorption*, begins to take place [8]. The difference between adsorption and absorption is that adsorption is the attraction between the outer surface of the sorbent particle and the sorbate species, whereas absorption is the retention of the sorbate into the structure of the solid sorbent (Figure 2).

Due to the multiples mechanism possibilities involved in the sorption process, mass transfer rates are difficult to treat, and the nature of the sorption process is also difficult to predict. Various models are available to characterize the nature, the kinetics and the equilibrium of the sorption process. Most of these models works based on the postulation of sorbents as homogeneous spheres; however, this hypothesis is not completely true [18].

How is the metal bound within the biosorbent? This question is fundamental for understanding the biosorption process. The complexity of natural sorbents implies that there are many ways for metal sorption. When the metal ion plays a role in the metabolism of the biomass, the transport across cell membranes is the most important mechanism. This is an *active interaction*, which is capable of affect the metabolism of living microorganisms such as bacteria, algae and others. The *passive interaction* does no affect the metabolism of living organisms and involves several kinds of sorption mechanisms for all natural sorbents [2].

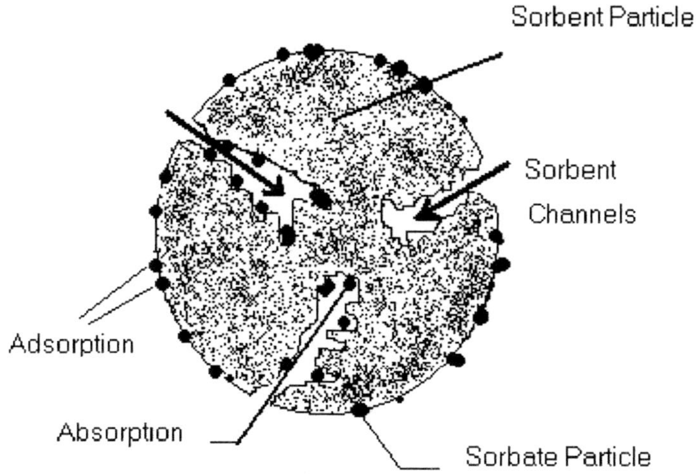


Figure 2: Adsorption and absorption process

A number of different metal binding mechanisms has been thought to be present in biosorption [38]:

- Chemisorption, by ion exchange, complexation, coordination, chelation
- Physical adsorption
- Microprecipitation

Redox reactions taking place in the biosorbent are also possible. Some of these mechanisms can act simultaneously to different degrees, depending on the biosorbent and the solution environment.

## 4. Sorption Isotherms

The sorption isotherm represents equilibrium between the concentration of a solute in solution and its concentration in the sorbent, at a given temperature [28]. The most appropriate method for assessing the sorbent capacity is the derivation of the whole sorption isotherm. This is very important because the isotherm can provide information on the sorption strength and the maximum sorption capacity of a given biosorbent. In addition, the knowledge of sorption isotherms is useful to evaluate to what extent a sorption system ca be improved, as well as helps us to foresee the conditions to work in open reactors and estimate the operating conditions for a more effective system. In general, sorption isotherms can exhibit different characteristic shapes, depending on the sorption mechanism [39].

The method to obtain the different points of the sorption isotherm consists of mixing different known amounts of the sorbent with a given volume of solution containing the sorbate at a known concentration. All the initial conditions – pH, concentration of the solution

– remain identical, and only the amount of the biosorbent added in each experiment varies. The mixtures are shaken uniformly to reach equilibrium (so, it is necessary a prior knowledge of the time needed to reach equilibrium, to be sure the concentration being measured is the equilibrium concentration). Once equilibrium has been reached, the sorption capacity can be obtained by using a mass equilibrium equation, and is expressed as [25]:

$$q_e = \frac{(C_o - C_e)\,V}{m} \tag{1}$$

$C_o$ (mg $L^{-1}$) and $C_e$ (mg $g^{-1}$) are the initial and equilibrium concentration, respectively; **V** is the volume of the solution (L) and **m** is the sorbent amount (g).

The sorption isotherm is constructed from the sorbed ($q_e$) and equilibrium ($C_e$) concentration values. An isotherm is said to be favorable if the sorption capacity increases rapidly with the concentration in equilibrium in the solution (convex form). Unfavorable isotherms have low sorption capacity at low concentrations in equilibrium (concave form). Figure 3 shows an example of an isotherm plot. The plateau region means saturation of the actives sites in the sorbent, and from this region the maximum sorption capacity of the sorbent can be determined by using mathematical models [40].

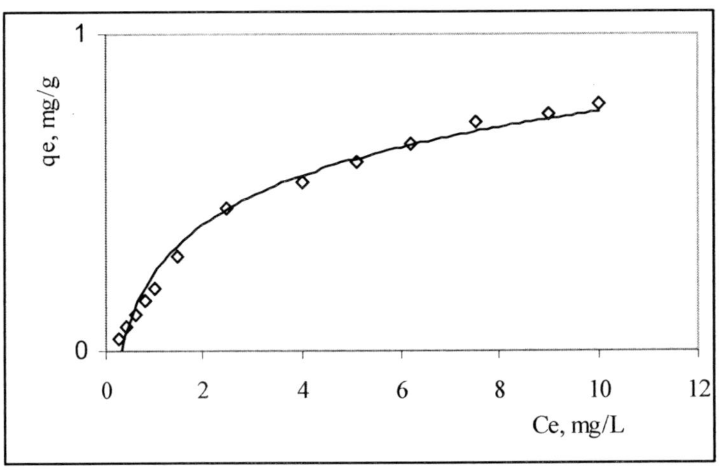


Figure 3: Sorption isotherm plot of the equilibrium metal retention in the solid sorbent *versus* the equilibrium metal concentration in the liquid phase

Sorption mathematical modeling is a very powerful tool for different tasks, and it is very useful for understanding the sorption process and essential for process design and optimization. Several equilibrium-based models have been used to describe the metal transfer between the solution and the solid phase – that is, the sorption process. Two, three and even four parameters isotherm models are available for modeling sorption data [41]. Two models have been most frequently used to describe the sorption isotherm - the Langmuir and the Freundlich models (two parameters). Simplicity, easy interpretation, physical meaning and easy linearization have been pointed out as the major reasons for the extensive use of these models. Moreover, there is no significant reason to use more complex models if a two parameters model can fit the data.

The Langmuir equation was developed in the 1990s to describe the sorption of gas molecules on planar surfaces [42]. This model was later extended to liquid systems, where species in solution are involved. The Langmuir model suggests that sorption occurs on homogenous surfaces by monolayer sorption (that is, sorption onto a surface with a finite number of identical sites), without interaction between sorbed molecules. The equation (2) describes this model.

$$q_e = \frac{q_m\, b\, C_e}{1 + b\, C_e} \quad (2)$$

where **b** (L mg$^{-1}$) is a constant related to the affinity of the binding sites, $\mathbf{q_e}$ (mg g$^{-1}$) and $\mathbf{C_e}$ (mg L$^{-1}$) are the sorbed metal ions on the biomass and the metal ions concentration in the solution at equilibrium, respectively. $\mathbf{q_m}$ (mg g$^{-1}$) is the maximum sorption capacity for a monolayer coverage – that is, the maximum amount of the metal ion per unit weight of biomass needed to form a complete monolayer on the surface. $\mathbf{q_m}$ represents a practical limiting sorption capacity when the surface is fully covered with metal ions and assists in the comparison of sorption performances. The sorption parameters ($\mathbf{q_m}$, **b**) can be determined from the linearised form by plotting $\mathbf{C_e / q_e}$ *vs* $\mathbf{C_e}$, according to equation (3) [43]:

$$\frac{C_e}{q_e} = \frac{1}{q_m\, b} + \frac{C_e}{b} \quad (3)$$

Probably the Langmuir isotherm model is the most widely applied model because it allows obtaining the maximum sorption capacity ($\mathbf{q_m}$) for given sorbent-sorbate system, at constant temperature. However, another characteristic of the particular sorption process can be obtained from the Langmuir isotherm equation: the *equilibrium parameter* or the *separation factor* ($\mathbf{R_L}$) [4]. This constant predicts the sorption efficiency of the process, and is given by equation (4):

$$R_L = \frac{1}{1 + b\, C_o} \quad (4)$$

In this expression, $\mathbf{C_o}$ is the initial concentration of the sorbate solution (mg L$^{-1}$) and **b** is the Langmuir isotherm parameter. Favorable sorptions are represented by $\mathbf{R_L} < 1$.

Freundlich presented the earliest known sorption isotherm equation in the 1900s. The Freundlich model proposes a monolayer sorption with a heterogeneous energetic distribution of active sites, and/or interactions between sorbed species – that is, multilayers sorption. This model is expressed by the equation below [44]:

$$q_e = k_F\, C_e^{1/n} \quad (5)$$

$\mathbf{k_F}$ (mg$^{1-1/n}$ g$^{-1}$ L$^{1/n}$) and **n** (dimensionless) represent the Freundlich constants characteristic on the system. $\mathbf{k_F}$ is an indicative of sorption capacity and **n** is related to sorption intensity. These parameters are empirical constants, and they depend on several factors. **n** is greater than one; if **1/n** approaches 1, the equation becomes linear. The linearised Freundlich expression is presented as below:

$$\log q_e = \log k_F + (1/n) \log C_e \qquad (6)$$

If **1/n** value is within 0.1 and 1.0 there is a favorable sorption of the sorbate on the given sorbent.

The Langmuir model assumes a sorption at specific homogeneous sites within the sorbent, and the Freundlich model deals with physicochemical sorption on heterogeneous surfaces. This heterogeneous system, which is characterized by the heterogeneity factor **1/n**, is not restricted to a monolayer formation. The $\mathbf{q_m}$ value in the Langmuir expression represents the monolayer sorption capacity, but the Freundlich constant $\mathbf{k_F}$ represents a magnitude relative to multiplayer sorption capacity. Therefore, several species may presenting interaction with solid sorbent by both monolayer and multiplayer formation.

## 5. Sorption Kinetics

Sorption kinetics shows large dependence on the physical and chemical characteristics of the biosorbent, which also influence the sorption mechanism. The knowledge of the kinetics of a sorption process is required for selecting optimum operating conditions for full-scale batch process, and gives useful information for designing batch biosorption systems.

The solute transfer process is assumed to proceed through a succession of steps, which are represented in the Figure 4:

(i) solute transfer from the bulk solution to the boundary film
(ii) solute transport from the boundary film to the surface of the sorbent (external diffusion)
(iii) transfer of the solute from the surface of the sorbent to the intraparticle active sites (intraparticle diffusion)
(iv) retention of the solute species on the active sites in the biosorbent.

The first and last steps – solute transfer to the boundary film and retention on the active sites – are thought non limiting as the stirring is sufficient to avoid a concentration gradient in the solution, and the sorption is seen as a *quasi* instantaneous mechanism. Therefore, external mass transfer resistance and intraparticle mass transfer resistance are the major controlling steps [45].

The external mass-transfer rate for single-species sorption on spherical particles may be expressed by:

$$\frac{\partial q}{\partial t} = \beta A (C - C_s) \qquad (7)$$

**q** is the solute concentration in the solid biosorbent (g kg$^{-1}$), **C** and $\mathbf{C_s}$ are the solute concentrations in the solution and on the particle surface (g m$^{-3}$), **A** is the surface area for the unit mass of biosorbent ($m^2$ $kg^{-1}$) and **β** is the external mass transfer coefficient (m $s^{-1}$). The value of **β** depends on the transport conditions around the particle [46].

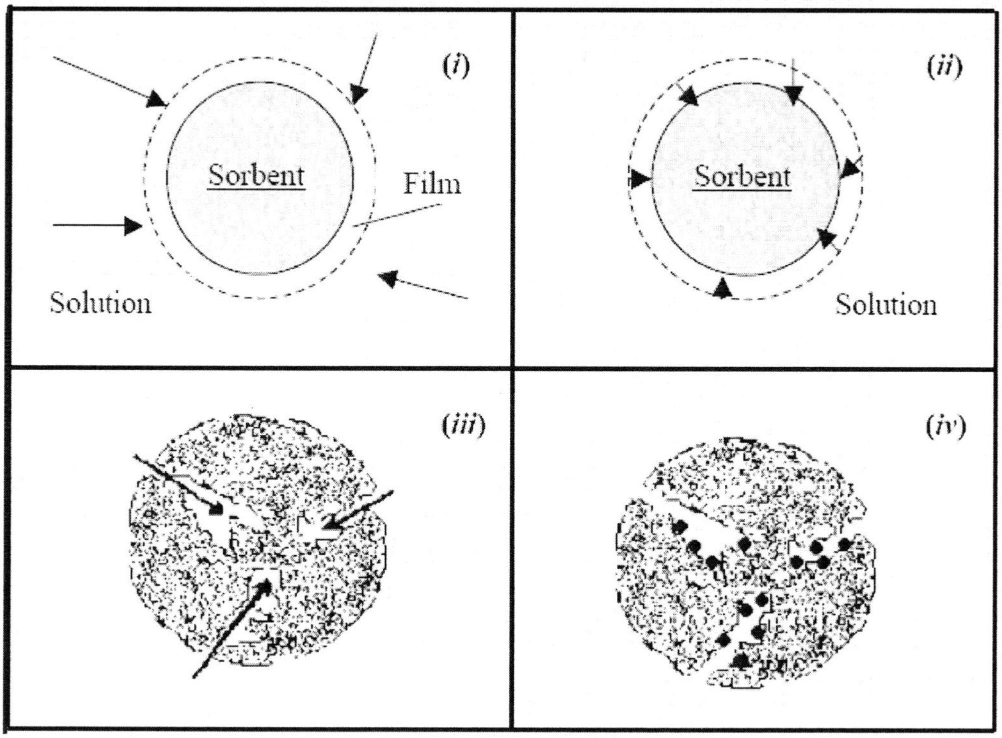


Figure 4: Mechanisms steps

The diffusion of a solute through the particle has been assumed to follow Fick's law. Obviously, in a rigorous quantitative treatment of complex systems, such as sorption systems, mathematical difficulties are often found. Therefore, simplified models have been widely used to assists in the sorption process understanding.

Some single models have been proposed to determine the effect of external mass transfer in sorption systems. Findon *et al.* proposed the boundary model, which assumes that the surface concentration of the metal ions (solute) is very low, that is, is negligible at $t$=0; as a consequence, intraparticle diffusion is also negligible [47]. The change in metal concentration with respect to time is related to the liquid-solid film mass transfer coefficient $\beta$ by the equation (8):

$$\frac{dC}{dt} = -\beta S (C - C_S) \quad (8)$$

**C** and $\mathbf{C_S}$ are the solute concentration in the solution at time **t** (g $m^{-3}$) and on the surface sorbent (g $m^{-3}$), respectively; $\beta$ is the mass transfer coefficient (m $s^{-1}$); and **S** is the specific outer surface of sorbent particles per unit of particle volume ($m^{-1}$) [45].

As t=0, $\mathbf{C_S} \rightarrow 0$ and **C** = **Co**, and equation (8) becomes:

$$[dC / dt]_{t=0} = -\beta S C_o$$

Than, when **t** > 0, $\mathbf{C_S} > 0$ and $\mathbf{C} < \mathbf{C_o}$ and equation (8) becomes:

$$\frac{d(C/C_o)}{t} = -\beta S \qquad (9)$$

Therefore, plotting $\mathbf{C / C_o}$ *vs* time, $\boldsymbol{\beta}$ can be determined from the slope as $\mathbf{t} \rightarrow 0$.

Intraparticle diffusion is characterized by the relationship between specific sorption (**q**) and the square root of time, according to the expression (10):

$$q_t = (K_i / m)\, t^{1/2} \qquad (10)$$

In the equation (10), $\mathbf{q_t}$ is the amount of solute sorbed at time **t** (mg $g^{-1}$), **m** is the mass of sorbent (g) and $\boldsymbol{K_i}$ is the initial intraparticle diffusion rate [18,48].

Urano and Tachikawa [49] proposed an intraparticle diffusion coefficient **D** based on the concentration in solids, assuming that the sorption rate is independent of the stirring speed, and external mass transfer is non-limiting. The intraparticle diffusion coefficient **D** can be determined by plotting the function log $[1 - (q / q_e)^2]$ *vs* time.

The isotherms described above take the assumption that sorption is constantly occurring at equilibrium, but this is not valid at all times with all systems. Therefore, kinetics must be used to calculate how fast the solute species are being sorbed on the solid sorbent. In general, non-equilibrium sorption occurs when sorption is slow. Rate constants determine how fast a given concentration of solute is sorbed onto a sorbent particle, but because of the difficulty in calculating these values, the equilibrium assumption is usually accepted [50].

The chemical kinetics describes reaction pathways along the time to reach the equilibrium, whereas chemical equilibrium data give no information about pathways and reaction rates. Although there are many factors that will influence sorption process, such as contact time, pH, temperature, nature and concentration of the solute, and sorbent concentration, a kinetic model is concerned only with the effect off observable parameters on the overall rate. These include initial nature and concentration of the solute, temperature, and amount of the solid biosorbent. In order to investigate the mechanism of sorption, a number of models have been applied to different metal ions-biosorbent systems [51].

The *first-order reversible kinetic model* is based on reversible reactions and equilibrium establishment between the solid and the aqueous phases [28, 52]. According the schematic representation of a metal ion **M** sorption process as given below, the first- order rate constant can be obtained as follows:

$$M_{aq} \underset{k_2}{\overset{k_1}{\leftrightarrows}} M_{sb}$$ (*aq* and *sb* are, respectively, the aqueous and the sorbent phase)

$$dC_{sb} / dt = - dC_{aq} / dt = k_1 C_{aq} - k_2 C_{sb} \qquad (11)$$

where $\mathbf{C_{aq}}$ (mg $L^{-1}$) and $\mathbf{C_{sb}}$ (mg $g^{-1}$) are the solute concentrations in the aqueous phase and on the sorbent, respectively, at any time, and $\boldsymbol{k_1}$ and $\boldsymbol{k_2}$ are the first-order rate constants. If $\mathbf{C_{aq}^{o}}$

and $\mathbf{C_{sb}^{o}}$ represent the initial metal ion concentration in the aqueous phase and in the sorbent, $\mathbf{C_{sb}}$ can be obtained from the mass balance expression:

$$C_{sb} = C_{sb}^{o} + C_{aq}^{o} - C_{aq} \tag{12}$$

And substituting equation (12) in equation (11), the rate expression becomes:

$$dC_{sb} / dt = - dC_{aq} / dt = (k_1 + k_2)\, C_{aq} - k_2\, (C_{aq}^{o} + C_{sb}^{o}) \tag{14}$$

At equilibrium, the forward and the reverse reaction rates are equal:

$$k_1\, C_{aq,e} = k_2\, C_{sb,e} \quad \text{and} \quad K = k_1 / k_2 = C_{sb,e} / C_{aq,e} \tag{15}$$

$\mathbf{C_{aq,e}}$ and $\mathbf{C_{sb,e}}$ represents the concentration of the solute, at equilibrium, in the aqueous phase and in the solid phase, and $\boldsymbol{K}$ is the equilibrium constant.

Rearranging equation (14), the reaction rate expression becomes:

$$dC_{sb} / dt = - dC_{aq} / dt = k_1\, \{\, [\, 1 + (k_2 / k_1)\,]\, C_{aq} - (k_2 / k_1)\, (C_{aq}^{o} + C_{sb}^{o})\, \} \tag{16}$$

Substituting $\boldsymbol{K}$ from equation (15) in equation (16),

$$- dC_{aq} / dt = k_1\, \{\, [\, (K + 1)\, K^{-1}\,]\, C_{aq} - K^{-1}(\, C_{aq}^{o} + C_{sb}^{o}\, )\} \tag{17}$$

From equations (12) and (15), the equilibrium constant $\boldsymbol{K}$ can be expressed as:

$$(K + 1)\, C_{aq,\,e} = C_{sb}^{o} + C_{aq}^{o} \tag{18}$$

Thus, equation (17) can be rewritten:

$$- dC_{aq} / dt = k_1\, (K + 1)\, K^{-1}\, (C_{aq} - C_{aq,e}) \tag{19}$$

and

$$\frac{dC_{aq}}{C_{aq} - C_{aq,\,e}} = K^{*}\, dt \tag{20}$$

The final expression can be obtained integrating equation (20) with respect to the limits $\mathbf{C_{aq} = C_{aq}^{o}}$ and $\mathbf{C_{aq} = C_{aq}}$, and $\mathbf{t} = 0$ and $\mathbf{t = t}$, respectively:

$$- \mathrm{Ln} \left[ \frac{C_{aq} - C_{aq,e}}{C_{aq}^{o} - C_{aq,e}} \right] = K^{*}\, t \tag{21}$$

$\boldsymbol{K^{*}}$ can be obtained from the slope of the plot $- \mathrm{Ln}\, [\, (C_{aq} - C_{aq,e}) / (C_{aq}^{o} - C_{aq,e})\, ]$ *vs* time. With this $\boldsymbol{K^{*}}$ value and the equilibrium rate constant $\boldsymbol{K}$ (equation (15)), the rate constants $\boldsymbol{k_1}$ and $\boldsymbol{k_2}$ can be also obtained from:

$$K^{*}\, K = k_1\, (K + 1) \qquad \text{and} \qquad k_2 = k_1 / K \tag{22}$$

The *pseudo-first-order model* is the first equation for a solid/liquid sorption process; the sorption data follows the Lagergren expression [53]:

$$\frac{dq}{dt} = K_P\,(q_e - q) \tag{23}$$

**q** and $\mathbf{q_e}$ are, respectively, the amount of solute on the sorbent at time **t** and at equilibrium (mg $g^{-1}$). $\boldsymbol{K_P}$ ($min^{-1}$) is the equilibrium rate constant for the pseudo-first-order sorption. Equation (23) can be integrated for the boundary conditions **t** = 0, $\mathbf{q_t}$ = 0 and **t** = **t**, $\mathbf{q_t} = \mathbf{q_t}$ :

$$\text{Log}\,(q_e - q) = \text{Log}\,q_e - (K_P / 2.303)\,t \tag{24}$$

From the slope of the plot Log ($q_e$ – q) *versus* time, $\boldsymbol{K_P}$ value can be obtained.

The *pseudo-second-order model* proposed by Ho and Mc Kay [54] is based on the hypothesis that biosorption follows the second order chemisorption. This kinetic model can be expressed as:

$$\frac{dq}{dt} = K_S\,(q_e - q)^2 \tag{25}$$

or

$$\frac{dq}{(q_e - q)^2} = K_S\,dt \tag{26}$$

Integrating equation (26) for the boundary conditions **t** = 0, **q** = 0 and **t** = **t**, **q** = **q**:

$$\frac{t}{q} = \frac{1}{K_S\,q_e^2} + \frac{1}{q_e} \tag{27}$$

Thus, plotting **t / q** *vs* **t** allows the determination of $\boldsymbol{K_S}$ (g $mg^{-1}$ $min^{-1}$) and $\mathbf{q_e}$. The initial sorption rate, **h** (mg $g^{-1}$ $min^{-1}$), can be determined from $\boldsymbol{K_S}$ value:

$$h = K_S\,q_e^2 \tag{28}$$

## 6. Sorption of Heavy Metal Ions onto Orange Bagasse: Kinetic, Thermodynamic and Equilibrium Data

In order to exemplify the potentialities of the natural sorbents for the metal ions removal from aqueous solutions – as in the effluent remediation – the sorption capacity of *orange bagasse* is shown.

Orange bagasse constitutes roughly 49% of the orange fruit and is a waste of orange juice industries. This waste contains pulp, skin, kusk and seeds of the fruit. In Brazil, which is the major orange producer in the world, several million tons of orange bagasse are generated annually. This causes a disposal problem, and the bagasse re-utilization is advantageous. Therefore, finding ways to use this agricultural by-product profitably will benefit orange juice industries, and seems to be an interesting alternative for its disposal.

Orange bagasse is obtained after pressing out the juice and removing the essential oils from the fruit. The bagasse can retain some of the oil after extraction, and this oil deteriorates by bacteria action, forming humic substances. The presence of these humic species increases the potential use of the orange bagasse as biosorbent for metal ions, since the humic substances present high sorption capacity for metals due to their negatively charged functional groups (such as phenolic, carboxylic and alcoholic hydroxyls).

Thus, considering all the orange bagasse characteristics, its sorption proprieties will be discussed in this text. As an application, the sorption of heavy metal ions – Cu(II), Ni(II), Pb(II) and Zn(II) - onto orange bagasse was evaluated. The sorption experiments were carried out in batch systems, and metal ions sorption was studied with respect to pH, ionic strength, initial metal ion concentration, biosorbent amount and time. The sorption equilibrium data were modelled by using the Langmuir and Freundlich isotherm models. Kinetics and thermodynamic parameters were also determined. Before initiating sorption studies, orange bagasse was first characterized by using spectroscopic techniques in order to know the main chemical groups (functional groups) present to bind the metal ions.

## 6.1. Characterization of Orange Bagasse – Techniques Used

In order to investigate the functional groups present in the orange bagasse-based biosorbent, and probably involved in the sorption process, a Bromem FT-IR 100 infrared spectrometer was used; the experiments were done using KBr discs to prepare the orange bagasse samples, and the spectral range varied from 400 to 4000 $cm^{-1}$. The thermal behaviour of the biosorbent was investigated by using a Shimadzu TGA 50 thermogravimetric analyser; *ca.* 10 mg of orange bagasse were heated up to 1000 °C at 10 °C $min^{-1}$, in an inert atmosphere (20 mL $N_2$ $min^{-1}$). The B.E.T. specific surface area of the biosorbent was determined by using a Quantachrome Nova 1200 instrument.

## 6.2. Orange Bagasse Characterization

FT-IR spectra allowed the identification of some of the functional groups present in the biomass (Figure 5) The broad band found in the region of 3400 $cm^{-1}$ indicated the presence of hydroxyl groups on the orange bagasse surface and sorbed water on the surface. The weak band at 2919 $cm^{-1}$ is due to asymmetric vibration of the $CH_2$ group and to stretching of the OH groups bound to methyl radicals. These groups have been usually associated to ligninic structures [55]. The bands located at 1748 and 1650 $cm^{-1}$ are characteristic of carbonyl group stretching from carboxylic acids and amides (probably from proteins). The band between 1100 and 1015 $cm^{-1}$ can be attributed to vibrations of the groups C-O-C from polysaccharides and C-O from phenol and alcohol. Therefore, the IR-spectrum of the orange bagasse biomass indicated the existence of groups capable to remove heavy metal ions from aqueous solutions; these functional groups are strongly affected by the pH of the solution.

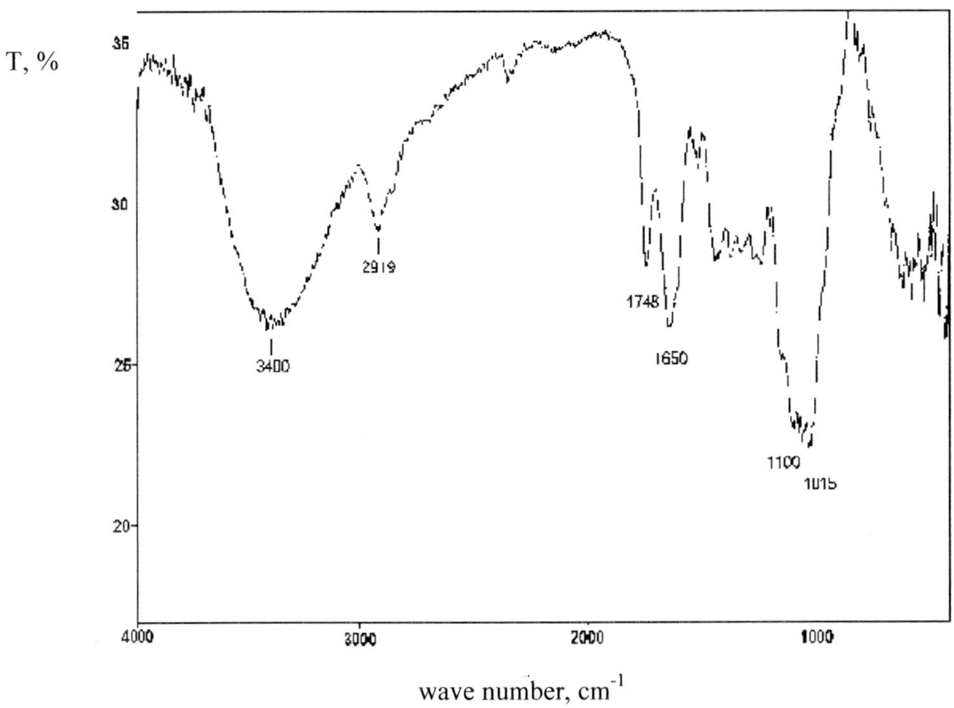


Figure 5: FT-IR spectrum of dried orange bagasse

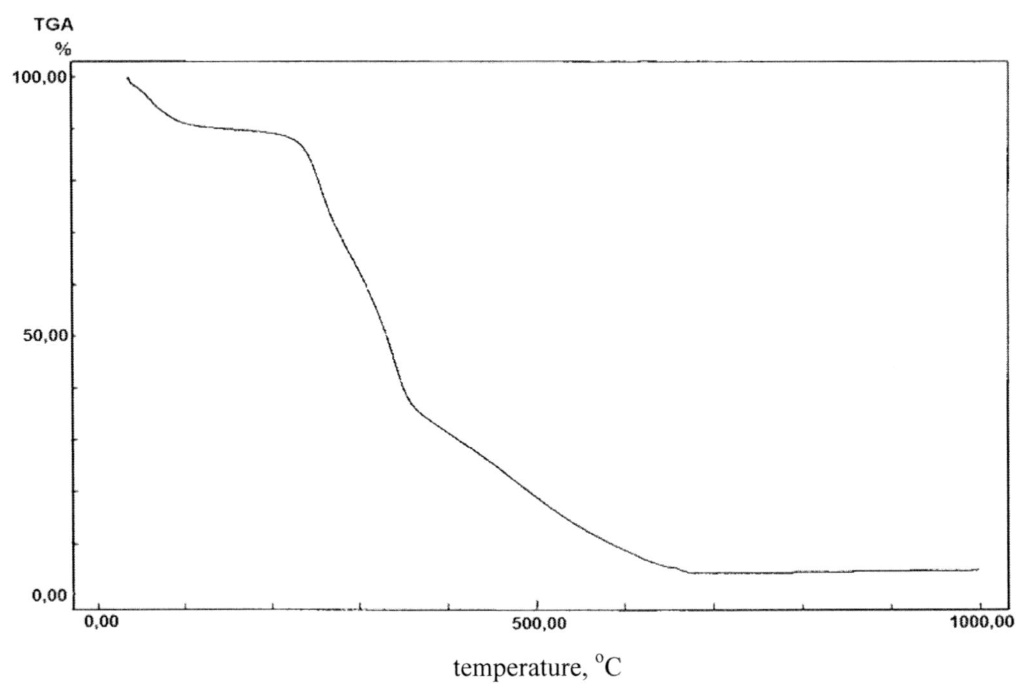


Figure 6: Thermogravimetric curve for the orange bagasse

The thermogravimetric curve (Figure 6) indicates the thermal behaviour of the orange bagasse; thus, a mass loss of *ca.* 9% by heating the biomass up to 225 °C is due to the moisture elimination. Beyond this temperature begins the biomass degradation. In the second pyrolysis step, varying the temperature between 225 and 360 °C, a higher mass loss (*ca.* 63%) was observed. This behaviour indicates cellulose and hemicellulose decomposition, as well as loss of the remaining sorbed water [56] The third step, between 360 and 665 °C, is related to lignin decomposition, wherein 96% of orange bagasse is decomposed. Finally, the plateau beyond 665 °C corresponds to the metallic oxides present in the biomass.

The specific surface area (BET method) of orange bagasse particles used in the sorption experiments was 5.62 $m^2$ $g^{-1}$. This value is comparable to the surface area of other biosorbents that showed high affinity for the heavy metal ions.

## 6.3. Biosorbent Preparation

*Citrus sinensis* oranges were used in this work. After the juice was pressed out (home juicer), the bagasse (pulp, skin, seeds, husks) was dried naturally, washed with distilled water, and dried in an oven at 70 °C for 4 hours. The dried material was then ground and sieved to obtain a uniform material (particle size 425 μm).

## 6.4. Sorption Parameters

Sorption experiments were performed in batch procedures at room temperature. A known amount of the biosorbent was contacted with a single metal ion solution with known concentration (20.0 mL), at the desired pH values. The flasks were shaken on a homemade horizontal shaker for the required time. The orange bagasse biomass was separated form the solution by a filtration step, and the residual metal ion concentration in the filtrate was determined by flame atomic absorption spectrometry. In the sorption kinetics experiments, samples were stirred for predetermined time intervals, and the metal removal was determined. Constant ionic strength was used throughout the experiments.

The variables related to the sorption behaviour of the metal ions Cu(II), Ni(II), Pb(II) and Zn(II) – pH, ionic strength, metal ion and biosorbent concentrations, contact time – were evaluated. The effect of the pH on the metal ions sorption onto the orange bagasse was studied at pH ranging from 1 to 7. pH influences the sorption sites dissociation as well as the solution chemistry of the heavy metals and their speciation [57]. The same behaviour was observed for all the metal investigated, and the sorption was more efficient above pH 4, where the biosorbent surface presents negative sites and the metals are present as cations (divalent cations). The poor removal efficiency under acidic conditions may be related to the presence of excess hydrogen ions, which compete with the metal cations for the sorption sites. The maximum removal of Cu(II), Ni(II), Pb(II) and Zn(II) was obtained at about pH 5.

Industrial effluents present different levels of a great variety of ionic species. The presence of these species can affect the desired removal due to a competition for the active sites [58]. Therefore, the effect of the ionic medium on the sorption efficiency was evaluated; the cation size, the anion nature and the ionic strength were studied. The ionic medium composition (nature of cation and anion) had a little influence on the sorption process and the most important parameter was the ionic strength (Table 1). The sorption efficiency decreased with the increasing of the ionic strength. This effect, probably, resulted from the hindrance of the heavy metal ions migration due to the excess of charged species, or from a competition for the active sites in the biomass. Pb(II) sorption showed no effect due to the ionic strength, and probably this can be attributed to the great affinity for the biomass.

**Table 1. Ionic effect on the sorption of heavy metal ions onto orange bagasse**

| Ionic medium ($mol\ L^{-1}$) | Removal (%) | | | |
|---|---|---|---|---|
| | Pb(II) | Cu(II) | Zn(II) | Ni(II) |
| LiCl 0.1 | 96.3 | 86.9 | 67.9 | 61.2 |
| NaCl 0.1 | 96.5 | 85.3 | 64.7 | 52.8 |
| KCl 0.1 | 96.5 | 85.5 | 61.4 | 52.3 |
| $NaNO_3$ 0.1 | 96.4 | 88.9 | 63.7 | 53.2 |
| $NaNO_3$ 1.0 | 93.5 | 66.8 | 25.2 | 18.2 |

Experimental conditions: biosorbent amount, 200 mg; pH, 5; contact time, 60 min; room temperature; Cu(II), 2mg $L^{-1}$; Ni(II), 5mg $L^{-1}$; Pb(II), 20mg $L^{-1}$; Zn(II), 1mg $L^{-1}$

All the heavy metal ions in this study form complexes with chloride; these complexes can be cationic (as $[PbCl]^+$), neutral (as $[PbCl_2]$) or anionic (as $[PbCl_3]^-$ or $[PbCl_4]^{2-}$). In the experimental conditions of this study the metal ions are mostly as free ion or cationic complexes. However, the small amount of anionic or neutral complexes does not interfere on the sorption efficiency as indicate the sorption efficiency values determined in the presence of nitrate, very similar to those determined in chloride medium. 0.1 mol $L^{-1}$ NaCl was the choice for the continuity of the experiments.

Increasing the amount of biosorbent from 25 mg to 800 mg, while keeping all other parameters constants, improved the sorption of the heavy metal ions. This is due to the availability of more active sites for metal ions removal. The maximum efficiency was observed for 400 mg of the orange bagasse, and a further increase in orange bagasse amount did not cause a significant effect in the sorption process; 400 mg of the biomass was used in the next experiments.

## 6.4. Sorption Kinetics

The data for the metal ions sorption on the bagasse orange showed that a contact time of 30 min was sufficient to achieve equilibrium for Cu(II), Ni(II), Pb(II) and Zn(II); the sorption did not change with further increase until 120 min. The sorption capacity increased with increasing contact time up to 30 min and most of the metal ions were removed by the orange bagasse in the first 5 min (Figure 7). In the other hand, the metal ion removal $q_e$ (mg $g^{-1}$) decreased with increasing orange bagasse amount. This can result from aggregation and/or agglomeration of biosorbent particles, which could prevent the sorption process. Besides, the sorption sites remain unsaturated in these conditions. As a result, a more economical removal of a given amount of metal ions can be carried out using small batches of sorbent, rather than in a single batch with large amounts of the biomass.

The sorption process occurred in two phases. The first one may be explained as a rapid sorption on the biomass surface through physical processes, or an ion-exchange process. Adsorption phenomena characteristically tend to attain instantaneous equilibrium, and many agricultural products act as natural ion exchangers; therefore, the initial fast-phase removal of metal ions by orange bagasse is within previous considerations. The rate of metal ions removal is of great significance for developing a biomass-based effluent-remediation treatment; the ability of orange bagasse to remove the metal ions within 30 min is an indicative of its potential as an efficient biosorbent.

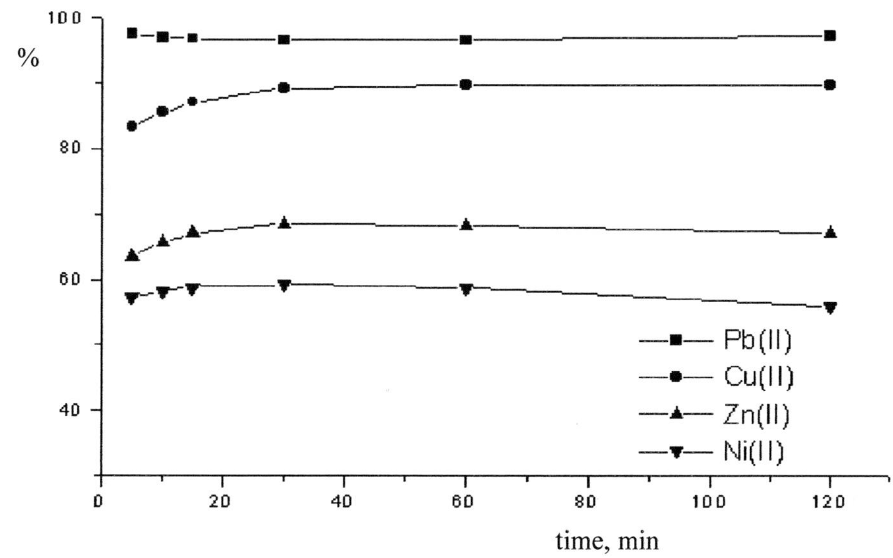


Figure 7: Time contact effect on the sorption efficiency (%) of orange bagasse. Biosorbent amount, 400 mg; pH, 5; room temperature; Cu(II), 2.0mg $L^{-1}$; Ni(II), 5.0mg $L^{-1}$; Pb(II), 20.0mg $L^{-1}$; Zn(II), 1.0mg $L^{-1}$

As it was said before, the sorption process occurs in different steps. The rate of the sorption depends on several parameters, such as structural properties of the biosorbent (*e.g.* porosity, surface specific area, particle size), metallic ions properties (*e.g.* ionic radius, coordination number), ion concentration in the solution and interactions between the metallic ions (sorbate) and active sites of the biosorbent [45]. Therefore, the kinetics data for Cu(II), Ni(II), Pb(II) and Zn(II) ions sorption onto orange bagasse were then investigated in order to increase sorption process understanding.

External mass transfer is related to the initial rate of sorption and characterized by the external mass transfer coefficient **β** [45]. According to equation (9), the plots $C_t$ / $C_o$ *versus* time give the external mass transfer coefficient values by tracing the slopes at $t$=0 (Figure 8). Table 2 shows the corresponding **β** values for the metal ions studied. These values mean that the resistance to the external mass transfer may be important for the sorption process onto orange bagasse particles.

Intraparticle diffusion (second step of the biosorption process) can be characterized by the intraparticle diffusion rate constant (***K**$_i$*) and intraparticle diffusion coefficient (***D**$_i$*), according to equation (10) and Urano and Tachikawa model [50]. ***K**$_i$* and ***D**$_i$* values are shown in Table 2. The values of **β** show that the external mass transfer from the bulk solution to the sorbent surface was rapid enough to consider this biomass as a sorbent for metal ions retention.

The plots ***q*** *vs* $t^{1/2}$ were linear within a certain extension, but did not pass through the origin. So, intraparticle diffusion is not the only rate-controlling step. Calculating the intraparticle diffusion coefficient Di further supported this thought. According to Michelsen [59], a $D_i$ value of the order of $10^{-11}$ cm$^2$ s$^{-1}$ is an indicative of a sorption process rate which is limited by the intraparticle diffusion step. Metal ions sorption onto orange bagasse presented $D_i$ values at around $10^{-6}$ – $10^{-5}$ cm$^2$ s$^{-1}$, thus confirming that the intraparticle diffusion is not the only rate-controlling step. Therefore, both boundary layer and intraparticle diffusion might be involved in the sorption process. Also, from these values it becomes clear that the external mass transfer of the heavy metal ions inside the biosorbent particles is most important than intraparticle diffusion. $D_i$ values are similar to metal ions diffusion coefficients in aqueous solutions, thus suggesting the presence of free water inside the sorbent particle.

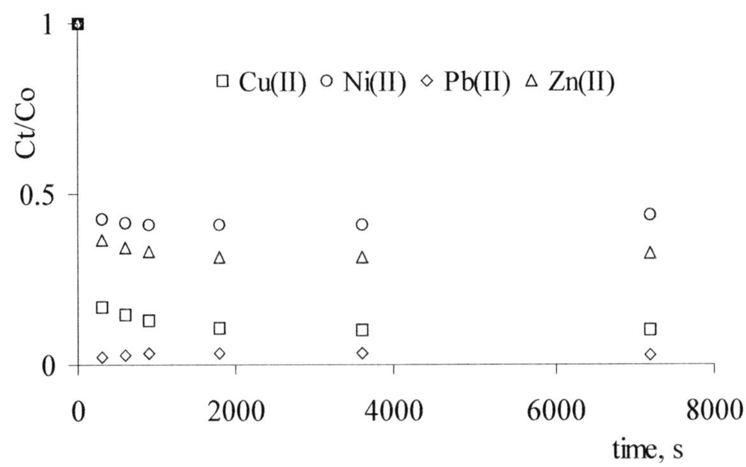


Figure 8: $C_t$ / $C_o$ plots as a function of time for metal ions sorption onto orange bagasse – external mass transfer determination. Experimental conditions: see Figure 7.

**Table 2: Kinetic parameters for metal ions sorption onto orange bagasse**

| | $K_i$, mg g$^{-1}$ s$^{-1/2}$ | $D_i$, cm$^2$ s$^{-1}$ | β, m s$^{-1}$ |
|---|---|---|---|
| Cu(II) | 0.0038 | 2.1x10$^{-6}$ | 4.9x10$^{-8}$ |
| Ni(II) | 0.0065 | 1.7 x10$^{-5}$ | 3.4x10$^{-8}$ |
| Pb(II) | 0.044 | * | 5.9x10$^{-8}$ |
| Zn(II) | 0.0014 | 1.2x10$^{-5}$ | 3.7x10$^{-8}$ |

*not determined

Pseudo first–order and second-order kinetics models were checked according equations (24) and (27) [60-63]. The experimental kinetic data did not fit well with the Lagergren pseudo first-order model, but they fitted well to the pseudo second-order model ($r^2$>0.999); plots of ***t/q*** *versus* ***t*** for metal ions - orange bagasse systems are shown (Figure 9). This pseudo second-order model assumes that the rate-limiting step may be a chemical sorption involving valence forces through sharing or exchange of electrons between biomass and metal ions. The theoretical $q_e$ values agree well with the experimental values, showing the applicability of this kinetic model.

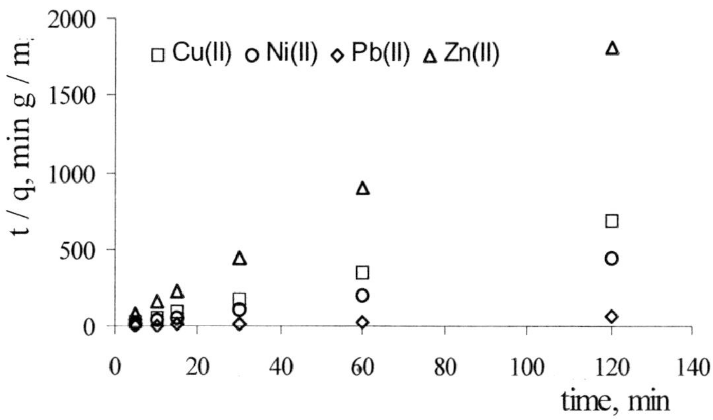


Figure 9: Pseudo second-order kinetic model plot for metal ions sorption onto orange bagasse. Experimental conditions: see Figure 7.

The second-order sorption constant $K_{II}$ and $q_e$ values were determined from the slopes and intercepts of the plots, and are shown in Table 3.

**Table 3: Pseudo second-order kinetic constants for heavy metal ions sorption onto orange bagasse**

| | $K_{II}$, g mg$^{-1}$ min$^{-1}$ | $q_e$, mg g$^{-1}$ |
|---|---|---|
| Cu(II) | 9.76 | 0.18 |
| Ni(II) | 46.4 | 0.29 |
| Pb(II) | 279 | 1.9 |
| Zn(II) | 258 | 0.066 |

## 6.5. Sorption Isotherms

The equilibrium sorption efficiency for Cu(II), Ni(II) and Zn(II) ions increased with an increase in the initial metal ion concentration up to *ca.* 15, 10 and 4 mg L$^{-1}$, respectively, and then almost did not change with further increase in the initial metal ion concentration (Figure 10), indicating sorbent saturation. Pb(II) ion sorption behaviour did not show a greatly concentration dependent profile and maximum efficiency was obtained even for low concentrations.

The sorption capacities of orange bagasse towards Cu(II), Ni(II), Pb(II) and Zn(II) ions were determined after finding the best experimental conditions – pH 5, 30 min of mixing 20.0mL of the metal ion solution and 400 mg of the biomass. A saturation condition of the surface of the biomass was obtained by increasing the metal ion concentrations. The sorption isotherms (***$q_e$*** *versus* ***$C_e$***) of the metal ions onto orange bagasse are shown in the Figure 11. Pb(II) and Zn(II) isotherms showed a convex shape - that is, the sorption process of Pb(II) and Zn(II) onto orange bagasse is favourable, and great amounts of these metal ions can be sorbed even from dilute solutions. Cu(II) and Ni(II) isotherms were linear, indicating that the sorbed amount is proportional to the metal ion concentration in the solution. These isotherms were linearised according to Langmuir and Freundlich models.

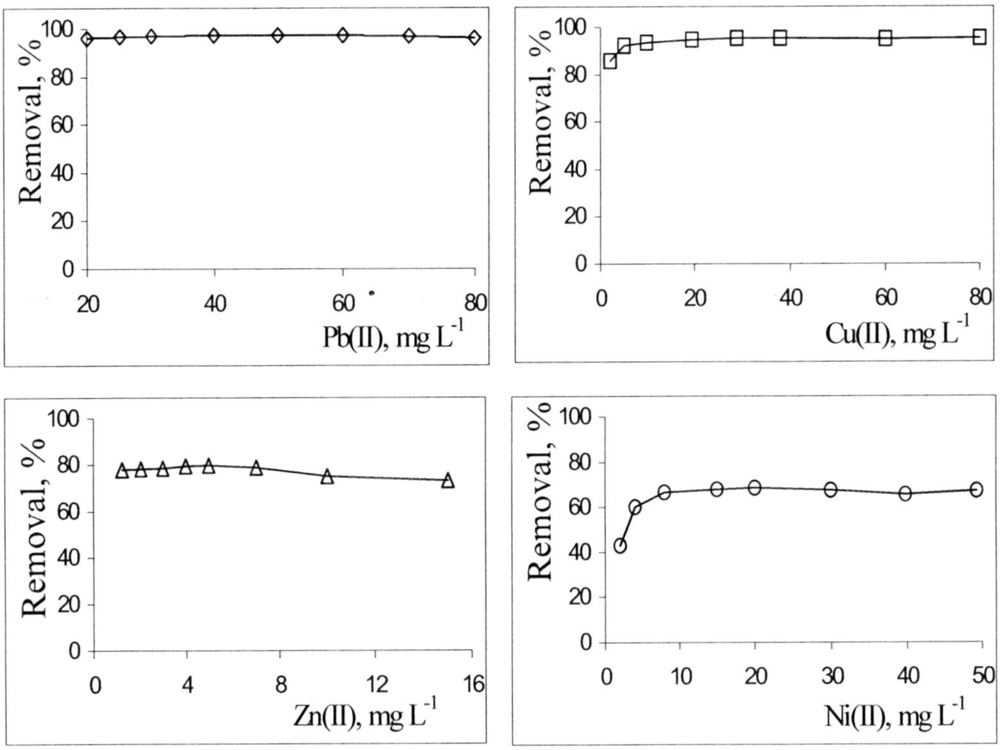


Figure 10: Sorption efficiency for different initial metal ion concentrations. Experimental conditions: see Figure 7; time of contact: 30 min.

Experimental data did not fit well to the Freundlich expression (equations (5) and (6)), but a very good fitness ($r^2 > 0.999$) was found to the Langmuir model (equations (2) and (3)). Consequently, the sorption process of metal ions retention on the orange bagasse follows the Langmuir model, where the metal ions are independently retained on a single type of binding site, and the sorption of a first metal ion does not affect the sorption of another one. $\mathbf{q_m}$ represents a practical limiting sorption capacity when the surface is completely covered with metal ions. This value assists in the evaluation of sorption performance, particularly when the sorbent did not reach its full saturation in the experiments [64].

The maximum sorption capacity $\mathbf{q_m}$ for this sorption system – heavy metal ion and orange bagasse – was determined for the metal ions studied; these values are shown in Table 4. The maximum amount of Pb(II) that can be sorbed, per gram of orange bagasse, *ca.* 67.6 mg compares favourably with the most of the $\mathbf{q_m}$ values reported with other types of biosorbents [65], suggesting a great affinity of orange bagasse towards Pb(II) ions. The $\mathbf{q_m}$ values obtained for the orange bagasse sorbent (Table 4) indicate that 1 kg of raw dried orange bagasse can treat 19 500, 14 600, 67 600 and 16 900 L of wastewater containing 1 mg $L^{-1}$ of Cu(II), Ni(II), Pb(II) and Zn(II), respectively.

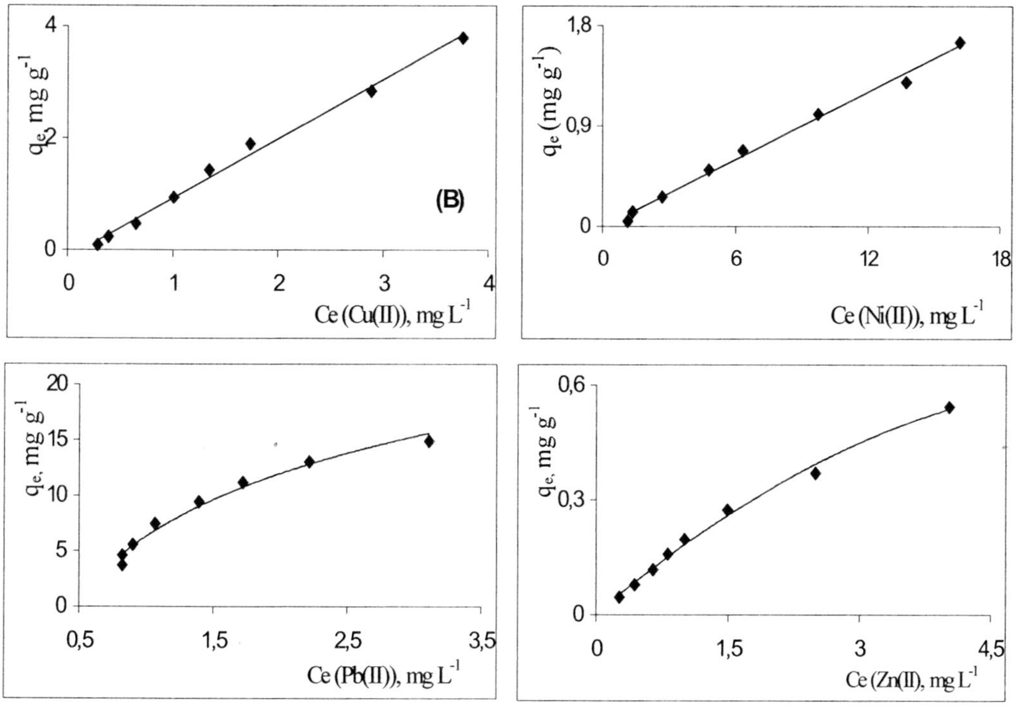


Figure 11: Sorption isotherm plots for the metal ions sorption onto orange bagasse. Experimental conditions: see Figure 10.

The monolayer coverage of the sorbent surface by the metal ion can be used to estimate the specific surface area ***S*** ($m^2$ / g biosorbent) of the orange bagasse (35):

$$S = q_m \, N \, A / M \tag{29}$$

**Table 4: Langmuir parameters for the heavy metal ion sorption onto orange bagasse**

| | qm, mg g-1 | b, L mg-1 |
|---|---|---|
| Cu(II) | 19.5 | 0.043 |
| Ni(II) | 14.6 | 0.0065 |
| Pb(II) | 67.6 | 0.025 |
| Zn(II) | 16.9 | 0.0075 |

$q_m$ is the Langmuir parameter (g metal ion / g biosorbent), ***N*** is the Avogadro number, ***A*** is the cross sectional area of metal ion ($m^2$) and ***M*** is the metal molecular weight (g / mol). The maximum specific surface area of orange bagasse for Cu(II), Ni(II), Pb(II) and Zn(II) binding were determined as 3.0 $m^2$ $g^{-1}$, 2.2 $m^2$ $g^{-1}$, 10.7 $m^2$ $g^{-1}$ and 3.4 $m^2$ $g^{-1}$, respectively. Comparison of these ***S*** values of sorbents for these metal ions shows that orange bagasse has a larger specific surface area than other natural sorbent, specially for Pb(II) [31,35,66].

The equilibrium parameter $\mathbf{R_L}$ values obtained from the Langmuir isotherm for the metal ions - orange bagasse systems, between 0.23 and 0.97, suggest that the sorption of Cu(II), Ni(II), Pb(II) and Zn(II) onto this biosorbent is feasible at the range of concentrations that was used in the sorption experiments.

Ion exchange may be the principal mechanism for the sorption of metal ions such as Cu(II), Ni(II), Pb(II) and Zn(II). The major components in the orange bagasse, as a vegetable biomass, have the ability of sequestering heavy metal ions. Therefore, it can be speculated that these lignic compounds act as the active ion exchange compounds. The models based on exchange reactions are the most common to describe the sorption process on natural sorbents [65,66]. Among these models, surface complex formation takes account of the effects related to the metal valence and ionic radius. The greatest sorption of Pb(II) onto orange bagasse can be attributed to its greater ionic radius and coordination sphere – that is, the coordination sphere of Pb(II) embraces a greater number of binding sites in the biosorbent surface. The sorption onto orange bagasse followed the sequence Pb(II) > Cu(II) > Zn(II) > Ni(II), ant this behaviour did not obey the ionic radius sequence. Thus, other kinds of chemical bindings can be responsible for the sorption onto the orange bagasse surface [26]. Covalent bonding can result form the sharing of free electron pairs between the surface oxygen atom and the metal atom, or the formation of an O---M bonding. Hydrogen bonding results from the long-range force between surface oxygen atom and the hydrogen atom of the hydrated metal ions. It has been reported that the more electronegative a metal is, the better its sorption is [25], but orange bagasse showed a metal ions sorption trend that is not fully similar to the metal ions electronegativity sequence. However, Pb(II) - the more electronegative among the ions studied - showed the greatest sorption capacity. Under the currently available experimental results, it was not possible to differentiate the exact nature of the mechanism involved in the metal ions sorption onto the orange bagasse biosorbent.

## 6.6. Thermodynamic Parameters

The thermodynamic parameters of the sorption process - entalphy ($\Delta H^o$), entropy ($\Delta S^o$) and free energy ($\Delta G^o$) - can be evaluated from the apparent equilibrium constant, $K_c'$, for the metal ion sorption onto orange bagasse. $K_c'$ was obtained from the relation between metal ion concentrations in the sorbent (Csorbent) and in the solution (Csolution), at equilibrium, according to equation (30):

$$K_c' = \frac{(\text{Csorbent})_e}{(\text{Csolution})_e} \qquad (30)$$

$\mathbf{K_c'}$ values determined at different metal ion initial concentrations and extrapolated to zero allow to calculate the thermodynamic equilibrium constant, $\mathbf{K_c^o}$ and the Gibbs free energy of the metal ion sorption process (68):

$$\Delta G^o = - R\,T\,\text{Ln}\,K_c' \qquad (31)$$

***$K_c$'*** can be plotted as a function of the temperature (***T***) according to equation (32), which allows the ***ΔH*** and ***ΔS*** from the slope and the intercept of the plot.

$$\mathrm{Ln}\, K_C' = \Delta H / RT + \Delta S / R \qquad (32)$$

The metal ion sorption onto orange bagasse was carried out at three temperatures (26, 40 and 60 °C) in order to determine the thermodynamic parameters for the sorption process. The time to reach the equilibrium at the different temperatures was found to be equal, thus indicating that the equilibrium time is not dependent on the temperature. The thermodynamic parameters for Cu(II), Ni(II), Pb(II) and Zn(II) sorption onto orange bagasse are shown in Table 5.

**Table 5: Thermodynamic parameters for metal ions sorption onto orange bagasse**

| | $K_C^o$ | $-\Delta G^o$, kJ mol$^{-1}$ | ΔH, J mol$^{-1}$ | ΔS, J K$^{-1}$ mol$^{-1}$ |
|---|---|---|---|---|
| Cu(II) | 10.4 | 5.82 | 2.72 | 19.7 |
| Ni(II) | 2.28 | 2.05 | 1.80 | 6.90 |
| Pb(II) | 44.2 | 9.42 | 36.5 | 33.4 |
| Zn(II) | 3.36 | 3.01 | 12.0 | 10.9 |

The negative values of $\Delta G^o$ indicate the spontaneous nature of the sorption process and confirm the feasibility of the process. $\Delta H^o$ values were positive, indicating the endothermic nature of the sorption reaction. However, the entropic contribution was even larger than the free energy of sorption ($\Delta H^o - T\Delta S < 0$). Therefore, the sorption of Cu(II), Ni(II), Pb(II) and Zn(II) onto orange bagasse is entropically governed [27]. The positive values of $\Delta S^o$ reflect the affinity of the sorbent for the metal ions, and suggest some structural change in metal and sorbent [67].

## 6.7. Metal ions Recuperation

Desorption studies indicated (Table 6) that HCl 0.1 mol L$^{-1}$ is capable of remove more than 60% of Cu((I), Ni(II) and Zn(II) ions from the orange bagasse; for Pb(II), the maximum recuperation (87%) was obtained with HCl 1.0 mol L$^{-1}$. Therefore, the metal ions can be recuperated from the biomass of orange bagasse without any special procedure.

**Table 6: Recovery of metal ions sorbed onto the orange bagasse**

| | Cu(II) | Ni(II) | Pb(II) | Zn(II) |
|---|---|---|---|---|
| HCl 0.1 mol L$^{-1}$ | 77.4 | 67.1 | 20.9 | 65.5 |
| HCl 1.0 mol L$^{-1}$ | 82.0 | 67.7 | 87.0 | 70.4 |
| $HNO_3$ 0.1 mol L$^{-1}$ | 75.8 | 66.9 | 15.2 | 67.5 |
| EDTA 0.1 mol L$^{-1}$ * | 67.2 | 58.6 | 83.0 | 45.5 |

* pH = 10

## 7. Conclusion

In the last decades, there has been great interest in natural sorbent materials for wastewater remediation, and many different examples have been reported. The potentialities of several different new materials, called as ***biosorbents***, have been shown for effluent treatment, exploring the affinity for the contaminant, and especially for the heavy metal ions removal. Although several biomaterials have been applied for effluent remediation, certainly a great variety can still be suggested for this task.

Due to the accumulated knowledge and to the greatly significant economic aspect some new biosorbents materials are currently poised for commercial utilization. Although much has been accomplished in the area of *low-cost sorbents*, much effort is still necessary. To this point, it seems that there is no limit to increasing the ***sorption science***, in order to provide deeper understanding of sorption phenomena and to support valuable application attempts. Since pollution seems inevitable men can wonder, "what we should do to minimize the pollution?" The world needs new methods and new technologies to clean and preserve waters, to weaken the environmental hazards, to assure the welfare to all the human being, no matter where he lives. Any contribution is welcome!

## 8. References

[1] Vieira, R. H. S. F.; Volesky, B. *Internatl Microbiol* 2000, 3, 17-24.
[2] Wilkins, E.; Yang, Q. *J Environ Sci Health* 1996, A31, 2111-2123.
[3] Lacour, S.; Bollinger, J. C.; Serpaud, B.; Chantron, P.; Arcos, R. *Anal Chim Acta* 2001, 428, 121-132.
[4] Hsieh, C. T.,; Teng, H. *Carbon* 2000, 38, 863-869.
[5] Ajmal, M.; Rao, R. A. K.; Anwar, S.; Ahmad, J.; Ahmad, R. *Biores Technol* 2003, 86, 147-149.
[6] Katsumata, H.; Kaneco, S.; Inomata, K.; Itoh, K.; Funasaka, K.; Masuyama, K.; Suzuki, T.; Ohta, K. *J Environ Manag* 2003, 69, 187-191.
[7] Volesky, B. *Hydrometallurgy* 2001, 59, 203-216.
[8] Dabrowski, A. *Adv Colloid Interf Sci 2000*, 93, 135-224.
[9] Bailey, S. E.; Olin, T. J.; Bricka, R. M.; Adrian, D. D. *Water Res* 1999, 33, 2469-2479.
[10] Volesky, B. *Trends Biotechnol* 1987, 5, 96-101.
[11] Bag, H.; Lale, M.; Turker, A. R. *Talanta* 1998, 47, 689-696.
[12] Volesky, B.; Prasetyo, I. *Biotechnol Bioeng* 1994, 43, 1010-1015.
[13] Ohtman, M. R.; Amin, A. M. *Biochem Eng* J 2003, 16, 361-364.
[14] Tsezos, M.; Volesky, B. *Biotechnol Bioeng* 1981, 23, 583-604.
[15] Tsezos, M. *Hydrometallurgy* 2001, 59, 241-243.
[16] Sa, Y.; Aktay, Y. *Biochem Eng* J 2002, 12, 143-153.
[17] Seki, h.; Suzuki, A. *J Coll Interf* Sci 2002, 249, 295-300.
[18] McKay, G. *Chem Eng Sci* 1984, 39, 129-138.
[19] Horsfall, M.; Abia, A. A. *Water Res* 2003, 37, 4913-4923.
[20] Ayyappan, R.; Sophia, C. A.; Swaminathan, K.; Sndhya, S. *Proc Biochem* 2005, 40, 1293-1299.

[21] Oliveira, E. A.; Montanher, S. F.; Andrade, A. D.; Nóbrega, J. A.; Rollemberg, M. C. *Proc Biochem* 2005, 40, 3485-3490.
[22] Kadirvelu, K.; Namasivayan, C.; Sivaraj, R. *Waste Manag* 2001, 21, 105-110.
[23] Kadirvelu, K.; Thamaraiselvi, K.; Namasivayan, C. *Biores Technol* 2001, 76, 63-65.
[24] Oliveira, E. A.; Montanher, S. F.; Rollemberg, M. C. *J Hazard Mater* 2005, 117, 207-211.
[25] Tay-Costodes, V. C.; Fauduet, I. I.; Porte, C.; Delacroix, A. *J Hazard Mater* 2003, 105, 121-143.
[26] Daifullah, A. A. M.; Girgis, B. S.; Gad, H. M. H. *Mat Letters* 2003, 57, 1723-1731.
[27] Özer, A.; Özer, D.; Özer, A. *Proc Biochem* 2004, 39, 2183-2191.
[28] Ho, Y. S.; Huang, C. T.; Huang, H. W. *Proc Biochem* 2002, 37, 1421-1430.
[29] Dakiki, M.; Khanmis, M.; Manasra, A.; Mer'eb, M. *Adv Environ Res* 2002, 6, 533-540.
[30] Mathialazon, T.; Viraraghavan, T. *J Hazard Mater* 2002,0B94, 291-303.
[31] Srivastava, S. K.; Singh, A. K.; Sharma, A. *Environ Technol* 1994, 15, 353-361.
[32] Tarley, C. R. T.; Arruda, M. A. Z. *Chemosphere* 2004, 54, 987-995.
[33] Marshall, W. E.; Wartelle, L. H.; Boler, D. E.; Johns, M. M; Toles, C. A. *Biores Technol* 1999, 69, 263-268.
[34] Lee, C. K.; Low, K. S.; Mah, S. *J. Adv Environ Res* 1998, 2, 351-359.
[35] Senthilkumaar, S.; Bharathi, S.; Nithyanandhi, D.; Subburam, U. *Biores Technol* 2000, 75, 163-165.
[36] Volesky, B.; Holan, Z. R. *Biotechnol Progr* 1995, 11, 235-250.
[37] McKay, G.; McConvey, I. F. *J Chem Technol Biotechnol* 1981, 31, 401-410.
[38] Volesky, B. *Hydrometallurgy* 2003, 71, 179-190.
[39] Bajpai, J.; Shrisvastava, R.; Bajpai, A. K. *Coll Surf A* 2004, 236, 81-90.
[40] Sheng, P. X.; Ting, Y. P.; Chen, J. P.; Hong, L. *J Coll Interf Sci* 2004, 275,131-141.
[41] Khan, A. R.; Ataulah, R.; Al-Haddad, A. *J Coll Interf Sci* 1997, 194, 154-165.
[42] Chong, K. H.; Volesky, B. *Biotechnol Bioeng* 1996, 49, 629-638.
[43] Agrawal, A.; Sahu, K. K.; Pandey, B. D. *Coll Surf 2004*, 237, 133-140.
[44] Ng, C.; Losso, J. N.; Marshall, W. E.; Rao, R. M. *Biores Technol* 2002, 85, 131-135.
[45] McKay, G.; Poots, V. J. P. *J Chem Technol Biotechnol* 1980, 30, 279-292.
[46] Singh, K. K.; Rastogi, R.; Hasan, S. H. *J Hazard Mater* 2003, 121, 51-58.
[47] Sag, Y.; Aktay, Y. *Sep Sci Technol* 2002, 37, 2801-2822.
[48] Ho, Y. S.; Wase, D. A. J.; Forster, C. E. *Environ Technol* 1996, 17, 71-77.
[49] Urano, K.; Tachikawa, I. I. *Ind Eng Chem Res* 1991, 30, 1897-1899.
[50] Masters, G. M. *Introduction to Environmental Engineering and Science*; Prentice Hall: Englewood, NJ, 1991; .
[51] Ho, Y. S.; McKay, G. *Water Res* 2000, 34, 735-742.
[52] Ho, Y. S.; McKay, G. *Adsp Sci Technol* 2002, 20, 797-815.
[53] Cheung, W.; Porter, J. F.; McKay, G. *Water Res* 2001, 35, 605-612.
[54] Ho, Y. S.; McKay, G. *Proc Biochem* 1999, 34, 451-465.
[55] Pasquali, C. E. L.; Herrera, H. *Thermoch Acta* 1997, 293, 39-46.
[56] Mansaray, K. G.; Ghaly, A. E. *Energy Sources* 1999, 21, 899-911.
[57] Espósito, A.; Pagnelli, F.; Veglio, F. *Chem Eng Sci* 2002, 57, 307-313.
[58] Schiewer, S.; Volesky, B. *Environ Sci Technol* 1997, 31, 2478-2485.
[59] Michelsen, L. D.; Gideon, P.D.; Pace, E. G.; Kutsal, KL. H. U S Dept Ind Off Wat res Technol; Bull 74; 1975.

[60] Donmez, G.; Aksu, Z. *Proc Biochem* 2002, 38, 751-762.

[61] Nuhoglu, Y.; Malkoc, E.; Giirses, A.; Campolat, N. *Biores Technol* 2002, 85, 331-333.

[62] Ho, Y. S.; Ng, J. C. Y.; McKay, G. *Sep Sci Technol* 2001, 36, 241-261.

[63] Low, K. S.; Lee, C. K.; Liew, S.C. *Proc Biochem* 2000, 36, 59-64.

[64] Kekinkan, O.; Goksu, M. Z. L.; Yuceer, A.; Basibuyuk, M.; Forster, C. E. *Proc Biochem* 2003, 39, 179-183.

[65] Shiewer, S.; Wong, M. H. *Chemosphere* 2000, 41, 271-282.

[66] Lee, S. M.; Davis, A. P. *Water Res* 2001, 35, 534-540.

[67] Donmez, G.; Aksu, Z. *Proc Biochem* 1999, 35, 135-142.

In: Hazardous Materials and Wastewater
Editor: Allison A. Lewinsky, pp. 79-125
ISBN 1-60021-257-3


*Chapter 3*

# REMOVAL OF HAZARDOUS METALS FROM WATER BY CLAYS

***Krishna G. Bhattacharyya*[1*] *and Susmita Sen Gupta*[2]**
[1] Department of Chemistry, Gauhati University, Guwahati, Assam, India
[2]Department of Chemistry, B N College, Dhubri, Assam, India

## ABSTRACT

Hazardous metal cations enter water through the natural geochemical route or the industrial wastes. These non-biodegradable and persistent metals accumulate in living tissues getting biomagnified in the food chain. The present work investigates the interaction of Cd(II), Cu(II), Ni(II) and Pb(II), with clay minerals, kaolinite, montmorillonite and their acid-activated forms. As an important constituent of soil, clay minerals play the important role of a scavenger of toxic materials from the soil. The clay-heavy metal interactions were studied under different conditions of pH, concentration of the metal ions, amount of clay, interaction time and temperature. The interactions were dependent on pH and the uptake was controlled by amount of clay and initial concentration of the metal ions. 2 g of clay (kaolinite, acid-activated kaolinite, montmorillonite, acid-activated montmorillonite) per liter of solution could remove 15.9 to 92.1 % Cd(II), 15.7 to 87.7 % Cu(II), 20.7 to 69.2 % Ni, and 20.9 to 90.8 % Pb(II) from a solution of initial concentration 50 mg/L at equilibrium at 303 K. Adsorption was in the order of kaolinite $<$ acid-activated kaolinite $<$ montmorillonite $<$ acid-activated montmorillonite. A detailed study was undertaken with respect to the kinetics, Freundlich and Langmuir isotherms, and the thermodynamics of the processes. The interactions were adsorptive in nature and were relatively fast. Equilibrium was approached within 180, 360, 240 and 180 min for Pb(II), Cu(II), Cd(II) and Ni(II) respectively. The rate processes did not match the pseudo first order kinetics of Lagergren and therefore, they were further investigated with the simple second order kinetics, Elovich equation, liquid film diffusion and intra-particle diffusion mechanisms, with the second order kinetic model giving the best results. The adsorption data fitted both Langmuir and Freundlich isotherms indicating that strong adsorptive forces were involved in the interactions and that the interaction sites were energetically non-uniform. Langmuir monolayer capacity was 11.5 to 31.4, 4.3 to 28.0, 6.8 to 30.7 and 7.1 to 21.3 mg/g respectively for the four clays. The thermodynamics of the rate processes showed that the adsorption of Pb(II) and Ni(II) was exothermic accompanied by decreases in entropy and Gibbs energy, but the adsorption of Cd(II) and Cu(II) was endothermic with an increase in entropy and an appreciable decrease in Gibbs energy. The results have established good potentiality for kaolinite, montmorillonite and their acid activated forms to be used as adsorbents for Cd(II), Cu(II), Ni(II) and Pb(II) ions from aqueous medium.

**Keywords:** kaolinite; montmorillonite; acid activated clays; adsorption kinetics; adsorption isotherm; temperature; enthalpy; metal ions

# INTRODUCTION

Rapid growth of industrialization and urbanization has been introducing a large number of pollutants into the environment multiplying the ecological problems. Water is particularly vulnerable to contamination with toxic chemicals discharged by various industries. Heavy metals in water are one such group of contaminants having serious consequences for aquatic life and human beings (Chiron et al., 2003). Hazardous and toxic metal cations reach surface and groundwater following the natural geochemical route as well as the industrial pathway. These are also harmful to the environment as a whole because of their non-biodegradable and persistent nature even at low concentration. The metals accumulate in living tissues and get biomagnified through the food chain (An et al., 2001). The metals of major concern are aluminium, arsenic, cadmium, chromium, cobalt, copper, iron, lead, manganese, mercury, nickel and zinc.

## Toxic Effects of Cu(II), Cd(II), Ni(II) and Pb(II)

Copper is primarily used as a metal or an alloy (e.g., brass, bronze, gun metal). Copper sulfate is used as a fungicide, algicide, and nutritional supplement. Copper particulates are released into the atmosphere by windblown dust; volcanic eruptions; and anthropogenic sources, primarily copper smelters and ore processing facilities. Copper is often found near mines, smelters, industrial settings, landfills, and waste disposal sites. In surface waters, copper exists either as free copper ions or, more likely, bound to suspended particulates.

One of the most commonly reported adverse health effects of copper is gastrointestinal distress. Vomiting, nausea and abdominal pain usually occur shortly after drinking beverages stored in copper or untinned brass containers. It is also irritating to the respiratory tract and coughing, sneezing, running nose, pulmonary fibrosis, and increased vascularity of the nasal mucosa have been reported in workers exposed to copper dust. Exposure to excessive levels

* Corresponding author. Tel: +91 361 2571529; fax: +91 361 2570599. *Email address*: krishna2604@sify.com (K G Bhattacharyya)

of copper can result in a number of adverse health effects including liver and kidney damage, anemia, immunotoxicity, and developmental toxicity and even death. There is strong evidence to suggest that Wilson's disease, Indian childhood cirrhosis, and possible idiopathic copper toxicosis are caused by an increased genetic susceptibility to copper toxicity (ATSDR, 2002).

The liver is also a sensitive target of Cu(II) toxicity. Liver damage (necrosis, fibrosis, abnormal biomarkers of liver damage) has been reported in individuals ingesting lethal doses of copper sulfate. Decreased hemoglobin and erythrocyte levels have been observed in workers exposed to airborne copper levels of 0.64–1.05 mg $m^{-3}$ (Finelli et al., 1981). The co-carcinogenic character of copper is also accepted.

Cadmium is released into the aqueous system from metal plating, smelting, mining, cadmium-nickel batteries, phosphate fertilizers, paint industries, pigments and alloy industries as well as from sewage (Kadirvelu and Namasivayam, 2003). Breathing air with lower levels of cadmium over long periods of time (for years) results in a build-up of cadmium in the kidney, which is the main target organ for cadmium toxicity (ATSDR, 1999a) but inhalation of high levels of cadmium can severely damage the lungs and may cause death. Prolonged exposure to cadmium has been shown to be responsible for making the bones fragile. Most ingested cadmium passes through the gastrointestinal tract without being absorbed (Kjellstrom et al., 1978). Case studies indicate that calcium deficiency, osteoporosis, or osteomalcia can develop in some workers after long-term occupational exposure to high levels of cadmium. Chronic cadmium toxicity has also been the cause of Japanese Itai–Itai disease (Mathialagan and Viraraghavan, 2002). The nervous system appears to be the most sensitive target of Cd(II) toxicity. New-born baby exposed to cadmium before birth have shown effects on behavior and learning. Adverse health effects due to cadmium such as lung insufficiency, bone lesions, and hypertension in humans are well documented (Yin and Blanch, 1989). An increased risk of lung cancer from cadmium exposure was also reported (Stayner et al., 1992).

Nickel is released into the environment during nickel mining and by industries that make or use nickel, nickel alloys, or nickel compounds, oil-burning power plants, coal-burning power plants, and trash incinerators. The higher nickel load on soils may increase the uptake by plants and lead to adverse effects on ecosystem and human health. Changes in faunal composition and soil fertility due to nickel pollution are observed in many places.

The respiratory tract is the most sensitive target of inhaled nickel toxicity. Chronic bronchitis, emphysema, and impaired lung function have been observed in nickel welders and foundry workers. The carcinogenic effect of nickel has been well documented in occupationally exposed individuals (ATSDR, 2003). Humans suffer from allergy due to nickel-containing diets and also due to exposure to nickel-containing materials. The immunotoxicity of nickel has been established in humans following inhalation, oral, and dermal exposure. Moreover, it was established that the male reproductive system might be a sensitive target of ingested nickel toxicity.

The main sources of lead pollution are lead smelters, battery manufacturers, paper and pulp industries, boat and ship fuels and ammunition industries. In addition, the production of television picture tubes, pigments, petroleum fuels, printing, glass industries, photographic materials, etc., also adds lead to the environment (Kiff, 1987). Low levels of Pb have been identified with anemia as it causes injury to the blood forming systems while high levels cause severe dysfunction of the kidneys, liver, the central and peripheral nervous system, and high blood pressure (ATSDR, 1999b). Hypertension has also been associated with lead exposure in the general population. At the typical levels to which individuals are exposed, lead can cross the placenta and damage developing fetal nervous systems. The most severe neurological effect of lead in adults is lead encephalopathy, which is a general term to describe various diseases that affect brain function. Lead exposure may cause weakness in fingers, wrists, or ankles.

Children are more sensitive to the effects of lead than adults. A child who swallows large amounts of lead may develop blood anemia, kidney damage, severe stomachache, muscle weakness, and brain damage. The lower IQ values and other neuropsychological deficits among the children exposed to higher lead levels have been well documented (Kotok et al., 1977).

Usually the toxic metals will be present in water at very low levels compared to the levels which may cause major damage as described above. However, even at very low levels, prolonged exposure and biomagnification may create enough problems for humans.

## Removal of Toxic Heavy Metals from Aqueous Environment

A number of technologies are accessible to curtail the problem of heavy metals in the environment. A few familiar methods are chemical precipitation, ion exchange, solvent extraction, reverse osmosis, etc. The removal of heavy metals by reverse osmosis is not very lucrative as the membrane is easily spoiled and needs frequent replacement. Moreover this technique requires a high operating pressure and some pre-treatment is often required. Chemical precipitation is not very suitable when the heavy metals are present in trace amounts and the large amount of sludge produced after treatment may cause side effects. Ion exchange has the advantage of allowing the recovery of metallic ions, but it is expensive and sophisticated. Solvent extraction or electrolytic processes are also available but they are considered to be cost-effective only for more concentrated solutions. The need for safe and economical methods for removing heavy metals, even in a very small concentration, from polluted wastewaters has resulted in the search for alternative techniques.

The process of adsorption has become one of the preferred methods for removal of toxic contaminants from water as it has been found to be very effective, economical, versatile and simple (Tran et al., 1999). Different conventional and non-conventional type of adsorbents have been tried for removal of various metal ions, viz., peat (for Cu(II) and Pb(II), Ho and McKay, 1999a), red mud (for Cr(VI) and Pb(II), Gupta et al., 2001), activated carbon (for Zn(II), Ramos et al. 2002), coconut coirpith (for Cd(II), Kadirvelu and Namasivayam, 2003), sewage sludge (for Cu(II), Pan et al., 2003), sawdust (for Cr(VI), Yu et al., 2003), chitosan (for Hg(II), Jeon and Holl, 2003), silica (for Cu(II) and Pb(II), Chiron et al., 2003), bone char (for Cu(II), Zn(II) and Cd(II), Ko et al., 2004), rice husk (for Cd(II) and Pb(II), Tarley et al., 2004), coconut shell charcoal (Cr(VI), Babel and Kurniawan, 2004), low-grade phosphate (for Zn(II) and Cd(II), Kandah, 2004), Azadirachta indica (Neem) leaf powder (for Pb(II), Bhattacharyya and Sharma, 2004; for Cr(VI) and Cd(II), Sharma and Bhattacharyya, 2004; 2005) polymetallic sea nodule (for As(III) and As(V), Maity et al., 2005), modified zeolite (for Pb(II) and Cd(II), Wingenfelder et al., 2005), etc.

## Soil Mineralogy

The minerals in soils can be classified into (a) Primary minerals, which are formed at elevated temperatures and inherited from the igneous and metamorphic rocks, and (b) Secondary minerals, those formed by low temperature reactions and inherited by soils from sedimentary rocks or formed in soils by weathering (Brewer, 1964; Dixon and Weed, 1977). Primary minerals make up the main part of sand (particles ranging from 0.05 – 1.00 mm in diameter) and silt (particles ranging from 0.002 – 0.05 mm in diameter) fractions of most soils. Primary minerals occur in the clay fraction (particles less than 0.002 mm in diameter) of little weathered soils, e.g., those derived from fine-grained rock flour derived by glacial grinding of rocks. Layer silicates of soils, play a prominent role in most soils in determining the physical properties such as plasticity and structure and chemical properties such as cation exchange, ion release, and ion fixation when fertilizer is added to soil (Stelly and Dinauer, 1968).The deposition of significant amounts of mica, vermiculite, and other layer silicates as well as quartz and carbonates occurs as aerosolic dusts. These are common in the arid regions of both hemispheres, transported by circumglobal westerly winds and deposited in rainfall on land and ocean bottom (Jackson et al., 1973). Micas occur almost universally in soil clays, primarily originating by inheritance from the parent rock from which the soil was formed. Vermiculite, which is a product of weathering of micas and possibly also of chlorites, occurs extensively in soils. Potassium in micas has been replaced predominantly by exchangeable calcium, aluminum, and magnesium and the interlayer space is expanded to yield a unit cell c-spacing of 14 Å or more so as to accommodate exchangeable interlayer cations and water (Ramon and Jackson, 1964). The minerals of the montmorillonite or smectites isomorphous series are freely expansible layer silicates responsible for soil plasticity. An environment abundant in the exchangeable cations, calcium and magnesium, silica rich, moist, and not excessively leached favors the formation of montmorillonite. Kaolinite occurs commonly in well-weathered soils.

## Structure of Kaolinite and Montmorillonite

The structure of kaolinite was first suggested by Pauling (1930). Kaolinite is a 1:1 layer silicate where the basic structure consists of a tetrahedral sheet of $SiO_4$ and an octahedral sheet with $Al^{+3}$ as the octahedral cation. Both the sheets combine to form a common layer such that the tips of the silica tetrahedra point towards the octahedral layer. The tetrahedral layer is inverted over the octahedral layer with the apical 'O' atoms being shared by the two.

In the layer common to the octahedral and tetrahedral sheets, two-thirds of the O-atoms are shared between Si and Al atoms. The remaining one-third of the sites in this layer consists of hydroxyl group coordinated to the octahedral Al atoms alone. Only two-thirds of the possible positions in the octahedral sheet are filled with Al, and there are three possible places of regular population of the octahedral layer with Al. The remaining one-third of the octahedral sites is vacant. The Al-atoms are placed in such a manner that any two Al-atoms are separated by two hydroxyl groups – one above and one below; making a hexagonal distribution in a single plane in the center of the octahedral sheet. The hydroxyl groups are placed directly against the centers of oxygen-hexagons of the basal plane of the tetrahedral layer (Theng, 1979).

Kaolinite, $(Si_4)^{IV}(Al_4)^{VI}O_{10}(OH)_8$, has the theoretical composition of $SiO_2$ 46.54 %, $Al_2O_3$ 39.50 %, $H_2O$ 13.96 % expressed in terms of the oxides. The formula indicates that there is no substitution of $Si^{4+}$ with $Al^{3+}$ in the tetrahedral layer and no substitution of $Al^{3+}$ with $Mg^{2+}$, $Zn^{2+}$, $Fe^{2+}$, $Ca^{2+}$, $Na^{+}$ or $K^{+}$ in the octahedral layer. Thus, the net layer charge of kaolinite is: [4 (+ 4)] + [4 (+ 3)] + [10 (- 2)] + [8 (- 1)] = 0, but in nature, kaolinite has a small net negative change arising from broken edges on the clay crystals. This negative charge, although small, is responsible for the surface not being completely inert. Some workers have also reported substitution of octahedral $Al^{3+}$ with $Fe^{2+}$ and/or $Ti^{4+}$ in kaolinite (Deer et al., 1985).

The structure of montmorillonite was given by Marshall (1935). Montmorillonite is composed of units made up of two silica tetrahedral sheets with a central alumina octahedral sheet. The tetrahedral and octahedral sheets combine such that the tips of the tetrahedra of each silica sheet and one of the hydroxyl layers of the octahedral sheet form a common layer. The atoms in this layer, which are common to both sheets, become oxygen instead of hydroxyl. It is thus referred to as a three-layered clay mineral with T-O-T layers making up the structural unit.

The silica-alumina-silica units are continuous in the 'a' and 'b' crystallographic directions and are stacked one above the other in the 'c' direction. In the stacking of these units, oxygen layers of each unit are adjacent to oxygen of the neighboring units. This causes a very weak bond and an excellent cleavage between the units. The outstanding feature of this clay structure is that water and other polar molecules can enter between the layers causing the lattice to expand in the 'c' direction. Thus, the 'c'axis dimension in montmorillonite varies from a minimum value of 9.6 A° when there is no polar molecules between the unit layers and above. The 'c'axis spacing also varies with the nature of the interlayer cations present between the silicate layers.

The formula for montmorillonite is, $(Si_{7.8}Al_{0.2})^{IV}(Al_{3.4}\ Mg_{0.6})^{VI}O_{20}(OH)_4$ and the theoretical composition without the interlayer material is $SiO_2$, 66.7%, $Al_2O_3$, 28.3%, $H_2O$, 5%. The above formula indicates that there is substitution for $Si^{4+}$ by $Al^{3+}$ in the tetrahedral layer and for $Al^{3+}$ by $Mg^{2+}$ in octahedral layer. Thus net layer charge of montmorillonite is: [7.8 (+4)] + [0.2 (+3)] + [3.4 (+3)] + [0.6 (+2)] + [20 (-2)] + [4 (-1)] = – 0.8 charge/unit cell. The resulting negative net charge is balanced by exchangeable cations adsorbed between the unit layers and around their edges.

## Role of Clay Minerals in the Environment

The clay minerals in soil play an important role in the environment by acting as a natural scavenger of pollutants. As water flows over soil or penetrates underground, clays in soil take up various pollutants from water through ion exchange and adsorption mechanisms. Clays are hydrous aluminosilicates broadly defined as those minerals that make up the colloid fraction ($< 2\mu$) of soils, sediments, rocks and water (Pinnavaia, 1983) and may be composed of mixtures of finer grained clay minerals and clay-sized crystals of other minerals such as quartz, carbonate and metal oxides. Usually the term clay is used in reference to material that becomes plastic when mixed with a small amount of water.

All clay minerals contain some exchangeable cations and anions on their surfaces, held through the process of adsorption. A few common cations and anions found on clay surface are $Ca^{2+}$, $Mg^{2+}$, $H^+$, $K^+$, $NH_4^+$, and $Na^+$, and $SO_4^{2-}$, $Cl^-$, $PO_4^{3-}$, and $NO_3^-$ in order of relative abundance. These ions can be exchanged with other ions relatively easily without affecting the clay mineral structure. Clays are highly valued for their adsorptive properties, which stem from their high surface area and their tendency to adsorb water in the interlayer sites (Naseem and Tahir, 2001). The high specific surface area, chemical and mechanical stability, layered structure, high cation exchange capacity (CEC), etc., have made the clays excellent adsorbent materials. Both Brönsted and Lewis type of acidity in clays (Tanabe, 1981) boosted the adsorption capacity of clay minerals to a great extent. The Brönsted acidity arises from $H^+$ ions on the surface, formed by dissociation of water molecules of hydrated exchangeable metal cations on the surface:

$$[M\,(H_2O)_x]^{n+} \rightarrow [M\,(OH)\,(H_2O)_{x-1}]^{(n-1)+} + H^+$$

The Brönsted acidity may also arise if there is a net negative charge on the surface due to the substitution of $Si^{4+}$ by $Al^{3+}$ in some of the tetrahedral positions and the resultant charge is balanced by $H_3O^+$ cations. The Lewis acidity arises from exposed trivalent cations, mostly $Al^{3+}$ at the edges, or $Al^{3+}$ arising from rupture of Si-O-Al bonds, or through dehydroxylation of some Brönsted acid sites.

The edges and the faces of clay particles can adsorb anions, cations, non-ionic and polar contaminants from natural water. These contaminants accumulate on clay surface and are retained through the processes of ion exchange, co-ordination, or ion-dipole interactions. Sometimes the pollutants can be held through H-bonding, van der Waals interactions or hydrophobic bonding. In other cases, the pollutants may be held through weak or strong adsorptive interactions. The strength of such interactions is determined by the particular structural and other features of the clay mineral.

The role of the clay minerals in the environment can be categorized into the following four types (Van Olphen, 1977):

(i) The clay mineral can behave as a strong Brönsted acid, due to interaction of cations with adsorbed or interlayer water molecules. The strength of the acid sites is determined by the type of cation, its position on the surface and the water content of the clay,

(ii) In absence of adsorbed water molecules (e.g. a calcined clay), the clay mineral behaves as a weak Lewis acid,

(iii) In presence of ions like $Fe^{3+}$ in octahedral positions or adsorbed oxygen, the clay mineral behaves as an oxidizing agent,

(iv) In presence of ions like $Fe^{2+}$ in octahedral positions the clay mineral behaves as a reducing agent.

Treatment of clay minerals with concentrated inorganic acids usually at high temperature is known as acid activation. Acid treatments of clay minerals are an important control over mineral weathering and genesis (Jackson and Sherman, 1952; Eberl et al., 1993). Such treatments can often replace exchangeable cations with $H^+$ ions and release $Al^{3+}$ and other cations out of both tetrahedral and octahedral sites, but leaving the $SiO_4$ groups largely intact

(Theocharis et al., 1988). It was reported that acid activation followed by thermal treatment increases the adsorbent capacity to a good extent (Rodrigues, 2003).

## Clay Minerals as Heavy Metal Scavengers

A good number of studies have been reported for metal ion removal by using clay minerals and their modified from. Mellah and Chegrouche, (1997) used natural bentonite for uptake of zinc from aqueous solution and reported the maximum Langmuir monolayer capacity as 52.91 mg/g (for an initial Zn(II) concentration of 300 mg/L) at 293 K. The bentonite-Zn(II) interactions were exothermic and favoured by high values of initial Zn(II) concentration and decreased particle size of clay adsorbent. Bentonite has a Langmuir monolayer capacity of 16.3 meq/1000g at 298 K for Sr(II) (Khan et al., 1995) with as much as 99% adsorption at pH 8.5. Thermodynamically the adsorption process was shown to be spontaneous supported by increase in entropy ($\Delta S$ = 0.139 kJ/K mol at 298 K) and decrease in Gibbs energy ($\Delta G$ = -10.69 kJ/mol at 298 K), despite being endothermic ($\Delta H$ = 30.62 kJ/mol at 298 K).

Alvarez-Ayuso and Garcia-Sanchez, (2003) used natural and Na-exchanged bentonites for the removal of Cr(III), Ni(II), Zn(II), Cu(II) and Cd(II) from aqueous solution. The authors reported that Langmuir model was most suitable in explaining the results and the maximum adsorption capacity was 49.8, 24.2, 23.1, 30.0 and 26.2 mg/g respectively for Cr(III), Ni(II), Zn(II), Cu(II) and Cd(II). The process, favoured by increasing pH, was shown to be very rapid with >95 % uptake in 30 min. In a similar work, Yavuz et al. (2003) used raw kaolin for removing Mn(II), Co(II), Ni(II) and Cu(II) from aqueous solution where the adsorption affinity was in the order of Cu(II) > Ni(II) > Co(II) > Mn(II) and the monolayer adsorption capacity was 10.78, 1.67, 0.92 and 0.45 mg/g respectively at 298 K. The interactions were endothermic were driven by entropy increase.

Another clay mineral, illite was used as an adsorbent for Cd(II) (Echeverria et al., 2002) and the process was found to be dependent on ionic strength, initial concentration of Cd(II) and pH. Cd(II) adsorption increased with pH reaching 100 % at pH 8.0. The interactions were endothermic in nature. Kara et al. (2003) used sepiolite as an adsorbent for Co(II) with the results indicating spontaneous, physical and entropically driven interactions. Strawn et al. (2004) reported the adsorption of Cu(II) by smectites (montmorillonite and bedellite) as a function of varying ionic strength and pH. The X-ray adsorption fine structure (XAFS) and electron paramagnetic resonance (EPR) spectroscopic experiments on Cu(II) adsorbed on clays revealed that the Cu-Cu linkages in the multinuclear complexes were 2.65 Å apart having a coordination number near one.

Along with the pure clay minerals, many modified clay minerals have also been tested for their effectiveness as adsorbents. Lin and Juang (2002) tried to modify the adsorptive properties of montmorillonite by introducing sodium dodecylsulfate before adsorbing Cu(II) and Zn(II). The modified clay had higher affinity towards the metal ions and the equilibrium was attained within 120 min. The interactions were endothermic, which followed pseudo-first order kinetics yielding a rate constant of $6.64 \times 10^{-4}$ and $3.14 \times 10^{-3}$ $min^{-1}$ at 298 K for Zn(II) and Cu(II) respectively. Adsorption of Cu(II) ions onto a 1:10 phenanthroline-grafted Brazilian bentonite was reported by De Leon et al. (2003). The modified clay showed a maximum uptake of Cu(II) at ~ 110 mg /g, which was almost 10 times more compared to the untreated bentonite. The desorption of Cu(II) from the modified clay revealed high uptake irreversibility and physical stability of the clay adsorbent either in strongly acidic or basic solutions. Kaolinite, montmorillonite and their poly(oxo zirconium) and tetrabutylammonium derivatives have been shown to be effective in removing Pb(II), Cd(II) and Ni(II) from aqueous solution (Sen Gupta and Bhattacharyya, 2005, 2006a, 2006b).

Use of acid-activated clays for metal ion adsorption has received attention recently (Bhattacharyya and Sen Gupta, 2006). The acid-activated bentonite has been used for removal of dyes and surfactants from tannery wastewater (Espantaleon et al. 2003) and for removal of vegetable extracts (Marsal et al. 2003).

The present work was undertaken to investigate the feasibility of using kaolinite, montmorillonite and their acid activated forms for removal of Cd(II), Cu(II), Ni(II) and Pb(II) from aqueous solution by adsorption under various environmental conditions in single batch systems.

# Materials and Methods

## Clay Adsorbents

Kaolinite, KGa-1b (K1) and Montmorillonite, SWy-2 (M1) were obtained from the University of Missouri-Columbia, Source Clay Minerals Repository, USA.

## Acid Activation

Acid activation of kaolinite and montmorillonite was carried out by treating with 0.25 M $H_2SO_4$ (E. Merck, Mumbai, India) by standard procedure (Espantaleon et al., 2003). 20 g of the clays (kaolinite K1 and montmorillonite M1) were refluxed with 200 ml of 0.25 M $H_2SO_4$ for 3 h. The resulting clays (acid activated kaolinite K2, acid activated montmorillonite M2) were centrifuged and washed with water several times till it was free of $SO_4^{2-}$ and dried at 383 K in an air oven until constant weight was attained. All the clays were calcined before using them as adsorbents at 773 K for 10 h.

### Characterization of the Adsorbents

#### *XRD Measurement*

XRD measurements were taken with Phillips Analytical X-ray spectrometer (PW 1710) using Cu Kα radiations.

#### *Surface Area*

The surface areas for the clay adsorbents were estimated utilizing the Sears' method (Sears, 1956). A sample containing 0.5 g of clay was acidified with 0.1 N HCl to pH 3.0 to 3.5. The volume was made up to 50 ml with distilled water after addition of 10.0 g of NaCl. The titration was carried out with standard 0.1 M NaOH in a thermostatic bath at 298 $\pm$ 0.5 K to pH 4.0, and then to pH 9.0. The volume, V, required to raise the pH from 4.0 to 9.0 was noted and the surface area was computed from the following equation

$$S\ (m^2/g) = 32\ V - 25 \qquad (1)$$

### *Cation Exchange Capacity*

The CEC of the clays was estimated by using the copper bisethylenediamine complex method (Bergaya and Vayer, 1997). 50 ml of 1 M $CuCl_2$ solution was mixed with 102 ml of 1 M ethylenediamine solution to allow for the formation of the $[Cu(en)_2]^{2+}$ complex . The slight excess of the amine ensures complete formation of the complex. The solution is diluted with water to one liter to give a 0.05 M solution of the complex. 0.5 g of a dry clay sample was mixed with 5 ml of the complex solution in a 100 ml flask, diluted with distilled water to 25 ml and the mixture was shaken for 30 min in a thermostatic water bath and centrifuged. The concentration of the complex remaining in the supernatant is determined by iodometric method. For this, 5 ml of the supernatant was mixed with 5 ml of 0.1 M HCl to destroy the $[Cu(en)_2]^{2+}$ complex and KI salt was added at 0.5 g per ml of solution. The mixture was titrated with 0.02 M $Na_2S_2O_3$ solution with starch as indicator. The CEC was calculated from the following formula:

$$CEC\ (meq/100\ g) = MSV(x\text{-}y)/1000m \qquad (2)$$

such that

M = the molar mass of the complex,
S = the concentration of the thio solution,
V = volume (ml) of the complex taken for iodometric titration,
m = the mass of adsorbent taken (g),
x = volume (ml) of thio required for blank titration (without the adsorbent), and
y = volume (ml) of thio required for the titration (with the adsorbent).

## Adsorbate Solutions

The adsorption experiments were conducted by using synthetic effluents containing Cu(II), Cd(II), Ni(II) and Pb(II) separately. The stock solution containing 1000 mg of Cu(II), Cd(II), Ni(II) and Pb(II) per liter was prepared by dissolving $CuSO_4.5H_2O$ (Qualigens, Mumbai, India), $Cd(NO_3)_2 \cdot 4H_2O$ (Qualigens, Mumbai, India) and $Ni(NO_3)_2.6H_2O$ (Qualigens, Mumbai, India) and $Pb(NO_3)_2$ (Glaxo, Mumbai, India) respectively in 1 L of double distilled water and were used to prepare the adsorbate solutions by appropriate dilution. The pH of the aqueous solution of Cd(II) as prepared was 5.5 while that of the aqueous solution of Cu(II), Ni(II) and Pb(II) as prepared was 5.7.

### *Batch Adsorption*

The adsorption experiments were carried out in 100 ml Erlenmeyer flasks by mixing clay and 50 mL aqueous solution of metal ions and agitating the mixture in a constant temperature water bath thermostat for a desired time interval. The mixture was centrifuged (Remi R 24, 20000 rpm) and the metal ions remaining unadsorbed in the supernatant liquid were determined with Atomic Absorption Spectroscopy (Varian SpectrAA 220, air-acetylene oxidizing flame, Cd(II): lamp current 4 mA, wavelength 228.8 nm, slit width 0.5 nm, optimum working range 0.02 – 3.0 μg/mL; Cu(II): lamp current 4 mA, wavelength 324.8 nm, slit width 0.5 nm, optimum working range 0.02 – 3.0 μg/mL; Ni(II): lamp current 4 mA, wavelength 232 nm, slit width 0.5 nm, optimum working range 0.02 – 3.0 μg/ml; Pb(II): lamp current 5 mA, 217.0 nm, slit width 1.0 nm, working range 0.1 – 30 μg/mL).

The pH of the adsorbate solutions was adjusted, if necessary, by adding 0.01 N NaOH or 0.01 N $HNO_3$ in drops. The following conditions were maintained for the different sets of experiments:

(i) *Effects of pH:*

Cd(II): Clay 2 g/L, Cd(II) 50 mg/L, temperature 303 K, interaction time 240 min, pH 1.0 to 10.0 at unit intervals

Cu(II): Clay 2 g/L, Cu(II) 50 mg/L, temperature 303 K, interaction time 360 min, pH 1.0 to 6.0 at unit intervals

Ni(II): Clay 2 g/L, Ni(II) 50 mg/L, temperature 303 K, interaction time 180 min, pH 1.0 to 10.0 at unit intervals

Pb(II): Clay 2 g/L, Pb(II) 50 mg/L, temperature 303 K, interaction time 180 min, pH 1.0 to 6.0 at unit intervals

(ii) *Kinetics:*

Cd(II): Clay 2 g/L, Cu(II) 50 mg/L, temperature 303 K, pH 5.5, time 20, 40, 60, 90, 120, 180, 240, 300, 360 min

Cu(II): Clay 2 g/L, Cu(II) 50 mg/L, temperature 303 K, pH 5.7, time 15, 30, 45, 60, 90, 120, 150, 180, 240, 360 min

Ni(II): Clay 2 g/L, Ni(II) 50 mg/L, temperature 303 K, pH 5.7, time 20, 40, 60, 90, 120, 150, 180, 240, 300, 360 min

Pb(II): Clay 2 g/L, Pb(II) 50 mg/L, temperature 303 K, pH 5.7, time 20, 40, 60, 90, 120, 150, 180, 240, 300, 360 min

(iii) *Effects of adsorbent amount:*

Cd(II): Cd(II) 50 mg/L, temperature 303 K, pH 5.5, time 240 min, Clay 2, 3, 4, 5, 6 g/L

Cu(II): Cu(II) 50 mg/L, temperature 303 K, pH 5.7, time 360 min, Clay 2, 3, 4, 5, 6 g/L

Ni(II): Ni(II) 50 mg/L, temperature 303 K, pH 5.7, time 180 min, Clay 2, 3, 4, 5, 6 g/L

Pb(II): Pb(II) 50 mg/L, temperature 303 K, pH 5.7, time 180 min, Clay 2, 3, 4, 5, 6 g/L

(iv) *Effects of adsorbate concentration:*

Cd(II): Clay 2 g/L, temperature 303 K, pH 5.5, time 240 min, Cd(II) 10, 20, 30, 40, 50 mg/L

Cu(II): Clay 2 g/L, temperature 303 K, pH 5.7, time 360 min, Cu(II) 10, 20, 30, 40, 50 mg/L

Ni(II): Clay 2 g/L, temperature 303 K, pH 5.7, time 180 min, Ni(II) 10, 20, 30, 40, 50 mg/L

Pb(II): Clay 2 g/L, temperature 303 K, pH 5.7, time 180 min, Pb(II) 10, 20, 30, 40, 50 mg/L

(v) *Isotherm*

Cd(II): Clay 2 g/L, temperature 303 K, pH 5.5, time 240 min, Cd(II) 10, 20, 30, 40, 50 mg/L

Cu(II): Clay 2 g/L, temperature 303 K, pH 5.7, time 360 min, Cu(II) 10, 20, 30, 40, 50 mg/L

Ni(II): Clay 2 g/L, temperature 303 K, pH 5.7, time 180 min, Ni(II) 10, 20, 30, 40, 50 mg/L

Pb(II): Clay 2 g/L, temperature 303 K, pH 5.7, time 180 min, Pb(II) 10, 20, 30, 40, 50 mg/L

(vi) *Thermodynamics:*

Cd(II): Clay 2 g/L, time 240 min, pH 5.5, temperature 303, 308, 313 K, Cd(II) 10, 20, 30, 40, 50 mg/L

Cu(II): Clay 2 g/L, time 360 min, pH 5.7, temperature 303, 308, 313 K, Cu(II) 10, 20, 30, 40, 50 mg/L

Ni(II): Clay 2 g/L, time 180 min, pH 5.7, temperature 303, 308, 313 K, Ni(II) 10, 20, 30, 40, 50 mg/L

Pb(II): Clay 2 g/L, time 180 min, pH 5.7, temperature 303, 308, 313 K, Pb(II) 10, 20, 30, 40, 50 mg/L

The amount, q (mg/g) of metal ions adsorbed per unit mass of the adsorbent and the extent of adsorption (in %) are given by the following expressions:

$$q = (C_o - C_t) / m$$

$$\% \text{ Adsorption} = (C_o - C_t) \times 100 / C_o$$

where $C_o$ (mg/L) and $C_t$ (mg/L) are the concentrations of the metal ions before and after adsorption for a time interval of t, and m (g) is the amount of clay adsorbent taken for 1 L of solution.

## Theoretical Foundation

Adsorption process involves an array of phenomena that can alter the distribution of solutes among the constituent phases and interfaces of subsurface system (Weber et al., 1991). Different types of inter-molecular forces involving permanent dipoles, induced dipoles and van der Waal's forces facilitate adsorbate- adsorbent interactions (Thomas and Crittenden, 1998). A number of parameters specific for a given adsorption process influence the system. Some of these are molecular mass, size, structure and polarity, concentration, steric form of adsorbate, surface area, the physicochemical nature of the surface, the availability of adsorbate ions, the physical size of adsorbent particles, etc. Sometimes, the extent of adsorption also influences system parameters like pH and temperature (Weber, 1985).

## Kinetics of Adsorption

The adsorption kinetics normally includes two phases: a rapid removal stage followed by a much slower stage before the equilibrium is established (Ho and McKay, 1999b). The rate law has three primary requirements (Ho and McKay, 1999c):

(i) Knowledge of all the molecular details of the reaction including the energetics and stereochemistry,
(ii) Interatomic distances and angles throughout the course of the reaction and,
(iii) The individual molecular steps involved in the mechanism.

In the present work, the adsorption kinetics was tested with respect to the following well known models:

(a) Pseudo first order kinetics based on the Lagergren equation (Lagergren, 1898; Ho, 2004):

$$dq_t / dt = k_1 (q_e - q_t) \quad (3)$$

where $k_1$ is the pseudo first order adsorption rate constant. The integrated form of the equation (3) for the boundary condition $t = 0$ to $t = t$ and $q_t = 0$ to $q_t = q_t$,

$$\ln (q_e - q_t) = \ln q_e - k_1 t \quad (4)$$

where $q_e$ and $q_t$ are the values of amount adsorbed per unit mass at equilibrium and at any time t is usually utilized. The values of $k_1$ can be obtained from the slope of the linear plot of log $(q_e-q_t)$ vs. t.

b) When the applicability of the first order kinetics becomes untenable (for example, if $q_e$ obtained on the basis of equation (4) is not the same as the one obtained experimentally), the second order equation (Ho and McKay, 1999d):

$$dq_t / dt = k_2 (q_e - q_t)^2 \quad (5)$$

is applied with a view to obtain $k_2$, the second order rate constant. The integrated form of the equation (5) under the boundary conditions of $q_t = 0$ at $t = 0$ and $q_t = q_t$ at $t = t$ in the linear form,

$$t/q_t = 1/(k_2 q_e^2) + (1/q_e) .t \quad (6)$$

makes it possible to obtain $q_e$ and $k_2$ from the plots of $t/q_t$ vs. t.

(b) As the Elovich Equation (Ho and McKay, 1998; Rudzinski and Panczyk, 2002) has been found useful in describing second order kinetics for energetically heterogeneous solid surfaces, this equation was also used in the form

$$dq_t / dt = \alpha \exp(-\beta q_t) \qquad (7)$$

Assuming $\alpha\beta t >> 1$, and $q_t = 0$ at $t = 0$ and $q_t = q_t$ at $t = t$, the linear form of the equation (7) is given by (Chien and Clayton, 1980):

$$q_t = \beta \ln(\alpha\beta) + \beta \ln t \qquad (8)$$

where $\alpha$ and $\beta$, known as the Elovich coefficients, represent the initial sorption rate (g $mg^{-1}min^{2}$) and the desorption coefficient (mg $g^{-1}min^{-1}$) respectively. The Elovich coefficients could be computed from the plots of $q_t$ vs. ln t.

(c) Due to rapid stirring in a batch adsorption study, adsorbate species can be transported from aqueous solution to the solid surface of adsorbent and there is a possibility of slow diffusion of adsorbate species into the interior of the pores of the adsorbent. The intra-particle diffusion (Weber and Morris, 1963) rate constant ($k_i$) is obtained from the equation:

$$q_t = k_i.t^{0.5} \qquad (9)$$

If the plots of $q_t$ vs. $t^{0.5}$ are linear and pass through the origin, then it is to be understood that intra-particle diffusion has a significant role in the adsorption process and the rate constant, $k_i$, could be obtained from the slope of the plots.

e) When the flow of the reactant from the bulk liquid to the surface of the adsorbent is the slowest process determining kinetics of the rate processes, the liquid film diffusion model (Boyd et al., 1947) given by the simple relation

$$\ln(1 - F) = - k_{fd}t \qquad (10)$$

could be the appropriate way to characterize the kinetics. The fractional attainment of equilibrium, F (= $q_t/q_e$) and the film diffusion coefficient, $k_{fd}$ are obtained from the plot of –ln (1 – F) vs. t, which should be linear with zero intercept.

## Adsorption Isotherm

The adsorption equilibrium is usually described by an isotherm equation whose parameters express the surface properties and affinity of the adsorbent at a fixed temperature and pH. An adsorption process is usually described by the following two widely used isotherms (Freundlich, 1906; Langmuir, 1918);

a) Freundlich isotherm: $q_e = K_fC_e^{n}$ (11)

where $C_e$ and $q_e$ are the equilibrium concentrations of the metal ions in the liquid phase and in the solid phase respectively, $K_f$ and n being Freundlich coefficients.

b) Langmuir isotherm: $C_e/q_e = 1/(bq_m) + (1/q_m)\, C_e$ (12)

where b and $q_m$ are Langmuir coefficients representing the equilibrium constant for the adsorbate-adsorbent equilibrium and the monolayer capacity respectively.

The linear Freundlich and Langmuir plots are obtained by plotting (i) log $q_e$ vs. log $C_e$ and (ii) $C_e/q_e$ vs. $C_e$ respectively, from which the adsorption coefficients could be evaluated. The Langmuir equation is also used to obtain $R_L$, the dimensionless equilibrium parameter or the separation factor (Hall et al. 1966), from the expression:

$$R_L = 1/\,(1+bC_e) \quad (13)$$

The shapes of the isotherms are indicated by $R_L$ values as follows (Ho, 2003):

| *$R_L$ value* | *Type of isotherm* |
|---|---|
| $0< R_L <1$ | Favourable |
| $R_L >1$ | Unfavourable |
| $R_L =1$ | Linear |
| $R_L =0$ | Irreversible |

## Thermodynamic Study

The thermodynamic parameters for the adsorption process, ΔH (kJ $mol^{-1}$), ΔS (J $K^{-1}$ $mol^{-1}$) and ΔG (kJ $mol^{-1}$), were evaluated using the equation (Thomas and Crittenden, 1998):

$$\Delta G = -\,RT \ln K_d \quad (14)$$

$$\Delta G = \Delta H - T\Delta S \quad (15)$$

Equation (14) and (15) can be expressed as:

$$\ln K_d = \Delta S/R - \Delta H/RT \quad (16)$$

where $K_d$, known as the distribution coefficient of the adsorbate, is equal to $q_e/C_e$ (L/g), T the temperature (K), and $R = 8.314 \times 10^{-3}$ kJ $K^{-1}$ $mol^{-1}$. The plot of ln $K_d$ vs. 1/T should be linear with the slope ($-\Delta H/R$) and the intercept ($\Delta S/R$) giving the values of ΔH and ΔS. The Gibbs energy change, ΔG, is then obtained from the relation, $\Delta G = \Delta H - T\Delta S$ at constant temperature. All these relations are valid when the enthalpy change remains constant in the range of temperature considered.

# Results and Discussion

## Adsorbent Characterization

### *XRD Study*

Acid treatment and calcination introduces some changes into the crystal structure of clay minerals. The intensity of the characteristic XRD peaks of the clay minerals are reduced on acidification and also the peaks become widened (Radojevic et al., 2002). This implies changes in the regular pattern of the clay structure along with partial destruction compared to the original. The dispersion and amorphization of the acid treated clay minerals are known to give rise to an increase in the very low angle diffraction intensities (Jozefaciuk and Bowanko, 2002) and a reduction in the corresponding tip widths (Diaz and Santos, 2001). The XRD patterns (Figure 1) in the present work on acid-activated and calcined clays confirm these observations as shown below:

(i) For the calcined, acid activated montmorillonite, the basal spacing expanded from 4.44 to 4.47 Å ($2\theta = 19.98^{o}$), which was accompanied by a decrease in intensity from 43.06 to 29.09 %. In case of the calcined, acid activated kaolinite, widening of the basal spacing was comparatively less prominent (from 4.45 to 4.46 Å; $2\theta = 19.92^{o}$) and the peak intensity changed from 23.14 to 21.32 % only.

(ii) The intensity of most of the XRD peaks of montmorillonite decreased sharply on acid treatment such that both octahedral and tetrahedral sites might have been affected drastically. This type of change is significantly less in case of acid treated kaolinite. Treatment with strong acid causes preferential release of octahedral Al ions from the kaolinite structure with formation of additional Al-OH and Si-OH bonds without disturbing the mineral structure (Suraj et al., 1998).

(iii) The relative intensity of a low angle peak occurring at $2\theta = 5.70^{o}$ increased from 1.28 (kaolinite) to 4.44 % (Acid activated kaolinite). Similar effects were however not very prominent in montmorillonite.

(iv) Acid activation of montmorillonite yielded two new peaks at 22.91 Å ($2\theta = 3.85^{o}$) and 12.49 Å ($2\theta = 7.06^{o}$), which were absent in the untreated montmorillonite. Appearance of new peaks indicates the formation of expansible phases and interlamellar expansion (Fanning et al., 1989).

(v) The tip width of the $19.98^{o}$ peak ($2\theta$) in montmorillonite (M1) reduced from 0.32 to 0.30 (acid activated montmorillonite, M2); but this feature has not been observed in acid activated kaolinite.

Thus, activation with 0.25M $H_2SO_4$ influences the structural properties of both kaolinite and montmorillonite, the latter being affected much more. Similar results have been found earlier with vermiculite, illite and kaolinite treated with hydrochloric acid (Jozefaciuk and Bowanko, 2002).

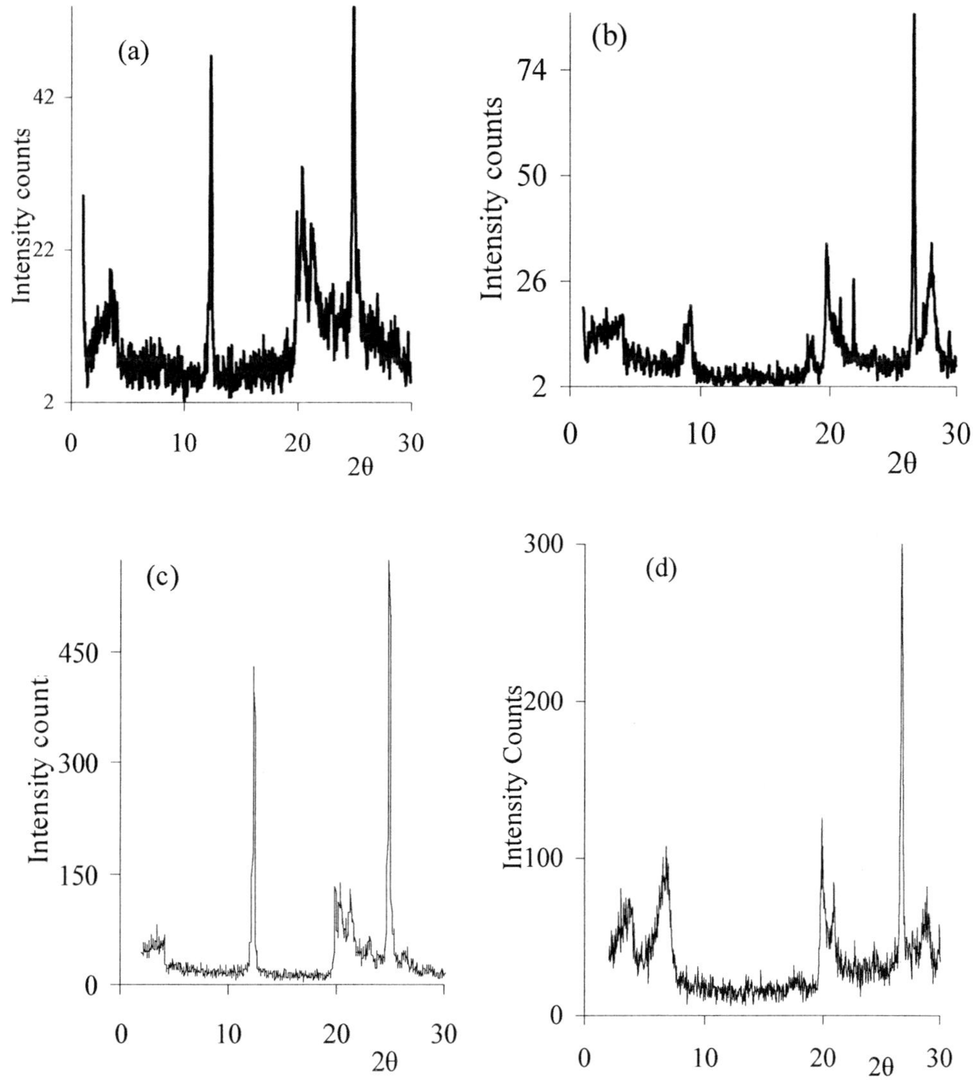


Fig.1 XRD spectra of (a) kaolinite (K1), (b) montmorillonite (M2), (c) acid-activated kaolinite (K2), (d) acid-activated montmorillonite (M2)

***Surface Area***

Activation of clay by acid treatment and subsequent calcination has been shown to be effective in limiting possible decomposition of the crystalline structure and in increasing the specific surface area (Sabah et al., 2002). The acid treatment opens up the edges of the platelets and as a consequence, the surface area and the pore diameter increase (Volzone et al., 1999).

In the present work, acid activation was responsible for increasing the specific surface area of kaolinite from 3.8 to 15.6 $m^2/g$ (> 400 %) and that of montmorillonite from 19.8 to 52.3 $m^2/g$ (> 250 %). Ravichandran and Sivasankar (1997) have shown that the specific surface area of montmorillonite (19.0 $m^2/g$) increased between 98.7 to 188.3 $m^2/g$ on treatment with HCl (0.1 to 0.7 M). Such high values of specific surface area were not achieved in the present work by treatment with 0.25 M $H_2SO_4$ acid. No report on the effect of acid treatment on the specific surface area of kaolinite could be found, but the untreated kaolinite was reported to have specific surface area of 5 to 25 $m^2/g$ (Volzone et al., 1999). Acid treatments thus had a positive impact on the surface area of the clay minerals.

### *Cation Exchange Capacity (CEC)*

Montmorillonite has a very large CEC compared to that of kaolinite, the values obtained being 11.3 meq/100g (kaolinite) and 153.0 meq/100g (montmorillonite) in agreement with the reported values (Grim, 1968). On acid treatment, the CEC of kaolinite and montmorillonite increased to 12.2 meq/100g and 341.0 meq/100g respectively. In a previous work, Ghosh and Bhattacharyya (2002) have reported CEC of 1.94 to 3.08 meq/100 g for some locally available impure kaolinite samples based on measurement by the conventional $BaCl_2$ method.

Clays contain Brönsted acid sites associated with the interlamellar region and the Lewis acid sites allied with the edge sites. When clay is heated (to ~ 373 K), most of the interlamellar water is removed leaving only 'one layer' of water and the Brönsted acidity increases markedly (Yahiaoui et al., 2003) to that of a very strong acid. The clay interlayer structure collapses if the temperature is raised to 473 – 573 K due to the removal of the structural water and the Lewis acidity increases at the expense of the Brönsted acidity. Further heating (calcination at ~ 700 K) results in complete dehydroxylation of the aluminosilicate lattice, while retaining the Lewis acidity.

The ion exchange capacity of clay minerals is attributed to structural defects, broken bonds and structural hydroxyl transfers (Radojevic et al., 2002), which are obviously in short supply in kaolinite than in montmorillonite. Acid treatment increases the total number of exchange sites marginally in kaolinite (CEC increase ~ 8 %) and quite profoundly in montmorillonite (CEC increase ~ 123 %). The treatment of the clays, montmorillonite in particular, with 0.25 M $H_2SO_4$ results in replacement of a large number of different cations with $H^+$ ions and on subsequent heating and calcination, dehydroxylation occurs leaving behind a large number of Lewis acid sites. Much of the increase in CEC in the present work is likely to be due to increase in Lewis acidity as the acid treated clay was calcined at 773 K before CEC measurement. Further, acid treatment of montmorillonite has been shown to create enhanced mesoporosity with important textural and structural changes (Rhodes et al., 1991). The CEC increase may be a complex process involving these effects as well.

## Adsorption of Heavy Metals

### *Effects of pH*

The pH of the aqueous solution is an important controlling parameter in the adsorption process. It is not possible to carry out adsorption experiments with Cu(II) and Pb(II) at pH > 6.0 and with Ni(II) at pH > 8.0 due to precipitation of the metals as the hydroxides, which introduces uncertainty into the interpretation of the results. The precipitation was very rapid for Cu(II) and Pb(II) at pH > 6.0 and the experiments had to be terminated at pH 6.0. The process of precipitation was slow for Ni(II) and the adsorption measurements could be carried out till pH 10.0. However, from blank experiments, it was observed that precipitation commenced at pH > 8.0 and therefore the large increase in the extent of adsorption after this pH had to be attributed primarily to onset of precipitation. Adsorption of Cd(II) could be studied over the pH range of 1.0 to 10.0 without any abrupt changes in $q_e$. The general trend in the extent of adsorption (%) as well as amount adsorbed ($q_e$) with increase in pH is positive. Figure 2 represents the variation of amount adsorbed ($q_e$) in different pHs for all the clay adsorbents. The significant feature of the plots is that the curve for the acid activated clay is almost parallel to the one for non-activated clay (whether kaolinite or montmorillonite). It can therefore be inferred that acid activation enhances the adsorption capacity uniformly independent of the change in pH.

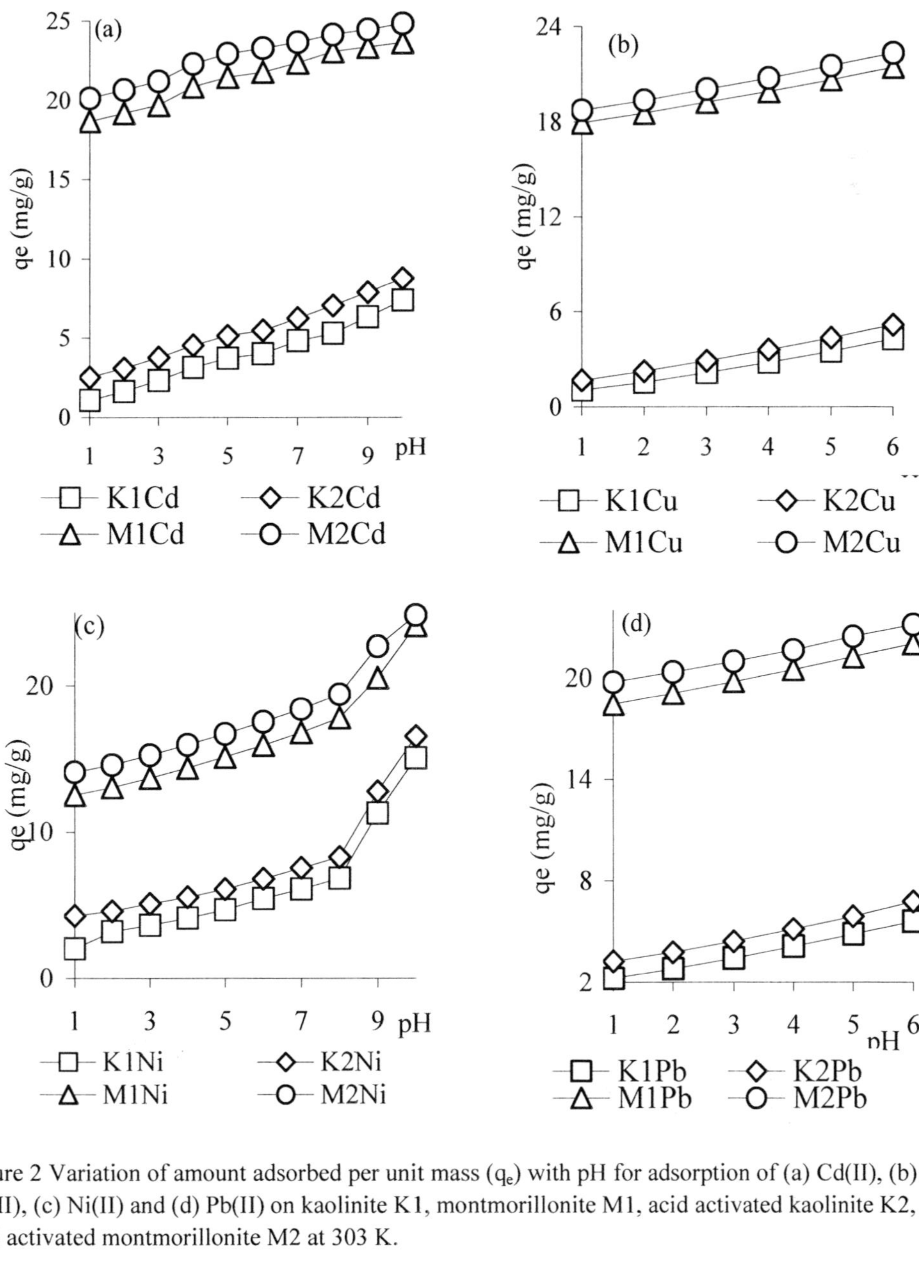


Figure 2 Variation of amount adsorbed per unit mass ($q_e$) with pH for adsorption of (a) Cd(II), (b) Cu(II), (c) Ni(II) and (d) Pb(II) on kaolinite K1, montmorillonite M1, acid activated kaolinite K2, and acid activated montmorillonite M2 at 303 K.

At very low pH, the number of $H_3O^+$ ions exceeds that of the metal ions several times and the metal ions can hardly compete with $H_3O^+$ ions for the binding sites on the clay adsorbents. The surface remains covered with $H_3O^+$ ions and the metal ions are compelled to remain in the liquid phase. With an increase in pH, the concentration of $H_3O^+$ ions decreases and some of the sites become available to the metal ions. As the acidity decreases, more and more $H_3O^+$ ions on clay surface are replaced by metal ions (Taty-Costodes et al., 2003). As the pH turns towards alkaline side, precipitation of the insoluble metal-hydroxides may add to higher metal ion removal from solution (Yu et al., 2001; Bayramglu et al., 2003). This happens at a comparatively lower pH (~6.0) for Cu(II) (Nuhoglu and Oguz, 2003) and Pb(II) (Jain and Ram, 1997) while the adsorption could be safely carried out to pH ~10.0 for Cd(II) (Mathialagan and Viraraghavan, 2002) and ~ for Ni(II) (Padmavathy et al., 2003).

The effect of pH on the adsorption of metal ions on clay adsorbents may be explained on the basis of aqua complex formation of the oxides present in the minerals. The oxygen atoms present on the clay surface interact with water in an acidic medium forming some aqua complexes (Mathialagan and Viraraghavan, 2002), which result in positive charge formation as follows:

$$-MO + H-OH \xrightarrow{H^+} M-OH_2{}^+ + OH^-$$

This surface positive charge is responsible for preventing metal ions from approaching the surface and explains the smaller extent of adsorption at low pH. In an alkaline medium, the clay surface becomes negatively charged favouring $M^{+2}$ uptake:

$$-MOH + OH^- \rightleftharpoons -MO^- + H_2O$$
$$-MO^- + M^{2+} \rightleftharpoons -M\text{–}O^- \cdots\cdots M^{2+}$$

Montmorillonite (2:1 clay) has high surface charges resulting from the spread of isomorphous substitution in tetrahedral and octahedral sheets, whereas 1:1 layered kaolinite has little isomorphous substitution (Chantawong et al., 2003). This accounts for its low capacity to adsorb cations. It is clear from Figure 2 that acid activation increases the number of sites responsible for adsorption of various metal-species and at any pH, the acid activated kaolinite or montmorillonite had a higher adsorption capacity compared to the non-activated clay. This can be seen from the following data:

| Adsorbate | pH | Amount adsorbed ($q_e$ mg/g) and extent of adsorption (%) | | | |
|---|---|---|---|---|---|
| | | Non-activated kaolinite | Acid-activated kaolinite | Non-activated montmorillonite | Acid-activated montmorillonite |
| Cu(II) | 2.0 | 1.6 (6.2) | 2.3 (9.0) | 17.9 (71.7) | 19.4 (77.4) |
| | 6.0 | 4.3 (17.1) | 5.2 (20.7) | 21.4 (85.7) | 22.3 (89.3) |
| Cd(II) | 2.0 | 1.6 (6.2) | 3.1 (12.4) | 19.2 (76.8) | 20.7 (82.6) |
| | 6.0 | 4.0 (16.0) | 5.5 (21.9) | 21.8 (87.1) | 23.3 (93.1) |
| Ni(II) | 2.0 | 3.2 (12.8) | 4.6 (18.4) | 13.1 (52.2) | 14.6 (58.3) |
| | 6.0 | 5.5 (21.4) | 6.8 (27.2) | 15.9 (63.7) | 17.5 (70.1) |
| Pb(II) | 2.0 | 2.8 (11.2) | 3.8 (15.1) | 19.1 (76.3) | 20.4 (81.4) |
| | 6.0 | 5.6 (22.4) | 6.8 (27.1) | 22.0 (88.1) | 23.2 (92.6) |

Note: The values in the parenthesis indicate the extent of adsorption (%)

## Effects of Interaction Time and Kinetics of Adsorption

The equilibrium for the interactions was studied by measuring adsorption capacities of the clays for the metal ions at various interaction times. For all the metal ions, adsorption was very fast initially and maximum uptake was recorded within 40 min for Pb(II), Cd(II) and Ni(II), and 45 min for Cu(II). As seen from Figure 3, the equilibrium was approached very slowly. Pb(II)-clay and Ni(II)-clay interactions reached equilibrium at ~180 min, Cd(II)-clay interactions at ~240 min, and Cu(II)-clay interaction at ~360 min. How the extent of adsorption (%) changes between 40 min and equilibrium time is shown in Table 1.

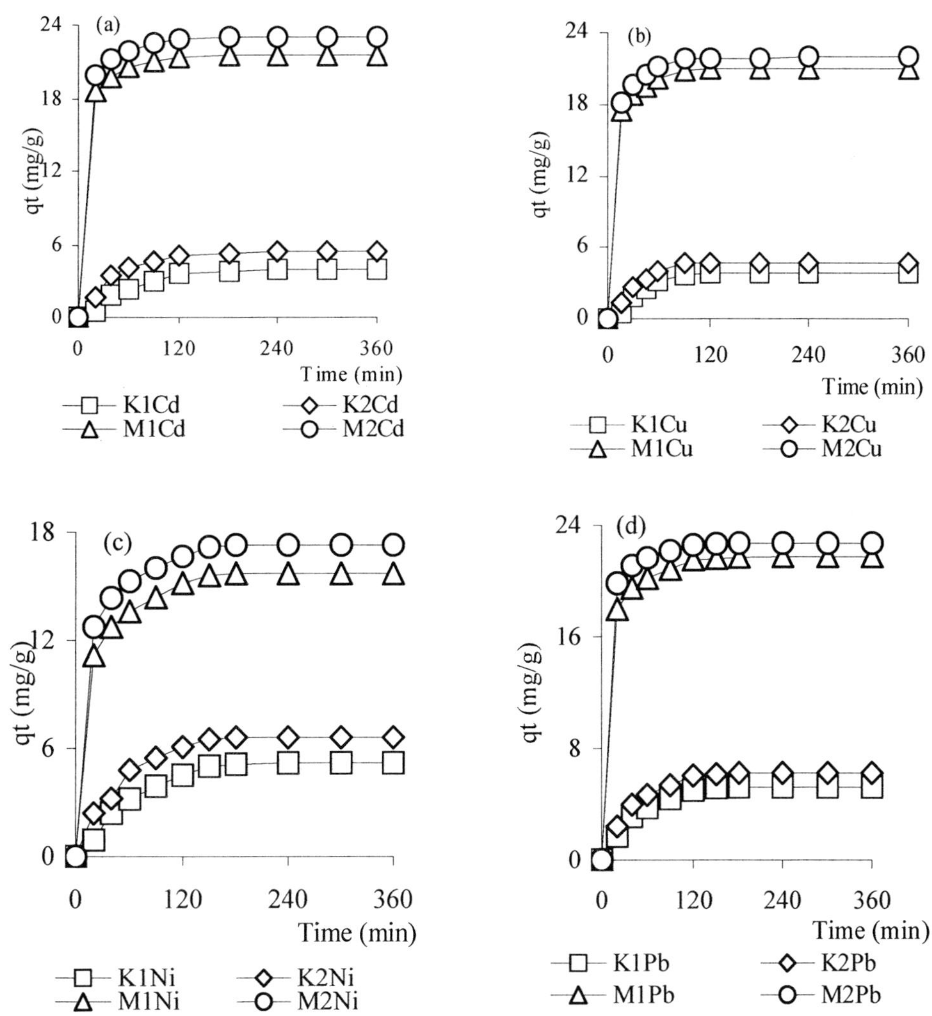


Figure 3 Influence of interaction time on amount adsorbed per unit mass ($q_e$) for adsorption of (a) Cd(II), (b) Cu(II), (c) Ni(II) and (d) Pb(II) on kaolinite K1, montmorillonite M1, acid activated kaolinite K2, and acid activated montmorillonite M2 at 303 K.

**Table 1. Typical data on influence of interaction time on extent of adsorption (%) of Cd(II), Cu(II), Pb(II) and Ni(II), on clay adsorbents at 40 min and at equilibrium time at 303 K (clay 2 g/L, initial metal concentration 50 mg/L, pH 5.5 for Cd(II), 5.7 for Cu(II), Pb(II) and Ni(II), K1 kaolinite, K2 acid activated kaolinite, M1 montmorillonite and M2 acid activated montmorillonite)**

| Metal ion | Clay | Equilibrium time (min) | Extent of adsorption (%) at | |
|---|---|---|---|---|
| | | | 40 min | Equilibrium |
| Cd(II) | K1 | 240 | 7.3 | 15.9 |
| | K2 | | 13.8 | 21.4 |
| | M1 | | 79.2 | 86.4 |
| | M2 | | 84.8 | 92.1 |
| Cu(II) | K1 | 360 | 10.0 | 15.7 |
| | K2 | | 13.5 | 19.1 |
| | M1 | | 77.9 | 84.1 |
| | M2 | | 81.6 | 87.7 |
| Ni(II) | K1 | 180 | 9.7 | 20.7 |
| | K2 | | 16.2 | 26.5 |
| | M1 | | 50.9 | 62.8 |
| | M2 | | 57.4 | 69.2 |
| Pb(II) | K1 | 180 | 12.5 | 21.0 |
| | K2 | | 15.9 | 25.0 |
| | M1 | | 77.9 | 86.9 |
| | M2 | | 84.5 | 90.8 |

During the experiment, initial and final pH values were measured which did not change significantly in all the cases. The acid treatment has resulted in a higher uptake of metal ions at the same interaction time. This can be attributed to increase in the number of equivalent adsorption sites on both kaolinite and montmorillonite following acid treatment.

The initial uptake rate for all the metals was very high as a large number of adsorption sites were available for adsorption at the onset of the process. As the sites are gradually filled up, adsorption becomes slow and the kinetics becomes more dependent on the rate at which the adsorptive is transported from the bulk phase to the actual adsorption sites (Yu et al., 2000). It is to be noted that at equilibrium, the net increase in metal ion uptake for kaolinite and acid activated kaolinite, and for montmorillonite and acid activated montmorillonite was almost equal. Acid activation thus influences the surface characteristics of both the clays almost to the same extent.

The rates of adsorption were first tested with the pseudo first order mechanism of Lagergren by plotting log ($q_e - q_t$) vs. t (min) (Figure 4). The regression coefficients and the first order rate constants obtained from these plots are given in Table 2. All the plots are linear showing that the interactions might be following first order mechanism and that the rate is dependent on initial concentration. However, linearity of the Lagergren plots does not establish a first order mechanism for the interactions (Ho and McKay, 2002) due to the inherent disadvantage of correctly estimating the equilibrium adsorption capacity, $q_e$. This was observed when $q_e$ values obtained from the Lagergren plots were compared with the experimental $q_e$ values (Table 3). The deviations between the experimental $q_e$ values and those obtained from Lagergren plots were very large as shown below:

(i) 83.8 to + 24.1 % for Cd(II),

(ii) 80.5 to + 15.4 % for Cu(II),

(iii) 27.2 to + 90.9 % for Ni(II), and

(iv) 65.5 to +160.6 % for Pb(II).

**Table 2. First and second order rate constants for adsorption of metal ions on clays at 303 K (clay 2 g/L, initial metal concentration 50 mg/L, pH 5.5 for Cd(II), 5.7 for Cu(II), Pb(II) and Ni(II), K1 kaolinite, K2 acid activated kaolinite, M1 montmorillonite and M2 acid activated montmorillonite)**

| Metal ion | Clay | Pseudo first order | | Pseudo second order | |
|---|---|---|---|---|---|
| | | $k_1 \times 10^2$ ($min^{-1}$) | r | $k_2 \times 10^2$ ($g\ mg^{-1} min^{-1}$) | r |
| Cd(II) | K1 | 1.9 | – 0.99 | 3.2 | + 0.99 |
| | K2 | 2.2 | – 0.99 | 4.0 | + 0.99 |
| | M1 | 1.9 | – 0.73 | 3.0 | + 0.99 |
| | M2 | 2.3 | – 0.99 | 11.1 | + 1.00 |
| Cu(II) | K1 | 4.1 | – 0.99 | 4.6 | + 0.99 |
| | K2 | 2.8 | – 0.96 | 12.8 | + 0.99 |
| | M1 | 4.1 | – 0.99 | 11.8 | + 0.99 |
| | M2 | 3.2 | – 0.98 | 15.8 | + 0.99 |
| Ni(II) | K1 | 2.6 | – 0.98 | 2.2 | + 0.99 |
| | K2 | 2.9 | – 0.98 | 2.5 | + 0.99 |
| | M1 | 3.1 | – 0.97 | 5.3 | + 0.99 |
| | M2 | 3.1 | – 0.96 | 5.9 | + 0.99 |
| Pb(II) | K1 | 3.1 | – 0.99 | 3.5 | + 0.99 |
| | K2 | 4.1 | – 0.95 | 4.1 | + 0.99 |
| | M1 | 2.7 | – 0.95 | 8.4 | + 0.99 |
| | M2 | 4.0 | – 0.97 | 11.2 | + 0.99 |

The first order kinetics is therefore inappropriate to explain the rate processes. Second order kinetic plots ($t/q_e$ vs. t) (Figure 5) had better linearity (r ~ + 0.99, Table 2) and appeared to give a better understanding of the interactions. Acid activation enhanced the second order rate constant for adsorption of the metal ions on both kaolinite and montmorillonite as shown below:

| Second order rate constant, $k_2$, in $g\ mg^{-1}\ min^{-1}$ | | | | |
|---|---|---|---|---|
| | Kaolinite | Acid-kaolinite | Montmorillonite | Acid-montmorillonite |
| Ni(II) | $2.2 \times 10^{-2}$ | $2.5 \times 10^{-2}$ | $5.3 \times 10^{-2}$ | $5.9 \times 10^{-2}$ |
| Cd(II) | $3.2 \times 10^{-2}$ | $4.0 \times 10^{-2}$ | $3.0 \times 10^{-2}$ | $11.1 \times 10^{-2}$ |
| Pb(II) | $3.5 \times 10^{-2}$ | $4.1 \times 10^{-2}$ | $8.4 \times 10^{-2}$ | $11.2 \times 10^{-2}$ |
| Cu(II) | $4.6 \times 10^{-2}$ | $12.8 \times 10^{-2}$ | $11.8 \times 10^{-2}$ | $15.8 \times 10^{-2}$ |

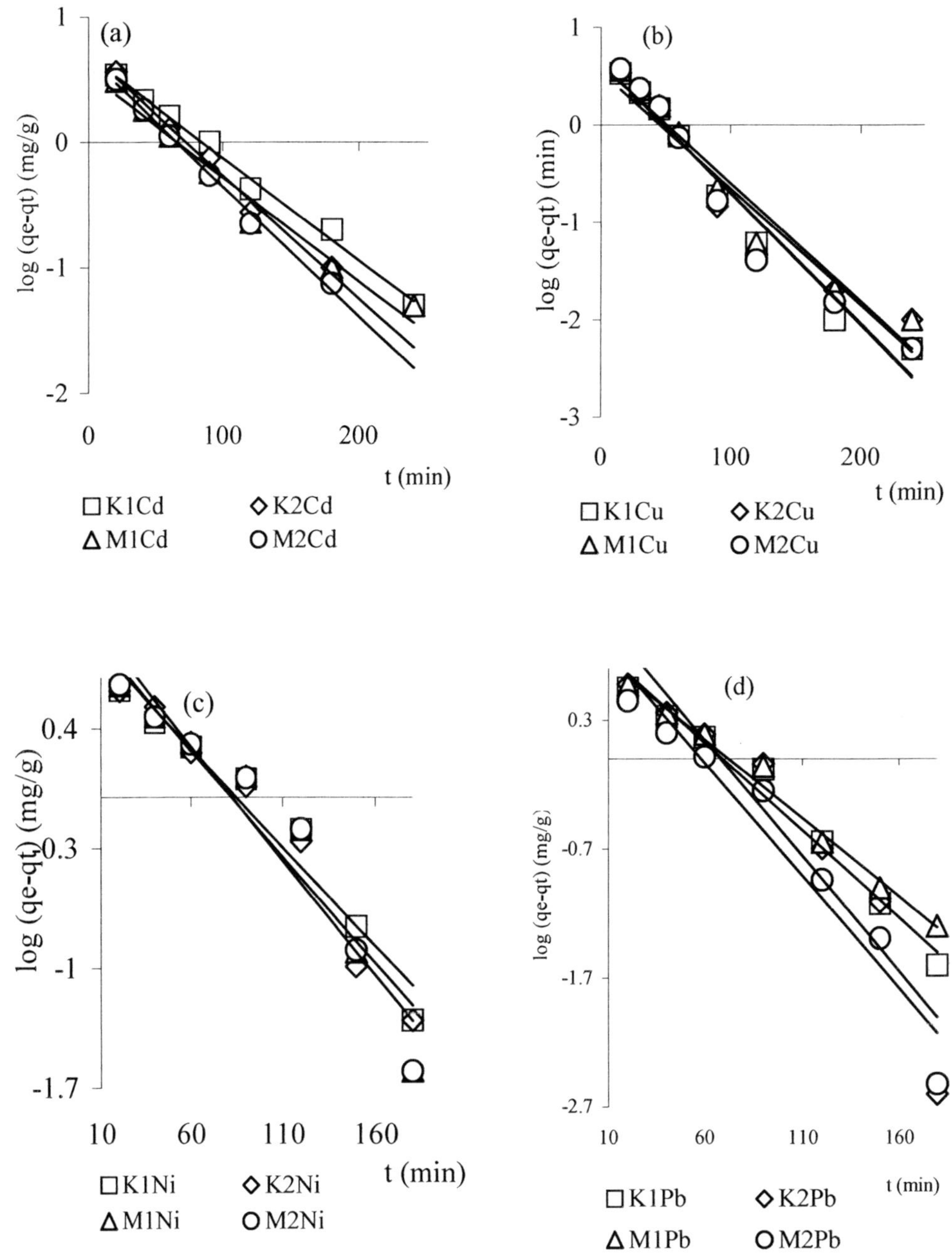


Figure 4 Lagergren pseudo first order plots for adsorption of (a) Cd(II), (b) Cu(II), (c) Ni(II) and (d) Pb(II) on kaolinite K1, montmorillonite M1, acid activated kaolinite K2, and acid activated montmorillonite M2 at 303 K.

**Table 3. Experimental and computed $q_e$ values from Lagergren and second order plots (Temperature 303 K, clay 2 g/L, initial metal ion concentration 50 mg/L, pH 5.5 for Cd(II), 5.7 for Cu(II), Pb(II) and Ni(II), K1 kaolinite, K2 acid activated kaolinite, M1 montmorillonite and M2 acid activated montmorillonite)**

| Metal ion | Clay | $q_e$ (mg/g) (Expt) | Lagergren Plot | | Second order plot | |
|---|---|---|---|---|---|---|
| | | | $q_e$ (mg/g) | Deviation (%) | $q_e$ (mg/g) | Deviation (%) |
| Cd(II) | K1 | 4.0 | 4.9 | + 24.1 | 4.0 | 0 |
| | K2 | 5.4 | 6.4 | +18.5 | 5.1 | -5.6 |
| | M1 | 21.6 | 3.5 | - 83.8 | 18.2 | -15.7 |
| | M2 | 23.0 | 4.7 | -79.5 | 23.4 | + 1.7 |
| Cu(II) | K1 | 3.9 | 4.5 | + 15.4 | 4.3 | +10.2 |
| | K2 | 4.8 | 3.5 | – 27.1 | 5.1 | + 6.3 |
| | M1 | 21.0 | 4.1 | – 80.5 | 21.3 | + 1.4 |
| | M2 | 21.9 | 4.4 | – 79.9 | 22.2 | + 1.4 |
| Ni(II) | K1 | 5.2 | 9.4 | + 80.8 | 6.9 | + 32.7 |
| | K2 | 6.6 | 12.6 | + 90.9 | 8.2 | + 24.2 |
| | M1 | 15.7 | 12.7 | – 19.1 | 16.5 | + 5.1 |
| | M2 | 17.3 | 12.6 | – 27.2 | 18.2 | + 5.2 |
| Pb(II) | K1 | 5.3 | 8.5 | +61.7 | 6.4 | +20.7 |
| | K2 | 6.3 | 16.3 | +160.6 | 7.3 | +15.8 |
| | M1 | 21.7 | 7.4 | -65.5 | 22.3 | +2.7 |
| | M2 | 22.7 | 10.2 | -55.5 | 23.3 | +2.6 |

Acid activation enhanced the second order rate constant sufficiently in case of Cu(II) adsorption on both kaolinite and montmorillonite and also for Cd(II) and Pb(II) adsorption on montmorillonite. The influence was minimal for Ni(II) adsorption on both kaolinite and montmorillonite as well as Cd(II) and Pb(II) adsorption on kaolinite. The affinity of the acid activated montmorillonite towards metal ions is definitely much more and the interactions take place rapidly compared to those in case of non-activated clays, and even activated kaolinite.

The experimental $q_e$ values and those obtained from the slopes of the second order plots now show much better agreement (Table 3). Some deviations still exist that might be due to the uncertainty inherent in obtaining the experimental $q_e$ values and also due to the actual process being not in conformity with simple first order or second order kinetics.

Interactions of Cu(II), Cd(II), Ni(II) and Pb(II) with kaolinite and montmorillonite yielded good, linear Elovich plots ($q_t$ vs. ln t, Figure 6) with correlation coefficients of ~ +0.91 to +0.99 (Table 4). The Elovich equation describes predominantly chemical adsorption on highly heterogeneous adsorbents, but the equation does not propose any definite mechanism for adsorbate-adsorbent interaction while predicting that metal ions are held strongly to the clay surface by chemisorptive bonds. With all the metal ions, acid activated montmorillonite had a much higher initial rate of uptake ($\alpha$), which might be due to very high surface area and CEC of acid activated montmorillonite compared to those of the other clay adsorbents used in this study.

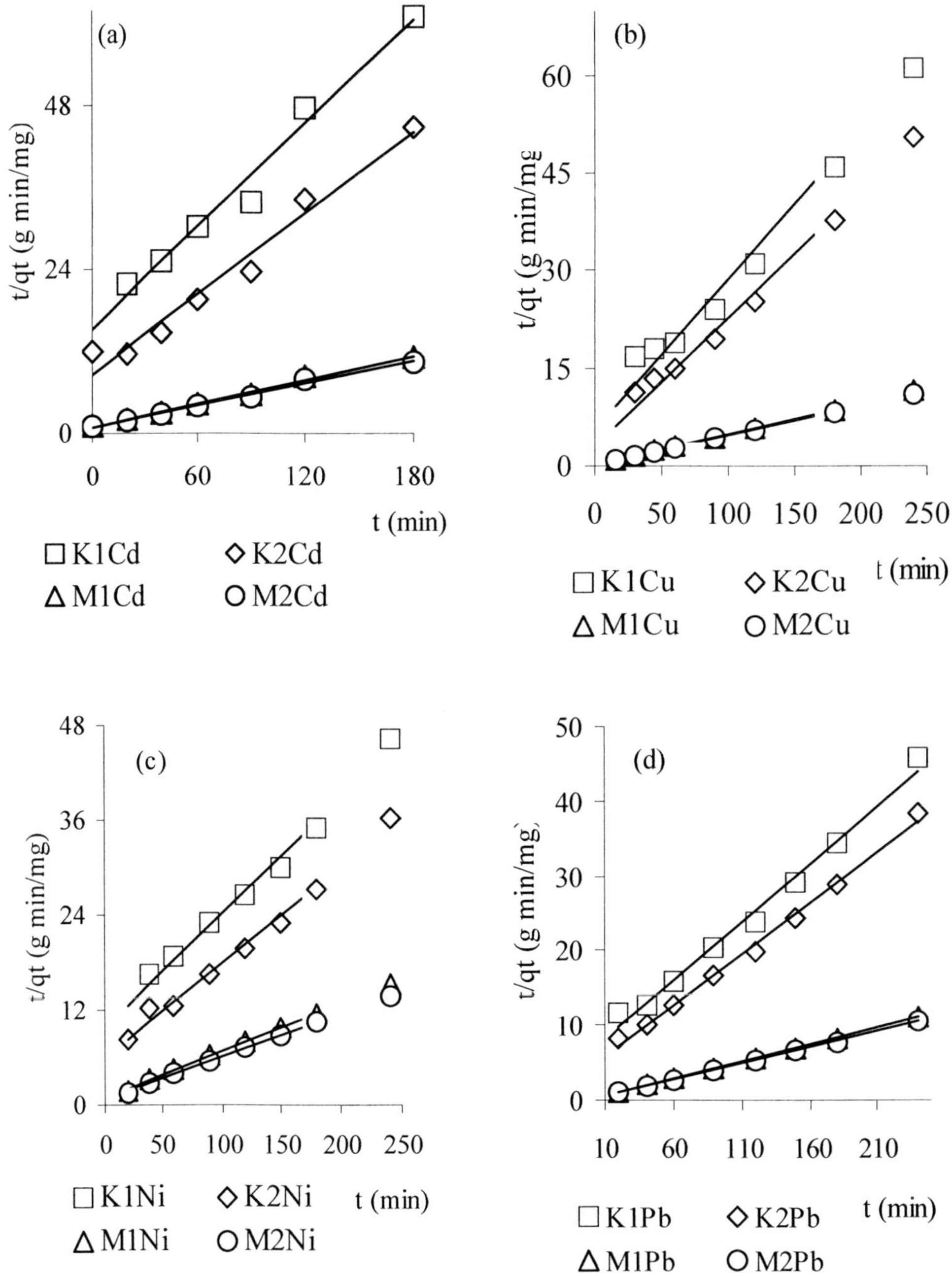


Figure 5 Second order plots for adsorption of (a) Cd(II), (b) Cu(II), (c) Ni(II) and (d) Pb(II) on kaolinite K1, montmorillonite M1, acid activated kaolinite K2, and acid activated montmorillonite M2 at 303 K.

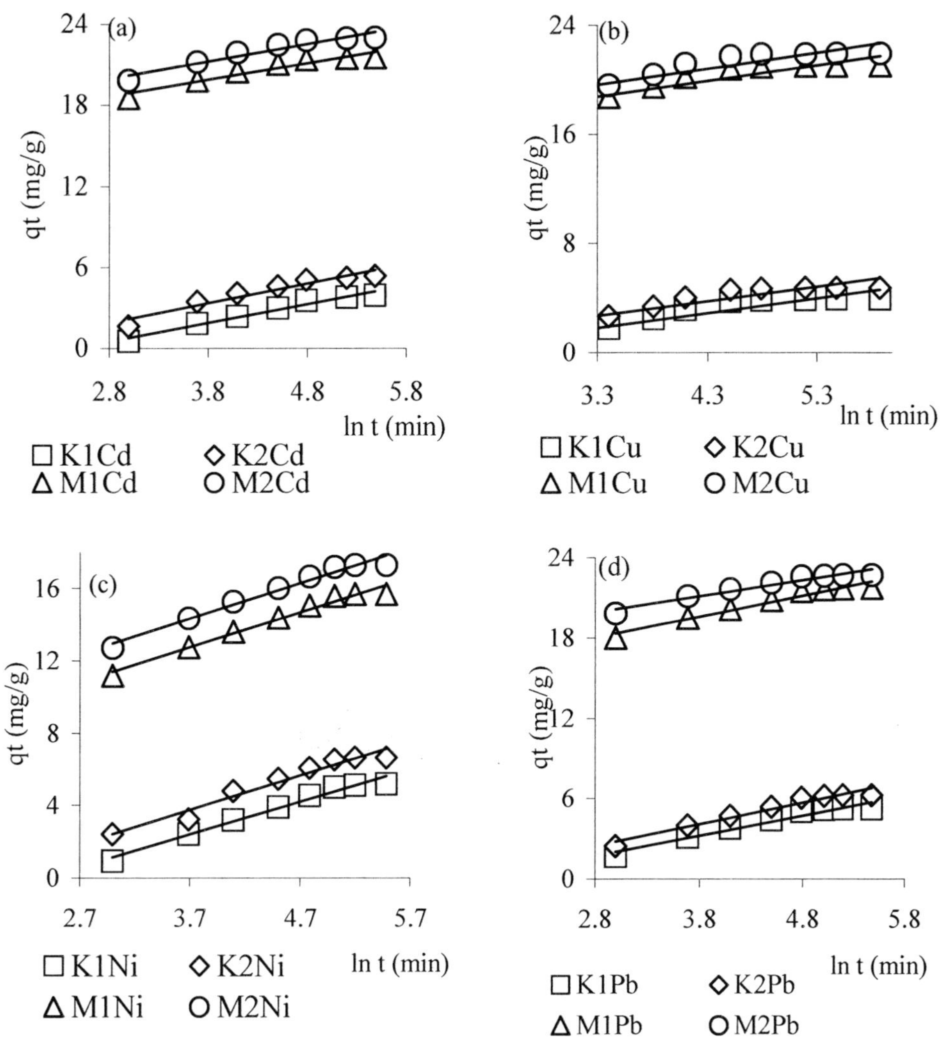


Figure 6 Elovich plots for adsorption of (a) Cd(II), (b) Cu(II), (c) Ni(II) and (d) Pb(II) on kaolinite K1, montmorillonite M1, acid activated kaolinite K2, and acid activated montmorillonite M2 at 303 K.

The plots for intra-particle diffusion ($q_t$ vs. $t^{0.5}$; Figure 7) were linear with regression coefficient, r, from +0.80 to +0.95 (Table 5), but the lines do not pass through the origin indicating that this model did not fit the adsorption process. Similarly, the liquid film diffusion model despite yielding good linear plots of $-\ln(1 - F)$ vs. t (Figure 8) (r ~ + 0.96 to + 0.99) was found to be inapplicable to the present work, as the plots did not have zero intercept. The rate constants and coefficients calculated from the liquid film diffusion model have been given in Table 6. Very small values of the intercepts might however point out some limited applicability for this model when transport of the metal ions from the liquid phase to the clay surface could well have control over the kinetics.

**Table 4. Elovich coefficients for adsorption at 303 K (clay 2 g/L, initial metal concentration 50 mg/L, pH 5.5 for Cd(II), 5.7 for Cu(II), Pb(II) and Ni(II), K1 kaolinite, K2 acid activated kaolinite, M1 montmorillonite and M2 acid activated montmorillonite)**

| Metal ion | Clay | Elovich coefficients | | |
|---|---|---|---|---|
| | | $\alpha$ (g mg$^{-1}$min$^{2}$) | $\beta$ (mg g$^{-1}$min$^{-1}$) | r |
| Cd(II) | K1 | $20.9 \times 10^{-4}$ | 1.3 | + 0.98 |
| | K2 | $36.8 \times 10^{-4}$ | 1.4 | + 0.96 |
| | M1 | $93.9 \times 10^{-4}$ | 1.4 | + 0.96 |
| | M2 | $299.3 \times 10^{-4}$ | 1.3 | + 0.96 |
| Cu(II) | K1 | $50.3 \times 10^{4}$ | 1.4 | + 0.97 |
| | K2 | $150.1 \times 10^{-4}$ | 1.1 | + 0.91 |
| | M1 | $371.0 \times 10^{-4}$ | 1.5 | + 0.97 |
| | M2 | $1851.0 \times 10^{-4}$ | 1.2 | + 0.91 |
| Ni(II) | K1 | $23.8 \times 10^{-4}$ | 1.8 | + 0.99 |
| | K2 | $99.9 \times 10^{-4}$ | 1.9 | + 0.98 |
| | M1 | $382.8 \times 10^{-4}$ | 1.9 | + 0.99 |
| | M2 | $2267.5 \times 10^{-4}$ | 2.0 | + 0.99 |
| Pb(II) | K1 | $14.8 \times 10^{-3}$ | 1.5 | + 0.98 |
| | K2 | $31.3 \times 10^{-3}$ | 1.6 | + 0.98 |
| | M1 | $29.4 \times 10^{-3}$ | 1.8 | + 0.98 |
| | M2 | $84.9 \times 10^{-3}$ | 1.2 | + 0.97 |

**Table 5. Intra-particle diffusion coefficients for adsorption at 303 K (clay 2 g/L, initial metal concentration 50 mg/L, pH 5.5 for Cd(II), 5.7 for Cu(II), Pb(II) and Ni(II), K1 kaolinite, K2 acid activated kaolinite, M1 montmorillonite and M2 acid activated montmorillonite)**

| Metal ion | Clay | Intra-particle diffusion | | |
|---|---|---|---|---|
| | | $k_i \times 10$ (mg g$^{-1}$ min$^{-0.5}$) | Intercepts | r |
| Cd(II) | K1 | 2.5 | +0.2 | + 0.92 |
| | K2 | 3.0 | +1.3 | + 0.90 |
| | M1 | 2.1 | +18.5 | + 0.89 |
| | M2 | 2.7 | +19.4 | + 0.90 |
| Cu(II) | K1 | 3.5 | -0.1 | + 0.91 |
| | K2 | 1.9 | +1.9 | + 0.80 |
| | M1 | 3.7 | +16.7 | + 0.91 |
| | M2 | 2.1 | +18.8 | + 0.80 |
| Ni(II) | K1 | 3.9 | -0.1 | + 0.95 |
| | K2 | 4.1 | +1.1 | + 0.94 |
| | M1 | 4.2 | +10.1 | + 0.95 |
| | M2 | 4.3 | +11.6 | + 0.95 |
| Pb(II) | K1 | 3.2 | +1.1 | + 0.93 |
| | K2 | 3.4 | +1.7 | + 0.92 |
| | M1 | 3.3 | +17.3 | + 0.93 |
| | M2 | 2.5 | +19.4 | + 0.91 |

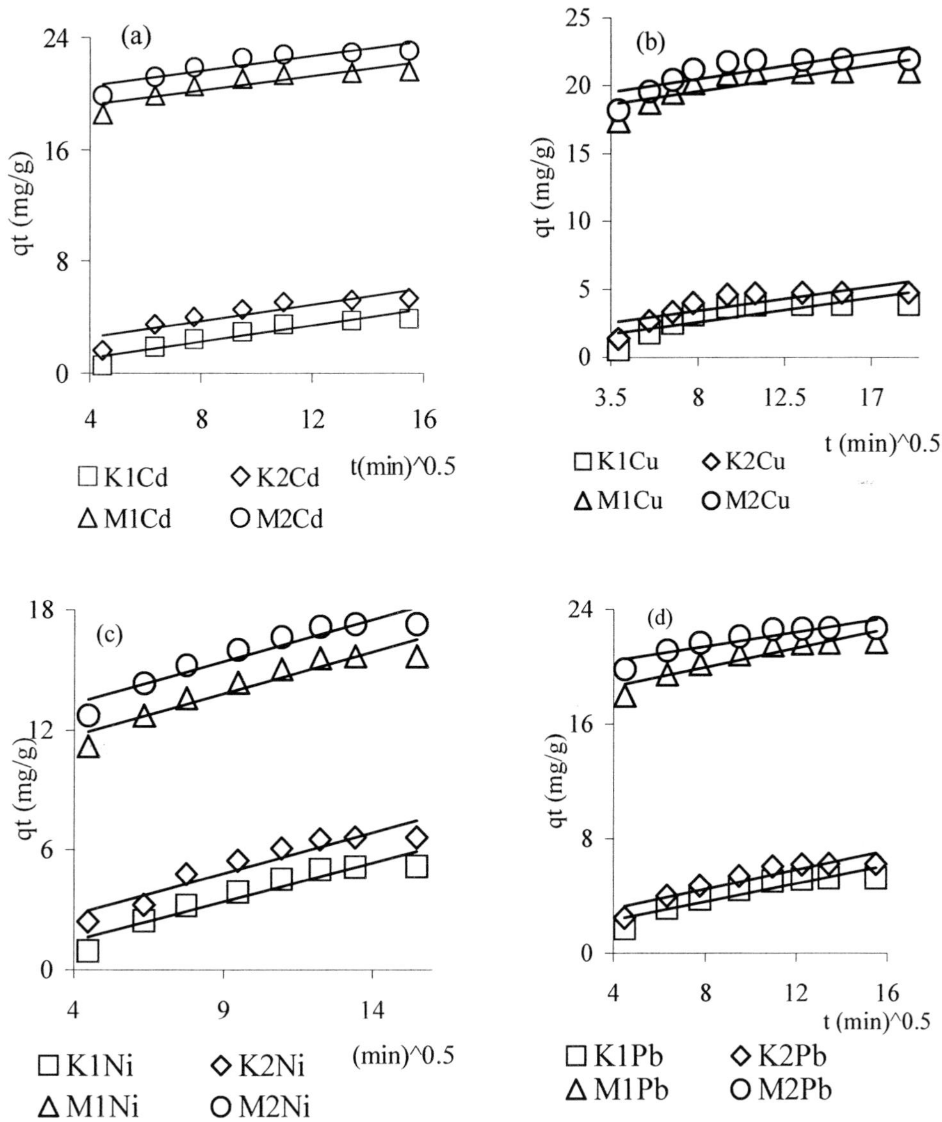


Figure 7 Intraparticle diffusion plots for adsorption of (a) Cd(II), (b) Cu(II), (c) Ni(II) and (d) Pb(II) on kaolinite K1, montmorillonite M1, acid activated kaolinite K2, and acid activated montmorillonite M2 at 303 K.

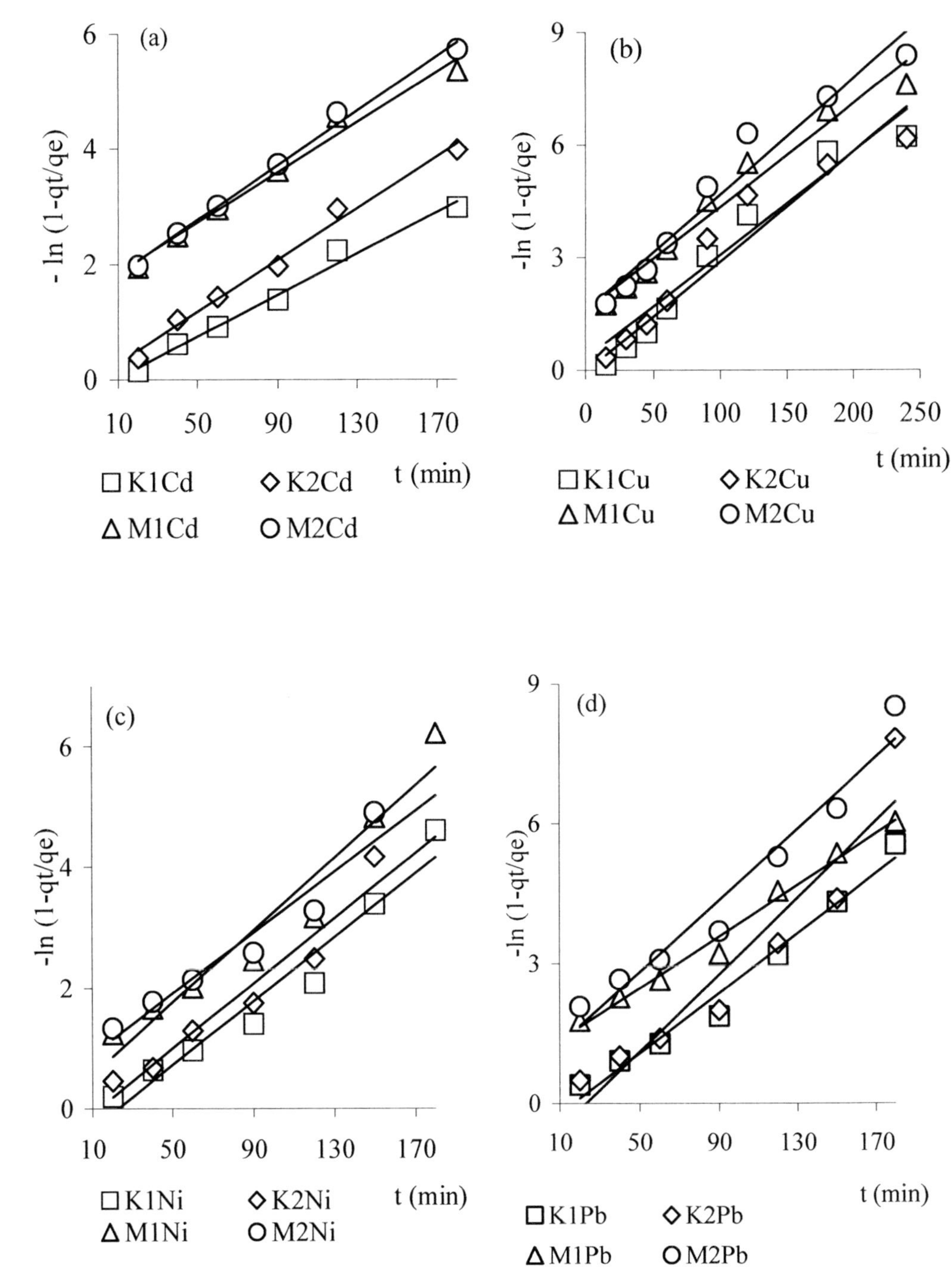


Figure 8 Liquid film diffusion plots for adsorption of (a) Cd(II), (b) Cu(II), (c) Ni(II) and (d) Pb(II) on kaolinite K1, montmorillonite M1, acid activated kaolinite K2, and acid activated montmorillonite M2 at 303 K.

**Table 6. Liquid film diffusion coefficients for adsorption at 303 K (clay 2 g/L, initial metal concentration 50 mg/L, pH 5.5 for Cd(II), 5.7 for Cu(II), Pb(II) and Ni(II), K1 kaolinite, K2 acid activated kaolinite, M1 montmorillonite and M2 acid activated montmorillonite)**

| Metal ion | Clay | Liquid film diffusion coefficients | | |
|---|---|---|---|---|
| | | $k_{fd}$ x $10^2$ ($min^{-1}$) | Intercepts | r |
| Cd(II) | K1 | 1.8 | -0.2 | + 0.99 |
| | K2 | 2.2 | +0.1 | + 0.99 |
| | M1 | 2.2 | +1.6 | + 0.98 |
| | M2 | 2.4 | +1.6 | + 0.99 |
| Cu(II) | K1 | 3.9 | -0.5 | + 0.99 |
| | K2 | 2.8 | +0.3 | + 0.96 |
| | M1 | 3.9 | +1.1 | + 0.99 |
| | M2 | 3.1 | +1.6 | + 0.98 |
| Ni(II) | K1 | 2.6 | -0.6 | + 0.98 |
| | K2 | 2.7 | -0.4 | + 0.97 |
| | M1 | 3.0 | +0.3 | + 0.98 |
| | M2 | 2.5 | +0.7 | + 0.97 |
| Pb(II) | K1 | 3.2 | -0.5 | + 0.99 |
| | K2 | 3.0 | -0.5 | + 0.99 |
| | M1 | 2.8 | +1.1 | + 0.99 |
| | M2 | 3.3 | +1.2 | + 0.99 |

The kinetics of Cd(II), Cu(II), Ni(II) and Pb(II) adsorption on both non-activated and acid-activated kaolinite and montmorillonite appears to be very complex in nature and although no kinetic model gives an absolute fit, the second order model gives better results and therefore, the interactions could be explained on the basis of a second order mechanism.

## Effects of Adsorbent Amount

The influence of clay amount (2, 3, 4, 5, and 6 g/L) on metal ion uptake is shown in Figure 9. The amount adsorbed per unit mass ($q_e$) showed a decrease for all the four clay adsorbents. The decrease in adsorption density may be due to two different reasons:

(i) A large amount of adsorbent effectively reduces the unsaturation of the adsorption sites and correspondingly, the number of such sites per unit mass comes down resulting in comparatively less adsorption at higher adsorbent amount, and
(ii) Higher adsorbent amount creates particle aggregation, resulting in a decrease in the total surface area and an increase in diffusional path length both of which contribute to decrease in amount adsorbed per unit mass (Sukla et al., 2002).

Figure 9 shows that acid-activated kaolinite maintained a higher level of adsorption than the non-activated kaolinite for different amounts. For montmorillonite, the amount adsorbed per unit mass decreased for both acid-activated and non-activated varieties, but there is only a little difference between the two. Thus, acid treatment did not significantly alter the number of adsorption sites per unit mass in montmorillonite, although the total adsorption on the basis of percentage uptake of metal ions increased. For example, the amount of Pb(II) adsorbed per unit mass of clay decreased from 5.2 to 3.2 mg/g, 6.3 to 3.5 mg/g, 21.6 to 8.3 mg/g and 22.7 to 8.4 mg/g for kaolinite, acid activated kaolinite, montmorillonite and acid activated montmorillonite respectively as the clay amount was changed from 2.0 to 6.0 g/L. The corresponding extent of adsorption increased from 20.9 to 38.3 %, 25.1 to 41.7 %, 86.5 to 99.1 %, and 90.7 to 99.9 % in the same order. Similar trends were observed in case of clay-Cu(II), clay-Cd(II) and clay-Ni(II) systems. The variation in the extent of adsorption (%) for all the systems is given in Table 7.

**Table 7. Extent of adsorption (%) of metal ions on different amount of clay adsorbents at 303 K (metal ions 50 mg/L, pH 5.5 for Cd(II), 5.7 for Cu(II), Pb(II) and Ni(II), time 240 min for Cd(II), 360 min for Cu(II), 180 min for Pb(II) and Ni(II), K1 kaolinite, K2 acid activated kaolinite, M1 montmorillonite and M2 acid activated montmorillonite)**

| Metal ion | clay (g/L) | Extent of adsorption (%) | | | |
|---|---|---|---|---|---|
| | | K1 | K2 | M1 | M2 |
| Cd(II) | 2 | 15.7 | 21.4 | 86.1 | 92.1 |
| | 3 | 20.8 | 26.6 | 91.0 | 95.2 |
| | 4 | 25.2 | 31.5 | 95.2 | 97.7 |
| | 5 | 29.5 | 36.1 | 98.2 | 99.6 |
| | 6 | 30.1 | 36.6 | 98.5 | 99.9 |
| Cu(II) | 2 | 15.2 | 18.9 | 83.8 | 87.5 |
| | 3 | 21.8 | 25.1 | 87.8 | 91.4 |
| | 4 | 27.2 | 30.7 | 91.4 | 94.9 |
| | 5 | 32.9 | 36.1 | 94.7 | 98.0 |
| | 6 | 37.1 | 39.2 | 97.7 | 99.3 |
| Ni(II) | 2 | 20.5 | 26.5 | 62.8 | 69.2 |
| | 3 | 26.4 | 30.1 | 67.9 | 74.5 |
| | 4 | 31.6 | 34.2 | 72.4 | 79.3 |
| | 5 | 36.5 | 38.9 | 76.2 | 83.4 |
| | 6 | 38.4 | 41.2 | 78.3 | 85.5 |
| Pb(II) | 2 | 20.9 | 25.1 | 86.5 | 90.7 |
| | 3 | 27.0 | 31.3 | 91.5 | 94.0 |
| | 4 | 32.5 | 37.0 | 95.4 | 97.2 |
| | 5 | 37.6 | 41.3 | 98.9 | 99.2 |
| | 6 | 38.3 | 41.7 | 99.1 | 99.9 |

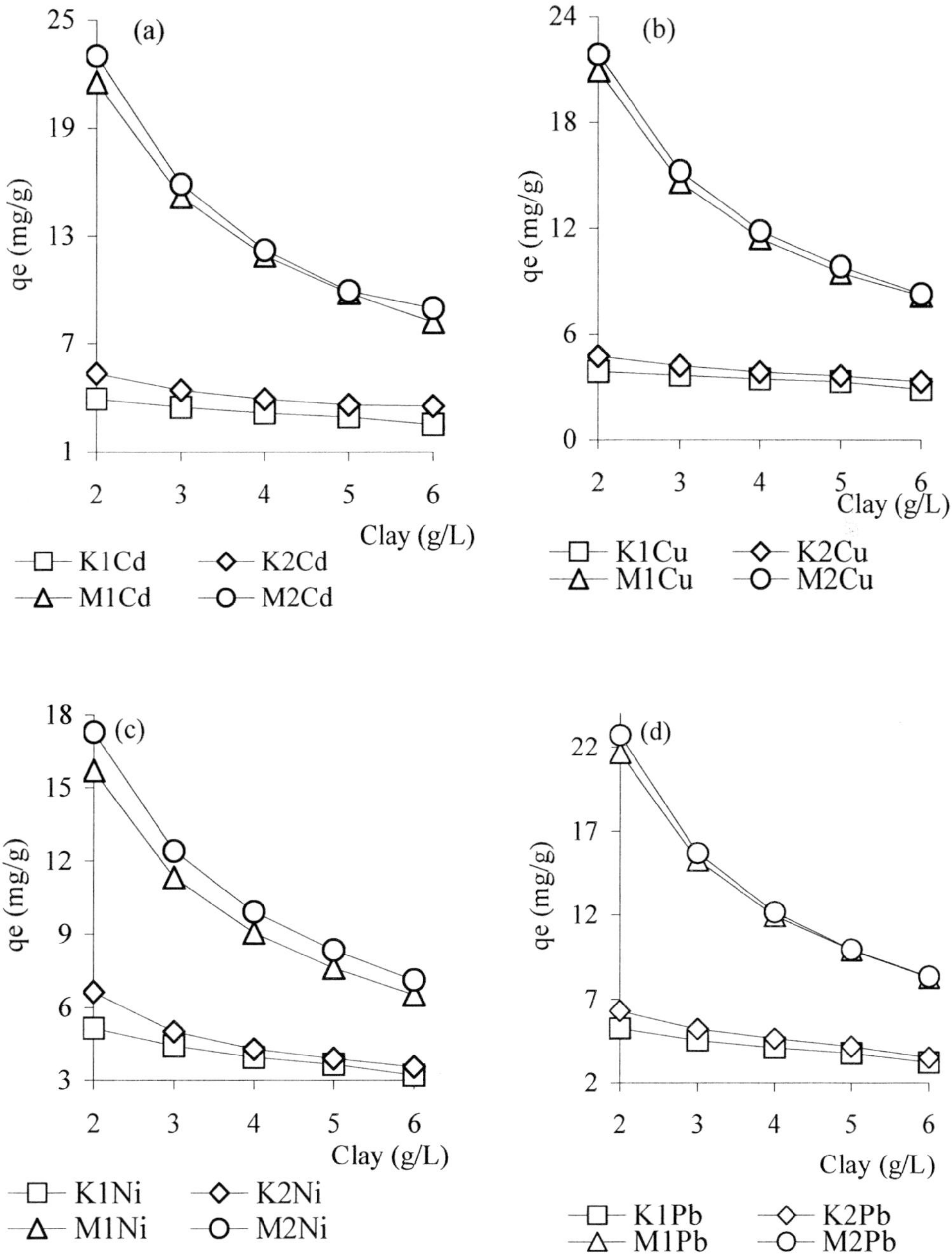


Figure 9 Influence of adsorbent amount (2, 3, 4, 5, 6 g/L) on uptake of (a) Cd(II), (b) Cu(II), (c) Ni(II) and (d) Pb(II) per unit mass of kaolinite K1, montmorillonite M1, acid activated kaolinite K2, and acid activated montmorillonite M2 at 303 K from 50 mg/L solution at equilibrium time.

## Effects of Initial Metal ion Concentration

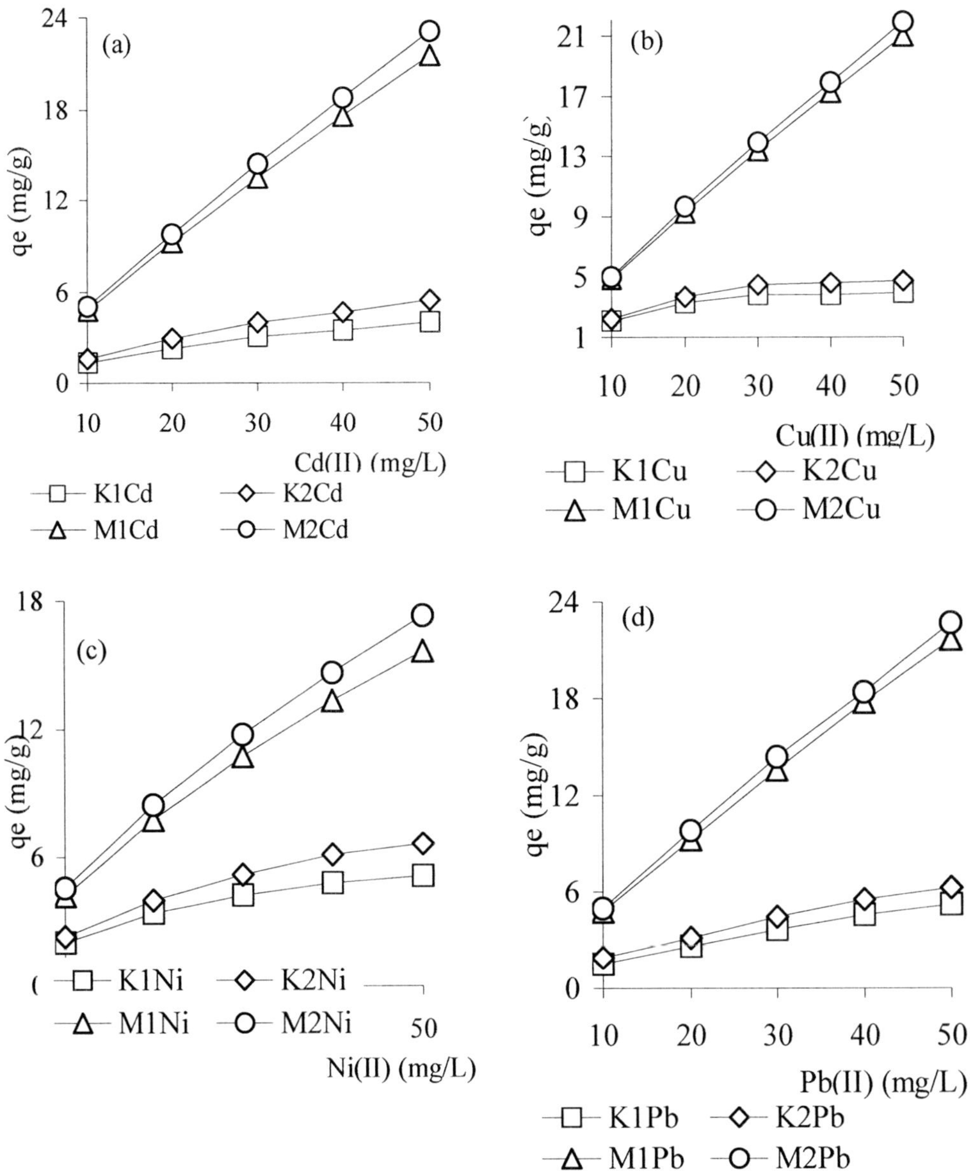


Figure 10 Influence of adsorbate concentration (10, 20, 30, 40, 50 mg/L) on uptake of (a) Cd(II), (b) Cu(II), (c) Ni(II) and (d) Pb(II) per unit mass of kaolinite K1, montmorillonite M1, acid activated kaolinite K2, and acid activated montmorillonite M2 at 303 K at equilibrium time with 2 g adsorbent per liter.

When the metal ion loadings were increased, the extent of adsorption (%) decreased, but the amount adsorbed per unit mass ($q_e$) showed an increasing trend (Figure 10). At low initial metal ion concentration, the ratio of the number of metal ions to the number of available adsorption sites is small and consequently the adsorption is independent of the initial concentration, but as the concentration of metal ions increases, the situation changes and the competition for adsorption sites becomes fierce (Ucun et al., 2003). At high concentration of metal, unit mass of the adsorbent is exposed to larger number of metal ions and progressively higher number of metal ions is taken up with the gradual filling up of the appropriate binding sites. This gives rise to an increase in $q_e$ although the net adsorption comes down.

The above can be seen from a comparison of the data for any of the systems. Considering the clay-Pb(II) system, when Pb(II) concentration was increased from 10 to 50 mg/L, the extent of adsorption decreased from 30.2 to 20.8 %, 38.0 to 25.0 %, 95.3 to 86.6 %, and 99.9 to 90.8 % respectively for kaolinite, acid-activated kaolinite, montmorillonite and acid-activated montmorillonite. Similar trends were observed for other clay-metal systems (Table 8). Such results are not uncommon and trends like these are found a large number of adsorption systems like Ni(II) on bagasse fly ash (Gupta et al., 2003) and Cu(II) on peat and pith (Ho and McKay, 2003).

**Table 8. Extent of adsorption (%) for different metal ion concentrations at 303 K (clay 2 g/L, pH 5.5 for Cd(II), 5.7 for Cu(II), Pb(II) and Ni(II), time 240 min for Cd(II), 360 min for Cu(II), 180 min for Pb(II) and Ni(II), K1 kaolinite, K2 acid activated kaolinite, M1 montmorillonite and M2 acid activated montmorillonite)**

| Metal ion | Concentration (mg/L) | Extent of adsorption (%) | | | |
|---|---|---|---|---|---|
| | | K1 | K2 | M1 | M2 |
| Cd(II) | 10 | 25.6 | 32.7 | 94.9 | 99.9 |
| | 20 | 22.9 | 30.3 | 92.2 | 98.4 |
| | 30 | 20.2 | 28.0 | 89.5 | 96.1 |
| | 40 | 17.3 | 25.5 | 87.9 | 94.2 |
| | 50 | 15.6 | 23.3 | 86.2 | 92.6 |
| Cu(II) | 10 | 40.8 | 44.2 | 96.6 | 99.9 |
| | 20 | 33.1 | 36.7 | 92.5 | 96.1 |
| | 30 | 25.1 | 29.4 | 88.9 | 92.5 |
| | 40 | 18.1 | 23.1 | 86.0 | 89.6 |
| | 50 | 15.4 | 18.9 | 83.9 | 87.5 |
| Ni(II) | 10 | 40.3 | 46.1 | 84.3 | 91.5 |
| | 20 | 34.2 | 40.0 | 77.8 | 84.7 |
| | 30 | 28.3 | 34.7 | 71.6 | 78.4 |
| | 40 | 24.1 | 30.6 | 66.7 | 73.1 |
| | 50 | 20.6 | 26.5 | 62.7 | 69.2 |
| Pb(II) | 10 | 30.2 | 38.0 | 95.3 | 99.9 |
| | 20 | 26.3 | 31.5 | 92.3 | 98.1 |
| | 30 | 24.4 | 29.6 | 90.4 | 95.8 |
| | 40 | 22.8 | 27.8 | 88.7 | 92.0 |
| | 50 | 20.8 | 25.0 | 86.6 | 90.8 |

## Adsorption Isotherm

The empirical Freundlich isotherm yields linear plots (r ~ + 0.91 to + 0.99) (Figure 11) and the values of the coefficients, n and $K_f$ (Table 9) indicate that the clays have good potential to be used as adsorbents for metal ions, Cd(II), Cu(II), Ni(II) and Pb(II). It is to be noted that Freundlich isotherm applies to adsorption on non-specific and heterogeneous sites on solid surfaces but no definite mechanism is proposed. The isotherm is valid for weak van der Waals type of adsorption as well as for strong chemisorption. The Freundlich adsorption capacity, $K_f$, was highest for the acid-activated montmorillonite and lowest for the non-activated kaolinite. The $K_f$ values are in the order of M2 (acid-activated montmorillonite) > M1 (non-activated montmorillonite) > K2 (acid-activated kaolinite) > K1 (non-activated kaolinite).

Langmuir isotherm is specific to strong monolayer chemisorption, which might be the case in the present work since the Langmuir plots (Figure 12) have good linearity (r ~ + 0.99) (Table 9). The Langmuir equilibrium coefficient, b, determines the direction in which the adsorbate-adsorbent equilibrium

Clay (solid phase) + Metal(II) (aqueous phase) = Clay Metal(II)

is shifted. Large values of 'b' ensure that the equilibrium is predominantly driven towards the right leading to formation of the adsorbate-adsorbent complex.

**Table 9. Freundlich and Langmuir coefficients for adsorption at 303 K (clay 2 g/L, initial metal concentration 50 mg/L, pH 5.5 for Cd(II), 5.7 for Cu(II), Pb(II) and Ni(II), time 240 min for Cd(II), 360 min for Cu(II), 180 min for Pb(II) and Ni(II), K1 kaolinite, K2 acid activated kaolinite, M1 montmorillonite and M2 acid activated montmorillonite)**

| Metal ion | Clay | Freundlich coefficients | | | Langmuir coefficients | | | |
|---|---|---|---|---|---|---|---|---|
| | | $K_f$ | n | r | $q_m$ | b | $R_L$ | r |
| Cd(II) | K1 | 0.4 | 0.6 | + 0.99 | 6. 8 | 32.3 | 0.0013 | + 0.99 |
| | K2 | 0.5 | 0.7 | + 0.97 | 11.0 | 24.1 | 0.0019 | + 0.99 |
| | M1 | 6.8 | 0.6 | + 0.99 | 24.5 | 29.6 | 0.0014 | + 0.98 |
| | M2 | 14.0 | 0.3 | + 0.99 | 30.7 | 19.3 | 0.0024 | + 0.98 |
| Cu(II) | K1 | 1.3 | 0.3 | + 0.91 | 4.3 | 19.9 | 0.0023 | + 0.99 |
| | K2 | 1.4 | 0.4 | + 0.96 | 5.6 | 136.6 | 0.0003 | + 0.99 |
| | M1 | 7.8 | 0.4 | + 0.99 | 25.5 | 43.7 | 0.0011 | + 0.98 |
| | M2 | 10.4 | 0.4 | + 0.99 | 28.0 | 122.9 | 0.0003 | + 0.98 |
| Ni(II) | K1 | 0.8 | 0.5 | + 0.99 | 7.1 | 70.1 | 0.0007 | + 0.99 |
| | K2 | 0.9 | 0.6 | + 0.99 | 9.9 | 57.1 | 0.0008 | + 0.99 |
| | M1 | 3.4 | 0.5 | + 0.96 | 21.1 | 137.7 | 0.0003 | + 0.99 |
| | M2 | 5.0 | 0.5 | + 0.99 | 21.3 | 237.9 | 0.0002 | + 0.99 |
| Pb(II) | K1 | 0.4 | 0.7 | + 0.99 | 11.5 | 20.7 | 0.0021 | + 0.99 |
| | K2 | 0.5 | 0.6 | + 0.99 | 12.3 | 27.3 | 0.0016 | + 0.98 |
| | M1 | 7.3 | 0.7 | + 0.97 | 31.1 | 31.0 | 0.0014 | + 0.98 |
| | M2 | 13.3 | 0.3 | + 0.99 | 31.4 | 15.8 | 0.0028 | + 0.96 |

(Units of $K_f$, $q_m$ and b are respectively $mg^{1-1/n}L^{1/n}\ g^{-1}$, $mg\ g^{-1}$, and $Lg^{-1}$.)

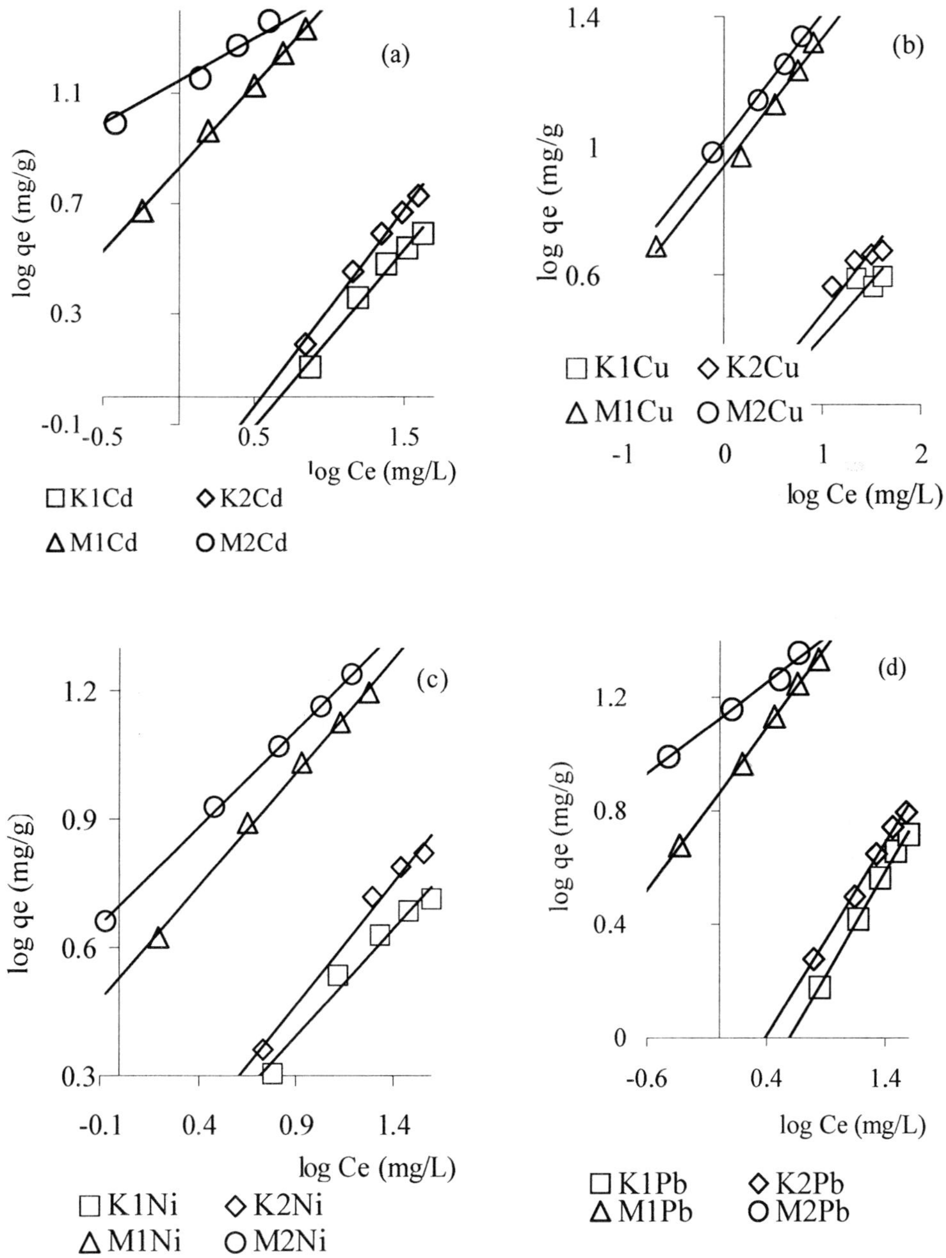


Figure 11 Freundlich isotherms for adsorption of (a) Cd(II), (b) Cu(II), (c) Ni(II) and (d) Pb(II) on kaolinite K1, montmorillonite M1, acid activated kaolinite K2, and acid activated montmorillonite M2 at 303 K.

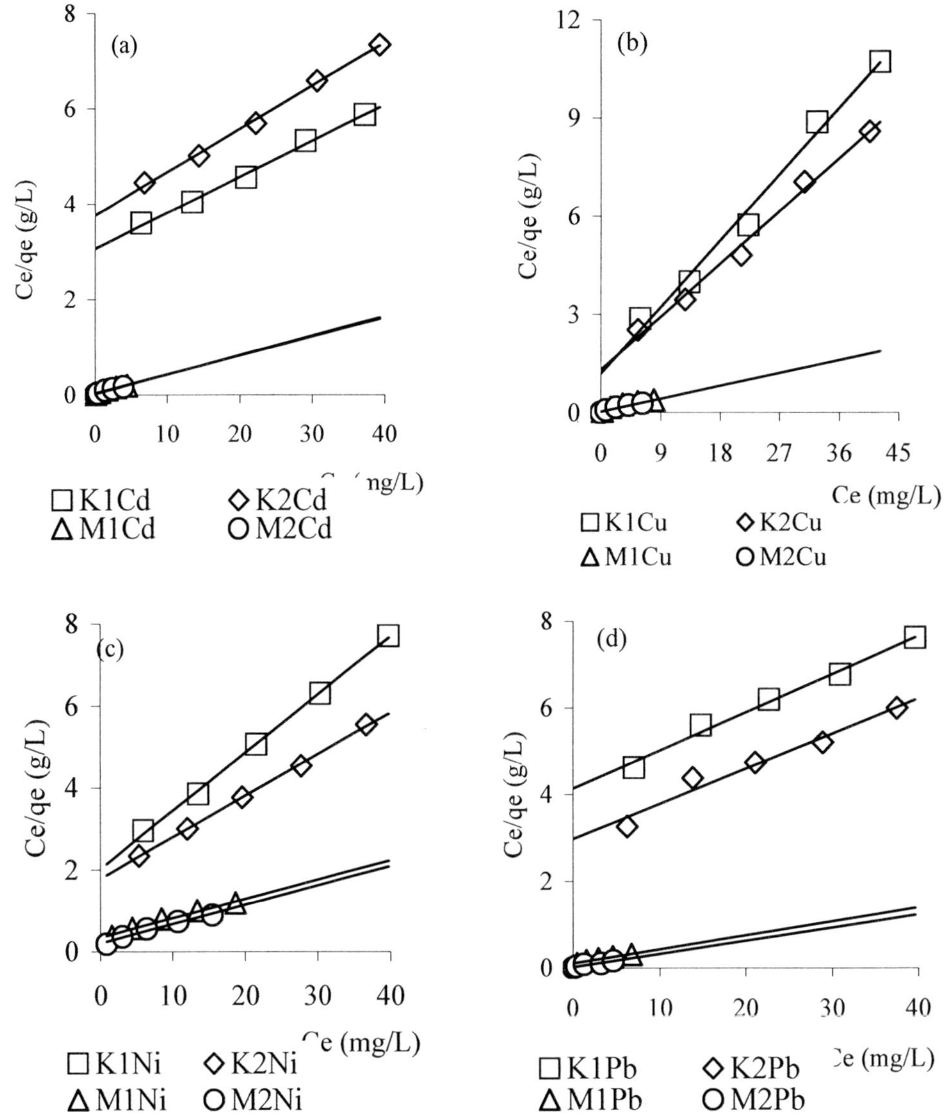


Figure 12 Langmuir isotherm for adsorption of (a) Cd(II), (b) Cu(II), (c) Ni(II) and (d) Pb(II) on kaolinite K1, montmorillonite M1, acid activated kaolinite K2, and acid activated montmorillonite M2 at 303 K.

From the values of Langmuir monolayer capacity, $q_m$, it is observed that acid activation did not have much influence on the values. This suggests that treatment with acid does not increase the number of adsorption sites to a large extent although the treatment influences the strength of the existing sites as revealed by the equilibrium coefficient (b) data. Langmuir monolayer capacity, $q_m$, is considerable for all the clays and therefore the clays can take up large amount of the metal ions. The $q_m$ values are 6.8 to 30.7 mg $g^{-1}$, 4.3 to 28.0 mg $g^{-1}$, 7.1 to 21.3 mg $g^{-1}$, 11.1 to 31.4 mg $g^{-1}$ for Cd(II), Cu(II), Ni(II) and Pb(II) respectively. The $q_m$ values follow the same order as the Freundlich adsorption capacity, i.e. M2 (acid-activated montmorillonite) > M1 (non-activated montmorillonite) > K2 (acid-activated kaolinite) > K1 (non-activated kaolinite). The separation factor, $R_L$ was << 1.0, indicating that the metal ions preferred to remain bound to the clay surface.

Values of the adsorption coefficients in similar ranges have been reported by other workers. The maximum adsorption capacity for Cu(II) on red mud was obtained as 19.72 mg $g^{-1}$ for a contact time of 48 h (R = 0.95) (Lopez et al. 1998). Lin and Juang, (2003) studied the adsorption of Cu(II) on montmorillonite modified with sodium dodecylsulfate and reported a Langmuir monolayer adsorption capacity of 0.016 mg/g. Taty-Costodes et al. (2003) reported $q_m$ between 8.45 to 22.22 mg $g^{-1}$, and between 6.72 to 15.27 mg $g^{-1}$ respectively for Pb(II) and Cd(II) adsorption on sawdust. Chen et al. (2001) have reported $q_m$ of 28.26 mg $g^{-1}$ (R = 0.99) and $K_f$ as 5.32 $Lg^{-1}$ (R = 0.96) for adsorption of Ni(II) on peat.

## Thermodynamic Studies

Adsorption of metal ions on clay adsorbents show different behaviour with respect to increase in temperature (Table 10).

**Table 10. Effect of temperature on extent of adsorption of metal ions on clays (clay 2 g/L, initial metal ion concentration 50 mg/L, pH 5.5 for Cd(II), 5.7 for Cu(II), Pb(II) and Ni(II), time 240 min for Cd(II), 360 min for Cu(II), 180 min for Pb(II) and Ni(II), K1 kaolinite, K2 acid activated kaolinite, M1 montmorillonite and M2 acid activated montmorillonite)**

| Metal ion | Temperature (K) | Extent of adsorption (%) | | | |
|---|---|---|---|---|---|
| | | K1 | K2 | M1 | M2 |
| Cd(II) | 303 | 15.6 | 21.4 | 86.2 | 92.1 |
| | 308 | 25.4 | 27.1 | 91.1 | 93.1 |
| | 313 | 21.3 | 23.3 | 89.7 | 92.6 |
| Cu(II) | 303 | 15.4 | 18.9 | 83.9 | 99.9 |
| | 308 | 20.4 | 23.6 | 86.9 | 89.9 |
| | 313 | 24.7 | 28.1 | 89.3 | 92.7 |
| Ni(II) | 303 | 20.6 | 26.5 | 62.7 | 69.2 |
| | 308 | 16.9 | 22.8 | 56.8 | 61.5 |
| | 313 | 13.0 | 18.1 | 51.3 | 55.5 |
| Pb(II) | 303 | 20.8 | 25.0 | 86.6 | 99.9 |
| | 308 | 13.3 | 22.2 | 79.0 | 84.3 |
| | 313 | 9.1 | 17.9 | 74.7 | 79.1 |

Cu(II) adsorption on clays, follows an endothermic path (Figure 13b). The amount adsorbed per unit mass of the adsorbents ($q_e$) as well as the extent of adsorption (%) increased appreciably in the temperature range 303 to 313 K. Cu(II) uptake changed from 3.9 mg/g (15.4 %) to 6.2 mg/g (24.7 %), 4.8 mg/g (18.9 %) to 7.0 mg/g (28.1 %), 21.0 mg/g (83.9 %) to 22.3 mg/g (89.3 %), and 21.9 mg/g (99.9 %) to 23.2 mg/g (92.7 %) for non activated kaolinite, acid activated kaolinite, non activated montmorillonite and acid activated montmorillonite respectively. Thus, adsorption of Cu(II) on clays has to overcome a small activation energy barrier and increasing energy supply makes it easier for Cu(II) to adsorb to the clay surface. Such activated adsorption following endothermic path has also been reported earlier (Shukla et al., 2002; Ho, 2003).

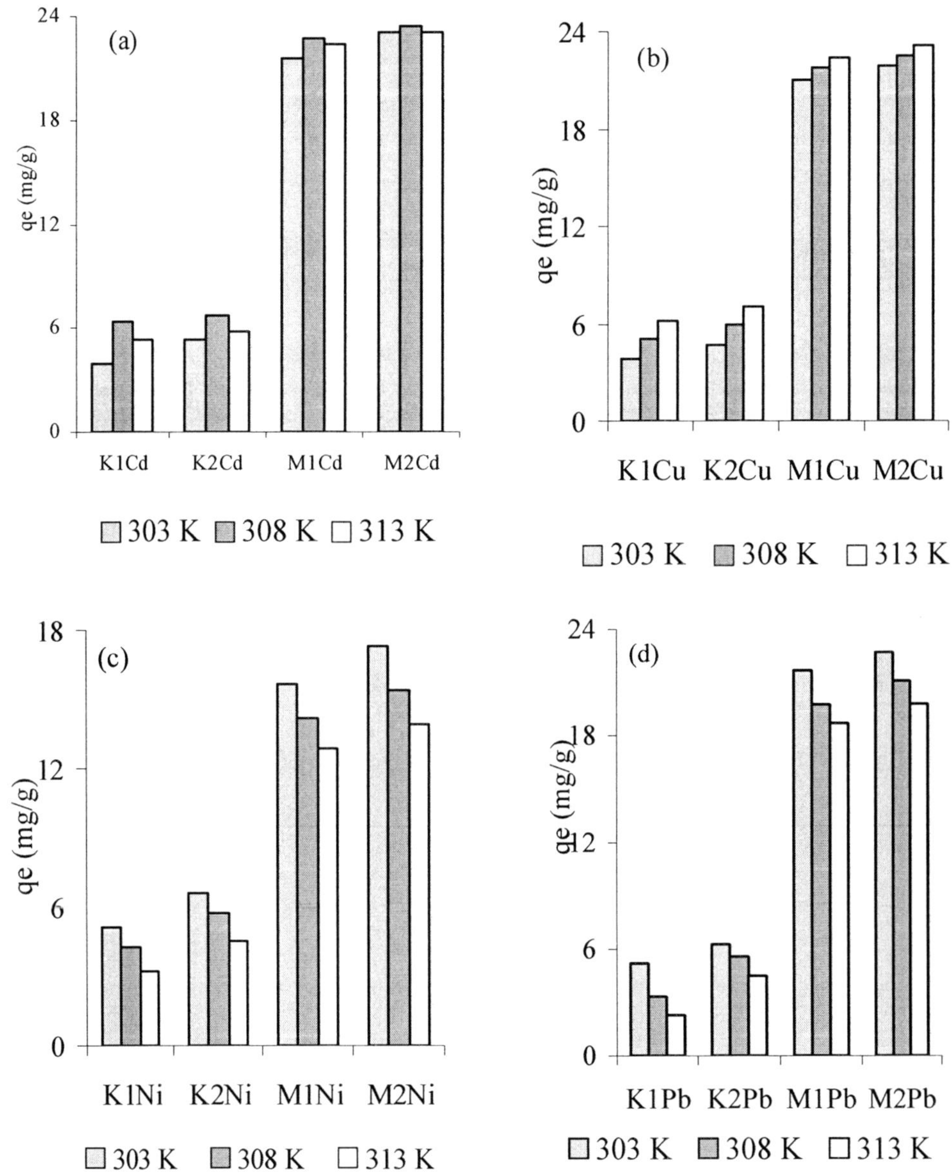


Figure 13 Effect of temperature on adsorption of (a) Cd(II), (b) Cu(II), (c) Ni(II) and (d) Pb(II) on kaolinite K1, montmorillonite M1, acid activated kaolinite K2, and acid activated montmorillonite M2.

The effect of increasing temperature was quite different in case of Cd(II), where increase of temperature from 303 to 308 K was found to enhance the amount of Cd (II) adsorbed per unit mass of the clays (as well as extent of adsorption, %), but the trend was reversed between 308 and 313 K (Figure 13a). The process is controlled by the adsorbate-adsorbent and adsorbate-adsorbate forces and it is clear from the results that the first one becomes weak in comparison to the latter as the temperature increases (Ajmal et al., 1998).

For both Ni(II) and Pb(II), increase in the temperature resulted in a decrease in the amount of metal adsorbed per unit mass of clay as well as the extent of adsorption (%) (Figure 13c and 13d), showing the interactions to be exothermic. For example, in the temperature range 303 K to 313 K, Ni(II) uptake changed from 5.2 mg/g (20.6 %) to 3.3 mg/g (13.0 %), 6.6 mg/g (26.5 %) to 4.5 mg/g (18.1 %), 15.7 mg/g (92.7 %) to 12.8 mg/g (56.8 %), and 17.3 mg/g (69.2 %) to 13.9 mg/g (55.5 %) for non activated kaolinite, acid activated kaolinite, non activated montmorillonite and acid activated montmorillonite respectively. Thus, the adsorbate ions have a tendency to escape from the solid phase to the bulk solution phase with the rise in temperature (Echeverria et al., 2003). The possible increase in the metal ion solubility and/or damage to the adsorption sites in the clay minerals due to increase in temperature may also be considered in accounting for the decrease in uptake of the metal ions at higher temperature (Othman and Amin, 2003). Moreover, rise in temperature and the excess energy supply promotes desorption. Such results are not uncommon for adsorption of metal ions from aqueous phase (Abou-Mesalam, 2003).

**Table 11. Thermodynamic data for adsorption of metal ion on clays (clay 2 g/L, initial metal ion concentration 50 mg/L, pH 5.5 for Cd(II), 5.7 for Cu(II), Pb(II) and Ni(II), time 240 min for Cd(II), 360 min for Cu(II), 180 min for Pb(II) and Ni(II), K1 kaolinite, K2 acid activated kaolinite, M1 montmorillonite and M2 acid activated montmorillonite)**

| Metal ion | Clay | $\Delta H$ (kJ mol$^{-1}$) | $\Delta S$ (J K$^{-1}$ mol$^{-1}$) | $-\Delta G$ (kJ mol$^{-1}$) | | |
|---|---|---|---|---|---|---|
| | | | | 303 K | 308 K | 313 K |
| Cd(II) | K1 | + 25.0 | + 66.3 | 20.0 | 20.4 | 20.7 |
| | K2 | + 8.6 | + 14.3 | 4.3 | 4.4 | 4.5 |
| | M1 | + 40.2 | + 147.5 | 44.6 | 45.4 | 46.1 |
| | M2 | + 25.0 | + 107.3 | 32.5 | 33.0 | 33.6 |
| Cu(II) | K1 | + 30.7 | + 86.8 | 26.2 | 26.7 | 27.1 |
| | K2 | + 32.7 | + 94.9 | 28.9 | 29.2 | 29.7 |
| | M1 | + 50.7 | + 180.7 | 54.7 | 55.6 | 56.0 |
| | M2 | + 45.8 | + 267.6 | 81.0 | 82.4 | 83.7 |
| Ni(II) | K1 | − 37.9 | − 118.2 | 41.9 | 42.6 | 43.3 |
| | K2 | − 33.7 | − 121.9 | 36.9 | 37.5 | 37.7 |
| | M1 | − 45.1 | − 146.4 | 44.4 | 45.1 | 45.9 |
| | M2 | − 56.9 | − 181.8 | 55.0 | 55.8 | 56.9 |
| Pb(II) | K1 | − 58.9 | − 209.7 | 63.5 | 64.5 | 65.6 |
| | K2 | − 31.5 | − 116.4 | 35.2 | 35.8 | 36.4 |
| | M1 | − 75.5 | − 235.9 | 71.4 | 72.6 | 73.8 |
| | M2 | − 83.4 | − 258.7 | 78.3 | 79.6 | 80.9 |

The thermodynamic adsorption parameters, $\Delta H$, $\Delta S$ and $\Delta G$, (Table 11), are computed from the plots of ln $K_d$ vs. 1/T. Typical ln $K_d$ vs. 1/T plots (clay 2 g/L, initial metal ion concentration 50 mg/L at natural pH of the metal ion solutions and at respective equilibrium time) are shown in Figure 14.

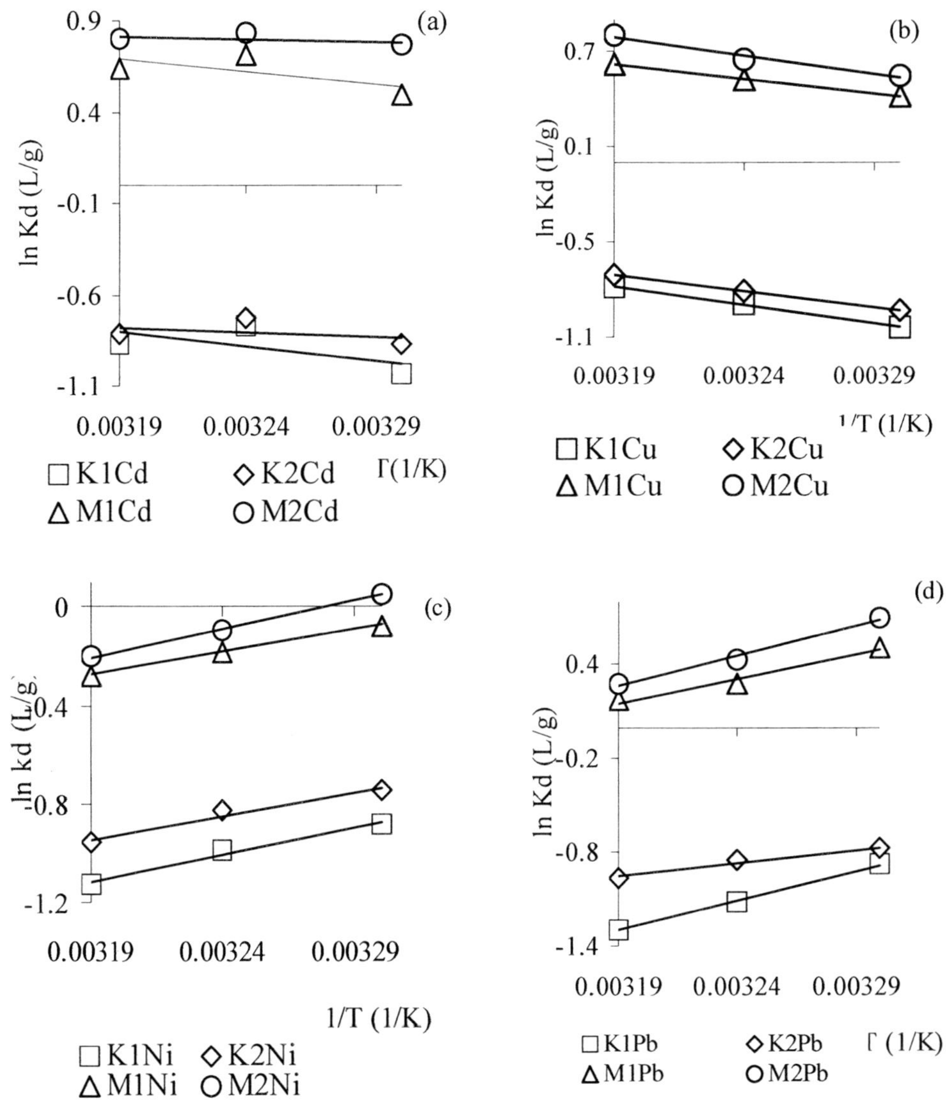


Figure 14 Plots of ln $K_d$ vs 1/T for adsorption of (a) Cd(II), (b) Cu(II), (c) Ni(II) and (d) Pb(II) (50 mg/L) on kaolinite K1, montmorillonite M1, acid activated kaolinite K2, and acid activated montmorillonite M2 (2 g/L, pH as prepared).

In case of Clay-Pb(II) interaction, the non-activated kaolinite and montmorillonite had mean adsorption enthalpy, ΔH of – 58.9 and – 75.5 kJ $mol^{-1}$, which changed to – 31.5 and – 83.4 kJ $mol^{-1}$ respectively after acid-activation. The exothermic enthalpy before acid-activation was higher for kaolinite than for montmorillonite with a clear indication that although the uptake might be more in case of montmorillonite, Pb(II) ions are held more strongly by kaolinite than montmorillonite. The situation changes after acid-activation and now ΔH has a higher magnitude for montmorillonite than that for kaolinite. The adsorption enthalpy increased for both the clays after acid activation, but the increase was much more for montmorillonite, implying that acid activation changed the adsorption sites in such a manner that Pb(II) was now held much more strongly to montmorillonite. The ΔH value for acid-activated montmorillonite is now almost in the region of chemisorption enthalpies. The trends are similar for exothermic adsorption of Ni(II) on the clays (kaolinite and acid-activated kaolinite: ΔH – 37.9 to – 33.7 kJ $mol^{-1}$, montmorillonite and acid-activated montmorillonite: ΔH –45.1 to –56.9 kJ $mol^{-1}$). The magnitude of the adsorption enthalpy, ΔH, indicates moderately strong bonding between the metal ions and the clay minerals (Gupta et al., 2001).

The endothermic enthalpy change (ΔH) associated with Cu(II) adsorption varied significantly for kaolinite and montmorillonite, but for the same clay, acid treatment showed only minor variation (kaolinite and acid-activated kaolinite: ΔH 30.7 to 32.7 kJ $mol^{-1}$, montmorillonite and acid-activated montmorillonite: ΔH 50.7 to 45.8 kJ $mol^{-1}$). Similar endothermic nature of interaction was observed in Cd(II) adsorption. The endothermic interactions decreased after acid activation in both kaolinite and montmorillonite.

Decrease in entropy during Ni(II) adsorption helps to stabilize the Ni(II)-clay adsorption complex. Mean ΔS values for clay-Ni(II) interaction are in the range of -181.8 J $K^{-1}$ $mol^{-1}$ (acid-activated montmorillonite) to -118.2 J $K^{-1}$ $mol^{-1}$ (non-activated kaolinite). Similarly, Pb(II)-adsorption on clays is accompanied by an entropy decrease and the mean values vary from –116.4 to – 258.7 J $K^{-1}$ $mol^{-1}$. Since stability is associated with an ordered arrangement, it is obvious that Ni(II) and Pb(II) are in a much more chaotic distribution in aqueous solution than when they are adsorbed on clays. The clay-Cu(II) interactions, on the other hand, were accompanied by an increase in entropy (+ 86.8 J $K^{-1}$ $mol^{-1}$ for kaolinite to + 267.6 J $K^{-1}$ $mol^{-1}$ for acid activated montmorillonite). The endothermic interactions were driven by entropy increase. The situation is similar for clay-Cd(II) interactions. Entropy increase accompanying the process illustrates increased randomness at the solid-solution interface indicating strong affinity of the adsorbent for Cu(II) and Cd(II).

Spontaneity of the adsorption processes is demonstrated by the decrease in Gibbs energy for all clay-metal systems.

Thermodynamic data on metal adsorption on clays are scarce. Yuvaz et al. (2003) have found that ΔH, ΔS and ΔG for adsorption of Cu(II) on Turkish kaolinite are 39.52 kJ $mol^{-1}$, 11.7 J $K^{-1}$ $mol^{-1}$ and 4.61 kJ $mol^{-1}$ respectively. Echeverria et al. (2003) have reported that ΔH, ΔS and ΔG for adsorption of Ni(II) on illite have values of +16.8 kJ $mol^{-1}$, + 58 J $mol^{-1}K^{-1}$ and –1.04 kJ $mol^{-1}$ respectively. ΔH, ΔS and ΔG for Cu(II) adsorption on surfactant-modified montmorillonite were reported as 7.05 kJ $mol^{-1}$, 9.09 J $K^{-1}$ $mol^{-1}$ and – 9.66 kJ $mol^{-1}$ respectively (Lin and Juang, 2002). All these values compare well with the values obtained in the present work.

## CONCLUSION

The following conclusions have been drawn from this study

- Both kaolinite and montmorillonite and their acid activated forms are capable of removing metal ions namely Cd(II), Cu(II), Ni(II) and Pb(II), from aqueous solution.
- Montmorillonite (2:1 clay) has high surface charges resulting from the spread of isomorphous substitution in tetrahedral and octahedral sheets, whereas 1:1 layered kaolinite has little isomorphous substitution. This accounts for its low cation

adsorbing capacity. Acid activation enhanced the adsorption capacity compared to the untreated clay minerals due to the increased surface area and pore volume.

- Adsorption increases with pH till the metal ions are precipitated out in alkaline solution.
- The rate of uptake of the metal ions by the clays is very high initially followed by a slow rate indicating entry of the metal ions into the interior of the adsorbent particles as the interactions proceed. The kinetics of clay-metal interactions were very complex, where different mechanisms of interactions based on pseudo first order and second order kinetics, intra-particle and liquid film diffusion processes might have some role. However, the processes were much more akin to second order kinetics.
- The uptake of metal ions depends appreciably on amount of clay adsorbents and initial concentration of metal ions. The amount adsorbed per unit mass decreased as the clay amount increased while the opposite trend was observed for increase in initial metal ion concentration.
- Both Langmuir and Freundlich isotherms yield good fits and the adsorption coefficients agree well with the conditions supporting favorable adsorption.
- All the metal-clay interactions are thermodynamically favorable. Clay-Pb(II) and clay-Ni(II) interactions were exothermic in nature whereas clay-Cu(II) and clay-Cd(II) were endothermic. The clay-metal interactions were spontaneous in all the cases supported by decrease in Gibbs energy.

## References

Abou-Mesalam, M.M. Colloids Surfaces *A Physicochem Eng Aspects.* 2003, 225, 85–94.

Ajmal, M.; Mohammad, A.; Yousuf, R.; Ahmad, A. *Indian J Environ Hlth.* 1998, 40, 15–26.

Alvarez-Ayuso, E.; Garcia-Sanchez, A. *Clays Clay Miner.* 2003, 51, 475–480.

An, H.K.; Park, B.Y.; Kim, D.S. *Water Res.* 2001, 35, 3551–3556.

ATSDR, (1999a). Toxicological profile for cadmium, Agency for Toxic Substances and Disease Registry, (http://www.atsdr.cdc.gov), U.S. Department of Health and Human Services, Atlanta, Georgia.

ATSDR, (1999b). Toxicological Profile for lead, Agency for Toxic Substances and Disease Registry, (http://www.atsdr.cdc.gov), U.S. Department of Health and Human Services, Atlanta, Georgia.

ATSDR, (2002). Draft toxicological Profile for copper, Agency for Toxic Substances and Disease Registry, (http://www.atsdr.cdc.gov), U.S. Department of Health and Human Services, Atlanta, Georgia.

ATSDR, (2003). Draft Toxicological profile for nickel, Agency for Toxic Substances and Disease Registry, (http://www.atsdr.cdc.gov), U.S. Department of Health and Human Services, Atlanta, Georgia.

Babel, S; Kurniawan, T.A. *Chemosphere.* 2004, 54, 951–967.

Bayramoglu, G.; Bekta, S.; Arica, M.Y. *J Hazard Mater.* 2003, B101, 285–300.

Bergaya, F.; Vayer, M. *Appl Clay Sci.* 1997, 12, 275–280.

Bhattacharyya, K.G.; Sharma, A. *J Hazard Mater.* 2004,113, 97–109.

Bhattacharyya, K.G.; Sen Gupta S. Colloids Surfaces *A Physicochem Eng Aspects.* 2006, 277, 191–200.

Boyd, G.E.; Adamson, A.W.; Myers, Jr., L.S. *J Am Che Soc.* 1947, 69, 2836–2848.

Brewer, R. *Fabric and mineral analysis of soils;* Wiley: New York, 1964; pp 470.

Chantawong, V.; Harvey, N.W.; Bashkin, V.N. *Water Air Soil Pollut.* 2003, 148, 111–125.

Chen, B.; Hui, C.W.; McKay, G. *Water Res.* 2001, 35, 3345–3356.

Chien, S.; Clayton, W.R. *Soil Sci Soc Am J* 1980, 44, 265–268.

Chiron, N.; Guilet, R.; Deydier, *E. Water Res.* 2003, 37 3079–3086.

De Leon, A.T.; Nunes, D.G.; Rubio, *J. Clays Clay Miner.* 2003, 51, 58–64.

Deer, W.A.; Howie, R.A.; Zussman, J. *An introduction to the rock-forming minerals;* ELBS Longman: England, 1985; pp 250–274.

Diaz, F.R.V.; Santos, P.D.S. *Quim Nova. 2001*, 24, 345–353.

Dixson, J.B; Weed, S.B.; Ed.; *Minerals in soil environments;* Madison: Soil Science Society of America, 1977; pp 948.

Eberl, D.D.; Velde, B.; McCormick, T. *Clay Minerals.* 1993, 28, 49–60.

Echeverria, J.; Indurain, J.; Churio, E.; Garrido, J. *Colloids Surfaces A Physicochem Eng Aspects.* 2003, 218, 175–87.

Echeverria, J.C; Churio, E.; Garrido, *J. Clays Clay Miner.* 2002, 50, 614– 23.

Espantaleon, A.G.; Nieto, J.A.; Fernandez, M.; Marsal, A. *Appl. Clay Sci.* 2003, 24, 105–110.

Fanning, D.S.; Keramidas, V.Z.; El-Desosky, M.A. In Minerals in soil environments; Dixon, J.B.; Weed, S.B.; Ed.; SSSA Book series; Soil Science Society of America, Medison: Wisconsin, 1989; pp 527–634.

Finelli V.N.; Boscolo, P.; Salimei, E. Heavy Met Environ Int Conf 4th, 1981; pp 475–478.

Freundlich, H. M. F. *J Phys Chem. 1906*, 57, 385–470.

Ghose, D; Bhattacharyya, K.G. *Appl Clay Sci.* 2002, 295–300.

Grim, R.E. Clay Mineralogy; McGraw Hill: New York, 1968.

Gupta, V.K.; Gupta, M.; Sharma, S. *Water Res.* 2001, 35, 1125–1134.

Gupta, V.K.; Jain, C.K.; Ali, I.; Sharma, M.; Saini, V.K. *Water Res.* 2003, 37, 4038–4044.

Hall, K.R.; Eagleton, L.C.; Acrivos, A.; Vermeulen, T. *Ind Eng Chem Fundam* 1966, 5, 212–223.

Ho, Y.S. *Water Res.* 2003, 34, 2323–2330.

Ho, Y.S., Scientometrics. 2004, 59, 171–177.

Ho, Y.S.; McKay, G. *Trans I Chem E.* 1998, 76B, 332–339.

Ho. Y.S.; McKay, G. *Adsorption.* 1999a, 5, 409–417.

Ho, Y.S.; McKay, G. *Water Res.* 1999b, 33, 578–584.

Ho, Y.S.; McKay, G. *Process Biochemistry.* 1999c, 34, 451– 465.

Ho, Y.S; McKay, G. *Trans I Chem E.* 1999d, 165–173.

Ho, Y.S.; McKay, G. *Adsorption Sci Technol.* 2002, 20, 797–815.

Ho, Y.S.; McKay, G. *Process Biochemistry* 2003, 38, 1047–1061

Jackson, M.L.; Sherman, G.D. *Advances in Agromony* 1952, 5, 219–318.

Jackson, M.L.; Gillette, D.A.; Danielsen, E.F; Blifford, I.H.; Bryson, R.A.; Syers, J.K. *Soil Sci.* 1973, 116, 135–145.

Jain, C.K.; Ram, D. *Water Res.* 1997, 34, 154–162.

Jeon, C.; Holl, W.H. *Water Res.* 2003, 37, 4770– 4780.

Jozefaciuk, G.; Bowanko, G. *Clays Clay Miner.* 2002, 50, 771–783.

Kadirvelu, A.; Namasivayam, C. *Advances Environ Res.* 2003, 7, 471– 478.
Kandah, M.I. *Sep Purif Technol* 2004, 35, 61–70.
Kara, M.; Yuzer, H.; Sabah, E.; Celik, M.S. *Water Res.* 2003, 37, 224–232.
Khan, S.A.; Rehman, R.; Khan, M.A. *Waste Manage.* 1995, 15, 641–650.
Kiff, R.F. *General inorganic effluents;* Longman: New York, 1987; Vol. 3, chapter 1.
Kjellstrom, T.; Borg, K.; Lind B. *Environ Res.* 1978. 15, 242–251.
Ko, D.C.K.; Cheung, C.W.; Choy, K.K.H; Porter, J.F; McKay, G. *Chemosphere.* 2004, 54, 273–281.
Kotok, D.; Kotok, R.; Heriot, T. *Am J Dis Child.* 1977, 131, 791–793.
Lagergren, S. Kungliga Svenska Vetenskapsakademiens. Handlingar. 1898, Band 24, No. 4, 1–39.
Langmuir, I. *J Am Chem Soc.*1918, 40, 1361–1403.
Lin, S-H.; Juang, R-S. *J Hazard Mater.* 2002, B92, 315–326.
Lopez, E.; Soto, B.; Arias, M.; Nunez, A.; Rubinos, D.; Barral, M.T. *Water Res.* 1998, 32, 1314–1322.
Maity, S.; Chakravarty, S.; Bhattacharjee, S.; Roy, B.C. *Water Res.* 2005, 39, 2579–2590.
Marsal, A.; Garcia-Espantaleon. A.; Ribosa, I.; Cot. J.; Manich, A. *J of the Society of Leather Chemist and Technologists.* 2003, 219–222.
Marshall, C.E. Z Krist. 1935, 91, 433– 449.
Mathialagan, T.; Viraraghavan, T. *J Hazard Mater.* 2002, 94, 291–303.
Mellah. A.; Chegrouche. S. *Water Res.,* 1997, 31, 621–629.
Naseem, R.; Tahir, S.S. *Water Res.* 2001, 35, 3982–3989.
Nuhoglu, Y.; Oguz, E. *Process Biochemistry.* 2003, 38, 1627–1631.
Othman, M.R.; Amin, A.M. *Biochemical Eng J.* 2003, 16, 361–364.
Padmavathy, V.; Vasudevan, P.; Dhingra, S.C. *Process Biochemistry.* 2003, 38, 1389– 1395.
Pan, S.C.; Lin, C.C.; Tseng, D.H. *Resources Conservation Recycling.* 2003, 39, 79–90.
Pauling, L. *Proc Natl Acad Sci.* 1930, 16, 578–582.
Pinnavaia, T.J. *Science.* 1983, 220, 365–371.
Radojevic, M.; Jovic, V.; Karaulic, D.; Vitorovic, D. *J Serb Chem Soc.* 2002, 67, 499–506.
Raman, K.V.; Jackson, M.L. *Clays Clay Min.* 1964. 12, 423–429.
Ramos, R.L.; Jacome, L.A.B.; Barron, J.M.; Rubio, L.F.; Coronado, R.M.G. *J Hazard Mater.* 2002, B90, 27–38.
Ravichandran, J.; Sivasankar, B. *Clays Clay Miner.* 1997, 45, 854–858.
Rhodes, C.N.; Franks, M.; Parkes, G.M.B.; Brown, D.R. *J Chem Soc Chem Comm.* 1991, 12, 804–807.
Rodrigues, M.G.F. *Ceramica,* 2003, 49, 146–150.
Rudzinski, W.; Panczyk, T. *Adsorption.* 2002, 8, 23–34.
Sabah, E.; Turan, M.; Celik, M.S. *Water Res.,* 2002, 36, 3957–3964.
Sears, G. *Anal. Chem.* 1956, 28, 1981–1983.
Sen Gupta, S.; Bhattacharyya, K.G. *Appl Clay Sci.* 2005, 30, 199–208.
Sen Gupta, S.; Bhattacharyya, K.G. *J Hazard. Mater.* 2006a, 128, 247–257.
Sen Gupta, S.; Bhattacharyya, K.G. *J Colloid Interface Sci.* 2006b, 295, 21–32.
Sharma, A. Bhattacharyya, K.G. *Adsorption* 2004, 10, 327–338.
Sharma, A. Bhattacharyya, K.G. *J Hazard. Mater.* 2005, 125, 100–110.

Shukla, A.; Zhang, Y-H.; Dubey, P.; Margrave, J.L.; Shukla, S.S. *J Hazard Mater*. 2002, 95 137–152.

Stelly, M.; Dinauer, R.C.; Ed.; Mineralogy in soil science and engineering; Madison: Soil Science Society of America, 1968; pp 106.

Strawn, D.G.; Palmer, N.E.; Furnare, L.J.; Goodell, C.; Amonette, J.E.; Kukkadapu, R.K. *Clays Clay Miner.* 2004, 52, 321–333.

Stayner, L.; Smith, R.; Thun M. In A quantitative assessment of lung cancer risk and occupational cadmium exposure; Nordberg, G.F.; Herber R.F.M.; Alessio, L.; Ed.; *Cadmium in the human environment: Toxicity and carcinogenicity. Lyon, International Agency for Research on Cancer* (IARC), 1992, 447–455.

Suraj, G.; Iyer, C.S.P.; Lalithambika, M., *Appl Clay Sci.* 1998, 13, 293–306.

Tanabe, K. In Catalysis Science and Technology; Anderson, J.R.; Boudart, M.; Ed.; *Solid acid and base catalysis;* Springer Verlag: New York, 1981; pp 231.

Tarley, C.R.T.; Arruda, M.A.Z., *Chemosphere*. 2004, 54, 987–995.

Taty-Costodes, V.C.; Faudue, H.; Porte, C.; Delacroix, A. *J Hazard Mater.* 2003, B105, 121–142.

Theng, B.K.G. *Formation and properties of clay polymer complexes;* Elsevier: New York, 1979; pp 1–12.

Theocharis, C.R.; Jacob, K.J.; Gray, A.C. *J Chem Soc Faraday Trans.*1988, 84, 1509–1516.

Thomas, W. J.; Crittenden, B. *Adsorption Technology and Design; Fundamental of adsorption equilibria;* Butterworth – Heinemann: Oxford, 1998; pp 31–63.

Tran, H.H.; Roddick, F.A.; O'Donell, *J.A. Water Res.* 1999, 33, 2992–3000.

Ucun, H.; Bayhan, Y.K.; Kaya, Y.; Cakici, A.; Algur, O.F. *Desalination*. 2003, 154, 233–238.

Van Olphen H. *An introduction to clay colloid Chemistry,* Wiley: New York, 1977.

Volzone, C.; Thompson, J.G.; Melnitchenko, A.; Ortiga, J.; Palethorpe, S.R. *Clays Clay Miner.* 1999, 5, 647–657.

Weber, W.J. Jr. In Adsorption technology: A step-by-step approach to process evaluation and application; Slejko, F.L.; Ed.; *Adsorption theory, concepts and models;* Marcel Dekker: New York, 1985; pp 1–36.

Weber, W.J. Jr.; Mc Ginley, P.; Katz, M.L.E. *Water Res.* 1991, 25, 499–528.

Weber, W.J.; Morris, J.C. *J Sanit Engg Div Am Soc Civ Engg.* 1963, 89, 31–63.

Wingenfelder, U.; Nowack, B.; Furrer, G.; Schulin, R. *Water Res.* 2005, 39, 3287–3297.

Yahiaoui, A.; Belbachir, M.; Hachemaoui, A. *Int J Molecular Sci.* 2003, 4, 548–561.

Yavuz, O.; Altunkaynak, Y.; Guzel, F. *Water Res*. 2003, 37, 948–952.

Yin, J.; Blanch, H.W. *Biotech Bioeng.* 1989, 34, 180–188.

Yu, B.; Zhang, Y.; Shukla, A.; Shukla, S.S.; Dorris, K.L. *J Hazard Mater.* 2000, B80, 33–42.

Yu, B.; Zhang, Y.; Shukla, A.; Shukla, S.S.; Dorris, K.L. *J Hazard Mater.* 2001, B84, 83–94.

Yu, L.J.; Shukla, S.S.; Dorris, K.L.; Shukla, A.; Margrave, J.L. *J Hazard Mater.* 2003, B100, 53–63.

In: Hazardous Materials and Wastewater
Editor: Allison A. Lewinsky, pp. 127-148
ISBN 1-60021-257-3


*Chapter 4*

# ECO-FRIENDLY TECHNOLOGIES FOR REMOVAL OF HAZARDOUS HEAVY METAL FROM WATER AND INDUSTRIAL WASTEWATER

***K. Kadirvelu and Jyotsna Goel***
Centre for Fire, Explosive and Environment Safety, Defense R and D Organization, TIMARPUR, Delhi, India

## ABSTRACT

The rapid development and changing technologies, industrial products and practices of the present day result in the generation of high quantum of hazardous heavy metal laden wastewater materials, and sludges, which need to be appropriately dealt with to prevent any adverse effects on public health and the natural environment system. Although source reduction and Waste minimization are the long term solutions for reducing or eliminating the release of toxic heavy metal ions to the environment, treatment technologies will continue to play a crucial role in controlling heavy metal discharges from current operations and in cleaning up the contaminated sites from previous operations.

The objective of this article is to overview the processes currently available for wastewaters contaminated with heavy metal ions and seeks to highlight the many of the significant contributions to the field of treatment of hazardous heavy metals in the environment.

This article reviews the applicable separation technologies relating to removal of heavy metals from industrial wastewater in order to present the state of the art in the field.

A number of specialized processes have been developed for the removal of heavy metals from waste discharges. Technologies for removal of hazardous heavy metal waste management can be illustrated into two major classes – chemical process technologies and bioprocess technologies. Chemical process technologies are of basically of two segments - conventional treatment technologies like chemical precipitation (including hydroxide, carbonate, or sulfide reagents) coagulation/flocculation, sedimentation, chelating complexation and filtration and alternative technologies, for end of the pipe treatment of hazardous heavy metal waste like adsorption, solvent extraction, ion exchange, electrochemical operations, membrane filtration and evaporation. Each technology is briefly described and typical operating conditions and technology performance are presented.

An attempt has been made to introduce engineers and scientist to the spectrum of practiced and emerging process technologies for removal of heavy metals from wastewaters.

In the present article the status of technologies adopted by the various industries and further requirement to meet the objective of treatment of hazardous heavy metal waste management have been assessed.

## HAZARDOUS HEAVY METALS IN THE WASTE STREAMS

### Heavy Metal Ion In Waste Streams

In the periodic table, the heavy metals constitute the Group III transition metals, the actinides series (uranium, neptunium, plutonium and americium), the lanthanide series, and three of the Group IV metalloids (arsenic, tellurium, selenium). The terms 'heavy metals' and 'transition metals' are often used interchangeably, although transition metals strictly refer to Group III elements.

In geological terms, trace elements are defined as those occurring at 1000 ppm or less in the earth crust. The trace metals may be divided into 'heavy', [with densities greater than $5g/cm^3$] and 'light [with densities less than $5g/cm^3$]. Trace metals with higher densities are termed as heavy metals.

Heavy metals exist in water-soluble as well as insoluble forms. Oxide, hydroxide, sulphide and carbonate salts of most heavy metals are not soluble or are sparingly soluble in water. Most common salts of heavy metal ions used in industrial operations are soluble in water, including chloride, sulphate and nitrates salts. As a dissolved salt, heavy metal ions are in cationic form such as $Cd^{2+}$, $Hg^{2+}$, $Pb^{2+}$, where exponent indicates valence oxidation state of the metal. Most of heavy metal cations are divalent ($2^+$), but some are monovalent (eg, silver $Ag^+$) or trivalent (eg. $Cr^{3+}$). Many heavy metals can be having multiple oxidation states (eg. Iron $Fe^{2+}$ or $Fe^{3+}$). Heavy metals can also exist in soluble form as oxymetal anions. The most prominent example is chromium, which can also exist as chromate $CrO_4^{2-}$ or dichromate $Cr_2O_7^{2-}$ with Cr oxidation state +6. Similar oxyanion states exist for arsenic, including arsenate ($AsO_4^{3-}$), where As has an oxidation state of +5, arsenite ($AsO_3^{3-}$), where As has oxidation state of +3, as well as selenium ($SeO_2^{3-}$, $SeO_4^{2-}$) [1].

### Toxicity of Heavy Metals

Many heavy metals, either in cationic or oxymetals anionic form are highly toxic. The toxicity of, industrially important metals ions have been encapsulated in Table 1.

Toxic effects of few of the heavy metals are detailed below.

## Mercury

Mercury is generally considered one of the most toxic metals found in the environment [2]. The adverse effects of mercury pollution have been reported from Minamata and Niigata, Japan, as Minamata disease in Japan [3]. Once mercury enters the food chain, larger accumulation of mercury takes place in human and animals. Mercury causes damage of the central nervous system and chromosomes, impairment of pulmonary functions, chest pain and dyspoea [4,5]. Even low concentration of mercury, may be considered potentially harmful to humans. Methyl mercury a compound involved in "Minamata Disease" is believed to inhibit enzyme activity in the cerebellum, which is responsible for neuron growth in early development stages [6]. The majority of reports of human intoxications are the results of methyl mercury or other organic compounds of mercury either working their way up the food chain, as in Minamata Bay, or being directly consumed due to mishandling of treated seed grain, as in Iraq. It is also possible however, for other forms of mercury to be converted to methyl mercury through microbial action, thereby increasing the possible sources of methyl mercury pollution. Chronic exposure to mercury leads to mental retardation. Clinically observable effects in adults occur at mercury concentration of 0.2 - 0.5 mg/ml in blood and body concentration of 0.5 - 0.8 mg/kg, or at concentration in hair as low as 15-20 mg/kg [7]. Due to mercury stability and high toxicity, even at low concentration, both direct (breathing) and indirect (food consumption) assimilation can cause health hazards to living beings. The later involves a complex passage through soil deposition, surface water enrichment, bio-assimilation and bioaccumulation by vegetation ending up in the food chain [8,9].

**Table 1: Sources and toxic effects of heavy metals**

| Metals | Sources | Effects and Significance |
|---|---|---|
| Arsenic | Mining By- Product, Pesticides, Chemical Waste | Toxic, Possibly Carcinogenic |
| Cadmium | Industrial discharge, mining waste, metal plating, water pipes | Replaces zinc biochemical, causes high blood pressure, kidney damage, destruction of testicular tissue and RBCs, toxicity to aquatic biota |
| Chromium | Metal plating, cooling tower water additive (chromate) normally found as Cr (VI) in polluted water | Essential trace elements; possibly carcinogenic as Cr (VI) |
| Copper | Metal plating, industrial and domestic waste, mining, mineral leaching | Essential trace elements; not very toxic to animals, toxic to plants and algae at moderate levels. |
| Lead | Industry, mining, plumbing, coal, gasoline | Toxic (anemia, kidney disease, nervous disorder), wild life destroyed |
| Mercury | Industrial waste, mining, pesticide, coal | Highly toxic |

## Lead

The toxicity and deleterious effects of lead are well documented. Its poisoning in human causes severe damages in kidney, reproductive system, liver, central nervous system or death [10]. Severe exposure to lead has been associated with sterility, abortion, still birth, and neonatal deaths [11], although there are no epidemiological data [12]. Women working in lead industries were found to have a higher rate of ovulatory dysfunction, and men with moderately increased lead absorption were found to have decreased fertility. Although lead is known to decrease thyroid function in man, the endocrine effects of lead are not well defined. Lead may also interfere with pituitary function in man, producing hypopitutarism in some individuals. Excessive oral ingestion may cause pathological changes in the pituitary-adrenal axis, decreasing metapyrone response and pituitary reserve [13,14]. Lead is a heavy metal poison which, by complexing with oxo-group enzymes, virtually affects all steps in the process of heam synthesis and porphyrin metabolism [15]. Lead is also toxic to aquatic organisms. Fish avoids the water containing 0.3mg/L lead and above 1 mg/L lead is lethal [16].

## Cadmium

Cadmium is one of several 'trace metals' existing in nature in small quantities that have no known nutritive value and are capable of producing a toxic effect. Cadmium ranks 67$^{th}$ of the 97 most common elements in the earth's crust and occurs in association with zinc (Zn). Cadmium is a potent inhibitor of several enzyme systems. Cadmium is highly toxic to some forms of life. Subleathal effects of cadmium toxicity have been reported from wild populations of fish Perch from a Swedish river contaminated with cadmium had a markedly enhanced lymphocyte count, slight anemia and changes in the concentration of potassium and magnesium in the blood. Residues of cadmium in the livers of exposed Perch were six to eight times higher than those of controls [17,18]. The harmful effects of cadmium include a number of acute and chronic disorders, such as `itai-itai' disease, renal damage, emphysema, hypertension, and testicular atrophy [19]. Some plant species also have some sensitivity to cadmium and have shown phytotoxic effects [20]. The no-effect level of cadmium administered at a single oral dose to adults has been estimated at 3 mg and the lethal doses from 350-8900 mg [21]. Cadmium may replace zinc in some enzymes, thereby altering the steric structure of the enzyme and impairing its catalytic activity [22]. Small amounts of cadmium (II) are believed to be harmful to man and have been associated with hypertensive diseases [23]. The International Agency for Research on Cancer (IARC) classified cadmium among chemicals, which are probably carcinogenic to man [24]. Cadmium is classified as a highly toxic non-essential metal that affects the action of enzymes and impedes respiration, photosynthesis and transpiration [25]. On a comparative basis, lead is neither as toxic nor as bioavailable as cadmium but it is more ubiquitous in the environment and acts as a cumulative toxin [26]. In the presence of other heavy metals in water has an appreciable effect on acidity [27] and the resultant decrease of pH is known to adversely effect fish stocks and vegetation [28].

## Sources of Heavy Metals to the Aquatic Environment

Sources of wastewaters are diverse, but the metals –related industries are the largest contributors. Primarily sources of wastewater containing heavy metals ions include mining operations, metal plating facilities (electroplating, electroless plating), metal cleaning and fabrication operations, electronic device manufacturing facilities, photo processing facilities, battery manufacturing and disposal sites, and nuclear weapons manufacturing and storage facilities etc. Non-metal related industries can also generate large volumes of aqueous wastes contaminated with low concentrations of heavy metal ions (below 1.00ppm). Prominent examples are coal – fired plants and pulp mills, where heavy metals indigenous to the raw materials (coal, wood, etc.) leach out during processing and ultimately find their way into wastewater streams.

Mining, metal plating industry, fertilizers, fungicidal sprays, and animal wastes are sources of copper to the aquatic environment [29]. Other sources include discharges from basic steel works and non-ferrous metals, motor vehicles and aircraft plating industries [30]: Effluent discharges from the paper and pulp, petroleum refining and printed circuit board also contribute copper to the water environment [31].

The major sources of mercury pollution in the aquatic environment are industries such as chloralkali, paint, pulp and paper, oil refining, electrical, rubber processing and fertilizer [32,33]. Most of the mercury used finds its way into the environment and it has been reported that about 0.1-0.2 kg of mercury is lost into the environment by chloralkali plant for every ton of chlorine produced [34]. Mercuric chloride and mercuric sulphate are used as catalysts, in the production of polyvinyl chloride and polyvinyl acetate [35].

Common sources of lead (Pb) in the environment are Pb-base pigments in paints, Pb-containing pesticides, discarded wet cell batteries, shooting ranges or water fowl hunting sites, putty, plumbing installations or repair sites, manufacturing fillers, untaxed whisky, and lotions. Other sources of Pb contamination have been lead pipes used in wells. Vegetation contamination resulting from heavy highway traffic or close proximity to a smelter has also been documented. Lead contamination of drinking water occurs as a result of corrosion and leaching from lead pipes, and Pb/Sn soldered joints associated with service lines commonly used in household plumbing [36]. Lead is introduced into natural waters from a variety of sources, such as storage batteries, lead smelting, tetraethyl lead manufacturing, mining, plating, ammunition, and ceramic and glass industries [10]. Anthropogenic sources of lead in the environment from the production of lead can be associated with (a) mining - milling operations, which include grinding, concentrating and transportation of the use, as well as disposed of the tailings and mine and mill water and (b) the smelter - refinery process and problems associated with storage, sintering, refining, atmospheric discharges and blowing dust [37]. In 1975, there was an estimated total of 19,225 tonnes of lead emitted to the atmosphere from sanitary sources. Of this, 400 tonnes were from primary lead smelting and 755 tonnes were from secondary smelting. In 1975, 1,42,000 tonnes of lead were emitted from mobile sources through gasoline [13].

Cadmium is introduced into bodies water from smelting [38], metal plating, cadmium - nickel batteries, phosphate, fertilizer, mining, pigments, stabilizers, alloy industries [39] and sewage sludge [40]. About one half of the total cadmium produced is used for plating purposes and the liquid wastes generated from these operations leads to contamination of ground water to the extent of 0.01 to 3.2 mg/L [41,42]. Significant quantities of nickel containing wastewater are introduced into water bodies, from the effluents of nickel-plating plants; silver refineries, zinc based casting industries and storage batteries [43]. Nickel has excellent corrosion resistance property and is used for alloy preparation. Lesser quantities of nickel are used in gasoline as additive and in ceramics, fungicides, pigment and in inorganic chemicals manufacturing industries etc. The recent users of nickel include nuclear power plants, gas turbine engines, cryogenic containers and pollution abatement equipments. Nickel present in the effluents of above stated industries pollutes water sources [44, 45].

Heavy metal containing wastewaters are either discharged into water bodies or onto land without any treatment, the latter forming the source of heavy metals to ground water. The extreme toxicity of heavy metals even in trace quantities, their tendency for bioaccumulation and transport in the food chain necessitated development of new methods for their removal from wastewaters.

## Chemical Process Technologies for Heavy Metal Removal

Various Chemical treatment methods have been developed for the removal and recovery of heavy metals from wastewaters. There are four major classes of chemical separation technologies: chemical precipitation, electrolytic recovery, adsorption/ ion exchange, and solvent extraction /liquid membrane seperation. These major classes involves various methods include chemical treatment with lime, caustic oxidation and reduction, ion – exchange, adsorption, reverse osmosis, solvent extraction, membrane filtration, electrochemical treatment and evaporative recovery. Chemical treatment methods for removal of heavy metal ions from waste are schematically illustrated in Figure 1. But most of these methods are often not economically viable especially when the effluents contain a large concentration of heavy metals. Most metal rich wastewater are treated by precipitation process. In this process, soluble metal ions are removed as insoluble metal hydroxide precipitates. But this process also has several disadvantages. The presence of aqueous organic ligands in wastewater can hinder metal hydroxide precipitation, which may result in residual metal concentrations that may no longer meet the increasingly stringent effluent discharge standards. The results are also not satisfactory with complex metals. Further, the major disadvantage with these treatment techniques is the production of sludge. As a result, an aquatic problem is changed into solid waste problem. Comparison of the major features, advantages and disadvantages of chemical processes technologies for removal of heavy metal ion from wastewater are provided in Table 2.

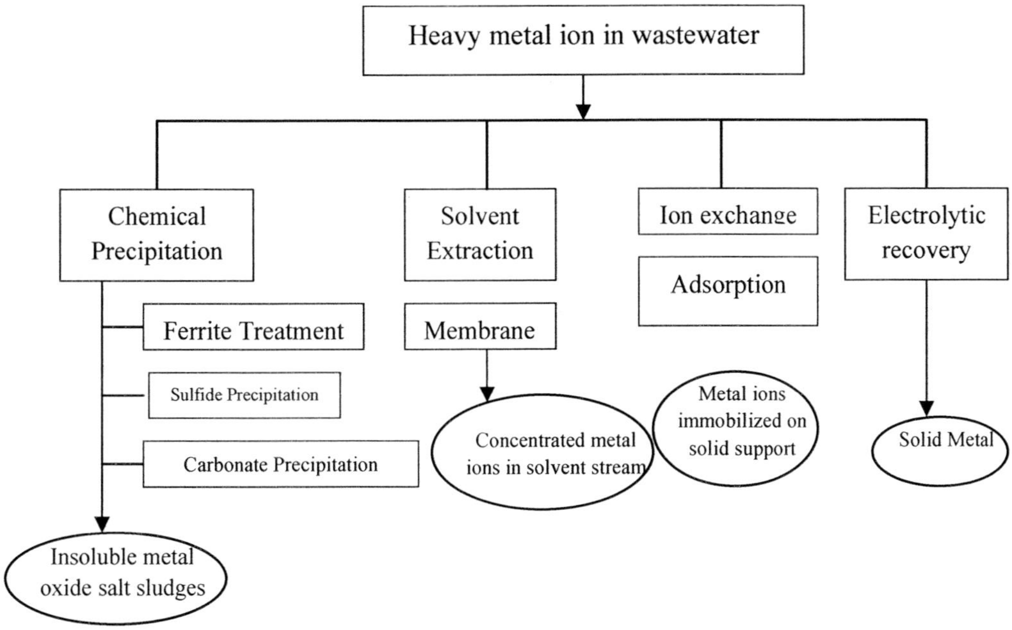


Figure 1: Available chemical treatment methods for hazardous heavy metal ion removal from wastewaters

**Table 2:Comparison of Chemical process Technologies for heavy metal ion removal**

| Process | Chemical /Energy Input | Major Advantage | Major Disadvantages |
|---|---|---|---|
| Chemical precipitation | Precipitant/flocculant, acid, base, mixing and fluid handling | Low metal concentration in the effluent are achieved | High chemical requirement, sludge disposal problem |
| Electrolytic Recovery | Electrical Energy | Lesser chemical consumption, recovery of pure metal is the added economic value | Energy intensive, high capital cost, reduced efficiency at dilute concentrations |
| Adsorption /Ion exchange | Fluid handling Unit/ regenerating solution | Highly effective for removing metal ions to a very low concentration | Chemical regeneration requirement, fouling and corrosion of plant, Disposal of exhausted adsorbent. |
| Solvent Extraction | Stripping solvent, makeup extraction solvent; fluid handling | Selective heavy metal removal, continuous concentrated metal solution recovery | Capital costs, Toxic solvent discharges |
| Membrane | Extractant for liquid – supported membrane; fluid handling | Selective heavy metal removal | Fouling and lesser durability of membranes |

## CHEMICAL PROCESS I: CHEMICAL PRECIPITATION, REDUCTION AND ELECRODEPOSITION

Chemical precipitation is the most common technique for the removal of heavy metals from water and wastewaters. The most frequently used chemicals for precipitation of metals are lime, caustic and sodium carbonates Precipitation with lime was found to be an effective and economical method for removing heavy metals from the discharge of a medium sized multi product company in Britain [48]. The lime also serves as an adsorbent for the removal of metal ions [49, 50]. The precipitation method has been used for the removal of iron, copper, zinc, tin, cadmium and nickel from the effluents of metal finishing industries [51] and for the removal of iron and aluminium in sewage water [52]. Sorg *et al.* [50] evaluated the use of lime for the removal of lead and cadmium from water and wastewaters and concluded that the effluents after treatment meet the requirements prescribed by Environmental Protection Agency (EPA), USA.

In the process of lime precipitation of heavy metals, caustic is added usually to reach the pH of minimum solubility. This process has limitation where multiple metals present have a pH range of minimum solubility. The disadvantage of sulfide precipitation is that the sludge does not thicken well and sulphides are a potential odor and health hazard. Co-precipitation is another process where many metals are adsorbed on alum or iron flocs but it has disadvantage of generating large quantities of sludge.

### Ferrite Treatment System

The electrolytic ferrite treatment system was devised by Japanese firm, Mitsubishi Petrochemical Co., [53]. In this process, iron dissolved in the waste steam by electrolysis of iron anodes supplies the hydroxide that coagulates the pollutants. After adding an alkali for pH control and a flocculant, the waste steam is fed to a thickener from which the sludge is withdrawn to a proprietary oxidation vessel. Aeration completes the formation of the crystalline ferrite precipitate, which is separated from liquid phase by magnetic or convection techniques.

### Sulfide Precipitation

Most heavy metals form stable sulfides; therefore, sulfide precipitation forms an excellent method for the removal of metals [54]. Sulfide precipitation gives lower residual metal concentrations than hydroxide precipitation. Further, metal sulfides usually settle faster and can be dewatered more readily than hydroxide sludges. Treatment with sulfide is most advantageous when used as a polishing step after conventional hydroxide precipitation or when very high percentage of metal removals is required.

Sulfide precipitation processes currently used for wastewater treatment fall into two broad categories (I) the soluble sulfide process and [II] the insoluble sulfide process. In the soluble sulfide process, the sulfide is added in the form of a water-soluble reagent, such as sodium sulfide.

## Carbonate Precipitation

The use of carbonate precipitation as a method for the removal of metals from wastewaters is very limited. Patterson et al. (55) used carbonate precipitation for cadmium (II) and lead (II) removal from electroplating effluents. At pH 7.5, the concentration of Lead (II) and cadmium (II) were reduced to 0.60 and 0.25 mg/L, respectively. Brantne and Cichon [56] compared pilot plant tests of carbonate precipitation with hydroxide precipitation of cadmium from an electric furnace blow down. At a sodium carbonate dosage of 650 mg/L and a pH ranging from 8.3 to 8.7, the total cadmium concentration was reduced from 1.37 mg/L down to 0.14 mg/L in clarified effluent. Soluble cadmium concentration was reduced from 0.79 mg/L to 0.1 mg/L and in another test from 1.2 mg/L to 0.04 mg/L with 'hydroxide precipitation comparable results could be obtained only at pH 10.5. Apart from low pH of precipitation, the relative filtration rates for carbonate precipitates were found to be approximately twice that of hydroxide precipitation.

## Chemical Reduction

The technique for treating a chromate waste by reduction to Cr (III) under acidic conditions, pH =3.0, followed by precipitation of Cr (III) hydroxide under slightly alkaline conditions is well known [57].

$$2\ HCrO4^- + 3\ HSO_3^- + 5\ H^+ \longrightarrow 2Cr^{3+} + 3\ SO_4^{2-} + 5H_2O \qquad [1.3]$$

$$HCrO_4^- + 3Fe^{2+} + 7H^+ \longrightarrow 3Fe^{3+} + 4H_2O \qquad [1.4]$$

The precipitation occurs according to the reaction

$$Cr^{3+} + 3OH^- \longrightarrow Cr\ (OH)_3 \qquad [1.5]$$

Another example of the application of reduction process is the use of sodium borohydride, which has been considered effective for the removal of mercury, cadmium, lead, silver and gold [58].

## Electro Deposition

Metal found in wastewater can be recovered by electrodeposition techniques using insoluble anodes. For example spent solutions resulting from sulphuric acid cleaning of Cu may be saturated with copper sulphate in the presence of residual acid. These are ideal for electrowinning where high quality cathode copper can be electrolytically deposited while free sulphuric acid is regenerated [58].

### Evaporation

Single stage or two stage evaporators are used chiefly to concentrate and recover valuable plating chemicals [59]. Recovery is accomplished by boiling off sufficient water from the collected rinse stream to allow the concentrate to be returned to the plating bath. Many of the evaporations in use also permit the recovery of the condensed steam or recycle as rinse water. Both capital and operational costs for evaporative recovery systems are high. Chemical recovery and water reuse values must offset these costs for evaporative recovery to become economically feasible. A closed hop system design for rinse and plating is most attractive.

## Chemical Processes II: Solvent Extraction and Membrane

Liquid-liquid extraction (also frequently referred to as solvent extraction) of metals from solutions on a large scale has experienced a phenomenal growth in recent years due to the introduction of selective complexing agents. In addition to hydrometallurgical applications, solvent extraction has gained widespread usage for waste reprocessing and effluent treatment.

Solvent extraction involves an organic and an aqueous phase. The aqueous solution containing metal or metals of interest is mixed intimately with the appropriate organic solvent and metal passes into the organic phase. In order to recover the extracted metal, the organic solvent is contacted with an aqueous solution whose composition is such that metal is stripped from the organic phase and is reextracted into stripping solution. The concentration of the metal in the strip liquor may be increased, often l0 to 100 times over that of the original feed solution. Once the metal of interest has been removed, the organic solvent is recycled either directly or after a fraction of it has been treated to remove impurities [60].

## Cementation

Cementation is the displacement of a metal ion from aqueous solution by a metal higher in the electromotive series. It offers an attractive possibility for treating any stream containing metallic ions. The fundamental principles of cementation have been known from time immemorial. Iron cementation in Launders, for example, is the oldest and most common technique for winning copper from its ore. Strictly speaking, the term cementation is restricted to cases where the displaced metal is precipitated from solution. However, the term may be extended to cover the reduction of any ionic species by contact with a metal of higher oxidation potential whether the reduced species is precipitated or not, for example, the reduction of ferric iron to the ferrous form by contact with metallic iron or zinc or the reduction of hexavalent chromium to the trivalent state by contact with iron [61,62].

Some common examples of cementation in wastewater treatment include the precipitation of silver from photo processing discharges, the precipitation of copper from printed etching solutions, and the reduction of Cr (VI) in chromium plating, and chromate inhibited cooling water discharges [60].

# MEMBRANE PROCESS

## Reverse Osmosis

Reverse osmosis is pressures driven membrane process in which a feed stream under pressure is separated into a purified permeate stream and a concentrated stream by selective permeation of water through a semi permeable membrane [63, 64]. Reverse osmosis enjoys wide spread popularity in the treatment of numerous diverse wastewaters. For example in the metal finishing industry, it is used for the recovery of plating chemicals from rinse water as well as purification of mixed wastewater to allow its reuse [65, 66]. In plating chemical recovery applications, reverse osmosis units separate the valuable metal salts from rinse solutions, yielding a concentrated metals solution, which can be recycled to the plating bath, a water of sufficient purity for use in rinsing.

Reverse osmosis has also been successfully demonstrated for the removal of Cr, Pb, Fe, Ni, Cu and Zn from vehicle wash rack water [67].

## Electro Dialysis

Electrodialysis is accomplished by placing cation and anion selective membranes alternatively across the path of an electric current. When the current is applied, the electrically attracted cations will pass through the cation exchanging membranes in one direction and the anions will pass through the anion exchanging membranes in the other direction. The result is that salinity decreases between one pair of membranes, increases between the next pair etc. Water can then pass through several such membranes until the required salinity is achieved [68].

Sorg *et al.* [50] reviewed the methods available for the removal of lead (II) and cadmium (II) from drinking water and wastewater. They concluded that though very little information was found in literature for Pb(II) and Cd (II) removal from water by electrodialysis, this technique would be as effective as reverse osmosis for the removal of these metal ions. The application of membrane process described is limited due to pretreatment requirements, primarily, for the removal of suspended solids. The methods are expensive and sophisticated, requiring a higher level of technical expertise to operate.

Dialysis involves the separation of solids by making use of unequal diffusion through membranes. The rate of diffusion is related to the concentration gradient, across the membranes. Dialysis has used to recover acids and metal salts and sodium hydroxide [68].

# CHEMICAL PROCESS III: ION EXCHANGE PROCESSES AND ADSORPTION

Ion exchange is a process in which ions on the surface of the solid are exchanged for ions of a similar charge in a solution with which the solid is in contact. Ion exchange can be used to remove undesirable ions from wastewater. Cations are exchanged for hydrogen or sodium and anions for hydroxide or chloride ions. The cation exchange on a hydrogen cycle can be illustrated by the following reactions; using in this example the removal of calcium ions, which are one of the ions ($Ca^{2+}$ and $Mg^{2+}$), which cause hardness of water:

$$H_2R + Ca^{2+} \rightleftharpoons CaR + 2H^+ \qquad [1.1]$$

where R, represents the resin.

The anion exchange can be similarly illustrated by the following reaction:

$$SO_4^{2-} + R\,(OH)_2 \rightleftharpoons SO_4R + 2OH^- \qquad [1.2]$$

When all the exchange sites have been replaced with calcium or sulphate ions, the resin must be regenerated. The cation exchanger can be regenerated by passing a concentrated solution of sodium chloride through the bed, while the anion exchanger which in this case is of hydroxide form, must be treated by a solution of hydroxide ions.

Ion exchange resins are available selectively for certain metal ions. The cation resins are mostly synthetic polymers containing an active ion group such as $SO_3H$. The natural materials such as zeolites can be used as ion exchange media [69]. The modified zeolites like zeocarb and chab carb have greater affinity for metals like nickel and lead [70]. The limitations on use of ion exchange for inorganic effluent treatment high cost of resins and the requirements for appropriate pretreatment systems.

Examples of selective removal of heavy metal ions by ion exchange are presented in Table 3. They include removal of Pb (II), Hg (II), Cd (II) Ni (II), V (IV, V), Cr (III, VI), Cu (II) and Zn (II) from water and industrial wastewaters by means various modern type of ion exchangers.

By ion exchange either all ions can be removed from a solution or substances are separated. Therefore, selective removal of ionic contamination and complete deionisation can be distinguished. The choice of between both depends mainly on the composition of solution and on the extent of decontamination required.

Hg (II) ions sorption on the ion exchanger Imac TMR (Styrenedivinylbenzene Copolymer) can be described by means of the following reactions:

$$2R\text{-}SH + Hg^{2+} \longrightarrow R\text{-}S\text{-}Hg\text{-}S\text{-}R + 2H^+ \qquad (21)$$

$$RSH + HgCl^+ \longrightarrow R\text{-}S\text{-}HgCl + H^+ \qquad (22)$$

where R is the styrenedivinylbenzene copolymer. After the Hg (II) ions sorption, the ion exchanger Imac TMR can be regenerated by means of the concentrated hydrochloric acids.

Metal ions and other cationic Impurities are removed on the cation exchanger whereas chromates on the anion exchangers:

$$nRSO_3H + M^+ \longrightarrow nRM^{n+} + nH^+ \qquad (23)$$

$$2ROM + CrO_4^{2-} \longrightarrow R_2CrO_4 + 2OH^- \qquad (24)$$

$$R_2CrO_4 + CrO_4^{2-} + H^+ \longrightarrow R_2Cr_2O_7 + OH^- \qquad (25)$$

The NaOH solutions are applied for anion exchanger regeneration. Further developments of conventional ion exchangers like polystyrenesulphonate cation exchangers and strongly basic anion exchangers were aimed at large exchange capacity and better chemical, thermal and mechanical stability. Kinetic capability of ion exchangers could be improved by means of macro porous ion exchangers developed in the nineteen seventies. The range of application of the ion exchange method was remarkably extended first by invention and application of new organic and inorganic ion exchangers.

**Table 3: Selective removal of heavy metal ions by ion exchange**

| Metal ion | Ion Exchangers | Composition and Functional groups |
|---|---|---|
| Pb (II) | Leawtit TP 207(Bayer, 2000) [71] | Chelating ion exchanger with functional iminodiacetate group |
| | Ethanotricarboxylate, 1,1-ddicarboxylate-2-etanophosphonate (Trochimczuk, 1999) [72] | Gel type Copolymer Of Vinyl benzene Chloride, Styrene and Divinyl Benzene |
| | Polyacrylamidoglycol Resin (Rivas et.al .1998) [73] | Amide, OH and carboxylic groups |
| | Clinoptilolite (Petruzzelli et. al.1999) [74] | Aluminosilicate structure |
| Hg (II) | Imac TMR | Styrenedivinylbenzene Copolymer |
| | Varian BTAM (Nike), [75]Varian BTKM, [76]Varian BSM [77], Ionac SR-3[78] | Containing nitrogen and sulphur functional group[22], Thiocarbamate group [23], Mercapathile group [24], Thiourea functional group [25] |
| | Dowex A-1[79] | Iminodiacetate group |
| Cd (II) | Dowex 50W-X4 | Polystyrenesulphonic cation exchanger |
| | Diphonix (Hubicki *et al.*1999) [80] | Sulphonic groups and diphosphonic groups |
| | KB-2 and KB-2-12P (Kauspediene et al., 1998) [81] | Polyacrylate carboxylic cation exchangers |
| | Phillipsite and Chabazite (Colella et al.1995) [82] | Natural zeolite from volcanic rocks of Central Italy |
| Ni (II) and V (IV, V) | Wofatit CA-20 (Halle et al. 1982) [83] | Macro porous carboxylic cation exchanger |
| | Chelex –100 (Leinonen and Lehto, 2000) [84] | Iminodiacaetae ion exchanger with Na/H/Ni in exchange system |
| | Sodium titnate, $Na_4Ti_9O_{20}$ (Lehto et.al, 1996) [85] | Layered material with exchangeable Na ions in the titanium oxide layer. |
| Cr (III, VI) | Lewatit MP-500A (Untea et.al, 2000) [86] | Basic anion exchanger |
| | Amberlite IR-120 (Drela, 1996) [87] | Polystyrenesulphone cation exchanger |

## Adsorption

Adsorption involves, in general, the accumulation (or depletion) of solute molecules at an interface (including gas-liquid interfaces, as in foam fractionation, and liquid-liquid interfaces, as in detergency). Adsorbent surfaces are often physically and/or chemically heterogeneous and bonding energies may very widely from one site to another. Adsorption involves accumulation of substance at an interface, which can either be liquid-liquid, gas liquid, gas-solid or liquid-solid. The material being adsorbed is termed the adsorbate and the adsorbing phase, the adsorbent. Chemical adsorption results in the formation of a monomolecular layer of the adsorbate on the surface through forces of residual valence of the surface molecules. Physical adsorption results from molecular condensation in the capillaries of the solid. In discussing the fundamentals of adsorption it is useful to distinguish between physical adsorption, involving only relatively weak intermolecular forces, and chemisorption, which involves, essentially, the formation of a chemical bond between the sorbate molecule and the surface of the adsorbent. The general features, which distinguish physical adsorption, chemisorption have been encapsulated in Table 4 [88].

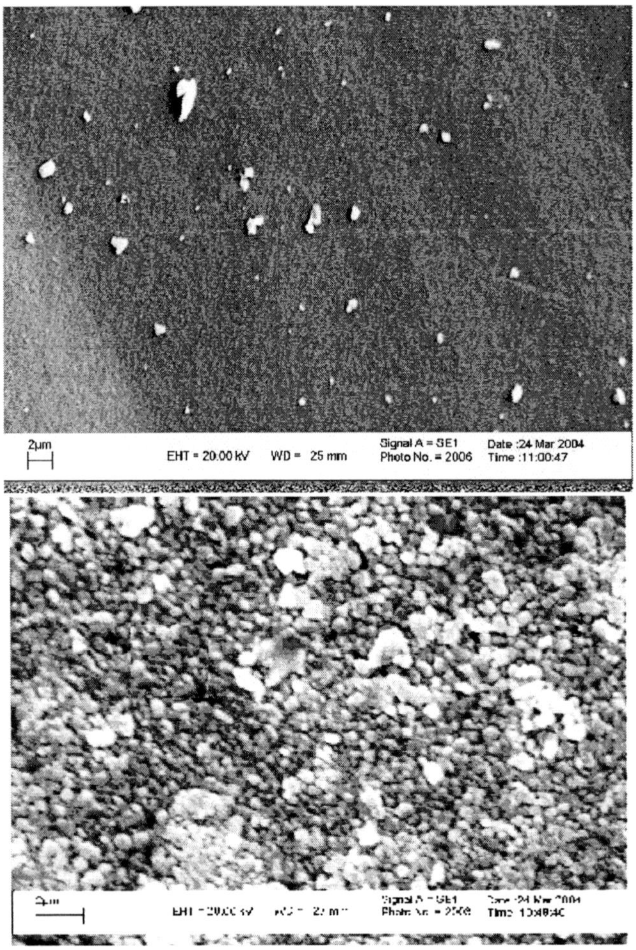


Figure 2 (a) SEM of Carbon Aerogel. Fig 2(b). SEM of carbon aerogel loaded with lead

## To Exemplify Adsorption process: Removal of Lead (II) from Aqueous Solution using Carbon Aerogel as Adsorbent

Hazardous heavy metal such as Pb (II), Hg (II) and Cd (II) can be removed from aqueous solution phase (industrial wastewaters) using carbon Aerogel as adsorbent [89, 90].

The nature and the presence of specific functional groups on the surface of the carbon Aerogel surface adsorbent impart certain characteristics that affect the adsorption process. It is found that that surface functional groups are formed during the carbon aerogel (adsorbent) activation process. Figure 2(a, b) shows SEM micrograph for carbon aerogel equilibrated with Pb (II). SEM of activated carbon surfaces before and after the adsorbent was equilibrated with the metal ion solution clearly shows the presence of Pb (II) ions.

Adsorption of heavy metals onto adsorbent is a pH dependent phenomenon. Extensive research investigations have been reported for adsorption of inorganic ions at different solution pH values. The dependence of adsorbent's adsorption capacity for inorganic ions removal from aqueous solution pH has been attributed to changes in surface chemical characteristics of adsorbent and species distribution of the inorganic ions varying with solution pH. In general the cationic species removal will increase with increasing solution pH, (Figure 3), while in the case of inorganic ion present as anionic species, removal generally increases with decreasing pH.

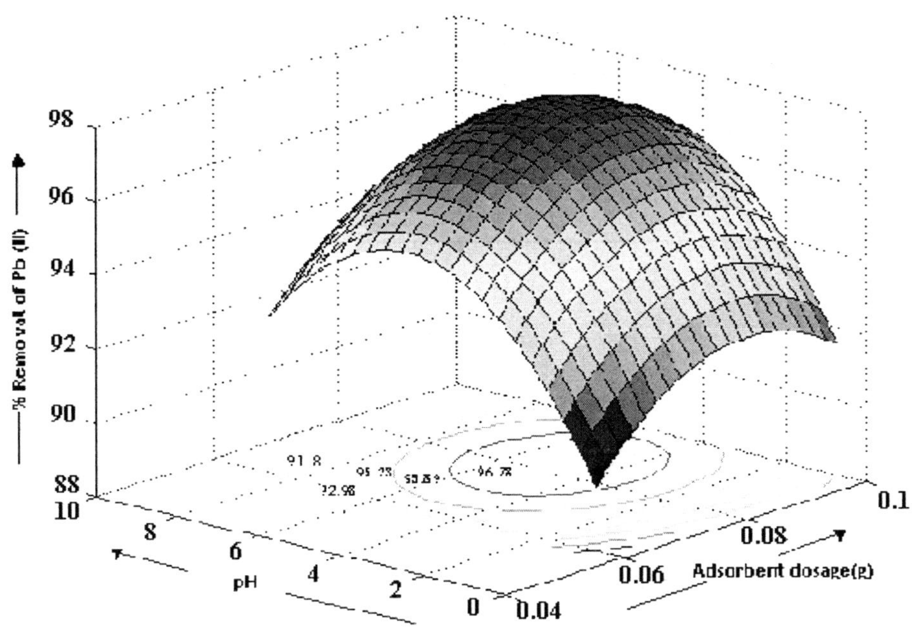


Fig.3. Plot for Percent removal of Pb (II) versus pH and adsorbent dosage using carbon aerogel as adsorbent

Specifically, pH affects the status of the outer hydration sheaths of the metal ion, metal speciation of inorganic ion in aqueous medium, complexation and solubility, and the electrochemical behavior of the carbon surface.

The adsorption capacity of carbon aerogel for Pb (II) removal was found to be 35mg of metal ion per gram of the adsorbent.

## ADSORBENTS FOR HAZARDOUS HEAVY METAL REMOVAL

Large number of adsorbents has been developed so far for the removal of inorganic. These include primarily activated carbon prepared from variety of materials of biological origin, industrial wastes, dead -biomass and organic wastes etc. Besides that several low cost adsorbents such as chitin, chitosin, clay, fly ash, peat, moss, zeolites, xanthates, seaweed/ algae/ alginate, lignin, bark/tannin rich material have also been explored for heavy metal containing wastewaters. Table 4 presents some of the highest adsorption capacity reported for various low-cost adsorbents.

**Table 4: Summary table of Maximum Reported Adsorption Capacities (mg/g)**

| Sorbent | Source | Cd (II) | Cr (III) | Cr (VI) | Hg (II) | Pb (II) |
|---|---|---|---|---|---|---|
| Bark | Randell *et al.* 1974a; Teles de Vasconcelos and Gonzalez Beca 1993;Masri *et al.*, 1974 Randell *et al.*, 1974a[91-94] | 32 | 19.45 | | 400 | 182 |
| CPEI cotton | Roberst and Rowland 1973[95] | | | | 1000 | |
| Chitin | Masri *et al.*1974 (Cd, Cr, Hg, Pb)[96] | 558 | 92 | | 100 | 796 |
| Chitosan | Pradas et al. 1994; Cadena *et al.*1990[97, 98] | 16.50 | | 57 | 1123 | 58 |
| Clay | Matis and Zouboulis, 1994; Niu *et al.* 1993[99,100] | 28 | | | | 116 |
| Dead biomass | Sharma and Foster, 1994 [101] | | | 43.0 | | |
| Leaf mould | Masri *et. al.* 1974; Srivastava *et. al.* 1994[96,102] | | | | 150 | 1587 |
| Lignin | Masri and Friedman, 1974(Cd, Cr, Hg and Pb)[96] | 87 | 17 | | 632 | 1587 |
| Modified wool | Low and Lee, 1991[102] | 46.65 | | | | |
| Moss | Tummavuori and Aho, 1980a;Kertman *et al.* 1993[103,104] | 5.058 | 76 | 43.9 | 16.2 | 135 |
| Peat | Tummavuori and Aho, 1980b;Kertman *et al.* 1993[103,104] | | | | | |
| Sawdust | Dikshit, 1989[105] | | | 16.05 | | 230 |
| Seaweed | Holan *et al.* 1993;Leusch *et al* 1995[106,107] | 215 | | | | 344 |
| Xanthate | Bricka and Hill 1989; Flynn *et. al*.1980[108,109] | 33.27 | 19.67 | | 1.149 | 18 |
| Coir pith carbon | Kadirvelu and Namasivayam 2000,Kadirvelu *et.al.* [110] | 94.0 | | | 154.0 | 264.0 |
| Fly ash | Apak *et al.*1998[111] | 198 | | | .073 | 444.0 |
| Red mud | Suemitsu *et al.* [112] | 67.0 | | | 7.1 | 165.0 |
| Peanut hull carbon | Periasamy, 1993[113] | 89.0 | | | 220.0 | 210.0 |
| Furnace Sludge | Lopez-Delgado *et al.*, 1998 [114] | 7.4 | | | | 68.0 |

Although a significant number of low cost adsorbent materials from sources, the commercial activated carbon (CAC) has still being used intensively up to date. Based on its size and shape, activated carbon is classified as following: Powder (PAC), granular (GAC), Fibre (ACF), and cloth (ACC). Due to the different sources of raw materials, the extent of chemical activation, and the physico-chemical characteristics; each type of activated carbon has its specific application with respect physicochemical properties, as well depending upon the inherent advantage and disadvantages in wastewater treatment.

**Table 4: Distinction between physical adsorption and chemisorption**

| Physical adsorption | Chemical adsorption |
|---|---|
| Low heat of adsorption (<2 or 3 times latent heat of evaporation.) | High heat of adsorption (> or 3 times latent heat of evaporation.) |
| Non-specific. | Highly specific. |
| Monolayer or multilayer.<br>No dissociation of adsorbed species. Only significant at relatively low temperatures. | Monolayer only.<br>May involve dissociation. Possible over a wide range of temperature. |
| Rapid, non-activated, reversible. | Activated, may be slow and irreversible. |
| No electron transfers although polarization of sorbate may occur. | Electron transfer leading to bond formation between sorbate and surface. |

## REFERENCES


[1] T. Stephengen, J.N. Lester. *Heavy metals in wastewater and sludge Treatment Processes. Vol I, Sources, Analysis, and Legislation*, CRC Press, Boca Raton, Fla. 31-64, (1987).

[2] N. Serpone, E. Borgarello, E. Pelizzeti.Photo reduction and photo degradation of inorganic pollutants :II Selective reduction and recovery of Au, Pt, Pb, Rh, Hg and Pb In: *Photocatalyis and environment*, Schiavello, M. (ed.) Kluwer Academic, p-57 (1988)

[3] F.W. Oehme. *Toxicity of heavy metals in the environment,* Part 2, Marcel Dekker Inc., New Marcell dekker Inc., New York and Basel, 624-628(1979)

[4] F.Berglund , M. Bertin, *Chemical Fallout*. Spring Field, IL. Thomas, (1969).

[5] WHO Environmental Health Criteria 101, *Methyl mercury*, Geneva: World Health Organisation. 68 –102 (1990).

[6] WHO Environmental Health Criteria, 118, *Inorganic mercury*. Geneva: World Health Organization, 68-112. (1991)

[7] J.K. Piotrowski, and M.J. Inskip, *Health effects of mercury*. Marc Report 24, Chelsea College, London (1984).

[8] L. Liberti, M. Natarnicola, V. Amicarelli, V. Campanaro, F. Roethel ,L. Swanson. Mercury removal with powdered activated carbon from flue gases at the cariano municipal solid waste incineration plant. *Waste Manage. Res.* 16, 183-189 (1998).

[9] M.A. Fernandez. Behavior of heavy metals in the combustion gases of urban waste incinerators. *Environ. Sci. and Technol.* 5, 1040-1047(1992).

[10] S. Manahan. *Environmental Chemistry*, Brooks/Colei California, USA (1984).

[11] R.A.Goyer, I.J. Chisolon. *Lead: in metallic contamination and human health,* Academic Press New York and London( 1972).

[12] K.S. Subramaniam, J.W. Cooner. Lead contamination of drinking water. *J Environ Sc Health,* Part A, 54, 29-33 (1991).

[13] WHO Environmental Health Criteria, *Lead*, Geneva: p.3 (1977).

[14] US Environmental Protection agency: *air quality Criteria for Lead*: USA (1977).

[15] N.N. Greenwood , A. Earnshaw. *Chemistry of the elements*, Pergamon press, Oxford (1989).

[16] O. Ravera. *Lead pollution in air and water*, FEG, Information Blatt, Nr. 21,46-54(1974).

[17] M.Buchaver. Contamination of soil and vegetation near zinc smelter by zinc, cadmium, copper and lead. *Environ. Sci. Technology.*7, 131-135 (1973)

[18] L. Friberg, M. Piscato, C.G.Nordberst, T. Kjellstrom. *Cadmium in the environments,* Springer, Berlin, Germany, (1979)

[19] C. P Huang,F.B. Ostovic. Removal of Cd (II) by activated carbon adsorption. *J. Environ. Eng. Div. ASCI.* 104, 863-878 (1978).

[20] R.J. Mahler, F. T. Bingham, A. L. Page and J. AS. Rayan. Cadmium enriched sewage sludge application to acids and calcareous soil: effects on soil and nutrition of lettuce, corn, tomato and Swies chard. *J. Environ. Qual.* 11, 694-701(1982).

[21] R. Lauwerys. *Criteria (dose /effect relationship) for cadmium*, CEC, Pergamon press, oxfords (1987).

[22] *Encyclopedia of environmental sciences*. S.PO. Parker (ed.) McGraw –Hill, New York p. 358 (1980).

[23] H.A. Schroeder, cadmium as a factor in hypertension, *J. Chron. Dis.,* 18, 647-656(1965).

[24] International agency for research on cancer (IARC) monographs on the evaluation of carcinogenic risk of chemicals to man, In: *cadmium and cadmium compounds* 11, 3974. (1976)

[25] B.J. Alloway ,D.C. Ayres. *Chemical principles of environmental Pollution*, Blackie, Glasgow, (1993).

[26] H. A. Waldron. *Metals in the environment*, Academic press, London, (1980).

[27] R.W. Batterbee, R.J. Flower, A.C. Stevewn son and B. Rippey. Lake acidifications in Galloway: a paleoecological test of competing hypotheses. *Nature*, 314: 350-352(1985).

[28] *Heavy metals and organochlorides. In: biology of fresh water pollution*, 3rd Edition, C.F. Mason ed. Longman, England, (1996).

[29] K. Pendias and H. Pendias. *Trace metals in soils and plants*, CRC Press, Boca Raton, Fl, 75-86(1992).

[30] J.G. Dean, F. L. Bosqui ,K. H. Lanouette. Removing heavy metals from wastewater. *Environ. Sci. Technol.* 6,518-522 (1972).

[31] R.C. Viashya, S.C. Prasad. Adsorption of Cu (II) on saw dust. *Indian J. Environ. Protection,* 11,284-289(1991).

[32] Kapoor ,T. Viraragahvan. Adsorption of mercury from waste water by fly ash. *Ads. Sci Technol.* 9,130 –147 (1994).

[33] Namasivayam, K. Periyasamy. Bicarbonate treated peanut hull carbon for mercury (II) removal from aqueous solution, *Water Res.* 27,1663-1668(1993).

[34] J.O. M. Bockris. *Environmental Chemistry*, Plenum Press, New York, (1977).

[35] G. Kaiser and G. Tolg. Anthropogenic compounds, In: *The handbook of environmental Chemistry*, O.Hutzinger (ed.) vol. 3, Part-A, Springer – Verlag, New York (1980).

[36] K. S. Subramaniam,J. W. Cooner. Lead contamination of drinking water. *J. Environ. Sci. Health,* Part A, 54, 29-33 (1991).

[37] L.W. Newland, K. A. Daum. Anthropogenic compounds, In: *The hand book of environmental chemistry*, O. Hutzinger (ed.) vol. 3, Part B , Springer -Verlag ,New York, p.22 (1982).

[38] M. J. Buchaver. Contamination of soil and vegetation near a zinc smelter for by zinc, cadmium, copper and lead. *Environ. Sci. Technol.,* 1, 131-135(1973).

[39] K. S. Low and C.K. Lee. Cadmium uptake by the moss claympers delesertii, *Biores. Technol.* 38, 1-6(1991).

[40] K. Bhattacharya, C. Venkobachar. Removal of cadmium (II) by low cost adsorbents. *J. Environ. Eng. Div. ASCE,* 140, 110 –122 (1984).

[41] S.D. Faust, O. A. Aly. Chemistry of water treatment, Ann. Arbor Science Book, Butter worth Publishers (1983).

[42] C.P.C. Poon. Removal of cadmium from wastewater, In: *cadmium in the environment*, H. Mislin and O. Reverva (eds.), Birkhas user basel, Switzerland, p. 46-55(1986).

[43] P.P. Viswakarma, K. P. Yadava ,V. N. Singh. Nickel (II) removal from aqueous solution by adsorption on fly ash. *Pertanika*, 12, 357-366(1989).

[44] *Encyclopedia of Environmental science* .S. P. Parker (ed.) 2nd Edition, McGraw Hill, New York, p. 354 (1980).

[45] S.A. Abbasi, R. Soni. Nickel in environment, *J. Indian Assoc. Environ. Management,* 17, 95-108 (1990).

[46] BIS, T*olerance limits for industrial effluents prescribed by Indian Standards Institution*, IS: 2490(Part I) (1981).

[47] U.S. EPA. *Control and Treatment Technology for the metal finishing industry: sulphide precipitation*, EPA 625,8-80-003, Cincinnati, OH (1980).

[48] Anderson, R. Clark. *Development of a pretreatment process for toxic metals removal, Effluent and water treatment convention*, Birmingham, Great Britain (1978).

[49] T. J. Sorg, M. Csanday,G. S. Logsdon. Treatment technology to meet the interim primary drinking water regulations for inorganics, *J. Am. Wat. Works Assoc.,* 70, 680-691(1978).

[50] G. Kakabadse. *Chemistry of effluents treatment*, Applied Science Publishers Ltd., London (1979).

[51] W. Stum, J.J. Morgan. *Aquatic chemistry*, John Wiley and Sons, New York (1970).

[52] Anon. Heavy metals removal by starch Xanthate, *Prod. Finishing,* 31, 72-74, (1978).

[53] J.G. Dean, F.L.Bosqui, K.H.Lanouette.Removing heavy metal from wastewaters. *Environ. Sc. Technol.* 6, 518-522(1972).

[54] J.W. Patterson, H. E.Allen, J.J. Scala. Carbonate precipitation of metal pollutants, *J. Wat. Pollut. Control Fed.* , 49, 2397-2410(1977).

[55] K.A.Brantner, E.J. Cichon. Heavy metal removal: comparison of alternative precipitation processes, Ind. Wast. *Proc. 13 th Mia. Atlantic Cont* Ann. Arbor Science, p.43-50 (1981).

[56] U.S. EPA, *Environmental Pollution control alternatives: Economics of waste water treatment alternatives for the electroplating industry,* EPA-625/5-79-016. Cincinnati, OH, 1979.

[57] M.J. Thomas, T.L. Thesis. Effects of selected ions on the removal of chrome (III) hydroxides. *J. Wat. Pollut. Control. Fed.* 48, 2032-2045(1976).

[58] R.F. Kiff.General inorganic effluents, In: *surveys in Industrial wastewater treatment-manufacturing and chemical industries.* D. Barnes, C.F.Forster and S.E. Hrudey (eds.), Vol. 3. Longman, NewYork (1987).

[59] U.S. EPA, *Control technology for metal finishing industry – Evaporators*, EPA – 625/8-79-002, Cincinnati, OH, (1977).

[60] T. Agelidies, K. Fytianos, G. Vasilikiotis, D. Janna Koudakis. Lead removal from wastewater by cementation utilizing a fixed bed of iron speres. *Envir. Pollution,* 50, 243-251(1988).

[61] Barnes, C.F. Forster, S.E. Hruday (eds.) Vol.3. *Chapter 1,* Longman, New York (1987).

[62] O.P. Case. *Metallic Recovery from wastewaters utilizing cementation*, EPA 670/2-74-008, U.S. Govt. Printing Office, Washington, DC (1974).

[63] H. Gooding. Reverse osmosis and ultrafiltrataion solve separations problems. *Chem. Eng.,* 92, 56-62(1985).

[64] P. Grampton. Reverse osmosis in the metal finishing industry. *Metal Finishing,* 80, 21-27(1982).

[65] M.K. Stiason. Emerging technologies for the treatment of electroplating wastewaters, Water –1978. *AICHE symp. Ser.* 75(190) 270-284 (1979).

[66] E.S.K. Chian. Renovation of vehicle wash rack wastewater for reuse, Water, 1975, *AICHE symp. Ser.* 71,87-92(1976)

[67] R.J.Keating , R. Dvorin, *Proc. 15 th Ind. Waste Conf. Porduer Univ.* 567 (1960).

[68] M.J. Thomas, T.L. Thesis. Effects of selected ions on the removal of chrome (III) hydroxides. *J. Wat. Pollut. Control. Fed.* 48, 2032-2045(1976).

[69] A.Groffman, S. Peterson ,D. Brookins. Removing lead from wastewaters using zeolite. *Wat. Environ. Technol.,* 4, 54-59(1992).

[70] Applegate. Membrane process. *Chem. Eng.* 91, 64-89 (1984).

[71] A.G.Bayer , Bayer AG Leverkusen (2000).

[72] W.A.Trochimczuk, *Europ. Polym. J.* 35,1457-1464 (1999).

[73] B.L Rivas, HA Maturana, S.Villegas, , E. Pereira, *Polym. Bull.* 40,721-728(1998).

[74] D.Petruzzelli, M.Pagano, G. Tiravanti, *Ind. Eng. Chem. Res.*34, 2612-2617(1995).

[75] B.A. Bolto, Pawlowski, L.E. and F.N. Spoon Ltd., London (1987).

[76] A.Warshawsky, S. Erlich-Rogozinski, *Microm. J.* 22,362-368(1977)

[77] K. Dorfne, Walter de Gruyter, Berlin – New York (1991).

[78] K.Inoue, K.Ohto , K. Yoshizuka R Shinubaru. Y., Babu, K Kina., *Bunseki Kagaku* 42, 725-731(1993).

[79] K Rengan. J. Radioanal. *Nucl. Chem.* 219, 211-215. (1997)

[80] Z. Hubicki, A. Jakowicz, A. Lodyga. Application of the ions from waters and sewages. In: Dabrowski, A. (Ed.). *Adsorption and environmental protection, Studies in surface science and catalysis*, Vol 120. Elsevier, Amsterdam- Lausanne – New York – Oxford – Shannon – Singapore - Tokyo.

[81] D.Kauspedience, J.Snukiskis, Gefeniene, A. J. Radioanal. *Nucl. Chem.* 229, 129-132(1998).
[82] Colella, M. De Gennaro, A. Langella, M. Pansini, Cadmium removal from wastewater using chabazide and phillipsite. In: ming, D.W., Mupton, F.A. (Eds.), *Natural Zeolites 93*. Brockport, New York, p.377.
[83] K. Halle, K. Fischwasser and B. Fenz, *Technol. Umweltschutz*, 25,1201032. (1982).
[84] Leinonen, H., Lehto *J.Rect. Funct. Polym.* 43,1-6(2000).
[85] J.Lehto, R., Harjula H. Leinonen, A. Paajanen, T. Laurila, Mononen, K., Saarinen, L. *J. Radio Nucl.Chem.*208, 435-443. (1996)
[86] Untea, E. Tudorache and C., *Florea Revista de Chimie* 51, 123-126(2000).
[87] Drela, Rudy *Met. Niezelaz*, 41, 321-323(1996).
[88] D.M. Ruthven. *Principle of adsorption and adsorption processes*. John Wiley and Sons, A Wiley Interscience Publication, New York, (1984).
[89] J.Goel, K.Kadirvelu, C.Rajagopal and V.K. Garg. *Carbon* 2004
[90] J. Goel, Ph.D. Thesis. *Heavy metal ion removal from contaminated water using column adsorption*. Guru Jambheshwar University, Hisar, Haryana, India, 2006.
[91] J.M.Randell, R. L Bermann, V. Garrett and A.C. Waiss, *J. Foorest Prod. J* 24,9,80-84(1974a)
[92] L.A. Teles de Vasconcelos and C.G. Gonzalez Beca, *Eur. Water Poll. Control* 4(1), 41-51(1994).
[93] M.S.Masri and M.Friedman, *J. Appl. Polymer Sci.* 18,2367-2377(1974).
[94] J.M. Randall, E. Hautala and A.C. Waiss Jr. (1974b) Removal and recycling of heavy metals ions from mining and industrial waste water streams with agricultural byproducts. *Proceedings of the fourth Mineral waste Utilization Symposium*, Chicago, IL, May 7-8 .1974,pp.329-334.
[95] E.J. Roberts and S.P. Rowland *Env.Sc Technol.* 7(6) 552-555(1973).
[96] M.S Masri., F.W. Reuter and M. Friedman *J. Appl. Poly. Sc.,* 18,6756-681 (1974).
[97] E.G. Prades, M.V Sanchez., F.C.Cruz , M.S. Viciana and M.F. Prez, *J. Chem. Tech. Biotechnol.* 59,289-295(1994).
[98] F.Cadena, R.Rizvi and R.W. Peters (1990) Feasibility studies for the removal of heavy metals from solution using tailored bentonite. In hazardous and industrial wastes, Proceedings of the twenty *–Second Mid-Atlantic Industrial Waste conference*, Drexel University, p-77-94.
[99] K.A.Matis and A.I. Zoublious *Biotech, Bioengg.* 44,354-360(1994).
[100] H. Niu, X.S. Xu and J.H. Wang, *Biotech. Bioengg.* 42,785-787(1993).
[101] D.C. Sharma and C.F. Forster, *Biores. Technol.* 49,31-40(1994).
[102] S.K.Srivastava, A.K. Singh and A. Sharma *Environ. Technol.*15, 353-361(1994).
[103] K.S. Low, C. K.Lee, F. Y. Cheong and G. S. Lim, *J. Environ. Sci. Health* A, 28 (8), 1689-1704 (1993).
[104] P. Tong, Y. Baba, Y. Adachi and K. Kawaza, *Chem Lett.* 9,1529-1532(1991).
[105] S.V.Kertman, G.M.Kertman and Z.H. Chibrikova., *J. Appl. Chem USSR,* 66(2), 465-466(1993).
[106] V.P. Dikshit, *Nat. Acad Sci. Lett.* 12(12), 419-421, (1989).
[107] Z.R. Holan, BN. Volesky and I. Prasetyo *Biotech, Bioengg.* 41(8), 819-825(1993).
[108] Leush, Z.R. Holan and B. Volesky, *J. Chem. Tech. Biotechnology*. 62, 279-288 (1995).

[109] R.M. Bricka and D.O. Hill (1989) Metal immobilization by solidification of hydroxide and xanthate sludges. *In environmental aspects of stabilisation and solidifications of hazardous and radioactive wastes ASTM STP 1033*, EDS P.L. Cote and T.M. Gilliam, pp-257-272. ASTM, Philadelphia.
[110] M Flynn. Jr., T.G. Carnahan and R.E. Lindastrom (1980) Adsorption of heavy metal ions by xanthated sawdust. . United States. Bureau of Mines.
[111] Kadirvelu K. and C.Namasivayam, *Advances in Environmental Research*, (2003)
[112] R. Apak, E. Tutem, M. Hugul and J. Hizal, *Water Res.*, 32,430-440(1998).
[113] R. Suemitsu, R. Venishi, I. Akashi and M. Nakana, *J. Appl. Poly. Sci.,* 31,75-83(1986).
[114] K.Periasamy, Adsorptive removal of heavy metals from synthetic indisustrial wastewater using peanut hull carbon. *Ph.D. Thesis*, Bharthiar University, Coimbatore, India, (1993).
[115] Lopez-Delgado, C. Perez and B.S.Girgis, *Water Res.*, 32,1169-1177(1998).

In: Hazardous Materials and Wastewater
Editor: Allison A. Lewinsky, pp. 149-199
ISBN 1-60021-257-3


*Chapter 5*

# AOP AS AN EFFECTIVE TOOL FOR THE MINIMIZATION OF HAZARDOUS ORGANIC POLLUTANTS IN COLORED WASTEWATER; CHEMICAL AND PHOTOCHEMICAL PROCESSES

***Natalija Koprivanac*** ***and Hrvoje Kusic*** [†]
Faculty of Chemical Engineering and Technology, University of Zagreb, Zagreb, Croatia

## ABSTRACT

The manufacturing and the application of organic dyes involve the production and the handling of many organic compounds hazardous to human health. Many of these substances are considered toxic, even carcinogenic. Over the past couple of decades, manufacturers and users of dyes have faced increasingly stringent legal regulations promulgated to safeguard human health and the environment. So, there is a clear need to treat dye wastewater prior to discharge into the primary effluent. The limitation of traditional wastewater treatment technologies (biological and physical), such as low rate, disability to degrade many of recalcitrant organic dyes and the production of secondary waste which demands further treatment, can be overcome by the utilization of advanced oxidation processes (AOPs). These wastewater treatment methods are considered as low- or even non-waste generation technologies. AOPs are based on the production of very reactive species, such as hydroxyl radicals, able to decolorize and to reduce recalcitrant colored wastewater loads due to the high oxidation power and the lack of selectivity of OH radicals towards a broad range of organic pollutants present in wastewater. Generally, AOPs can be broadly classified concerning the way of OH radicals generation into chemical, photochemical, photocatalytic, mechanical and electrical technologies. This research deals with the application of chemical and photochemical AOPs for the minimization of recalcitrant colored pollutants present in wastewaters. The comparative

[*] nkopri@fkit.hr
[†] hkusic@fkit.hr

study of several processes, $Fe^0/H_2O_2$, $UV/Fe^0/H_2O_2$, $UV/O_3$ and $UV/O_3/H_2O_2$, was performed through an experimental research of decolorization and mineralization of a reactive azo dye C.I. Reactive Blue 137, as a model wastewater pollutant. Applied processes were optimized according to their process parameters, $Fe^0$ dosages, initial $H_2O_2$ dosages and initial pH values of treated solutions. The influence of initial organic dye concentration, as well as the addition of solid particles, synthetic zeolites, on the process effectiveness was also investigated. Studied AOPs were evaluated on the basis of their eco-effectiveness, by the means of color ($A_{610}$), aromaticity ($UV_{280}$), TOC and AOX value decrease, and their cost-effectiveness as well.

# INTRODUCTION

## Colored Wastewater

Dyes make our world beautiful, but they bring pollution. Synthetic dyes are present in many spheres of our everyday life and their application is continuously growing. Organic synthetic dyes have been widely used as colorants in different industries such as textile, paper, color photography, pharmaceutical, food, cosmetic, electric... [1]. More than 0.7 million tons of organic synthetic dyes are produced annually worldwide. In additon, over 10,000 different dyes and pigments have been applied in those industries. Studies indicated that approximately 15 % of produced synthetic dyes per year have been lost during manufacturing and processing operations that involve the production and handling with many organic compounds hazardous to human health. Wastewaters originated from dye production and application industries present a very serious environmental problem because of the aesthetic nature due to the fact that the coloration is visible even in a low dye concentration. Although the presence of color in water might not appear to have a negative effect other than aesthetic, this is not actually the case. The color can adsorb and reflect sunlight entering the water. Consequently, bacteria can not grow sufficiently in the water and that can have a disastrous effect on the food chain [2]. But even more important, many substances in such wastewaters are considered toxic, and even carcinogenic and mutagenic [3]. Nowadays, the mostly used group among all dyes are reactive synthetic dyes, mostly applied for dyeing of natural fibbers such as cotton and wool. Due to the high solubility and hydrolysis as a side effect of such types of dyes, they can often cause loaded colored effluents [4]. Almost 70 % of all reactive dyes are of the azo type, while other are mainly of anthraquinonic and phtalocyanine types [5]. Furthermore, commercial reactive azo dyes are intentionally designed to resist biodegradation, i.e. the wastewater originated from reactive dye processes is characterized by poor biodegradability, passing unaffected through conventional treatment systems and discharging into the environment [6-8]. Like other colored wastestreams, those from reactive azo dye manufacturing and utilization industries pose a major threat to the surrounding ecosystems due to the documented health hazards caused by the toxicity and the potentially carcinogenic nature of such organic pollutants [9]. The group of most dangerous contaminants in such wastewaters are aromatic amines which may cause methemoglobinemia, and they are generated as by-products of partial degradation of reactive dye wastewater treatment or as impurities present in commercial azo dyes due to their usage in production process. The amines oxidize the heme iron of hemoglobin from Fe(II) to Fe(III), blocking the oxygen binding. This results in characteristic symptoms like cyanosis of lips and nose, weakness and dizziness. [10]. A number of azo dyes have been shown to be animal carcinogenous, and analysis based on structure-activity relationships suggests that many more azo dyes are likely to have a carcinogenic potential. Reductive cleavage to carcinogenic aromatic amines, which are known to act via genotoxic mechanisms, is in all likelihood an important mechanism of azo dye carcinogenicity [7]. Since 20 years ago some intermediates for azo reactive dye manufacturing, such as: *o*-aminoazotoluene, *p*-aminoazobenzene, *o*-anisidine, 2-amino-4-nitrotoluene, benzidine, 2-naphthylamine, *o*-toluidine, etc, were become to EPA's "black list" of hazardous chemicals that resulted with their restriction. But today, many amines are still in use, resulting with their presence in wastewater originated from azo reactive dye manufacturing and application industries.

## Wastewater Treatment Methods

Due to the above mentioned facts, manufacturers and users of reactive dyes have faced, over the past two decades, increasingly stringent legal regulations promulgated to safeguard human health and the environment [5]. To comply those requirements, the new technologies have been designed, specifically, to analyze and remove color and priority pollutants from wastewater effluents and the methodology of the cleaner production has been applied to circumvent pollution problems by eliminating their source [11].

Generally, wastewater treatment methods can be broadly classified into biological, physical (i.e. mechanical) and chemical [3, 12-14]. Biological treatment methods are widely applied for the treatment of municipal and industrial wastewaters. Although biological treatment methods have many advantages (economical acceptance, simplicity of treatment facility), they can not be successfully applied for the treatment of colored wastewater containing reactive dyes due to their poor biodegradability caused by recalcitrant nature [8, 14]. Reactive dyes are developed and designed to resist fading upon exposure to light, water and oxidizing agent [15, 16]. They are characterized by low biomass adsorbability and they are not degradable by aerobic conditions in conventional biological wastewater treatment facilities. Recently, some microorganism cultures, which are able to decolorize and mineralize some of reactive azo dyes, were developed [17-19]. However, biological processes by itself are often inappropriate due to the low rates of biological degradation. Physical, also known as mechanical, treatment methods such as: filtration, sedimentation, adsorption, coagulation/flocculation etc., like their name states, include only physical/mechanical processes, without chemical transformations of present contaminants, for wastewater treatment. Therefore, they generally transfer waste components from the liquid phase to a sludge phase. That makes them insufficiently effective and moreover, very expensive, due to the requested further treatment of generated secondary waste or due to the regeneration of inactive adsorbents. Although there are many examples in the literature about textile wastewater treatment by activated carbon adsorption with reported process effectiveness for wide range of textile dye types, the relatively low adsorption capacity of activated carbon for some of reactive dyes was shown [20]. Reactive dyes can be also removed from wastewaters by adsorption on various natural quaternized organic materials such as cellulose, sugarcane bagasse, rice husk, and coconut husk, alternatively cheap adsorbents. But here, too, no successful regeneration has been reported [21]. Reactive dye removal by adsorption onto mezoporous minerals such as sepiolites and zeolites was also studied, but successful removal was limited due to the low adsorption capacities of these natural minerals [20]. Removal by synthetic zeolites is possible but it depends on pore size and capacity level of applied zeolites [22-24]. Coagulation/flocculation process using Fe(III) and Al(III) salts, or polymeric flocculant Levafloc-R showed high efficiency for decolorization of wastewater containing reactive dyes [25-27]. The main problems of this relatively cheap and easily to maintain treatment method are related with the generation of large amounts of sludge as secondary waste that requires further treatment and disposal. Chemical treatment is based on the electron transfer, and most widespread application found processes using chlorine, ozone and potassium permanganate. The main disadvantage of chlorine usage for wastewater treatment is the production of chlorinated hydrocarbons that are in most cases even more toxic and carcinogenic than parent contaminants and therefore the usage of chlorine is restricted in many countries [28, 29]. Besides that, the usage of chlorine as an oxidant for wastewater treatment is characterized by its low selectivity and high consumption. The application of pure ozone oxidation is limited by high selectivity and slow kinetics. Although ozone was shown as very efficient for decolorization of colored wastewaters, it rarely produces complete mineralization to $CO_2$ and $H_2O$ [30]. The usage of potassium permanganate as rather expensive oxidative agent is widespread due to the easiness of its application. But the main problem is caused due to the production and the precipitation of manganese oxide which than should be removed from the process equipment. From the previous statements concerning wastewater treatment methods it could be concluded that common treatment methods suffer from various limitations and that higher efficient treatment methods need to be found.

## Advanced Oxidation Processes

Advances in chemical treatment of wastewater resulted with the development of a number of alternative chemical technologies; so-called Advanced Oxidation Processes (AOPs). In last two decades, the interest of not only world scientific community for the development and the application of AOPs as wastewater treatment methods was rapidly increased. The rapid increase of interest of research on AOPs resulted with numerous scientific publications, but also with their numerous applications on full-scale treatment facilities [31]. The main advantage of AOPs over other wastewater treatment processes; biological, physical-mechanical and classical chemical is their pronouncedly destructive nature that results with the mineralization of organic contaminants present in wastewater. AOPs are considered as low- or even none-waste generation technologies [32, 33], and generate short-lived chemical species with high oxidation power. At the first place, that are hydroxyl radicals, (HO•), which Glaze et al. [34] pointed out as main oxidation species in AOPs. Owing to high reactivity (Table 1) and unselectivity (Table 2) of HO•, AOPs are considered to be promising methods for the treatment of hazardous toxic organic pollutants in aqueous solutions [35]. AOPs generate OH• in sufficient quantities to oxidize the majority of organics present in the effluent water [36]. However, some of the simplest organic compound such as acetic, maleic and oxalic acid, as well as acetone, chloroform and tetrachloroethane can not be oxidized by OH radicals [37].

**Table 1. Standard redox potential of some oxidant species [35]**

| Oxidant | Redox potential, $E°$, V |
|---|---|
| Flour | 3.03 |
| Hydroxyl radical | 2.80 |
| Atomic oxygen | 2.42 |
| Ozone | 2.07 |
| Hydrogen peroxide | 1.77 |
| Permanganate ion | 1.67 |
| Chlorine | 1.36 |
| Chlorine dioxide | 1.27 |

**Table 2. Chemical species those are able to be oxidized by hydroxyl radicals [36]**

| Acids | formic, gluconic, lactic, malic, propionic, tartaric |
|---|---|
| Alcohols | benzyl, *tert*-butyl, ethanol, ethylene glycol, glycerol, isopropanol, methanol, propenediol |
| Aldehydes | acetaldehyde, benzaldehyde, formaldehyde, glyoxal, isobutyraldehyde, trichloroaldehyde |
| Aromates | benzene, chlorobenzene, chlorophenol, PCBs, phenol, hydroquinone, catechol, benzoquinone, *p*-nitrophenol, toluene, xylene, trinitrotoluene |
| Amines | aniline, cyclic amines, diethylamine, dimethylforamine, EDTA, propanediamine, *n*-propylamine |
| Dyes | azo, anthraqiunone, triphenylmethane |
| Ethers | tetrahydrofuran |
| Ketones | dihydroxyacetone, methylethylketone |

AOPs show high flexibility in their practical application due to the fact that it can be used either separately or in combination, or in combination with other classical wastewater treatment methods. Furthermore, another great advantage of AOPs over classical wastewater treatment methods is the possibility of conduction at ambient conditions, i.e. atmospheric pressure and room temperature. Such so-called ambient AOPs can be used for the treatment of low loaded wastewaters (Figure 1.) [38]. For the treatment of highly loaded wastewaters, even up to value of COD (chemical oxygen demand) ecological parameter of 200 $gL^{-1}$, other processes generating also OH radicals can be used. Some authors consider those processes as AOPs, although these processes require pressure and temperature highly over ambient conditions, so they can be named as non-ambient AOPs. Such processes are wet air oxidation, supercritical wet air oxidation and other hydrothermal oxidation processes that requires pressures and temperatures even above 1 MPa and 150 °C, respectively [33, 35, 38].

Regarding to the value of TOC (total organic carbon) content, AOPs are suitable for the treatment of wastewaters loaded in range of 100-1000 mgC $L^{-1}$ [39]. Ambient AOPs can be broadly classified concerning the way of OH radical generation. Generally, OH radical can be generated by chemical, electrical, mechanical or radiation energy. Therefore, AOPs can be classified into chemical and catalytic, photochemical and photocatalytic, mechanical and electrical processes (Figure 2.).

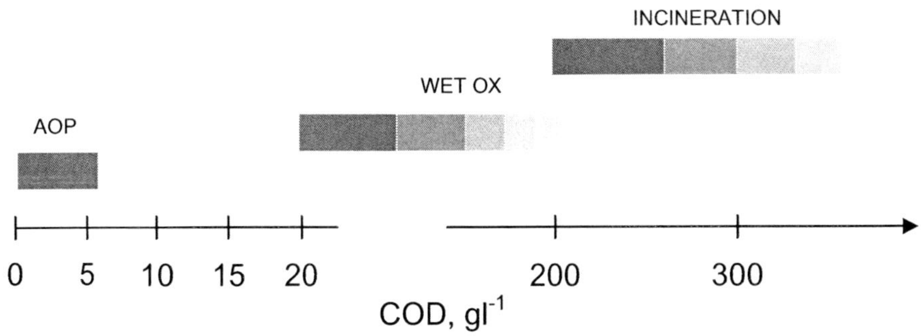


Figure 1. Suitability of water treatment technologies according to COD contents [38]

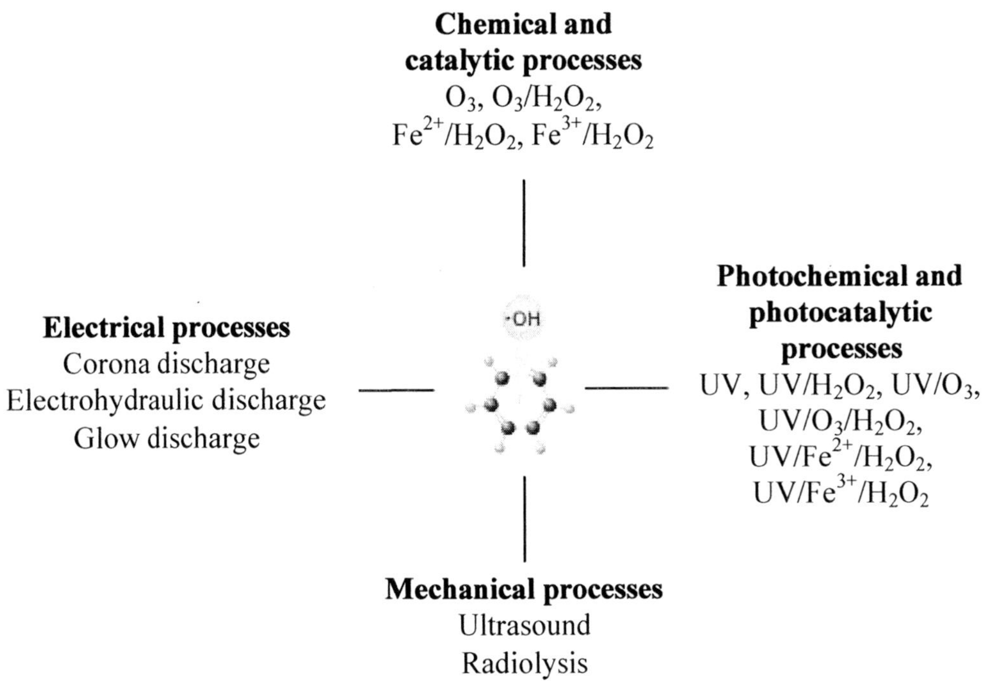


Figure 2. Schematic of advanced oxidation processes classification

Chemical processes involve the application of ozone and/or hydrogen peroxide, while a subcategory of this type of AOPs can be named catalytic processes that involve usage of some powerful catalyst (e.g. iron or cupper ions) in combination with hydrogen peroxide to produce OH radicals, so-called Fenton type processes. Photochemical and photocatalytic processes involve application of UV or solar irradiation in combination with some powerful either oxidant (ozone and/or hydrogen peroxide) or photocatalyst (e.g. $TiO_2$, ZnO, etc.). OH radical can be also produce under influence of mechanical (e.g. ultrasound process, radiolysis) or electrical (e.g. electrohydraulic discharge and non-thermal plasma processes) energy [31-40].

# Colored Wastewater Treatment by AOPs: A Review of Recent Studies

According to the number of AOPs investigations directed to the treatment of colored wastewater, the most studied processes are Fenton type processes, then UV-based technologies, ozone-based technologies, etc. This part of chapter contains the more detailed review of AOPs applied for the treatment of colored wastewater investigated in our laboratory, and the less detailed overview of other AOPs.

## Fenton Type Processes

All processes that involve catalytic reaction between hydrogen peroxide and metal homogeneous ions, in the most cases $Fe^{2+}$ ions, can be named as Fenton type processes. The reactivity of this system was first observed by Fenton in 1876 [41, 42]. Its utility was not recognized immediately, but 1930s when its mechanism based on OH radicals was proposed, the number of investigations concerning the application of Fenton reagent for the degradation of organic compounds dissolved in water is continuously growing. The application of Fenton process is based on the reaction that involves the oxidation of $Fe^{2+}$ ions to $Fe^{3+}$ ions with $H_2O_2$ and the subsequent generation of OH radicals (1):

$$Fe^{2+} + H_2O_2 \rightarrow Fe^{3+} + HO\bullet + HO^- \quad (1).$$

Such generated OH radicals could then react with organic matter present in water. Although reaction shown by equation (1) is often referred as Fenton reaction [43], and presents the key step in the Fenton process, other important reactions also occur. The following set of reactions describes the occurrence of Fenton catalytic cycle:

$$Fe^{2+} + HO\bullet \rightarrow Fe^{3+} + HO^- \quad (2)$$

$$Fe^{3+} + H_2O_2 \rightarrow Fe^{2+} + HO_2\bullet + H^+ \quad (3)$$

$$Fe^{2+} + HO_2\bullet \rightarrow Fe^{3+} + HO_2^- \quad (4)$$

$$Fe^{3+} + HO_2\bullet \rightarrow Fe^{2+} + O_2 + H^+ \quad (5)$$

If the organic matter, i.e. organic pollutant, is present in the system, the next reactions are also involved in chain radical mechanism:

$$RH + HO\bullet \rightarrow H_2O + R\bullet \quad (6)$$

$$R\bullet + Fe^{3+} \rightarrow Fe^{2+} + products \quad (7)$$

By-products of above reaction (7) could be further degraded by radical mechanism to the complete mineralization. Besides Fenton process, $Fe^{2+}/H_2O_2$, that involves the application of ferrous salts, mostly ferrous sulphates, as a source of iron catalyst for Fenton reaction (1) [44-49], there is a number of studies that investigated the application of so-called Fenton "like" processes for the degradation of pollutants in wastewater. There are three types of Fenton "like" processes. The first group of processes considers the usage of ferric salts instead of ferrous salts as catalysts for the incitation of Fenton reaction (1) [48-51]. Next group of processes considers the usage of heterogeneous Fenton type catalysts such as iron powder, iron-oxides, iron-ligands, or iron ions doped in zeolites, pillared clays or resins, instead homogenous ferrous ions obtained from the dissolution of added ferrous salts [48, 49, 52-62]. Third group are processes that use other metal ions, e.g. cupper, manganese or cobalt, instead of ferrous ions in Fenton reaction [44, 63]. The primary benefits of Fenton type processes are their ability to convert a broad range of pollutants to harmless or biodegradable products and the fact that their relatively cheap reagents are safe to handle and are environmentally benign [43]. Even though these systems offer a cost effective source of HO•, their efficiency is limited with a couple of disadvantages: (a) the need for the removal of remaining iron ions after the treatment and (b) a limited yield of reaction process due to the formation of stable $Fe^{3+}$-complexes [37, 43]. Those limitations can be overcome by the usage of heterogeneous Fenton-type catalysts which can lower the final concentration of iron ions in the bulk after the treatment, while by the assistance of UV irradiation the formed $Fe^{3+}$-complexes can be destroyed, thus allowing $Fe^{3+}$ ions to participate in the Fenton catalytic cycle [60, 64]. The oxidation power of Fenton type processes depends of several operating parameters such as iron concentration, source of iron catalysts, or catalyst if other metal ions are used instead of iron ions, $H_2O_2$ concentration, applied catalysts/hydrogen peroxide ratio (Fenton reagent), temperature, pH and treatment time [65]. According to the literature [65], a minimal threshold concentration of ferrous ions that allows the reaction to proceed within the reasonable period of time ranges between 3-15 $mgL^{-1}$, regardless of the concentration of organic pollutant. However, iron levels <25-50 $mgL^{-1}$ can require excessive reaction times (10-24 hours). Typical Fenton catalyst/$H_2O_2$ ratios range between 1 : 5 to 1 : 25 [65, 66], while it is markedly proven that the optimal pH range for Fenton process is between 2.8 and 3.0 [43-54].

As it was mentioned above, Fenton type processes are the most studied processes for the treatment of colored wastewater. Numerous studies dealt with the application of Fenton process $Fe^{2+}/H_2O_2$ for the decolorization of colored wastewater including those that contain reactive azo dyes. Xu et al. [44] investigated the efficiency of $Fe^{2+}/H_2O_2$ for the decolorization and the mineralization of 20 organic dyes including 3 reactive dyes and they obtained rather high decolorization, >96 %, while mineralization extents after 120 min of treatment ranged between 72-79 % TOC removal. Lucas and Peres [45] achieved bleaching between 83 and 96 % depending on the initial dye concentration. Meric et al. [46] applied $Fe^{2+}/H_2O_2$ process for the decolorization and the detoxification of three reactive dyes with the initial dye concentrations of 100 and 200 $mgL^{-1}$. The overall decolorization obtained in that study was >98 %, while COD removal yielded between 92 and 95 %. In our previous study [49], Fenton process was applied for the degradation of two reactive dyes (azo and anthraquinonic type), and the bleaching >95 % was achieved in cases of both reactive dyes, while final extents of partial mineralization, 34.3 and 72.1 % of TOC removal, depended on dye structure. Neamtu et al. [51] obtained >99 % of color removal for both studied reactive dyes with the initial concentrations of 100 $mgL^{-1}$. Hsueh et al. [67] investigated the efficiency of $Fe^{2+}/H_2O_2$ using the low concentration of ferrous salts for the decolorization of reactive dye model solution, and they obtained similar results as in previous cases, rather high decolorization, >96 %, and <20 % of TOC removal after a one-hour treatment. From the above presented results of several studies applying $Fe^{2+}/H_2O_2$ process for the degradation of reactive dyes in wastewater, it can be observed that rather high decolorization was achieved in all cases, while according to the other monitored ecological parameters, i.e. TOC or COD, in all studies the mineralization of colored wastewater was only partial. Similar results were obtained by using ferric salts instead of ferrous salts [49, 67]. Several studies investigated application of heterogeneous catalysts for Fenton reaction and obtained similar results. Decolorization >98 % and mineralization extents of 45.3 and 46.9 % TOC removals for C.I. Reactive Blue 137 and C.I. Reactive Blue 49 model wastewaters, respectively, were obtained in our previous study by using iron powder instead of homogenous ferrous ions [49]. Tang and Chen [54] applied iron powder/$H_2O_2$ system for the treatment of colored wastewater containing C.I. Reactive Red 120, and obtained color removal >90 %. They pointed out iron powder/$H_2O_2$ system as effective for dye removal owing to the Fenton reaction that occurs in the bulk and on the iron powder surface. Several studies include the application of ferrous or ferric ions doped in zeolites or resins, as well as the usage of ferrioxalate, for the decolorization of model wastewaters containing reactive dyes [51, 60, 62, 69, 70]. By these heterogeneous Fenton type processes rather high dye degradation was estimated, and mostly owing to the occurrence of Fenton reaction on the surface of heterogeneous catalysts resulting with lower concentration of iron ions in the bulk after the treatment. But, in most of those studies the reactivity of Fenton type processes were improved by UV irradiation or by non ambient conditions (so-called wet oxidation process). Presence of UV light in Fenton type processes could give them several benefits. One of benefits is an additional source of OH radicals besides primary source throughout Fenton mechanism. Therefore, OH radicals could be generated in photo-Fenton processes from the photolysis of hydrogen peroxide, equation (8), and from the reduction of ferric ions to ferrous ions, equation (9). Furthermore, UV light could provide the avoidance of breaking the Fenton catalytic cycle due to the formation of stable $Fe^{3+}$-complexes between free $Fe^{3+}$ ions and some aliphatic acids formed as by-products of dye degradation, shown by equation (10) [44, 45, 51, 62, 64, 68].

$$H_2O_2 \xrightarrow{h\upsilon} 2HO\bullet \tag{8}$$

$$FeOH^{2+} \xrightarrow{h\upsilon} Fe^{2+} + HO\bullet \tag{9}$$

$$Fe^{3+}\left(L^-\right) \xrightarrow{h\upsilon} Fe^{2+} + L\bullet \tag{10}$$

Also, the presence of UV irradiation could enable these processes to achieve the complete mineralization due to the degradation of some OH radical persistent by-products, such as oxalic and acetic acid, by UV light [64, 69-71]. Throughout all these studies it was found that the decolorization of reactive dye wastewater by Fenton type processes was rather high and fast. On the other hand, only partial degradation/mineralization of formed organic by-products after initial cleavage of dye chromophores indicate high susceptibility of dye chromophores to OH radicals in the comparison to the remained organic part of colored wastewater. However, Fenton type processes could be proposed as one of successful treatment methods for colored wastewaters containing reactive dyes.

## UV-based Processes

Under the term "UV-based processes" are considered all processes that apply UV light either for the degradation of organic pollutants or for the initiation of oxidation mechanisms by the irradiation of some powerful oxidants or photo-catalysts, throughout then will be provided degradation of organic pollutants present in water. Hence, UV-based processes could be classified into photochemical processes, where UV light is used for the photolysis of some powerful oxidants such as $H_2O_2$ and/or $O_3$, and photocatalytic, where UV light is used for the irradiation of some powerful photocatalyst such as $TiO_2$, ZnO etc.

### *UV Photolysis*

Since 1666 when Isaac Newton observed the diffraction of white light beam when passing through a prism, the investigations regarding UV light and its properties were began. At the begging of 19$^{th}$ century, the radiant energy beyond two ends of the spectra of visible lights was discovered. One of those is identified as infrared and another as ultraviolet region. Furthermore, it was shown that invisible chemically active irradiation beyond violet end of spectrum were the subject of laws of interference. All those observations, along with further investigations, indicated to the fact that irradiation with visible (VIS), infrared (IR) or ultraviolet (UV) light has characteristics of the same electromagnetic irradiation, but they differ regarding to its frequency, and, what was discovered latter, pertaining energy [35, 72].

**Table 3. Radiation type and pertaining energy [35]**

| Radiation | Wavelength, nm | Energy range, kJ Einstein$^{-1}$ |
|---|---|---|
| IC | >780 | <155 |
| VIS | 780 – 400 | 155 – 300 |
| UV-A | 400 – 315 | 300 – 377 |
| UV-B | 315 – 280 | 377 – 425 |
| UV-C | 280 – 100 | 425 – 1198 |

Therefore, UV radiation comprises energies from about 300 kJ Einstein$^{-1}$ (UV-A radiation, 1 Einstein = 1 mol of photons), up to 1200 kJ Einstein$^{-1}$ (vacuum UV). Table 3. summarizes the wavelength and the pertaining energy of different types of irradiation. For oxidation purposes, UV-C radiation is the most often in use, although in the literature could be found the application of other types of UV radiation [68, 73-75]. The most common wavelength of UV-C radiation is 254 nm that could be achieved by low-pressure vapor mercury lamp invented by Hewitt at 1901 [35]. The primary usage of UV radiation in the past was for disinfection, but with the development of reaction mechanism, UV radiation nowadays found the usage for oxidation purpose as well. At the room temperature, most molecules reside in their lowest-energy electronic state, i.e. "ground state". When molecules were exposured to UV radiation, they transferred to state with higher energy, i.e. "excited sate". The energy difference between the "ground" and the "excited" state depends on absorbed radiation $h\nu$, where $h$ and $\nu$ are Planck constant and the frequency of absorbed radiation, respectively. The molecule in the "excited state" has very short lifetime ($10^{-9}$ do $10^{-8}$ s), after which it returns to "ground state" by one of several mechanisms (fluorescence, phosphorescence...) or decomposes to yield a different molecule, resulting with photochemical reaction. Simple mechanism of direct photolysis is given below [35]:

$$M + h\upsilon \xrightarrow{k_a} M^* \tag{11}$$

$$M^* \xrightarrow{k_b} M \tag{12}$$

$$M^* \xrightarrow{k_c} produkti \tag{13}$$

UV radiation is always in use in combination with some powerful oxidant or photocatalyst. The efficiency of its separate use depends on the lot of limitation such as: (a) water solution should be treated in a way to achieve the highest possible UV transmission, i.e. turbidity should be as low as possible, (b) too high concentration of OH radicals could inhibit mineralization reaction of organic contaminant present in water, (c) water solution should be free of heavy metals and oils, and (d) the costs of UV radiation are higher than those for dark Fenton type processes. However, several authors investigated in their comparative studies the efficiency of UV radiation for the treatment of colored wastewater containing reactive dyes. Bali et al. [76] obtained only negligible decolorization, 2 %, of C.I. Reactive Black, initial concentration 100 $mgL^{-1}$, after two hours treatment by UV photolysis. Furthermore, Behnajady et al. [77] achieved after a one-hour treatment only 18.5 % of color removal with initial dye concentration of 30 $mgL^{-1}$. In our previous studies, in the case of C.I. Reactive Red 45 with initial concentration of 80 $mgL^{-1}$ again poor color removal, 30 %, and the mineralization degree of 3 % of TOC removal were achieved [78]. Treating model dye solutions with lower initial dye concentrations, 20 $mgL^{-1}$, significantly higher color removals of 87 and 98 % were achieved, whilst only 17 and 6 % of dye solutions mineralized [52, 79]. Such low decolorization in the case of higher dye concentration could be the fact that although UV radiation itself has ability to destroy organic molecules [35], the efficiency of direct photolysis of organic dyes decay has been proved to be difficult and it depends on the dye's reactivity and photosensitivity, and more important, solution transmission is very low in the case of higher dye concentration [80]. Moreover, most of the commercially used dyes are usually designed to be light resistant.

***Photochemical Processes***

These processes combine UV radiation and the addition of some powerful oxidant, $H_2O_2$ and/or $O_3$, in the system. The application of $UV/H_2O_2$ process for water purification is widely investigated AOP due to the fact that involves the simplest way of OH radical generation, through direct photolysis of $H_2O_2$. For its successful application it is necessary to determine the type and the initial concentration of pollutants in water, as well as the presence of other organic and inorganic compounds, so-called "scavengers", that could inhibit or even stop the treatment process. This process is in use for the removal of micro- and macro-pollutants from drinking water, for the treatment of organic toxic compounds present in lower concentration in ground waters, for the treatment of smaller volume of highly recalcitrant pollutants in order to achieve their detoxification and faster degradation, for the control of exhaust gases in the case of volatile organic compounds (VOCs)... [81]. Quantity of energy that emits by direct photolysis of $H_2O_2$ under UV radiation is very high, and theoretically two OH radicals could be generated per absorbed energy quant, equation (8) [82]. In practice, the highest quantum yield for generation of OH radicals is 0.5 mol of $H_2O_2$ per Einstein [35, 83]. That value is called the basic energy of direct photolysis of $H_2O_2$. By using higher energy than basic, formed OH radicals can be recombined and in that way $H_2O_2$ is producing. According to the literature [84, 85], besides reaction (8), other important reactions should be also taken into account in UV/$H_2O_2$ process:

$$H_2O_2 \Leftrightarrow HO_2^- + H^+ \tag{14}$$

$$HO\bullet + H_2O_2 \rightarrow HO_2\bullet + H_2O \tag{15}$$

$$HO\bullet + HO_2^- \rightarrow HO_2\bullet + HO^- \tag{16}$$

$$HO_2\bullet + HO\bullet \rightarrow H_2O + O_2 \tag{17}$$

Above reactions present "scavenger" mechanism of $H_2O_2$ (14) or OH radicals (15-17) negatively influencing the overall process efficiency. Important process parameters of UV/$H_2O_2$, besides UV lamp characteristics and reactor performances, are pH of solution and the initial concentration of $H_2O_2$. One of limitations of UV/$H_2O_2$ process that should be taken into consideration is the presence of iron and potassium salts in treated water resulting with reduction of UV radiation, and that could be avoided by adjusting pH to the value where those salts can precipitate. Further limitation is related to the large quantities of suspended particles resulting with increased turbidity and that can be solved by the filtration as the pretreatment of such wastewaters. Besides $H_2O_2$, ozone is also very used as oxidant in photochemical processes. Moreover, ozone is even better oxidant than $H_2O_2$ due to the significantly higher value of molar absorption coefficient at 254 nm, typical value for UV-C radiation (3300 $M^{-1}cm^{-1}$>>>18.6 $M^{-1}cm^{-1}$). Also, the rate of ozone photolysis is almost 1000 times higher than that of $H_2O_2$ [86]. UV/$O_3$ process is based on the fact that by the decomposition of ozone under UV radiation two OH radicals are generated which rather form $H_2O_2$ than react with organic matter present in water, shown by equation (18):

$$O_3 + H_2O \xrightarrow{h\upsilon} H_2O_2 + O_2 \tag{18}$$

Furthermore, such formed $H_2O_2$ decomposes under UV radiation to OH radicals that react with organic matter present in water, equation (8). This oxidative system contains three components that could generate OH radicals and/or directly oxidize organic matter: UV radiation, $H_2O_2$ and $O_3$. Accordingly, there are several ways to generate OH radicals: (a) photolysis of formed $H_2O_2$ or (b) reaction between formed $H_2O_2$ and ozone, so-called peroxone reaction. Also, there are several mechanisms for the degradation of organic pollutants in water: (a) direct photolysis, (b) OH radical attack generated from different sources and (c) direct ozone attack [87]. Process efficiency is limited by concentration of ozone and other factors mentioned above in the case of $UV/H_2O_2$ process. There is also combination of these two binary systems ($UV/H_2O_2$ and $UV/O_3$) the $UV/O_3/H_2O_2$ process. This process could enable to achieve the complete mineralization of organics present in water due to the enhanced generation of OH radicals throughout added $H_2O_2$ in the system [36, 78].

In the literature could be found many studies concerning $UV/H_2O_2$ process and colored wastewaters, but application of $UV/O_3$ and $UV/O_3/H_2O_2$ processes are not so widely investigated, particularly for the treatment of wastewater containing reactive dyes. Mohey El-Dein et al. [88] achieved complete bleaching of C.I. Reactive Black 5 with initial dye concentration of 500 $mgL^{-1}$ at pH 7 and 140 mM $H_2O_2$ solution after 150 minutes of treatment, while after 240 minutes mineralization reached almost 90 % of removed ecological parameter DOC (dissolved organic carbon). In our previous paper, time required for complete mineralization of C.I. Reactive Red 45 (80 $mgL^{-1}$) by $UV/H_2O_2$ process was 600 minutes, while complete bleaching was accomplished after 60 minutes of treatment [78]. Muruganandham and Swaminathan [89] achieved between 93.5 and 61.0 % of decolorization, whilst 62.9 and 24.6 % of aromatic products degraded after 150 minutes, applying $UV/H_2O_2$ process for the treatment of C.I. Reactive Orange 4 with various initial dye concentrations (0.1-0.5 mM). Bali et al. [76] decolorized C.I. Reactive Black (100 $mgL^{-1}$) model solution completely by $UV/H_2O_2$ process, pH 7 and initial $H_2O_2$ of 100 mM, after 60 minutes. In our previous study [79] where $UV/H_2O_2$ process was applied for the treatment of C.I. Reactive Blue 137 model solution with 20 and 80 $mgL^{-1}$ initial dye concentrations, complete decolorization was achieved after 10 and 50 minutes, respectively. Colonna et al. [90] reported in their study of decolorization of several dyes, including one reactive, by $UV/H_2O_2$ process complete bleaching of C.I. Reactive Blue 19 after 110 minutes and that ratio $[H_2O_2]/[dye]$ plays important role in the oxidation efficiency of applied system. That was also confirmed by Ince [91] and he named that ratio “effective $H_2O_2$ level”. Shu and Chang [92] reported significantly faster decolorization of studied dyes by $UV/O_3$ than by $UV/H_2O_2$. Similar results are obtained in our previous study [78], where significantly faster decolorization and correspondingly higher mineralization of C.I. Reactive Red 45 was obtained by $UV/O_3$ and $UV/O_3/H_2O_2$ than by $UV/H_2O_2$ process. It was established that mineralization by $UV/O_3/H_2O_2$ is four time faster than that by $UV/H_2O_2$ process. As the important process parameter of $UV/O_3$ and $UV/O_3/H_2O_2$ processes is pointed out pH of solution and it was established that optimal pH is at neutral or weak basic conditions [78, 93].

### ***Photocatalytic Processes***

Heterogeneous photocatalysis is based on the photonic excitation of a solid, which renders it more complex. The term photocatalysis may designate several phenomena that involve photons and catalyst, while this part of the chapter will consider only semiconductor photocatalysis. Photocatalytic activity of $TiO_2$ is based on its semiconductor properties. Radiation by photons, which have higher transfer energy, of such semiconductor leads to generation of electron-hole pairs [94]:

$$TiO_2 \xrightarrow{h\upsilon} h_{vb}^{+} + e_{cb}^{-} \quad (19)$$

Holes in valence band ($h_{vb}^+$) are strong oxidants, while electrons in conductance band ($e_{cb}^-$) act as reductants. Holes in valence band reacts with $H_2O$ or hydroxyl ions on the surface and producing OH radicals:

$$H_2O + h_{vb}^+ \rightarrow HO\bullet + H^+ \tag{20}$$

$$h_{vb}^+ + HO^- \rightarrow HO\bullet \tag{21}$$

Electrons react with dissolved oxygen producing superoxid radical ion, $O_2\bullet^-$, or its protonated form, perhydroxyl radical, $HO_2\bullet$, equations (22) and (23):

$$O_2 + e_{cb}^- \rightarrow O_2\bullet^- \tag{22}$$

$$O_2\bullet^- + 2H^+ \Leftrightarrow 2HO_2\bullet \tag{23}$$

In water, two $HO_2\bullet$ can be recombined if their concentration allow them to react significantly yielding $H_2O_2$ and $O_2$. It follows photocatalytic reduction of $H_2O_2$ by scavenging an electron from the conduction band where OH radicals are generated [95]:

$$H_2O_2 + e_{cb}^- \rightarrow HO\bullet + HO^- \tag{24}$$

The efficiency of $TiO_2$ photocatalysis could be improved by the addition of $H_2O_2$, but the addition of optimal $H_2O_2$ dosage should be taken into account. When it is in excess, it decrease process effectiveness. Photocatalysts, i.e. semiconductors, are comprised of microcrystalline or nanocrystalline particles and they are used in a form of thin film or as powder dispersion [96, 97]. Besides $TiO_2$ as the mostly known, used and studied photocatalyst, the usage of alternative photocatalysts, such as ZnO, CdS, $SnO_2$, is investigated as well [71, 94-101]. One of the main operating parameter of photocatalytic process is the dosage of photocatalyst.

In the studies concerning the application of photocatalytic processes for reactive dye wastewater degradation, $TiO_2$ is mostly used photocatalyst. Bizani et al. [102] investigated the efficiency of UV/$TiO_2$ process for the degradation of two commercial reactive dyes (initial concentrations 50 $mgL^{-1}$) and achieved the complete decolorization between 30 and 120 minutes of treatment depending on dye molecular structure and the type of $TiO_2$ photocatalyst. Goncalves et al. [97] investigated optimal pH range for reactive dye decolorization by UV/$TiO_2$ process and they reported the highest rate of C.I. Reactive Orange 4 decolorization at strong basic condition. Muruganandham and Swaminathan [103] investigated the application of other photocatalysts such as $SnO_2$, ZnO, CdS and $Fe_2O_3$ for decolorization of C.I. Reactive Yellow, and they reported that $SnO_2$, CdS and $Fe_2O_3$ have negligible activity on RY14 decolorization and degradation. In our previous paper [104], two effective photocatalysts, $TiO_2$ and ZnO, are compared in the case of C.I. Reactive Red 45 model solution decolorization and mineralization. The optimal conditions for both processes are established and it was found out that there are no significant differences between these two photocatalysts for decolorization of model solution at established optimal conditions for each process. Similar findings are reported in the study Akyol et al. [105] where complete color removal of Remazol Red was obtained after 35 minutes treatment by UV/$TiO_2$ and UV/ZnO processes.

## Ozone-based Processes

Ozone is main component in many oxidation processes assembled under the term ozonation processes. In these processes ozone is applied either alone ($O_3$ process) or with the addition of oxidant, e.g. $H_2O_2$ ($O_3$/$H_2O_2$ process), UV radiation (explained in above subchapter), catalyst, activated carbon, ultrasound etc. Ozone is inorganic molecule constituted by three atoms of oxygen. It is present in nature in upper atmosphere in the form of stratospheric layer around the earth, and it is formed by the photolysis of diatomic oxygen and further recombination of atomic and diatomic oxygen, shown by equations (25) and (26) [35]:

$$O_2 \xrightarrow{h\upsilon} 2O\bullet \tag{25}$$

$$O\bullet + O_2 \rightarrow O_3 \tag{26}$$

Ozone can be generated artificially in ozone generators. There is two ways of generating ozone by ozone generators: (a) the cleavage of oxygen molecules under the influence of strong electrical field and (b) the same mechanism like in nature, the photolysis of oxygen, but induced artificially. The mechanism of both ways can be described by above reactions (25) and (26). Ozone was discovered in 1840, while its molecular structure as triatomic oxygen was established 30 years latter. The first usage of ozone was reported at the half of $19^{th}$ century where ozone was used as disinfectant in many water treatment plants and hospitals. Therefore, at 1856 the first usage of ozone as disinfectant for operating rooms in hospital over Europe was reported [106, 107]. First usage of ozone for the purification of drinking water in Monaco and Germany was performed at 1860 [107, 108]. Ozone application as oxidant for water purification was retained through $20^{th}$ century, and its significant increase was noticed at 1970s when the production of trihalomethanes and other organohalogenated hydrocarbons were identified during the water treatment by chlorine [35]. Ozone is very reactive reagent with redox potential of 2.07 V (Table 2), either in liquid or gas phase. That high reactivity owes to electronic configuration; ozone can be presented as a hybrid in four different molecular resonance structures that give ozone characteristics of an electrophilic, dipolar or even nucleophilic agent. Hence, ozone molecule could react with organic compounds under two different mechanisms: direct and indirect. Direct mechanism involves organic compound degradation by molecular ozone, occurring at acidic pH range. The main reactions involved in direct mechanism are the reaction of addition to the unsaturated part of the hydrocarbon molecule and electron transfer. Rather high oxidation/reduction potential enables ozone to react with many organic, but also inorganic compounds [35].

$$O_3 + 2H^- + 2e^- \rightarrow O_2 + H_2O \quad (27)$$

Degradation of organics by indirect ozone mechanism occurs at basic conditions and involves generation of OH radicals and their further reaction with present organic compounds in water. Hydroxyl radicals are generated by the reaction of ozone with hydroxyl ions present in water [109]:

$$O_3 + H_2O \xrightarrow{OH^-} HO\bullet + O_2 + HO_2\bullet \quad (28)$$

Mechanism of ozone decomposition in water depends on the presence of chemical species that can initiate, promote and/or inhibit its decomposition. The most accepted ozone decomposition mechanism is given in Figure 3.

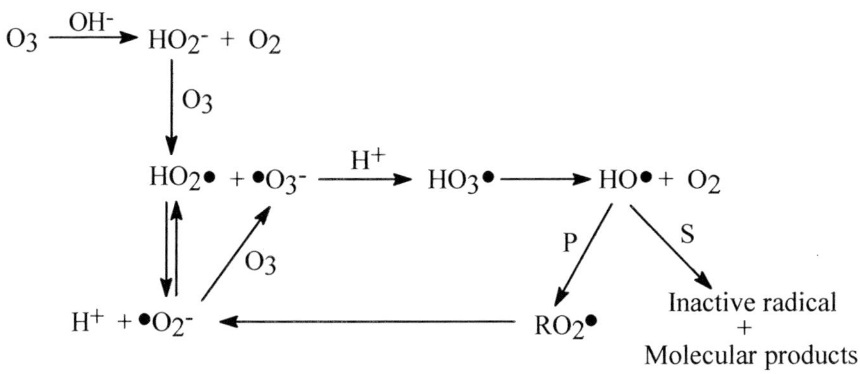


Figure 3. Scheme of ozone decomposition mechanism in water. P = promoter (e.g. ozone, methanol), S = scavenger or inhibitor (e.g. *t*-butanol, carbonate ion), I = initiator (e.g. hydroxyl ion, perhydroxyl ion) (adapted by Beltran [35])

According to the presented mechanism of ozone decomposition in water, it can be seen that ozonation can classified in AOPs, when the degradation of organics occurs by indirect mechanism, and in classical chemical treatment methods, when direct mechanism is dominant in the degradation of organics in water. One of the limiting factors of ozonation influencing overall process effectiveness presents the efficiency of ozone mass transfer from gas to liquid phase. One of possible solutions gave Mathews et al. [110] by the application of *in situ* ozone generator with porous electrode. They obtained 90 % of C.I. Reactive Blue 19 decolorization in two times shorter treatment period than by the conventional ozone generator. Besides $O_3$ process, either direct or indirect, one of widely investigated process from the group of ozone-based processes is $O_3/H_2O_2$, so-called peroxone process or perozonation [111]. Efficiency of that process is based on the fact that the presence of $H_2O_2$ could enhance ozone reactivity in bulk phase. Moreover, some authors consider degradation of organics by direct mechanism negligible in the case of $O_3/H_2O_2$ process [111, 112]. Hydroxyl radicals are generated by reactions of ozone and perhydroxyl ion, $HO_2^-$, which is present in water by the dissociation of $H_2O_2$. The simple overall mechanism of OH radical generation in peroxone process is shown by equation (29) [112]:

$$H_2O_2 + O_3 \rightarrow 2HO\bullet + 3O_2 \quad (29)$$

Degradation kinetic of organic pollutants by $O_3$ process, either direct or indirect mechanism, is reported as first order or "pseudo"-first order kinetic, while in the case of peroxone process, degradation kinetics follows second order due to the predominant degradation mechanism by OH radicals.

Many investigations are available in the case of ozonation as treatment methods for colored wastewaters containing reactive dyes. Chu and Ma [30] investigated direct and indirect $O_3$ process for the degradation of several dyes including three reactive. They reported dye decolorization by indirect mechanism significantly faster than that by direct mechanism. Wu and Wang [113] reported the complete decolorization of C.I. Reactive Black 5 (0.5 $gL^{-1}$) in 40 minutes of treatment by $O_3$ process with rate of ozone generation 26.1 $mgL^{-1}min^{-1}$, and their results revealed that rate of ozone transfer increased with increase in dye concentration, applied ozone dose and temperature. Peralta-Zamora et al. [114] obtained very fast complete decolorization of C.I. Reactive Blue 19, in <5 minutes, but degradation of dye monitored in UV region, i.e. degradation of aromatics formed after the initial cleavage of dye chromophores, was significantly slower. Furthermore, Arslan-Alaton et al. [115] investigated the application of $O_3$ process for the treatment of simulated dyebath effluent containing several different reactive dyes at different pH region: acidic, neutral and basic. Results of their study showed that decolorization of simulated dyebath is very fast at all three pH regions, yielding with 100 % color removal in couple of minutes, and that degradation of aromatics, monitored at 280 nm, was somewhat slower than decolorization, and that overall extent of aromatics degradation depends on operating pH value. Mineralization of solution was also monitored, and significantly lower extents of partial mineralization were achieved than those of decolorization and aromatic degradation. They obtained similar findings in previous study too [116]. Similar large difference between decolorization and mineralization extents were obtained in our previous comparative study of $O_3$, $O_3$/zeolite ($NH_4$ZSM5 or HY type) and $O_3/H_2O_2$ processes for degradation of C.I. Reactive Blue 137 model solution [117]. Model dye solution was completely decolorized at all initial dye concentrations (20-150 $mgL^{-1}$), while only partial mineralization was obtained in all cases. The somewhat higher mineralization extents were obtained by $O_3/H_2O_2$ than by $O_3$ process. Zeolites did not affected decolorization rate, only slightly due to changing of pH values of solution, but somewhat higher TOC removals were obtained in cases with zeolite addition than without. Explanation could be found in the adsorption of aromatic by-products with appropriate molecular structure to penetrate in zeolite pores. Besides the improvement of $O_3$ process by $H_2O_2$ addition, some authors combined $O_3$ and ultrasound processes in order to improve degradation of reactive dye solution [118]. From the presented results it can be seen that decolorization of colored wastewater can be easily achieved by ozonation processes, but dye degradation in order to get complete mineralization is very hardly to achieve.

## High Voltage Electrical Discharge Processes

Basic principles of processes including electrical discharge, so-called corona discharge, in liquid and gas phase, are that chemical and physical processes could be induced by strong electrical field, and therefore these processes could be consider as electrical AOPs. It is proved that by applying electrical discharge in liquid phase intensive UV irradiation and various active species (•OH, •H, •O, $HO_2^-$, $O_2^{\bullet -}$, $H_2O_2$, etc.) are considered to be produced [39, 40, 119-124]. By the presence of oxygen in the system, ozone and its ozone-related radical species can be formed. This process can be effective for the treatment of biological microorganisms and dissolved chemicals in liquid phase [125]. Application of the processes based on strong electric field in water or organic solution is investigated in the past due to importance in electrical transmission processes and they practical usage in biology, chemistry and electrochemistry. Recently, the application of reactors involving electrical discharge in liquid or above the surface, for the treatment of drinking and wastewaters is widely investigated. Nowadays, these reactors are in use, but only in laboratory scale. There are many types of corona reactors according to the configurations of their high voltage and ground electrodes. The most common reactor type has both electrodes, high voltage and ground, in liquid phase [40]. Generally, corona is type of electrical discharge that is formed when DC, AC or pulsed electric potential is applied between two non-uniform electrodes where a dielectric medium is placed between the electrodes. Corona onset begins when the discharge initiates in the dielectric medium. That manifests as thin violet layer which surrounds the high voltage electrode in liquid phase. That layer contains of very short streamers. Initial voltage of corona discharge depends on the type of applied electrodes and physical properties of dielectric material between the two electrodes. As the corona discharge voltage increases, the streamers become larger and corona discharge is formed. Further increasing of the voltage leads to the sparkover, i.e. the production of direct current channel that bridges two electrodes [119, 125]. As it was mentioned before, pulsed corona discharge is applied in gas and liquid phase. In gas phase corona, the mostly used electrode geometry is "plate-to-plane" and "wire-to-cylinder", while in liquid phase the most studied geometries are "point-to-plane" and "plate-to-plane" (Figure 4.) [40].

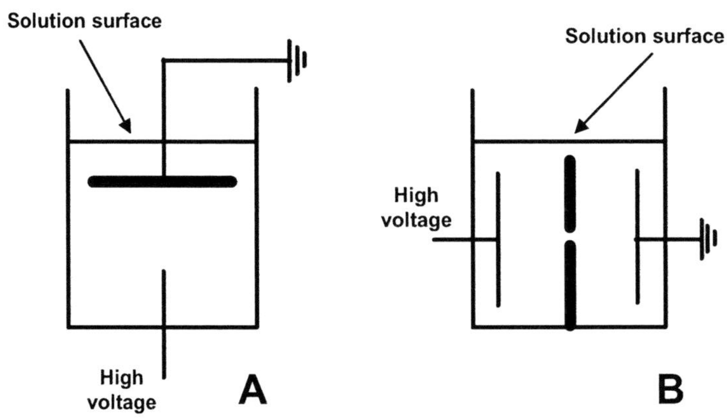


Figure 4. Most studied electrode geometry in liquid corona discharge reactors: (A) point-to-plane and (B) plate-to-plane

Pulsed corona discharge produce repetitive (60 Hz), fast raising high voltage pulses with short lifetime (μs). Application of short high voltage pulses (200-1000 ns) consider presence of non-thermal conditions, that means electrons in plasma having a higher mean energy ($T_e$

>> 1 eV), than the other constituents in liquid ($T_{H2O}$ < 0.1 eV) [121, 126]. On this way, the energy loss for the migration of ions, chemical species with less mobility than electrons and which have no influence to generation of radicals, is minimized as high as possible [120]. Among all radical species generated in corona when electrons with high energy collide with water molecules, equations (30) and (31), OH radicals present main species responsible for degradation of organics.

$$H_2O + e^-* \rightarrow HO\bullet + H\bullet + e^- \quad (30)$$

$$H_2O + e^-* \rightarrow H_2O^+ + 2e^-* \quad (31)$$

$$H_2O + H_2O^+ \rightarrow H_3O^+ + HO\bullet \quad (32)$$

Such formed radical species could react between themselves resulting with the production of $H_2O_2$, $O_2$, or $H_2$.

$$H\bullet + H\bullet \rightarrow H_2 \quad (33)$$

$$HO\bullet + HO\bullet \rightarrow H_2O_2 \quad (34)$$

$$HO\bullet + H\bullet \rightarrow H_2O \quad (35)$$

Degradation of organics could be performed by: (a) direct reactions with highly reactive species (OH radicals), (b) indirect reactions over radicals formed from stabile molecules ($H_2O_2$), and (c) direct reactions with stabile molecules.

### *Hybrid Corona Reactors*

These types of corona reactors involve simultaneous application of non-thermal plasma above liquid surface and direct high voltage electrical discharge in liquid phase. That is achieved by the combination of different electrode configuration and/or by changing electrode geometry of standard corona reactors which use “point-to-plane” geometry. There are two hybrid corona reactors, so-called hybrid-series and hybrid-parallel (Figure 5.). Hybrid-series involves “point-to-plane” geometry and configuration of electrodes where high voltage electrode is in liquid phase, and ground electrode is placed in gas phase just above the liquid solution. That enables the formation of conventional streamer like corona discharge in liquid and the formation of intensive plasma channels near to liquid surface. Hybrid-parallel corona reactor involves electrode geometries “point-to-plane” in liquid phase and “plane-to-plate” in gas phase. That reactor configuration combines mostly used electrode geometries in liquid and gas phase, i.e. on that way is combined classical corona discharge in liquid and gas.

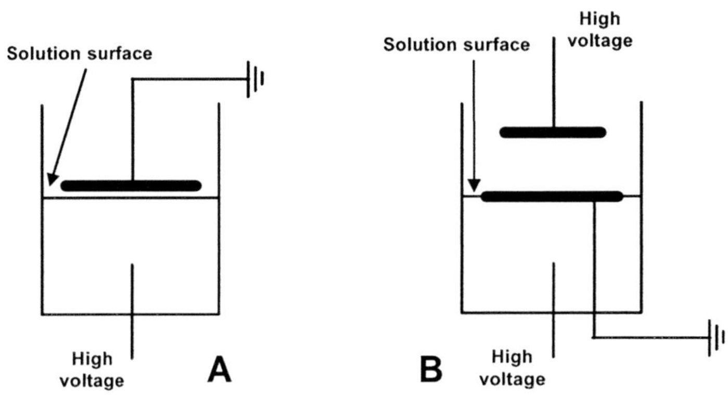


Figure 5. Configuration of hybrid corona reactors: (A) series and (B) parallel

Both hybrid reactor types are very interested due to improvements concerning standard liquid corona type reactor (Figure 4.). It is reported that hybrid reactors, besides other earlier mentioned radical, ionic and molecular species, include generation of ozone in gas phase and its transfer to liquid phase, improving degradation of organics in liquid [127]. Besides explained corona discharge, there are other types of discharges in water such as glow, barrier, glidarc, but they are not discussed in this subchapter.

There are not too many studies directed to the investigation of corona discharge reactors application for the treatment of colored wastewater, particularly containing reactive dyes. Tri Sugiarto et al. [128] investigated the application of corona discharge reactor, "plane-to-plate" geometry, for decolorization of three dyes (initial dye concentrations of 10 $mgL^{-1}$). They reported complete decolorization at neutral pH after 120 minutes treatment and dependence of decolorization rate on pH value. They obtained the highest decolorization rate at basic conditions. Sunka et al. [129] investigated new type of corona discharge reactor, "pin hole" geometry, for decolorization of C.I. Reactive Blue 137 (50 $mgL^{-1}$). In first case where only corona discharge was used for decolorization of reactive dye solution, very low color removal was achieved after 30 minutes of treatment, while the addition of Fe ions in the system improved color removal and in 5 minutes of treatment the complete color removal was obtained. Similar findings were obtained in study by Loncaric Bozic et al. [130] treating C.I. Reactive Blue 49 in the same reactor type. One of our earliest paper related to corona discharge dealt with the application of corona discharge reactor with "point-to-plane" electrode geometry for decolorization of model solution of four different dyes (20 $mgL^{-1}$), including two of reactive type [131]. Decolorization extents ranged between 17-25.6 % after 60 minutes of treatment, while by the addition of Fe ions in the system, complete decolorization was achieved after 30 minutes. Our two new papers related to corona are directed to the application of hybrid corona reactors for decolorization of reactive dyes [132, 133]. C.I. Reactive Blue 137 (20 $mgL^{-1}$) was decolorized up to 72.9 and 85.0 % regarding reactor type. Both papers also investigated the influence of synthetic zeolites on decolorization by corona discharge and it was found that zeolites could improve color removal in hybrid-parallel corona reactor. Decolorization effectiveness was decreased by increasing initial dye concentration, resulting with color removal in the case of 80 $mgL^{-1}$ initial dye concentrations between 34 and 48 % regarding dye structure, type of reactor, the addition of zeolites or not, and type of added zeolite. From the presented results of corona discharge processes for decolorization of colored wastewaters it can be seen that these type of processes could be efficient, but only when dye loading in wastewater is not so high.

## Other AOPs

This part of subchapter contains less detailed review of ambient AOPs that are not studied in our laboratory, such as: ultrasound or ultrasonic irradiation, radiolysis of water and electrochemical processes.

### *Ultrasound*

The term sonochemistry describes all chemical processes in which ultrasound irradiation is involved. The main premise of ultrasound process as wastewater treatment method is generation of free radicals, such as hydrogen and hydroxyl, upon the action of ultrasonic waves in liquid. The applied frequency ranges from 20-40 kHz; predominantly used in large scale applications, so-called "power ultrasound", up to 1 MHz; frequency used in so-called "diagnostic ultrasound" [134]. Ultrasound produces the chemical effects through several different physical mechanisms and the most important nonlinear acoustic process for sonochemistry is cavitation. Cavitation can be generated when large pressure differentials are applied in flowing liquid (hydrodynamical cavitation), or by means of an electrochemical transducer, piezoelectrical or magnetostrictive, in contact with the fluid (acoustical cavitation) [135]. Electrohydraulic cavitation in liquid causes bubbles. These bubbles start very small, and grow in diameter and implode under the periodic variations of the pressure field of the ultrasonic waves. The rapid implosion of the eventually instable gas bubbles causes adiabatic heating of the bubble vapor phase. It is assumed that creates drastic local conditions: temperatures of about 5500 °C inside the bubble and 2100 °C in the liquid that surrounds the cavity, and pressure up to 100 MPa within the collapsing cavity. Such conditions are limited to a very small region and the heat produced during cavitation is dissipated very quickly (heating and cooling rates greater than 109 K/s). As a result, the surrounding liquid remains at the ambient temperature. Organic compounds are highly degraded in such an environment and inorganic compounds can be oxidized or reduced. Water irradiation using ultrasound causes decomposition of the water molecules into extremely reactive radicals HO• and H•, shown by equation (36) [119, 136].

$$H_2O \xrightarrow{ultrasound} H\bullet + \bullet OH \tag{36}$$

Several authors investigated ultrasound as treatment process for colored wastewater degradation. Tezcanli-Guyer and Ince [137] investigated ultrasound, 520 kHz, for the treatment of four organic dyes including two of reactive type (initial concentrations 20-60 $mgL^{-1}$), and reported decolorization after two hours treatment between 81-99 % depending on dye structure, while degradation of formed aromatics after four hours treatment ranged between 11-93 %. Rehorek et al. [138] applied power ultrasound, 850 kHz, at different input powers for the treatment of several organic dyes. They obtained complete decolorization of all studied dyes in 1-15 hours of treatment regarding input power of ultrasound, and they reported that ultrasound is able to degrade organic dyes to non-toxic products.

***Water Radiolysis***

Irradiation of dilute aqueous solutions by high energy ionizing radiation, ranged from keV to MeV, results with the excitation and the ionization of water molecules leading to the production of radical species. The types of radiation may be classified by the source of production: (a) the decay of radioactive nuclei ($\alpha$-, $\beta$- and $\gamma$-radiation), (b) beams of accelerated charged particles (electrons, protons and deutrones) and (c) short-wavelength electromagnetic radiation (X rays). OH radicals and hydrated electrons, $e_{aq}^-$, can be considered as predominant species formed by water radiolysis. Besides these two species, water radiolysis produces also hydrogen atoms, hydrogen peroxide (predominant specie in the presence of $O_2$), hydroxyl ions, etc. The mostly investigated radiation sources for degradation of organics are $\gamma$-radiation and electron beam [119, 139-141]. Dajka et al. [140] reported the role of $e_{aq}^-$ in decolorization of C.I. Reactive Black. Zhang et al. [141] investigated $\gamma$-radiation for decolorization of three different dyes and reported that, depending on dye structure, different gamma irradiation dose are required to achieve the complete decolorization.

***Electrochemical Processes***

These processes can occur by a direct electron transfer reaction to (reduction) or from (oxidation) the present organic pollutant, or by a chemical reaction of the pollutant with previously electrogenerated species. The mechanism is generally viewed as a direct anodic oxidation of organic pollutant involving its reduction by direct electron transfer from organic molecule to the electrode to form a radical cation that readily deprotonates, equation (37):

$$RH \xrightarrow{electrolysis} R\bullet + H^+ + e^- \qquad (37)$$

Nowadays, various types of electrodes are in use: $Pt/IrO_2$, Ti/Pt, $SnO_2/Ti$, $Ti/RuTiO_2$, etc [119, 142]. Catanho et al. [143] used $Ti/RuTiO_2$ electrode for decolorization of C.I. Reactive Red 198 and reported 75 % color removal after 2 hours treatment. Awad and Abo Galwa [144] investigated different conductive electrolytes for decolorization of colored wastewater and obtained the highest decolorization in the case of NaCl. Carneiro et al. [145] investigated different types of electrochemical processes for degradation of C.I. Reactive Blue 4 and reported the as the best process $Ti/SnO_2/SbO_x/RuO_2$ electrode in $Na_2SO_4$ 0.2 M at pH 2.2 and potential of +2.4 V, where complete decolorization and 60 % mineralization were obtained.

## Study of C.I. Reactive Blue 137 Wastewater; Treatment by Several AOPs

After a review of available and currently studied AOPs, two types of processes were chosen for our research directed to degradation of C.I. Reactive Blue 137 as pollutant in model wastewater. Due to the very high effectiveness and fastness of Fenton type processes, but also considering secondary treatment of solution due to iron ions and counter ions, Fenton type process using iron powder, $Fe^0/H_2O_2$, was one of studied treatment processes. It was concluded that the assistance of UV light to $Fe^0/H_2O_2$ could offer significant benefit to process effectiveness due to the achievement of complete mineralization, as it was mentioned above in "Fenton type processes" subchapter. Another group of chosen processes include those combining ozone and UV light, $UV/O_3$ and $UV/O_3/H_2O_2$, due to the conductance at neutral or at weak basic conditions that excludes after-treatment solution conditioning like in the case of Fenton type processes, and also due to constant ozone supply enabling to constant OH radical generation. The influence of synthetic zeolites on chosen AOPs was also investigated in the framework of this research. Synthetic zeolites were shown positive effect when added to some of AOPs for the treatment of colored wastewater in our previous studies [104, 117, 132, 133, 146, 147]. Generally, synthetic zeolites, are three-dimensional, microporous, crystalline solids with well defined structures that contain aluminum, silica, and oxygen in a regular framework. Zeolites have void pore space that can host cations, water or other molecules. They have the ability to act as catalysts for chemical reactions which take place within the internal cavities. Furthermore, synthetic zeolites may act as ion exchangers, because the loosely-bound nature of the extra-framework metal ions allows the exchange of other types of metals when present in aqueous solution. Shape-selective properties of zeolites are also the basis for their use in molecular adsorption [148-150]

## Objectives

The aim of the study was to find the optimal treatment process for colored wastewater containing reactive dye C.I. Reactive Blue 137. The comparative study involved several processes, $Fe^0/H_2O_2$, $UV/Fe^0/H_2O_2$, $UV/O_3$ and $UV/O_3/H_2O_2$, and it was performed through an experimental research of decolorization and mineralization of a reactive azo dye C.I. Reactive Blue 137, as a model wastewater pollutant. Applied processes were optimized according to their process parameters, $Fe^0$ dosages, initial $H_2O_2$ dosages and initial pH values. The influence of initial organic dye concentration as well as the addition of solid particles, synthetic zeolites, on the process effectiveness was also investigated. Studied AOPs were evaluated on the basis of their eco-effectiveness, by the means of color ($A_{610}$), aromaticity ($A_{280}$), TOC and AOX value decrease, and their cost-effectiveness as well.

## Materials and Methods

Synthetic wastewater containing organic model pollutant was prepared using two different concentrations, 20 and 80 $mgL^{-1}$, of the commercial dye Cibacron Marine P2R-01, C.I. Reactive Blue 137 (RB137) obtained from Ciba-Geigy, and deionezed water; pH 7, conductivity less than 1 $\mu S\ cm^{-1}$. Other chemicals ($H_2O_2$, 30 %; sulfuric acid, >95 %; and sodium hydroxide, p.a.) were supplied by Kemika, Croatia.

Series of experiments were conducted in order to find an optimal $Fe^0/H_2O_2$ ratio. At the beginning of the experiment, pH was adjusted at 3 using 25 % sulphuric acid, which was followed by the addition of iron powder and $H_2O_2$. Iron powder dosages were 28 and 56 $mgL^{-1}$ that corresponded to the concentrations of 0.5 and 1.0 mM, respectively. Concentrations of hydrogen peroxide were varied to give molar ratios 1 : 5, 1 : 10, 1 : 20, 1 : 30, 1 : 40, and 1 : 50. Reaction mixture (V=250 mL) was continuously stirred at room temperature in an open batch system with magnetic stirring bar and was treated for two hours, while dye concentration and TOC values were measured at the end of the each experiment to established decolorization and mineralization extents.

All further experiments were carried out in a 0.8 L batch glass annular water-jacketed photoreactor, used in our previous studies [48, 52, 59, 78, 79, 87]. The value of incident photon flux at 254 nm, $7.36\times10^{-6}$ Einstein $s^{-1}$, was calculated on the basis of hydrogen peroxide actinometry experiments [151]. In experiments that required the addition of iron powder, $Fe^0/H_2O_2$ and $UV/Fe^0/H_2O_2$, optimized dosages of iron powder and $H_2O_2$ were added accordingly to the results of experiments explained in previous paragraph ($Fe^0/H_2O_2$ optimization study part). In experiments where UV irradiation was required, $UV/Fe^0/H_2O_2$, $UV/O_3$ and $UV/H_2O_2/O_3$, the UV lamp was switched on. In the experiments involving ozone, $UV/O_3$ and $UV/H_2O_2/O_3$, ozone was introduced into the reactor. Ozone was generated from pure oxygen, >99.9 %, by introducing it into the ozone generator, MIC Systems Inc., Valdosta, Georgia, USA to produce ozone. The outlet gas mixture from the ozone generator, i.e. mixture of ozone and oxygen, was fed into the reactor through a sintered glass plate diffuser located in the bottom half of the reactor. The inlet gas flow into the ozone generator was fixed at 0.15 $Lmin^{-1}$. The rate of ozone generation in reactor inlet stream, $7.9\times10^{-3}$ $gmin^{-1}$, was determined iodometrically [152]. In experiments of $UV/O_3$ and $UV/H_2O_2/O_3$ processes optimization, initial pH values and hydrogen peroxide concentration ranged from 3 to 11, and 0 to 10 mM, respectively, until the achievement of maximal mineralization extent. Initial pH was adjusted with the addition of NaOH or $H_2SO_4$. Added quantities of $H_2SO_4$, NaOH and $H_2O_2$ as an additional oxidizing agent, were negligible in comparison to the total volume of treated reaction mixture. The experiments were conducted by adjusting one variable, while others were held constant.

The total volume of the treated solution was 0.5 L in all cases, while the mixing of the solution was provided by both magnetic stirring and peristaltic pump at a flow rate of 0.1 $Lmin^{-1}$. Experiments were carried out at 25 (+/- 0.2) °C. The duration of each experiment was one hour. Samples were taken periodically from the reactor (2, 5, 10, 15, 20, 30, 40, 50 and 60 min) and thereafter immediately analyzed. In experiments where the influence of synthetic zeolites on process effectiveness was investigated, at the beginning of experiments different amounts of zeolites were added (0.4 $gL^{-1}$ of NH4ZSM5 in the case of $Fe^0/H_2O_2$ process and 1.0 $gL^{-1}$ in the cases of $UV/O_3$ and $UV/H_2O_2/O_3$ processes), accordingly to our previous studies [117, 146]. In these experiments, samples were centrifuged before the analysis in order to remove solid particles of zeolites from the solution. All experiments were repeated at least two times to give reproducibility of the experiments within 5 %, and averages are reported.

The initial pH values were measured by handylab pH/LF portable pH/conductivity-meter, Schott Instruments GmbH, Mainz, Germany. The decolorization and degradation of RB137 solution was monitored by a Perkin Elmer Lambda EZ 201 UV/VIS spectrophotometer, USA, by scanning complete UV and VIS spectra. The extent of RB137 mineralization was determined on the basis of measurements of total organic carbon (TOC), using TOC analyzer; TOC-$V_{CPN}$ 5000 A, Shimadzu, Japan, and absorbable organic halides (AOX), performed by Organic Halide Analyzer, DX-2000, Dohrmann.

## Results and Discussion

### *Fenton Type Processes*

First group of investigated AOPs were addressed to Fenton type process. As it was mentioned earlier, the oxidation power of Fenton type processes, particularly those performed in the dark; strongly depends on process parameters such as iron catalyst concentration, concentration of $H_2O_2$, pH, temperature, and treatment time. Therefore, it was necessary to performe laboratory investigations, so-called "jar tests", of those parameters in order to find optimal for chosen system Fenton type process/organic pollutant [65]. In the research "jar tests" were directed to find optimal concentration of iron powder and optimal $Fe^0/H_2O_2$ ratio, while other process parameters were in the range reported as optimal in the literature: pH 3, room temperature, and a one-hour treatment time [44, 45, 49, 51, 52, 54, 66]. The results of $Fe^0/H_2O_2$ process parameters for the treatment of RB137 model solution with initial concentration of 20 $mgL^{-1}$ are summarized in Figure 6. It can be seen that through complete investigated range of $Fe^0/H_2O_2$ ratio at both iron powder dosages rather high decolorization of RB137 colored wastewater, >95 %, was achieved. Obtained decolorization extents were below visibility limit, <1 $mgL^{-1}$ of remained dye in the solution [2]. During the same treatment period, only partial mineralization was obtained in all cases, with the highest 40.9 % of TOC removal in the case of $c(Fe^0)$=1.0 mM and 1 : 20 $Fe^0/H_2O_2$ ratio. It can be concluded that decolorization of RB137 was not significantly influenced by investigated process parameters, but at the same time, mineralization showed different results. Almost the same system behavior could be observed at both investigated iron powder dosages and the same $Fe^0/H_2O_2$ ratio. Mineralization of RB137 increased with increasing the $Fe^0/H_2O_2$ ratio, but only to the certain point. That can be explained with the increasing of OH radical concentration in the bulk, resulting with an increase of mineralized part of RB137. According to the results of TOC removal, further increase over obtained optimal ratio showed negative effect to the process effectiveness. That can be explained with the fact that $H_2O_2$ when is in excess, act as OH radical scavenger, resulting with the lowering of RB137 mineralization [44, 45, 47-49, 65]. Besides that effect, the creation of an inert oxidative film on the surface of iron powder at higher concentration of $H_2O_2$ should also be taken into account. That inert film could disable the leaching of $Fe^{2+}$ ions from iron powder surface in the bulk, as well as the occurrence of Fenton reaction on the iron powder surface [54].

After determining the optimal iron powder dosage and $Fe^0/H_2O_2$ ratio, the kinetic of RB137 degradation with $Fe^0/H_2O_2$ and $UV/Fe^0/H_2O_2$ processes was investigated. In Figure 7. the results of RB137 degradation with both processes are presented. By monitoring complete UV/VIS spectra, it was tried to get an overview on the degradation of complete dye molecular structure.

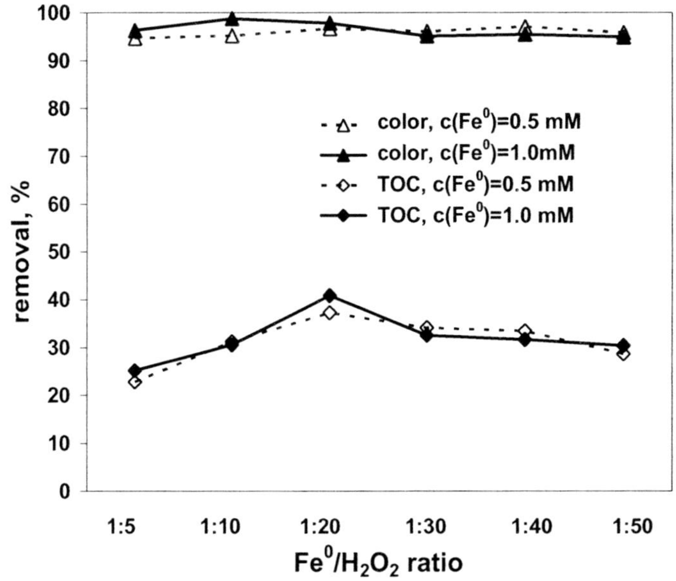


Figure 6. Influence of iron powder dosages and $Fe^0/H_2O_2$ ratio on color and TOC removal in Fenton "like" process for degradation of RB137 at pH 3

By full black line the chromatogram of RB137 through complete UV/VIS spectra was presented. Three characteristic peaks can be observed. First one absorbing in the VIS region between 480-680 nm with maximum wavelength at 610 nm corresponds to RB137 chromophore. Next two peaks absorb in UV region, first between 260-350 nm with maximum wavelength at 280 nm corresponding to the absorbance of aromatic part of dye molecule and second between 200-260 with maximum wavelength at 230 nm corresponding to triazine ring in dye molecular structure [45, 72, 153, 154]. Comparing degradation of dye chromophores, i.e. the disappearance of peak absorbing at $\lambda_{max}$=610 nm, it can be observed that decolorization is much slower with $Fe^0/H_2O_2$ than $UV/Fe^0/H_2O_2$ process. Therefore, complete decolorization of RB137 by $UV/Fe^0/H_2O_2$ process was achieved after 5 minutes of treatment, while in the case of $Fe^0/H_2O_2$ process, the complete bleaching, i.e. zero absorbance of RB137 model solution at 610 nm was observed barely after 20 minutes treatment process. The band of RB137 in UV region with two peaks at 230 and 280 nm was interfered through almost complete treatment time by absorbance of added $H_2O_2$ [35, 155]. It can be observed that the monitoring of dye degradation in UV region, as well as degradation of aromatic structures and triazine ring was partially disabled due to the interference by $H_2O_2$. But, it can be seen that new peak, produced by the interference of triazine ring peak with the absorbance of added $H_2O_2$, is lowering much faster in the case of $UV/Fe^0/H_2O_2$ than $Fe^0/H_2O_2$ process. That might be due to the much faster consumption of added $H_2O_2$, but also due to much faster degradation of triazine rings, by process combined with UV light. That is in accordance with the literature where is reported that $UV/Fe^0/H_2O_2$ process consumed $H_2O_2$ much faster than $Fe^0/H_2O_2$ process [48]. Furthermore, triazine rings have very high molar absorption coefficient at 230 nm, ≈40000 $M^{-1}cm^{-1}$, that makes them very unstable under UV-C irradiation [72]. Degradation of aromatic structures at $\lambda_{max}$=280 nm is also much faster by $UV/Fe^0/H_2O_2$ than by $Fe^0/H_2O_2$ process. Although that peak was also interfered by the addition of $H_2O_2$, in the case of $UV/Fe^0/H_2O_2$ process a significant decrease of that peak during the treatment period from 2.-20. minute could be observed, meaning that aromatic structures in RB137 molecules were degraded. After 30. minute of treatment time, the appearance and the constant increasing of new peak in UV region with maximum wavelength at 290 nm can be observed. That peak could correspond to the formation of organic acids as products of degradation of aromatic compounds in model solution that absorb in UV region at 220 and 290 nm [71, 153]. That is also confirmed by the increasing of peak at 220 nm in treatment period between 40. and 60. minute.

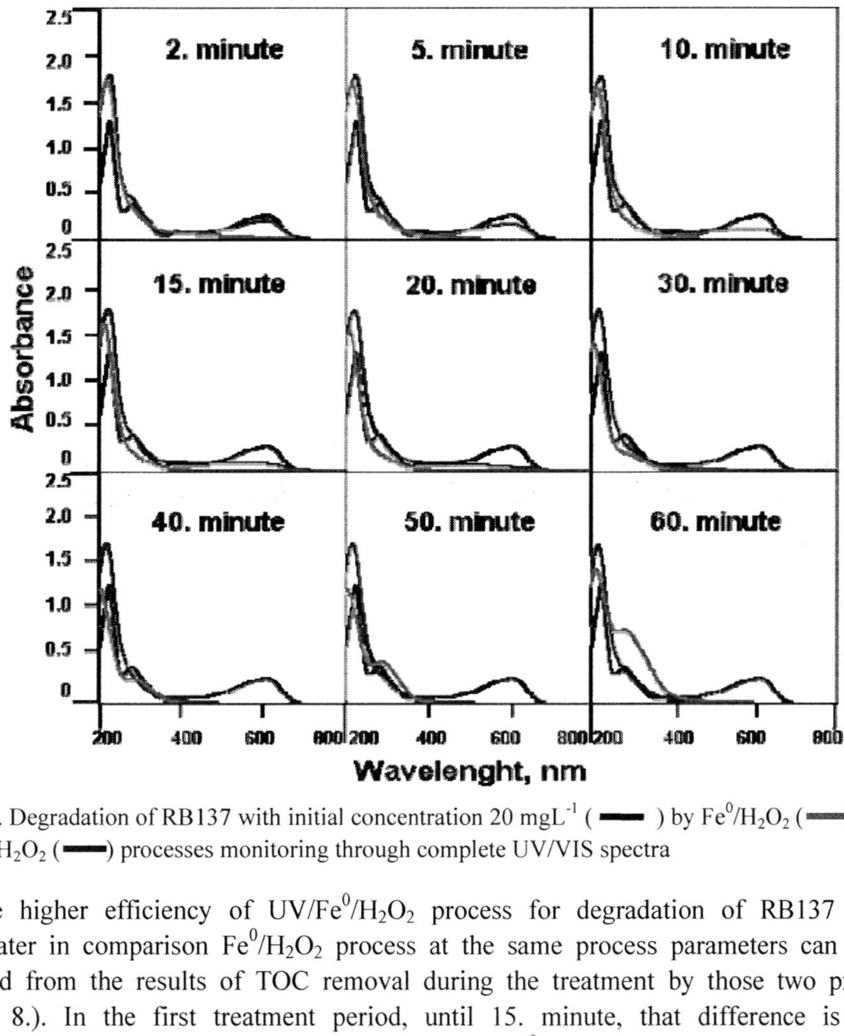


Figure 7. Degradation of RB137 with initial concentration 20 $mgL^{-1}$ ( ▬ ) by $Fe^0/H_2O_2$ ( ▬ ) and $UV/Fe^0/H_2O_2$ ( ▬ ) processes monitoring through complete UV/VIS spectra

The higher efficiency of $UV/Fe^0/H_2O_2$ process for degradation of RB137 colored wastewater in comparison $Fe^0/H_2O_2$ process at the same process parameters can be also observed from the results of TOC removal during the treatment by those two processes (Figure 8.). In the first treatment period, until 15. minute, that difference is not so pronounced. But after 15. minute, in the case of $Fe^0/H_2O_2$ process strong inhibition in mineralization of RB137 model solution can be observed. On the other hand, the inhibition of $UV/Fe^0/H_2O_2$ process occurred after 40. minute of process. Inhibition and possible complete stopping of dark Fenton type process could be due to the complete consumption of $H_2O_2$, or due to the breaking of Fenton catalytic cycle by the formation of $Fe^{3+}$-complexes, mostly carboxylates [64]. Inhibition of photo-Fenton type processes could be caused by complete consumption of $H_2O_2$ and in that case degradation of organics will be continued only by their direct photolysis. Further reason for the inhibition of photo-Fenton type process might be due to the formation of organics acids resistant to degradation by OH radicals, such as acetic and oxalic [37], which can be degraded under UV light but with lower rate than other organic content by OH radicals. In this case it is more likely that the inhibition of both processes was not caused by the complete consumption of $H_2O_2$, due to the results of rather high remained $H_2O_2$ concentration in bulk, >20 %, obtained in our previous study at the same process conditions, but with other organic pollutant [48].

Hence, it can be assumed that the inhibition of dark process, $Fe^0/H_2O_2$, occurred due to the formation of $Fe^{3+}$-complexes resulting with a lack of OH radicals in bulk produced through bulk Fenton reaction, while the mineralization efficiency of $UV/Fe^0/H_2O_2$ process was inhibited by the formation of large amounts of OH radical attack resistant organics acids. That is in accordance with the observation from Figure 7., where the appearance of new peaks occurred after 30. minute of treatment process.

The possible improvement of dye degradation efficiency of dark Fenton type process was tried to get by the addition of synthetic zeolite $NH_4ZSM5$. In our previous study [146], that type of zeolite showed positive effect on TOC removal by $Fe^0/H_2O_2$ process when added in lower dosages, <0.5 $gL^{-1}$. In Table 4 the comparison of $Fe^0/H_2O_2$, $Fe^0/H_2O_2/NH_4ZSM5$ and $UV/Fe^0/H_2O_2$ processes concerning the final decolorization, degradation of chloro-triazine rings and mineralization of RB137 model solution (20 $mgL^{-1}$) after a one-hour treatment is presented. It can be observed that all three compared processes showed very high decolorization efficiency, >99 % color removal in all cases. It should be noted that so-called "blank-tests" were performed in order to investigate adsorption of dye molecules on zeolites or possible penetration in zeolites pores. No changes of dye spectra monitored through complete UV/VIS region were obtained, indicating no dye adsorption by $NH_4ZSM5$ zeolite. Comparing mineralization efficiency of organic content by all three investigated processes, it can be observed that partial mineralization extents were obtained in all cases.

**Table 4. Degradation efficiency of RB137 model solution (20 $mgL^{-1}$) by $Fe^0/H_2O_2$, $Fe^0/H_2O_2/NH_4ZSM5$ and $UV/Fe^0/H_2O_2$ processes**

| process | $Fe^0/H_2O_2$ | $Fe^0/H_2O_2/NH_4ZSM5$ | $UV/Fe^0/H_2O_2$ |
|---|---|---|---|
| color removal, % | 100 | 99.2 | 100 |
| TOC removal, % | 40.8 | 51.5 | 83.5 |
| AOX removal, % | 42.2 | 42.5 | 90.5 |

The increase of TOC removal for 10.3 % by $Fe^0/H_2O_2/NH_4ZSM5$ process in comparison to $Fe^0/H_2O_2$ could be contributed to the addition of solid particles. But, it is more likely that synthetic zeolite acted as adsorbents for molecule of appropriate size to be adsorbed/penetrated on/in zeolite. No effect of zeolite addition to other monitored parameters, color and AOX removal, and the findings of our previous study where scanned IR spectrograms of zeolites after the treatment with $Fe^0/H_2O_2/NH_4ZSM5$ showed new peaks which corresponded to possible dye degradation by-products [146], indicating to above mentioned assumption. Moreover, Kawai and Tsutsumi [157] reported, and the same was confirmed in our previous studies [59, 158], that $NH_4ZSM5$ zeolite could adsorb phenol. If phenol could be adsorbed, it can be assumed that other mono-substituted benzenes could also be removed from the aqueous solution by adsorption on/in $NH_4ZSM5$ zeolite. Furthermore, Feng et al. [156] proposed degradation mechanism of reactive dye with similar structure like

RB137. They reported that after the initial cleavage of azo bonds, in the bulk are remained mono- and di-substituted benzenes and naphtholes and triazine rings. Moreover, Stylidi et al. [71] detected phenol and other mono-substituted benzenes as by-products of azo dye degradation by OH radicals. In the favor of adsorption of only mono-substituted benzenes speaks also the fact that AOX removal was almost the same in the case with and without zeolite addition, indicating inability of $NH_4ZSM5$ zeolite for adsorption of triazine rings. The highest removal of all parameters; color, TOC and AOX was obtained when $Fe^0/H_2O_2$ process is combined with UV irradiation, resulting with 105 and 115 % improvement in TOC and AOX removal efficiency.

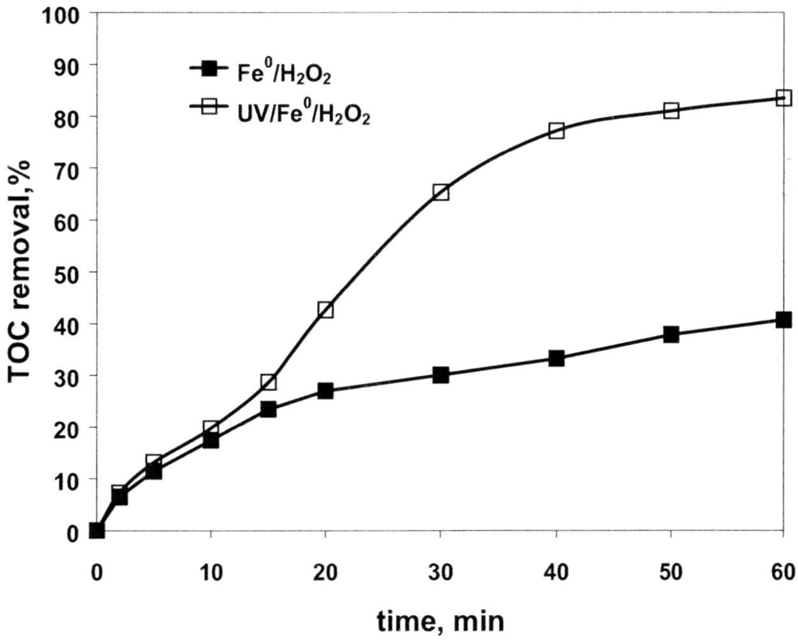


Figure 8. Mineralization of RB137 with initial concentration 20 $mgL^{-1}$ by $Fe^0/H_2O_2$ and $UV/Fe^0/H_2O_2$ processes monitoring decrease of TOC ecological parameter

The efficiency of $Fe^0/H_2O_2$ and $UV/Fe^0/H_2O_2$ processes was also tested on colored wastewater containing 80 $mgL^{-1}$ of RB137 (Figure 9.).

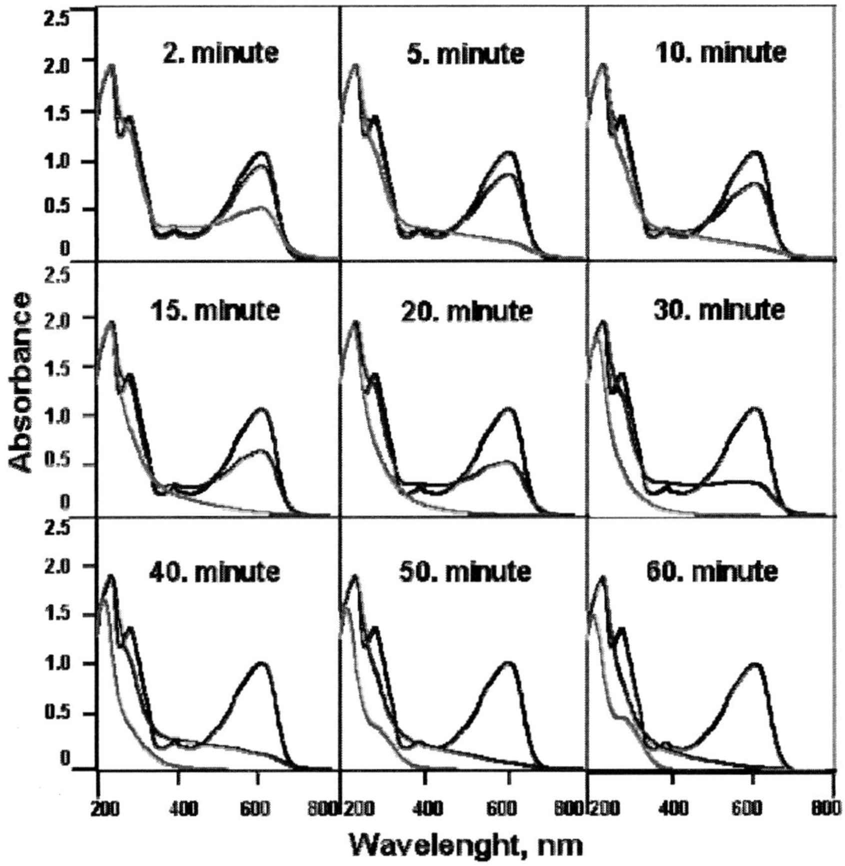


Figure 9. Degradation of RB137 with initial concentration 80 $mgL^{-1}$ (▬) by $Fe^0/H_2O_2$ (▬) and $UV/Fe^0/H_2O_2$ (▬) processes monitoring through complete UV/VIS spectra

Again, three specific peaks can be observed: dye cromophore peak at 610 nm, aromatic peak at 280 nm and triazine ring peak at 230 nm. The peaks in UV region were also interfered by absorbance of added $H_2O_2$, but not in such high manner like in the case with the initial dye concentration of 20 $mgL^{-1}$. $UV/Fe^0/H_2O_2$ process again showed rather high decolorization efficiency of colored wastewater; the complete bleaching was achieved between 20. and 30. minute of treatment, while in the case of $Fe^0/H_2O_2$ process complete bleaching was not achieved even after 60 minutes treatment. From Figure 9. a rather fast degradation of aromatics, i.e. disappearance of peak at 280 nm, can be observed even more clearly than in case of lower initial dye concentration due weaker interference of that peak by $H_2O_2$ absorbance. On the other hand, in the case of $Fe^0/H_2O_2$ process the lowering of that peak was much slower than in the case of $UV/Fe^0/H_2O_2$ process. Also, an appearance of new peak with maximum wavelength at 290, corresponding to the formation of aliphatic by-products, was observed after 40. minute of treatment in the case $UV/Fe^0/H_2O_2$ process.

Mineralization of RB137 solution (80 $mgL^{-1}$) by those two processes is showed in Figure 10. Again, the pronounced difference in process effectiveness can be observed after 10. minute of treatment, where the inhibition of $Fe^0/H_2O_2$ process can be observed.

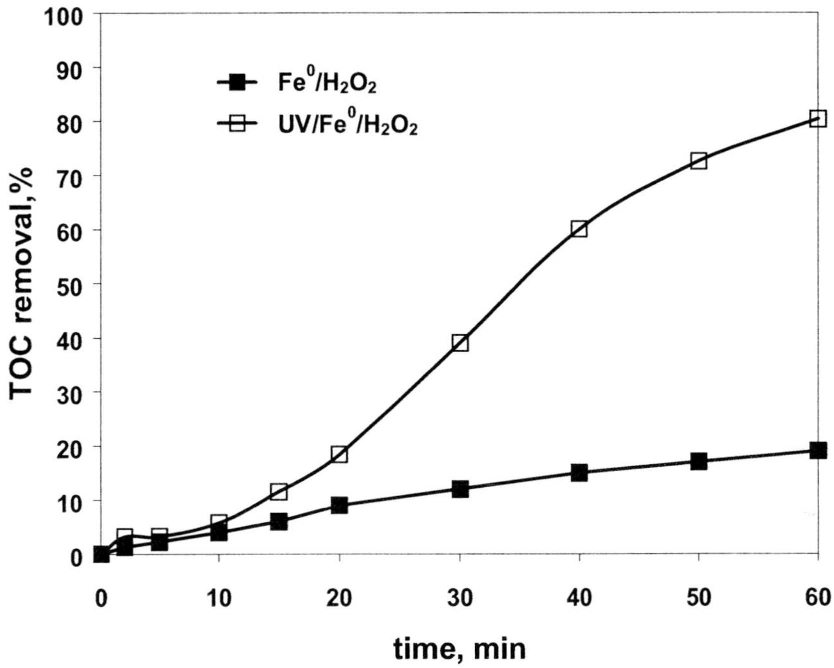


Figure 10. Mineralization of RB137 with initial concentration 80 mgL$^{-1}$ by $Fe^0/H_2O_2$ and $UV/Fe^0/H_2O_2$ processes monitoring decrease of TOC ecological parameter

In the case of $UV/Fe^0/H_2O_2$ process strong inhibition like in the case of lower initial dye concentration is not observed, but it is obvious that the rate of mineralization process is lowering by increasing of mineralization degree of model solution. It should be noted that final achieved mineralization extents; 19.1 and 80.3 % of TOC removal by $Fe^0/H_2O_2$ and $UV/Fe^0/H_2O_2$ process, respectively, are not four times lower than those obtained in the case of lower dye concentration, 40.8 and 83.5 %. Moreover, overall removed organic content expressed in mgC per L are much higher in the case of higher dye concentration: 4.9 >> 2.6 in the case of $Fe^0/H_2O_2$ and 20.4 >> 5.3 in the case of $UV/Fe^0/H_2O_2$ process. Such a large difference in the overall amount of mineralized dye might be due to the fact that besides OH radicals some other radical species, produced by the destruction of chromophoric parts and aromatic rings of the dye molecules and dye by-products, may be involved in further chain reactions of the dye degradation mechanism. Namely, with higher dye concentration, after initial cleavage of the chromophoric parts of the dye molecules, more radical species originated from chromophores and aromatic by-products destruction are present in the bulk and may participate in further dye degradation/mineralization. A similar effect is known from the literature. Chen and Pignatello [47] reported the formation and the involvement of organic radical species in phenol degradation mechanism by OH radicals. Furthermore, the simple mechanism proposed for the UV radiation chemistry is very similar. When UV radiation is incident upon the dye molecule, the radiation can be adsorbed, promoting the molecule to an excited state which can be involved in further degradation of dye molecules via radical mechanism [35, 159].

***UV-based Processes***

Next group of investigated AOPs were photochemical processes: $UV/O_3$ and $UV/O_3/H_2O_2$. Like in Fenton type processes, the efficiency of UV-based processes strongly depends on process parameters such as: light intensity, reactor configuration, pH, type of oxidant, concentration of oxidant. Therefore, it was necessary to find optimal parameters for those which are changeable in our case; pH and initial $H_2O_2$ concentration (when added), while other process parameters, such as ozone feed concentration, light intensity, were held constant. In Figure 11. final color removals, TOC removals after 30 and 60 minutes treatment, and final pH values after the treatment of RB137 model wastewater (20 $mgL^{-1}$) by $UV/O_3$ process through initial pH range between 3 and 11 are presented. According to the complete color removal, achieved in <10 minutes of treatment, through entire pH investigated, it can be concluded that bleaching is not influenced by initial pH value. On the other hand, presented results of TOC removal after 30 and 60 minutes treatment were influenced by initial pH value, particularly at strong basic conditions.

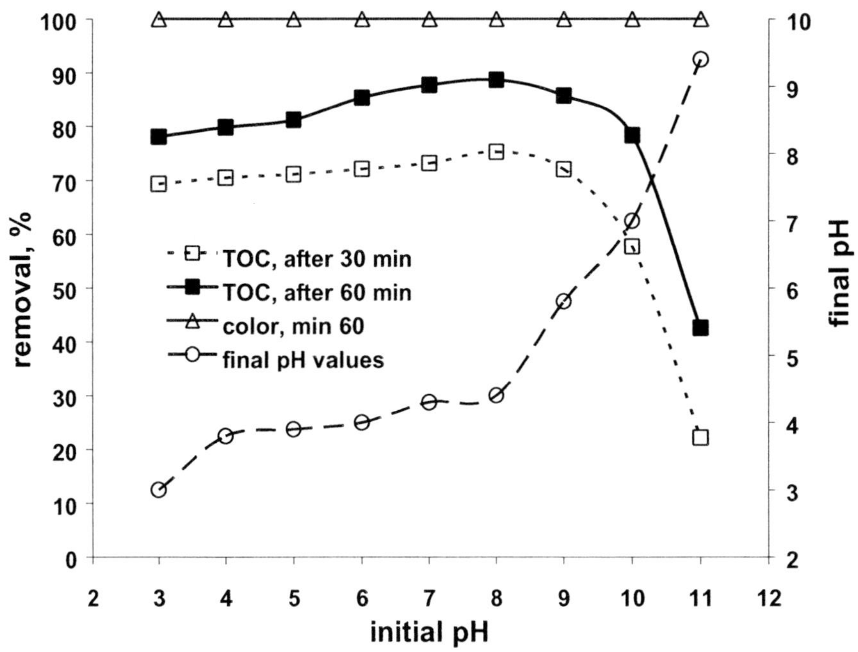


Figure 11. Influence of the initial pH value on the decolorization and mineralization efficiency, and final pH values after a one-hour treatment by $UV/O_3$ process.

The optimal pH area for the mineralization of RB137 is found to be between 6-9, i.e. at neutral and weak basic conditions, with the highest TOC removal, 88.7 %, achieved after 60 minutes at pH 8. It can be seen that the mineralization process is much faster in first 30 minutes of treatment process than in next 30 minutes. That can be attributed to the formation of large quantities of organic acids, including those resistant to OH radical degradation, in the last 30 minutes of treatment with significantly lower mineralization rate. The decrease in pH value after the treatment in comparison to initial value corresponds to the formation of acid by-products, mostly aliphatic [92].

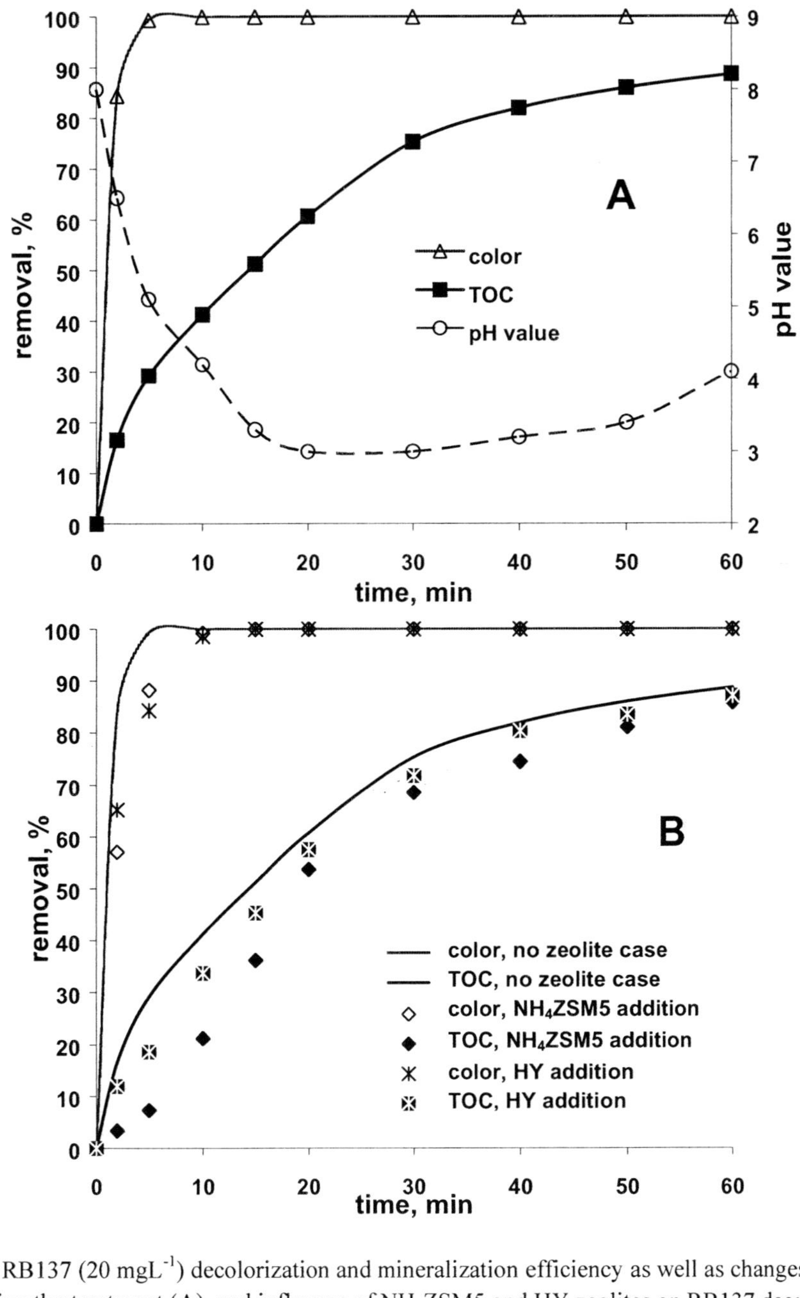


Figure 12. RB137 (20 $mgL^{-1}$) decolorization and mineralization efficiency as well as changes of pH values during the treatment (A), and influence of $NH_4ZSM5$ and HY zeolites on RB137 degradation efficiency (B) by $UV/O_3$ process (pH 8)

After optimizing $UV/O_3$ process regarding initial pH value, the kinetic of RB137 model solution (20 $mgL^{-1}$) decolorization and mineralization, as well changes of pH value during the treatment by $UV/O_3$ process (pH 8) were monitored (Figure 12.A). It can be seen that complete bleaching was achieved even after 5 minutes of treatment, while the rate of RB137 mineralization was decreased when treatment process progressed. That is particularly pronounced in last 30 minutes of treatment time. As it was mentioned above, the reason for such system behavior could be the formation of large quantities of organic acids, mostly aliphatic, as by-products after initial cleavage of azo bonds in dye molecules, and subsequent formation of aromatic by-products, and thereafter their degradation [71, 160]. Some of formed acids are recalcitrant to degradation through OH radial degradation mechanism [37]. However, they can be destroyed by UV irradiation, but slower than other organic content by OH radicals.

Additional reason for the slowing down of mineralization process could be found in the fact that mineralization efficiency of $UV/O_3$ process is slightly lower in strong acidic pH region than at neutral and weak basic pH (Figure 11.). Formation of acid by-products is also confirmed by large drop in pH value. As treatment process progressed, pH value started slightly to increase, that can be attributed to the mineralization of acid by-products. In Figure 12.B are presented results of experiments with the addition of synthetic zeolites $NH_4ZSM5$ and HY to $UV/O_3$ process. These two types of zeolites were added to the oxidative system combining UV irradiation and ozone due to the observed positive effect of those zeolite types on decolorization of model wastewaters containing reactive dyes in hybrid-parallel corona reactor, which involves, besides classical corona discharge in liquid and gas phase, UV irradiation in liquid, as well as ozone formation in gas and its transfer to liquid phase [132, 133]. It should be noted, that like in the case of $NH_4ZSM5$ zeolite, series of experiments for the determining adsorption characteristics of HY zeolite for RB137 removal from aqueous solution were conducted showing no dye adsorption/penetration on/in zeolite. From Figure 12.B it can be seen that the addition of synthetic zeolites showed slight negative influence to the rate of RB137 decolorization and mineralization. That can be explained by the fact that efficiency of processes based on UV irradiation strongly depends on solution turbidity, i.e. solution transparency and clearness [35, 59].

The efficiency of $UV/O_3$ process (pH 8) for decolorization and mineralization of RB137 was investigated at higher dye loading of model wastewater, 80 $mgL^{-1}$ (Figure 13.). It can be seen that complete bleaching was achieved after 20 minutes of treatment, while final extent of partial mineralization after a one-hour treatment is 64.4 % TOC removal. Trend of the curve that connects points of TOC removal extents measured during the treatment is almost linear, i.e. very different than that in the case of lower initial dye concentration. Although drop in pH value indicate the formation of acidic by-products, the mineralization of RB137 model solution was not slowed, indicating that the most of formed by-products are not resistant to OH radical attack. But, with further process progressing over 60 minutes, the slowing down of mineralization process could be expected, due to the fact that Hoigne and Bader [162] pointed out that organic acids with low reaction rate constants under UV irradiation and OH radical resistant, such as acetic and oxalic, always accumulate as final products when any type of reactive organic aqueous solutes are degraded by ozone or OH radicals in water. Hence, the overall oxidation of organic material to $CO_2$ and $H_2O$ will be delayed whenever the oxidations lead to oxalic or acetic acid as intermediates, which take part at acids conditions. Again, like in the case of lower dye loading, the influence of synthetic zeolites to $UV/O_3$ process efficiency was investigated (Figure 13.B). It can be seen that zeolite addition caused slight delay in decolorization of RB137 model wastewater (80 $mgL^{-1}$). That can be explained by earlier mentioned increase in the solution turbidity, that decrease UV irradiation efficiency for ozone photolysis resulting with slower generation of OH radicals.

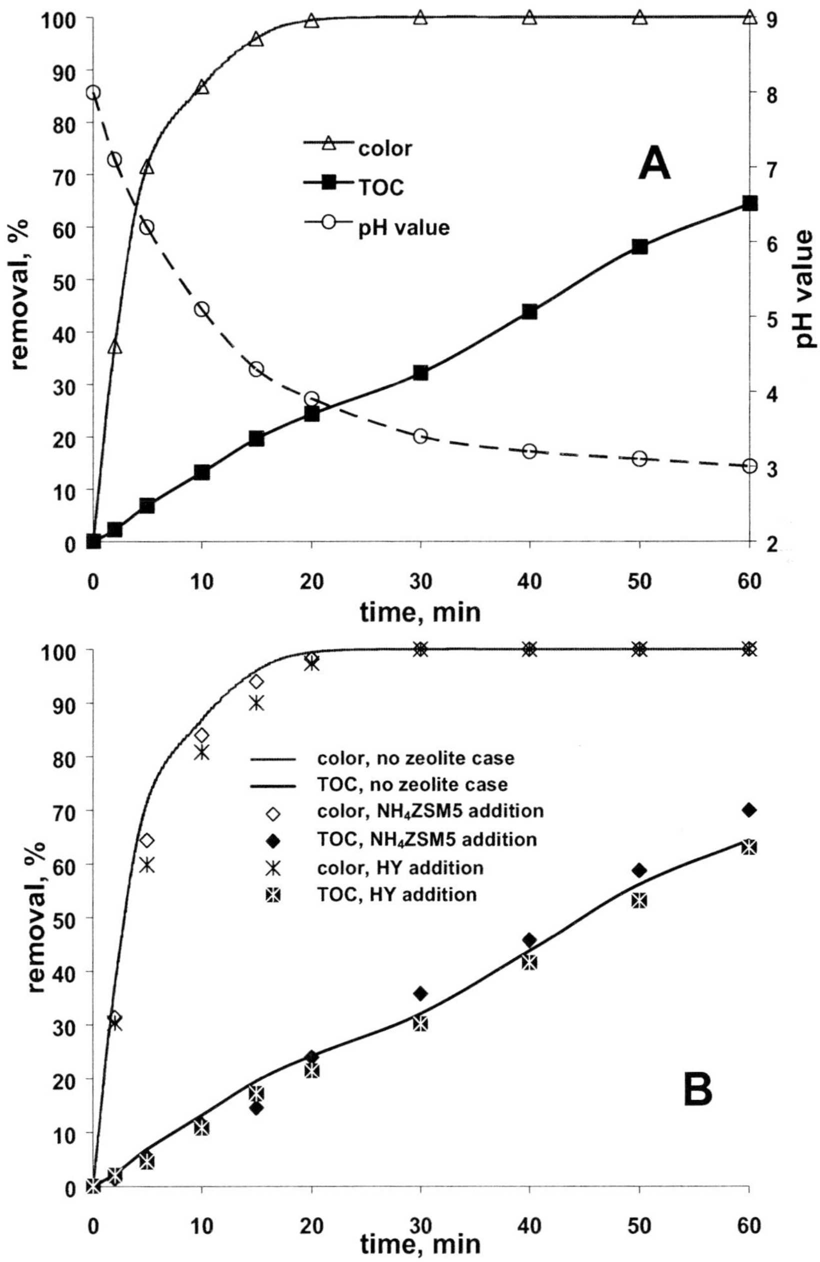


Figure 13. RB137 (80 $mgL^{-1}$) decolorization and mineralization efficiency as well as changes of pH values during the treatment (A), and influence of $NH_4ZSM5$ and HY zeolites on RB137 degradation efficiency (B) by $UV/O_3$ process (pH 8)

On the other hand, the addition of $NH_4ZSM5$ zeolite showed slight positive effect to TOC removal in treatment period after 30. minute, but that was not the case with HY zeolite. If addition of zeolites negatively influence decolorization, it can be assumed that negatively effects the mineralization too, due to the same reason; increase in system turbidity. But, such different system behavior concerning the addition of different type of zeolite could be explained with the adsorption/penetration of organic molecules formed as dye degradation by-products with appropriate size on/in zeolite $NH_4ZSM5$. Namely, these two zeolites are different according to pore size and structure, Si/Al ratio, surface area, etc. ZSM5 type zeolite has a pentasyl structure which is characterized by parallel channels with pores sizes 0.53-0.56 nm with crossed channels with pores sizes 0.51-0.55 nm, while Y zeolite has a basic cubic structure with pores sizes 0.74 nm [148-150]. As it was proposed above in the case of $Fe^0/H_2O_2/NH_4ZSM5$, $NH_4ZSM5$ zeolite showed ability to adsorb mono-substituted benzenes. According to the results presented in Figure 13.B, where TOC removal raised over that obtained in the non-zeolite case, it can be assumed that at particular treatment time, formed mono-substituted benzenes as by-products of dye degradation adsorbed on zeolites or even penetrated in zeolite pores. Such system behavior in process using ozone is already observed in our previous study [117].

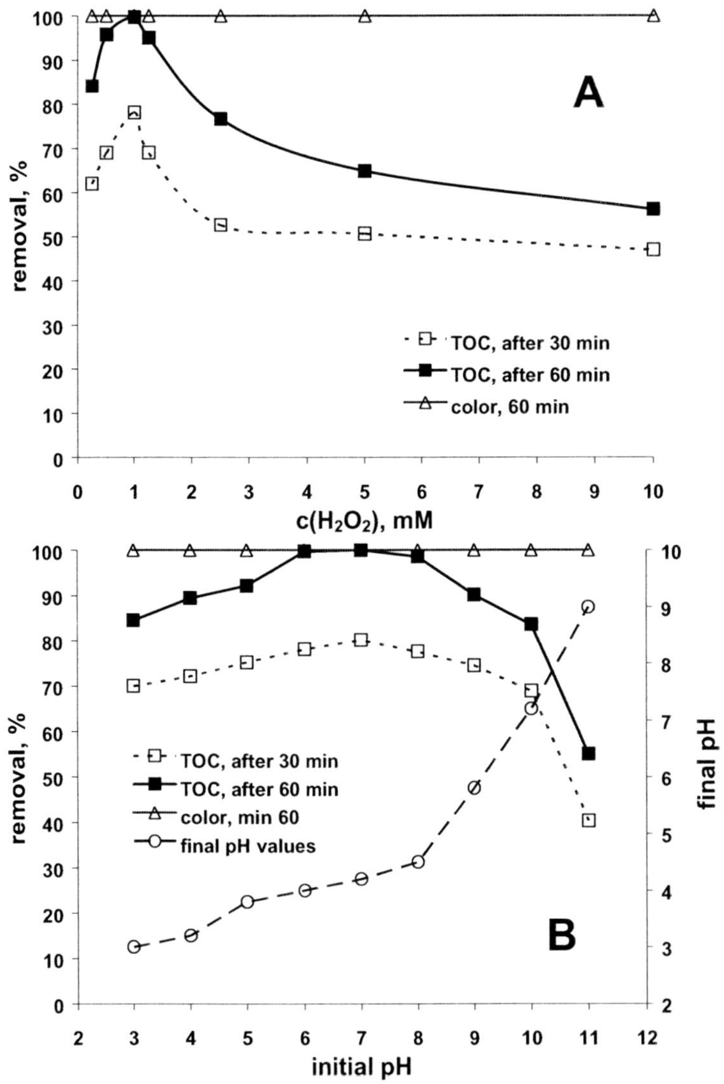


Figure 14. Influence of the initial $H_2O_2$ concentration (A), and initial pH value (B) on the decolorization and mineralization efficiency, and final pH values after a one-hour treatment by $UV/O_3/H_2O_2$ process.

In the same manner like $UV/O_3$ process, $UV/O_3/H_2O_2$ process was optimized regarding initial pH value, but also regarding to the other important changeable process parameters; initial $H_2O_2$ concentration. The results of optimization study are presented in Figure 14. Firstly, $UV/O_3/H_2O_2$ process was optimized concerning initial $H_2O_2$ concentration by changing that parameter and keeping others constant, including initial pH too, at zero pH charge of RB137 model wastewater (pH 6). It can be seen that the complete color removal was achieved through entire range of initial $H_2O_2$ investigated (0.25-10 mM). Moreover, that is achieved in very short treatment period, <10 minutes. Hence, it can be concluded that initial $H_2O_2$ concentration does not influence to much on decolorization efficiency of $UV/O_3/H_2O_2$ process. But, as it can be seen, initial $H_2O_2$ concentration strongly influence the mineralization efficiency. By increasing the initial $H_2O_2$ concentration, RB137 mineralization efficiency of $UV/O_3/H_2O_2$ process also increased, but only to the certain level, where further increase of $H_2O_2$ concentration negatively influences to the system effectiveness. Explanation for such system behavior is given as follows: slight enhancement of mineralization extents in the first part of the range of initial $H_2O_2$ concentration studied (0.25-1.0 mM), could be contributed to an increase of OH radical concentration in the bulk according to $H_2O_2$ photolysis mechanism, shown by equation (8). The decrease of RB137 degradation efficiency by $UV/O_3/H_2O_2$ process with further increasing of the initial $H_2O_2$ concentration could be the result of $H_2O_2$ bulk concentration higher than 1.0 mM. When hydrogen peroxide is in excess, it behaves as a scavenger, consequently lowering the hydroxyl radicals concentration and decreasing the overall mineralization efficiency of applied process [78, 87]. Similar system behavior was observed during the optimization of $H_2O_2$ concentration in Fenton type process (Figure 6.). By comparing mineralization extents of RB137 model wastewater by $UV/O_3$ at pH 6 (Figure 11.) and by $UV/O_3/H_2O_2$ at the same pH with optimal concentration of $H_2O_2$ (Figure 14.A), it can be seen that almost 20 % higher TOC removal was obtained by $UV/O_3/H_2O_2$ process.

In Figure 14.B the optimization of $UV/O_3/H_2O_2$ process regarding to the initial pH value is presented. That was performed in the way that pH values were changed from 3-11, while initial $H_2O_2$ concentration was held constant at 1.0 mM, found as optimal in the previous set of experiments (Figure 14.A). Again, complete bleaching was achieved in very short treatment period through complete range of pH investigated. According to mineralization extents achieved after 30. and 60. of minute treatment, the optimal pH region for degradation of RB137 by $UV/O_3/H_2O_2$ process was established. It can be observed similar system behavior like in the case of $UV/O_3$ process (Figure 11.); a significant decrease in process effectiveness at strong basic conditions, and the highest mineralization efficiency at neutral and weak basic pH region. But it should be pointed out that the addition of $H_2O_2$ to $UV/O_3$ system improved mineralization efficiency through complete investigated pH range. Moreover, the complete mineralization was achieved after a one-hour treatment in the pH region between 6 and 8. Due to such system behavior, optimal pH was found out according to the mineralization extents obtained after 30 minutes treatment time. On that way, it was found the highest mineralization of RB137, 80.1 % TOC removal, by $UV/O_3/H_2O_2$ process ($c(H_2O_2)$=1.0 mM) at pH 7. It should be noted that the trend of curve that connects points of final pH values measured after a one-hour treatment by $UV/O_3/H_2O_2$ process (Figure 14.B) is very similar to that obtained by $UV/O_3$ (Figure 11.).

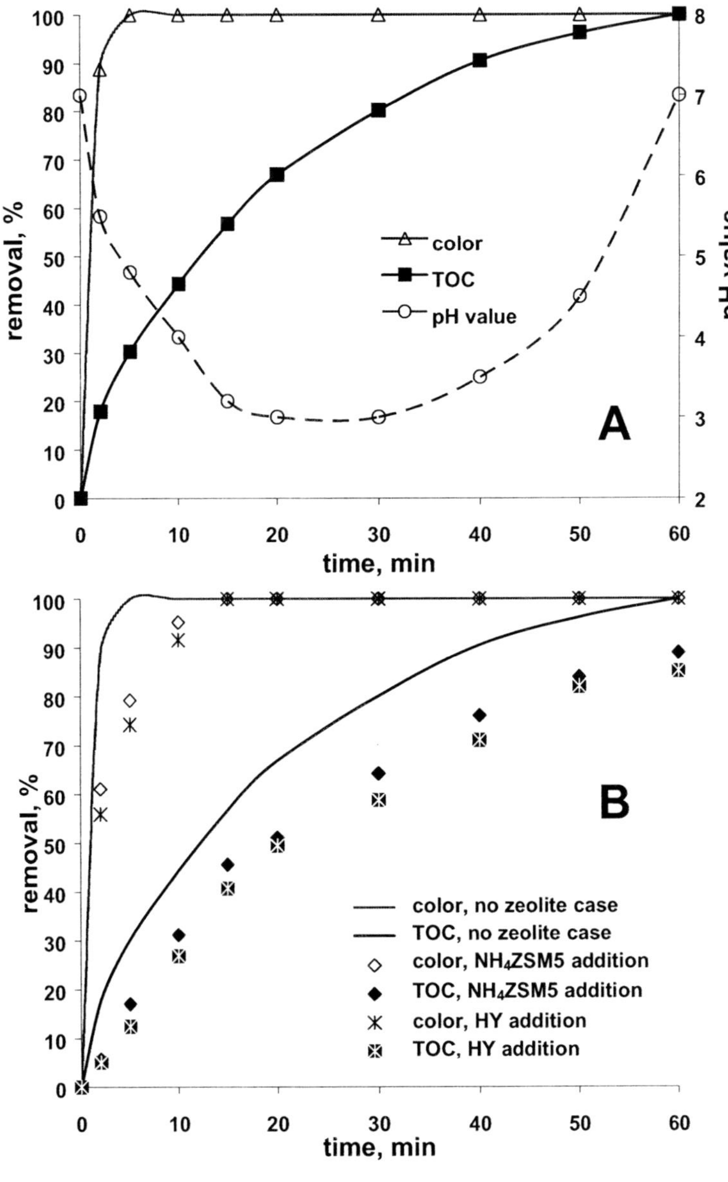


Figure 15. RB137 (20 $mgL^{-1}$) decolorization and mineralization efficiency as well as changes of pH values during the treatment (A), and influence of $NH_4ZSM5$ and HY zeolites on RB137 degradation efficiency (B) by $UV/O_3/H_2O_2$ process (pH 7 and $c(H_2O_2)$=1.0 mM)

Degradation kinetics of RB137 (20 $mgL^{-1}$) by $UV/O_3/H_2O_2$ process at optimal process conditions; pH 7 and $c(H_2O_2)$=1.0 mM, is showed in Figure 15.A. It can be seen that complete bleaching was achieved after 5 minutes of treatment. The same system behavior, like in the case of the treatment of RB137 (20 $mgL^{-1}$) by $UV/O_3$ process, slowing down of mineralization process with the progress of treatment process can be observed. However, complete mineralization was achieved. Decrease in the rate of mineralization is due to the earlier mentioned reason, production of large quantities of organic acids with lower rate of degradation. The drop in pH during the treatment time corresponds to the formation of acid by-products, while the achievement of complete mineralization resulted with reaching the initial pH 7 at the end of the process. Like in the case of UV/$O_3$/zeolite system for degradation of RB137 (20 mgL$^{-1}$) (Figure 12.B), the same system behavior could be observed in the case of UV/$O_3$/$H_2O_2$/zeolite (Figure 15.B). Moreover, negative influence of synthetic zeolites $NH_4$ZSM5 and HY addition is even more pronounced to RB137 decolorization and mineralization efficiency.

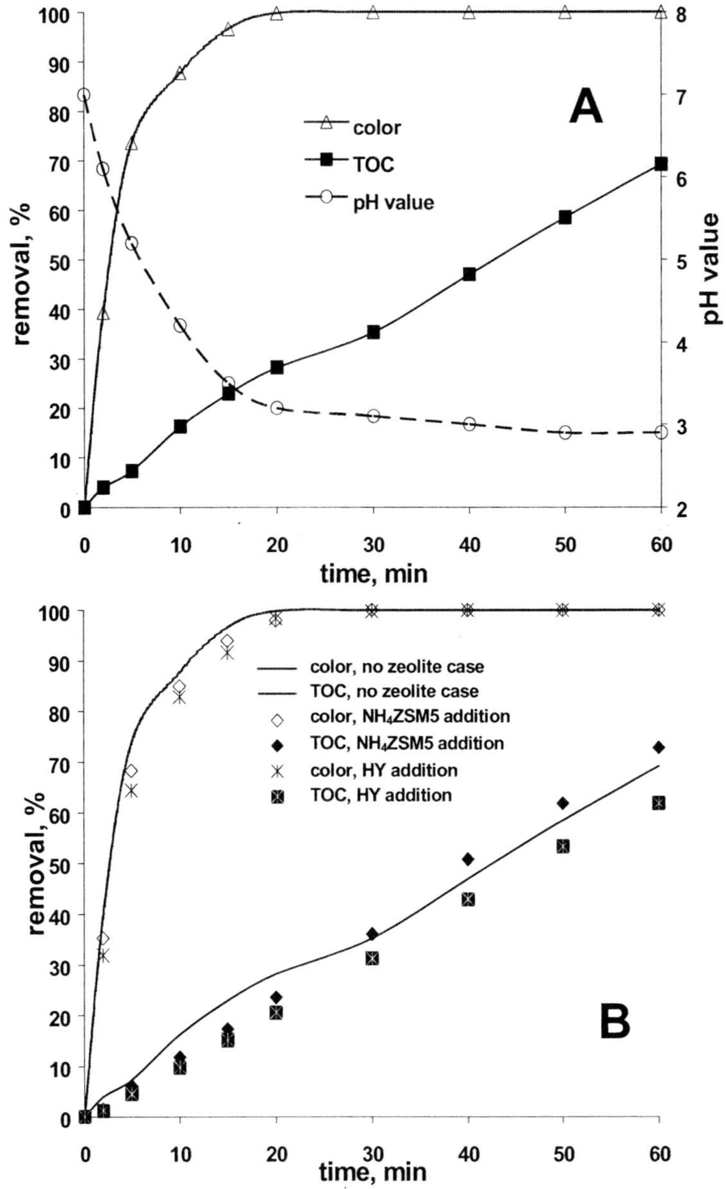


Figure 16. RB137 (80 mgL$^{-1}$) decolorization and mineralization efficiency as well as changes of pH values during the treatment (A), and influence of $NH_4$ZSM5 and HY zeolites on RB137 degradation efficiency (B) by UV/$O_3$/$H_2O_2$ process (pH 7 and c($H_2O_2$)=1.0 mM)

The $UV/O_3/H_2O_2$ process efficiency was tested for decolorization and mineralization of RB137 model wastewater with higher dye loading (Figure 16.). In all three cases: non-zeolite (Figure 16.A), with the addition of $NH_4$ZSM5 and with addition of HY zeolite (Figure 16.B) the very similar system behavior like that showed in the case of $UV/O_3$ process (Figure 13.) can be observed.

Complete bleaching with non-zeolite case was achieved after 20 minutes of treatment, trend of the curve which connects points of measured TOC removals is almost linear, zeolite addition caused delay in decolorization of RB137, while $NH_4$ZSM5 zeolite addition showed positive effect to TOC removal after 30. minute of process. But it should be pointed out that final achieved mineralization, 69.3 % TOC removal, by $UV/O_3/H_2O_2$ process is somewhat higher than that obtained by $UV/O_3$ process, 64.4 % TOC removal.

### *AOPs Cost Estimation*

The efficiency of studied AOPs in laboratory scale: $Fe^0/H_2O_2$, $UV/Fe^0/H_2O_2$, $UV/O_3$ and $UV/O_3/H_2O_2$, for the treatment of RB137 model wastewater was estimated from the point of view of required operating costs. The evaluation was performed on the basis of the obtained results in the case of higher dye loading in colored model wastewater. It should be noted that evaluation of only operating costs was performed. The overall costs are represented by the sum of the capital costs, the operating costs and the maintenance. For a full-scale system these costs will, besides the nature and the concentration of pollutants, strongly depend on the flow rate of the effluent and the configuration of the reactor [38]. In order to evaluate the operating costs for proposed processes, the costs of used reagents valid for Croatian market were taken into account (Table 5).

**Table 5. Cost of reagents used in applied processes**

| reagent | basis | cost (US$) |
|---|---|---|
| $Fe^0$ | 100 g | 13.62 |
| $H_2O_2$, 30 % | 1 L | 37.59 |
| $O_2$, >99.5 % | 100 L | 0.23 |
| electricity | 1 kWh | 0.16 |

Note: costs are given in US$ and the values are valid for Croatian market

The main estimation results are values of ratio *mineralization extent/process operating costs* (ME/POC). Besides costs estimation on the basis of mineralized organic content, the cost for decolorization below visibility limit, <1 $mgL^{-1}$ that corresponds to decolorization below 1.25 % in our case, are also showed. From the point of view of the most suitable processes for decolorization of RB137 model wastewater below visibility limit, it can be seen that processes combining ozone and UV irradiation, $UV/O_3$ and $UV/O_3/H_2O_2$, are double cheaper than processes using iron powder, and more important, decolorization is even faster. Moreover, by $Fe^0/H_2O_2$ process the complete decolorization was not achieved even after a one-hour treatment.

From the ratio ME/POC point of view, it can be seen that processes using UV irradiation are much more efficient, with ME/POC ratios ranged between 3.20-3.45 US$ $g^{-1}$, than that one conducted in dark. Moreover, UV irradiation could provide complete mineralization, which is not the case when only degradation over OH radicals is used. However, processes combining UV and ozone seem to be more appropriate due to that successful maintenance of $UV/Fe^0/H_2O_2$ process required strong acid conditions, pH 3. That could increase operating costs of treatment process, but also includes post-treatment as necessary due to the strong conditioning to the neutral pH, which also causes additional costs. Furthermore, post-treatment after the treatment by $UV/Fe^0/H_2O_2$ process includes removal of non-reacted iron powder and remaining iron ions from the bulk. On the contrary, $UV/O_3$ and $UV/O_3/H_2O_2$ processes were maintained at neutral or weak basic conditions, and after the complete mineralization, the pH values will be the same as at the beginning of treatment process. That excludes additional costs for the strong conditioning of solution before and after the treatment. The cost of processes using UV irradiation could be drastically reduced if solar light is used instead artificial UV.

**Table 6. Cost evaluation for all applied processes (processes conducted in laboratory scale)**

| process | time required for decolorization <1mgL$^{-1}$ (min) | operational cost for decolorization <1mgL$^{-1}$ (US$) | final TOC removal (%) | costs (US$ g$^{-1}$) |
|---|---|---|---|---|
| $Fe^0/H_2O_2$ | >60 | 0.046* | 19.0 | 9.37 |
| $UV/Fe^0/H_2O_2$ | <30 | 0.055 | 80.3 | 3.20 |
| $UV/O_3$ | <20 | 0.025 | 64.4 | 3.45 |
| $UV/H_2O_2/O_3$ | <20 | 0.028 | 69.3 | 3.31 |

* indicates treatment time over 60 minutes

## CONCLUSION

This chapter is dedicated to the investigation of highly efficient and destructive methods for the minimization of hazardous liquid waste containing reactive dyes as parent molecules and other contaminants present in such wastewaters. Besides an overview of recent publications concerning worldwide literature about AOPs for colored wastewaters treatment given in this chapter, a large part of our scientific investigations up to now, related to this subject, is presented too. The chapter describes different AOPs and the possibility of their usage for the treatment of colored wastewater containing reactive dyes in order to achieve colorless and non-toxic wastewater effluents at first, and, if it is possible, to reach significant minimization of overall organic content. In the research part of this chapter, several AOPs; $Fe^0/H_2O_2$, $UV/Fe^0/H_2O_2$, $UV/O_3$ and $UV/O_3/H_2O_2$, are investigated on degradation of C.I. Reactive Blue 137, azo reactive dye, as model pollutant in colored wastewater. It was shown that studied AOPs are very efficient for the bleaching of RB137 model solution. That is also reported through the review part of this chapter, where chemical and photochemical processes shown as the most efficient among all AOPs. Those findings are very important due to the fact that the present color in wastewaters could cause serious problems to aquatic life in recipient waters. Moreover, after decolorization less complex structures are formed that can be degraded by microorganisms, which is not the case when reactive dyes are present in water as parent molecules. Although AOPs seem to be much more expensive to start than common commercial treatment methods such as filtration, adsorption and biodegradation, they are highly efficient for the degradation of recalcitrant pollutants, such as reactive dyes. Moreover, evaluated values of ratio *mineralization extent/process operating costs* (ME/POC) for $UV/Fe^0/H_2O_2$, $UV/O_3$ and $UV/O_3/H_2O_2$ processes in the case of RB137 model solution treatment ranged between 3.20-3.45 US$ per gram of mineralized organic content. That seems reasonable if complete mineralization considers, i.e. complete removal of toxic organic substances from wastewater by destructive mechanism of AOPs.

## ACKNOWLEDGEMENT

We would like to acknowledge financial support from the Ministry of Science, Education and Sport, Republic of Croatia, Project # 0125-018. We are gratefully acknowledged for the support from the National Science Foundation (USA) (INT-0086351) for founding the ozone generator and synthetic zeolites.

## REFERENCE

[1] Zheng, Z; Levin, RE; Pinkham, JL; Shetty, K. Decolorization of polymeric dyes by a novel penicillium isolate. *Process Biochemistry,* 1999 34, 31-37.

[2] Zollinger, H. *Colour Chemistry – Synthesis, Properties and Applications of Organic Dyes and Pigments.* New York: VHC Publishers; 1987.

[3] Riefe, A; Freeman, HS. *Environmental Chemistry of Dyes and Pigments.* New York: John Wiley and Sons; 1996.

[4] Phillips, D. Environmentally friendly, productive and reliable: priorities for cotton dyes and processes. *Journal of Society of Dyers and Colourists,* 1996 112, 183-186.

[5] Hunger, K. *Industrial Dyes; Chemistry, Properties, Application.* Weinheim: Wiley-VCH; 2002.

[6] Allen, SJ; Khader, KYH; Bino, M. Electrooxidation of dyestuff in waste waters. *Journal of Chemical and Biochemical Technology,* 1999 62, 111-117.

[7] Wang, C; Yediler, A; Lienert, D; Wang, Z; Kettrup, A. Toxicity evaluation of reactive dyestuffs, auxiliaries and selected effluents in textile finishing industry to luminescent bacteria Vibrio fischeri. *Chemosphere,* 2002 46 (2), 339-344.

[8] Gottlieb, A; Shaw, C; Smith, A; Wheatley, A; Forsythe, S. The toxicity of textile reactive azo dyes after hydrolysis and decolourisation. *Journal of Biotechnology,* 2003 101(1), 49-56.

[9] Nillson, R; Nordlinder, R; Wass, U. Asthma, rhinitis, and dermatitis in workers exposed to reactive dyes. *British Journal of Industrial Medicine,* 1993 50, 65-70.

[10] Arpe, H-J. Ullmann's Encyclopedia of Industrial Chemistry: Volume A2, Amines, Aliphatic to Antibiotics. 5th Edition, Weinheim: Wiley-VCH; 1985.

[11] Kelshaw, P. The future for West European dyestuff manufacturers. *Journal of Society of Dyers and Colourists,* 1998 114, 35-37.

[12] Sincero, AP; Sincero, GA. *Physical-Chemical Treatment of Water and Wastewater.* New York: CRC Press, IWA Publishing; 2003.

[13] Droste, RJ. *Theory and Practice of Water and Wastewater Treatment.* New York: John Wiley and Sons; 1997.

[14] Forgacs, E; Cserhati, T; Oros, G. Removal of synthetic dyes from wastewaters: a review. *Environmental International,* 2004 30, 953-971.

[15] Pagga, U; Brown, D. The degradation of dyestuffs: part ii behavior of dyestuffs in aerobic biodegradation tests. *Chemosphere*, 1986 15, 479–491.

[16] Seshadri, S; Bishop, PI; Agha, AM. Anaerobic/aerobic treatment of selected azo dyes in wastewater. *Waste Management,* 1994 15, 127–137.

[17] Lourenço, ND; Novais, JM; Pinheiro, HM. Effect of some operational parameters on textile dye biodegradation in a sequential batch reactor. *Journal of Biotechnology,* 2001 89, 163-174.

[18] Georgiou, D; Hatiras, J; Aivasidis, A. Microbial immobilization in a two-stage fixed-bed-reactor pilot plant for on-site anaerobic decolorization of textile wastewater. *Enzyme and Microbial Technology,* 2005 37 (6) 597-605.

[19] Ambrósio, ST; Campos-Takaki, GM. Decolorization of reactive azo dyes by Cunninghamella elegans UCP 542 under co-metabolic conditions. *Bioresource Technology,* 2004 91, 69-75.

[20] Ozdemir, O; Armagan, B. Turan, M; Celik, MS. Comparison of the adsorption characteristics of azo-reactive dyes on mezoporous minerals. *Dyes and Pigments,* 2004 62, 49-60.

[21] Karcher, S; Kornmüller, A; Jekel, M. Screening of commercial sorbents for the removal of reactive dyes. *Dyes and Pigments,* 2001 51, 111-125.

[22] Wang, S; Li, H; Xu, L. Application of zeolite MCM-22 for basic dye removal from wastewater. *Journal of Colloid and Interface Science,* 2006 295, 71-78.

[23] Benkli, YE; Can, MF; Turan, M; Çelik, MS. Modification of organo-zeolite surface for the removal of reactive azo dyes in fixed-bed reactors. *Water Research,* 2005 39, 487-493.

[24] Armaan, B; Turan, M; Cęlik, MS. Equilibrium studies on the adsorption of reactive azodyes into zeolite. *Desalination*, 2004 170, 33-39.

[25] Papić, S; Koprivanac, N; Lončarić Božić, A. Removal of reactive dyes from wastewater using Fe(III) coagulant. *Journal of Society of Dyers and Colourist,* 2000 116, 352-358.

[26] Koprivanac, N, Lončarić Božić, A, Papić, S. Cleaner production processes in the synthesis of blue anthraquinone reactive dye. *Dyes and Pigments,* 2000 44, 33-40.

[27] Papić, S; Koprivanac, N; Lončarić Božić, A; Meteš A. Removal of some reactive dyes from synthetic wastewater by combined Al(III) coagulation/carbon adsorption process. *Dyes and Pigments,* 2004 62, 293-300.

[28] White, GC. Handbook of Chlorination and Alternative Disinfectants. New York: John Wiley & Sons; 1999.

[29] La Grega, MD; Buckingham, PL; Evans, JC. *Hazardous Waste Management.* New York: McGraw-Hill; 1994.

[30] Chu, W; Ma, C-W. Quantitative prediction of direct and indirect dye ozonation kinetics. *Water Research,* 2000 34 (12), 3153-3160.

[31] Parsons, SA; Williams, M. Introduction. In: Parsons SA, editor. *Advanced oxidation processes for water and wastewater treatment.* London: IWA Publishing; 2004; 1-6.

[32] Gogate, PR; Pandit, AB. A review of imperative technologies for wastewater treatment I: oxidation technologies at ambient conditions. *Advances in Environmental Research,* 2004 8, 501-551.

[33] Gogate, PR; Pandit, AB. A review of imperative technologies for wastewater treatment II: hybrid methods. *Advances in Environmental Research,* 2004 8, 553-597.

[34] Glaze, WH; Kang, JW; Chapin, DH. The chemistry of water treatment processes involving ozone, hydrogen peroxide and ultraviolet irradiation. *Ozone Science and Engineering,* 1987 9, 335–352.

[35] Beltran, FJ. Ozone-UV radiation-hydrogen peroxide oxidation technologies. In: Tarr MA, editor. *Chemical degradation methods for wastes and pollutants - environmental and industrial applications.* New York: Marcel Dekker; 2003; 1-75.

[36] Rodriguez, M. Fenton and UV-VIS based advanced oxidation processes in wastewater treatment: Degradation, mineralization and biodegradability enhancement. PhD Thesis. Barcelona: University of Barcelona, 2003.

[37] Bigda, RJ. Consider Fenton's chemistry for wastewater treatment. *Chemical Engineering Progressing,* 1995 91, 62-66.

[38] Andreozzi, R; Vaprio, V; Insola, A; Marotta, R. Advanced oxidation processes (AOP) for water purification and recovery. *Catalysis Today,* 1999 53, 51-59.

[39] Grymonpre, DR; Finney, WC; Clark, RJ; Locke BR. Hybrid gas-liquid electrical discharge reactors for organic compound degradation, *Industrial and Engineering Chemistry Research,* 2004 43, 1975-1989.

[40] Locke, BR; Sato, M; Sunka, P; Hoffmann, MR, Chang; JS. Electrohydraulic discharge and nonthermal plasma for water treatment. *Industrial and Engineering Chemistry Research,* 2006 45 (3), 882-905.

[41] Fenton, HJH. On a new reaction of tartaric acid. *Chemical News,* 1876 33, 190.

[42] Fenton, HJH. Oxidation of tartaric acid in presence of iron. *Journal of Chemical Society,* 1894 65, 899-910.

[43] Tarr, MA. Fenton and modified Fenton methods for pollutant degradation. In: Tarr MA, editor. *Chemical degradation methods for wastes and pollutants - environmental and industrial applications.* New York: Marcel Dekker; 2003; 1-77.

[44] Xu, X-R; Li, H-B; Wang, W-H; Gu, J-D. Degradation of dyes in aqueous solutions by the Fenton process. *Chemosphere*, 2004 57, 595-600.

[45] Lucas, MS; Peres JA. Decolorization of the azo dye Reactive Black 5 by Fenton and photo-Fenton oxidation. *Dyes and Pigments,* 2006 71, 236-244.

[46] Meric, S; Selcuk, H; Gallo, M; Belgiorno V. Decolourisation and detoxifying of Remazol Red dye and its mixture using Fenton's reagent. *Desalination*, 2005 173, 239-248

[47] Chen, R; Pignatello, JJ. Role of quinone intermediates as electron shuttles in Fenton and photoassisted Fenton oxidations of aromatic compounds. *Environmental Science Technology,* 1997 31, 2399-2406.

[48] Kušić, H; Koprivanac, N; Lončarić Božić, A; Selanec, I. Photo-assisted Fenton type processes for the degradation of phenol: a kinetic study. *Journal of Hazardous Materials,* 2006 *in press.*

[49] Kusic, H; Loncaric Bozic, A; Koprivanac, N. Fenton type processes for minimization of organic content in coloured wastewaters: Part I: processes optimization. *Dyes and Pigments,* 2006 *in press.*

[50] Gallard, H; De Laat, J. Kinetics of oxidation of chlorobenzenes and phenyl-ureas by Fe(II)/$H_2O_2$ and Fe(III)/$H_2O_2$. Evidence of reduction and oxidation reactions of intermediates by Fe(II) and Fe(III). *Chemosphere*, 2001 42, 405-413.

[51] Neamtu, M; Yediler, A; Siminiceanu, I; Kettrup, A. Oxidation of commercial reactive azo dye aqueous solutions by the photo-Fenton and Fenton-like processes. *Journal of Photochemistry and Photobiology A: Chemistry,* 2003 161, 87–93.

[52] Kusic, H, Koprivanac, N; Srsan, L. Azo dye degradation using Fenton type processes assisted by UV irradiation: A kinetic study. *Journal of Photochemistry and Photobiology A: Chemistry,* 2006 181 (2-3), 195-202.

[53] Bergendahl, JA; Thies, TP. Fenton's oxidation of MTBE with zero-valent iron. *Water Research,* 2004 38, 327-334.

[54] Tang, WZ; Chen, RZ. Decolorization kinetics and mechanism of commercial dyes by $H_2O_2$/iron powder system. *Chemosphere*, 1996 32 (5), 947-958.

[55] Costa, RCC; Lelis, MFF; Oliveira, LCA; Fabris, JD; Ardisson, JD; Rios, RRVA; Silva, CN; Lago, RM. Novel active heterogeneous Fenton system based on $Fe_{3-x}M_xO_4$ (Fe, Co, Mn, Ni): The role of $M^{2+}$ species on the reactivity towards $H_2O_2$ reactions. *Journal of Hazardous Materials,* 2006 129, 171-178.

[56] Moura, FCC; Araujo, MH; Costa, RCC; Fabris, JD; Ardisson, JD; Macedo, WAA; Lago, RM. Efficient use of Fe metal as an electron transfer agent in a heterogeneous Fenton system based on $Fe^0$/$Fe_3O_4$ composites. *Chemosphere*, 2005 60, 1118-1123.

[57] He, J; Ma, W; Song, W; Zhao, J; Qian, X; Zhang, S; Yu JC. Photoreaction of aromatic compounds at α-FeOOH/$H_2O$ interface in the presence of $H_2O_2$: evidence for organic-goethite surface complex formation. *Water Research,* 2005 39, 119-128.

[58] Phu, NH; Hoa, TTK; Tan, NV; Thang HV; Ha, PL. Characterization and activity of Fe-ZSM-5 catalysts for the total oxidation of phenol in aqueous solutions. *Applied Catalysis, B: Environmental, 2001* 34, 267-275.

[59] Kušić, H; Koprivanac, N; Selanec, I. Fe-exchanged zeolite as the effective heterogeneous Fenton-type catalyst for the organic pollutant minimization: UV irradiation assistance, *Chemosphere*, 2006 *in press*

[60] Catrinescu, C; Neamtu, M; Yediler, A; Macoveanu, M; Kettrup, A. Catalytic wet peroxide oxidation of an azo dye, Reactive Yellow 84, over Fe-exchanged ultrastable Y zeolite. *Environmental Engineering and Management Journal,* 2002 1, 177-186.

[61] Catrinescu, C; Teodosiu, C, Macoveanu, M; Miehe-Brendlé, J; Le Dred, R. Catalytic wet peroxide oxidation of phenol over Fe-exchanged pillared beidellite. *Water Research,* 2003 37, 1154-1160.

[62] Ly, X; Xu, Y; Ly, K; Zhang, G. Photo-assisted degradation of anionic and cationic dyes over iron(III)-loaded resin in the presence of hydrogen peroxide. *Journal of Photochemistry and Photobiology, A: Chemistry,* 2005 173, 121-127.

[63] Verma, P; Baldrian, P; Nerud, F. Decolorization of structurally different synthetic dyes using cobalt(II)/ascorbic acid/hydrogen peroxide system. *Chemosphere*, 2003 50, 975-979

[64] Kavitha, V; Palanivelu, K. The role of ferrous ion in Fenton and photo-Fenton processes for the degradation of phenol. *Chemosphere*, 2004 55, 1235-1243.

[65] www.h2o2.com

[66] Arslan, I; Balcioglu, IA. Degradation of Remazol Black B dye and its simulated dyebath wastewater by advanced oxidation processes in heterogeneous and homogeneous media. *Coloration Technology,* 2001 117, 38-42.

[67] Hsueh, CL; Huang, YH; Wang, CC; Chen; CY. Degradation of azo dyes using low iron concentration of Fenton and Fenton-like system. *Chemosphere*, 2005 58 (10), 1409-1414.

[68] Carneiro, PA; Pupo Nogueira, RF; Zanoni, MVB. Homogeneous photodegradation of C.I. Reactive Blue 4 using a photo-Fenton process under artificial and solar irradiation. *Dyes and Pigments*, 2006 *in press*.

[69] Hsueh, C-L; Huanga, Y-H; Wang, C-C; Chena, C-Y. Photoassisted Fenton degradation of nonbiodegradable azo-dye (Reactive Black 5) over a novel supported iron oxide catalyst at neutral pH. *Journal of Molecular Catalysis A: Chemical,* 2006 245, 78–86.

[70] Muruganandham, M; Swaminathan, M. Decolourisation of Reactive Orange 4 by Fenton and photo-Fenton oxidation technology. Dyes and Pigments, 2004 63, 315-321.

[71] Stylidi, M; Kondarides, DI; Verykios, XE. Pathways of solar light-induced photocatalytic degradation of azo dyes in aqueous $TiO_2$ suspensions. *Applied Catalysis B: Environmental,* 2003 40, 271-286.

[72] Stefan, MI. UV photolysis: background. In: Parsons SA, editor. *Advanced oxidation processes for water and wastewater treatment.* London: IWA Publishing; 2004; 7-48.

[73] Legrini, O; Oliveros, E; Braun, AM. Photochemical process for water treatment. *Chemical Review,* 1993 93 (2), 671-698.

[74] Arslan, I; Balcioglu, IA; Bahnemann, D. Advanced chemical oxidation of reactive dyes in simulated dyehouse effluents by ferrioxalate–Fenton/UV-A and $TiO_2$/UV-A processes. *Water Research,* 2001 37, 3061-3069.

[75] Muruganandham, M; Swaminathan, M. Solar driven decolourisation of Reactive Yellow 14 by advanced oxidation processes in heterogeneous and homogeneous media. *Dyes and Pigments,* 2007 72 (2), 137-143.

[76] Bali, U; Catalkaya, E; Sengul F. Photodegradation of Reactive Black 5, Direct Red 28 and Direct Yellow 12 using UV, UV/$H_2O_2$ and UV/$H_2O_2$/$Fe^{2+}$: a comparative study. *Journal of Hazardous Materials,* 2004 B114, 159–166.

[77] Behnajady, MA; Modirshahla, N; Fathi H. Kinetics of decolorization of an azo dye in UV alone and UV/$H_2O_2$ processes. *Journal of Hazardous Materials,* 2006 133 (1-3), 226-232.

[78] Peternel, I; Koprivanac, N; Kusic, H. UV-based processes for reactive azo dye mineralization. *Water Research,* 2006 40 (3), 525-532.

[79] Kušić, H; Koprivanac, N; Lončarić Božić, A; Papić, S; Peternel, I; Vujević, D. Reactive dye degradation by a photochemical AOP – development of a kinetic model for UV/$H_2O_2$ process. *Chemical and Biochemical Engineering Quarterly,* 2006 *in press*.

[80] Neppolian, B; Choi, HC; Sakthivel, S; Arabindoo, B; Murugesan, V. Solar/UV-induced photocatalytic degradation of three commercial textile dyes. *Journal of Hazardous Materials,* 2002 B89, 303-317.

[81] Tuhkanen, T. UV/$H_2O_2$ process. In: Parsons SA, editor. *Advanced oxidation processes for water and wastewater treatment.* London: IWA Publishing; 2004; 86-110.

[82] Hunt, JP; Taube, H. The photochemical decomposition of hydrogen peroxide, Quantum yields, transfer and fractionation effects. *Journal of American Chemical Society,* 1952 74, 5999-6002.

[83] Eckenfelder, WW. *Industrial Water Pollution Control.* 2nd Edition. New York: McGraw-Hill; 1989.

[84] Crittenden, C; Hu, S; Hand, DW; Green, SA. A kinetic model for $H_2O_2$/UV process in a completely mixed batch reactor. *Water Research,* 1999 33 (10), 2315-2328.

[85] De, AK; Chaudhuri, B; Bhattacharjee, S; Dutta, BK. Estimation of $OH^•$ radical reaction rate constants for phenol and chlorinated phenols using UV/$H_2O_2$ photo-oxidation. *Journal of Hazardous Materials,* 1999 64 (1), 91-104.

[86] Guittonneau, S; De Laat, J; Duguet, JP; Bonnel, C; Dore, M. Oxidation of parachloronitrobenzene in dilute Aqueous solution by $O_3$ + UV and $H_2O_2$ + UV: A comparative study, *Ozone Science and Engineering,* 1990 12, 73-94.

[87] Kusic, H; Koprivanac, N; Loncaric Bozic, A. Minimization of organic pollutant content in aqueous solution by means of AOPs: UV- and ozone–based technologies. *Chemical Engineering Journal,* 2006 *in press*

[88] Mohey El-Dein, A; Libra, JA; Wiesmann U. Mechanism and kinetic model for the decolorization of the azo dye Reactive Black 5 by hydrogen peroxide and UV radiation. *Chemosphere*, 2003 52, 1069–1077.

[89] Muruganandham, M; Swaminathan M. Photochemical oxidation of reactive azo dye with UV–$H_2O_2$ process. *Dyes and Pigments,* 2004 62, 269–275.

[90] Colonna, GM; Caronna, T; Marcandalli, B. Oxidative degradation of dyes by ultraviolet radiation in the presence of hydrogen peroxide. *Dyes and Pigments,* 1999 41, 211-220.

[91] Ince, N. Critical effect of hydrogen peroxide in photochemical dye degradation. *Water Research,* 1999 33, 1080-1084.

[92] Shu, H; Chang, M. Decolorization effects of six azo dyes by $O_3$, UV/$O_3$ and UV/$H_2O_2$ processes. *Dyes and Pigments*, 2005 65**,** 25-31.

[93] Wu, C-H; Chang, C-L. Decolorization of Reactive Red 2 by advanced oxidation processes: Comparative studies of homogeneous and heterogeneous systems. *Journal of Hazardous Materials,* 2006 B128, 265–272.

[94] Mills, A; Lee, S-K. Semiconductor photocatalysis. In: Parsons SA, editor. *Advanced oxidation processes for water and wastewater treatment.* London: IWA Publishing; 2004; 137-166.

[95] Pichat, P. Photocatalytic degradation of pollutants in water and air:basic concepts and applications. In: Tarr MA, editor. *Chemical degradation methods for wastes and pollutants - environmental and industrial applications.* New York: Marcel Dekker; 2003; 77-119.

[96] Carneiro, PA; Osugi ME; Sene, JJ; Anderson, MA; Boldrin Zanoni, MV. Evaluation of color removal and degradation of a reactive textile azo dye on nanoporous $TiO_2$ thin-film electrodes. *Electrochimica Acta,* 2004 49, 3807–3820.

[97] Goncalves, MST; Pinto, EMS; Nkeonye, P; Oliveira-Campos, AMF. Degradation of C.I. Reactive Orange 4 and its simulated dyebath wastewater by heterogeneous photocatalysis. *Dyes and Pigments,* 2005 64, 135-139.

[98] Reutergradh, LB; Iangpashuk, M. Photocatalytic decolorization of reactive azo dye: a Comparison between $TiO_2$ and CdS photocatalysis, *Chemosphere*, 1997 35, 585-596.

[99] Daneshvar, N; Salari, D; Khataee, AR. Photocatalytic degradation of azo dye Acid Red 14 in water on ZnO as an alternative catalyst to $TiO_2$. *Journal Photochemistry and Photobiology A: Chemistry,* 2004 162, 317-322.

[100] Sakthivel, S; Neppolian, B; Shankar, MV; Arabindoo, B; Palanichamy, M, Murugesan, V. Solar photocatalytic degradation of azo dye: Comparison of photocatalytic efficiency of ZnO and $TiO_2$. *Solar Energy Materials and Solar Cells,* 2003 77, 65-82.

[101] Vinogdopal, K; Wynkoop, DE; Kamat, PV. Environmental chemistry on semiconductor surfaces: Photosensitized degradation of a textile azo dye, Acid Orange 7, on the $TiO_2$ particles using visible light. *Environmental Science and Technology,* 1996 30, 1660-1666.

[102] Bizani, E; Fytianos, K; Poulios I; Tsiridis V. Photocatalytic decolorization and degradation of dye solutions and wastewaters in the presence of titanium dioxide. *Journal of Hazardous Materials,* 2006 *in press.*

[103] Muruganandham, M; Swaminathan M. $TiO_2$–UV photocatalytic oxidation of Reactive Yellow 14: Effect of operational parameters. *Journal of Hazardous Materials*, 2006 *in press.*

[104] Peternel, I; Koprivanac, N; Loncaric Bozic, A; Kusic, H. Comparative study of UV/$TiO_2$, UV/ZnO and UV/Fenton processes for the organic reactive dye in aqueous solution. *Waste Management,* 2006 *submitted*

[105] Akyol A; Yatmaz, HC; Bayramoglu, M. Photocatalytic decolorization of Remazol Red RR in aqueous ZnO suspensions. *Applied Catalysis B: Environmental,* 2004 54, 19–24.

[106] www.medicina.hr

[107] www.tifm.com/uvbozone

[108] Langlais, B; Reckhow, DA; Brink, DR. *Ozone in water treatment; application and engineering.* Boca Raton: Lewis Publishers; 1991.

[109] Tomiyasu, H; Fukutomi, H; Gordon, G. Kinetics and mechanism of ozone decomposition in basic aqueous solution. *Inorganic Chemistry,* 1985 24 (19), 2962-2966.

[110] Mathews, AP; Panda, KK; Ananthi, S; Padmanabhan, K. Mass transfer and oxidation kinetics in an in situ ozone generator. *Water Science and Technology,* 2004 49, 13-18.

[111] Arslan-Alaton, I; Dogruel, S; Baykal, E; Gerone, G. Combined chemical and biological oxidation of penicillin formulation effluent. *Journal of Environmental Management,* 2004 73 (2), 155-163.

[112] Glaze, W; Kang, J. Advanced oxidation processes. Test of a kinetic model for the oxidation of organic compounds with ozone and hydrogen peroxide in a semi batch reactor, *Industrial and Engineering Chemistry Research,* 1989 28, 1580-1587.

[113] Wu, J; Wang, T. Ozonation of aqueous azo dye in semi-batch reactor. *Water Research,* 2001 35 (4), 1093-1099.

[114] Peralta–Zamora, P; Kunz, A; Gomes de Moreas, S; Pelegrini, R; de Campos Moleiro, P; Reyes, J; Durán, N. Degradation of reactive dyes I. a comparative study of ozonation, enzymatic and photochemical processes. *Chemosphere*, 1999 38 (4), 835-852.

[115] Arslan-Alaton, I; Bacioglu, IA; Bahnemann, DW. Advanced oxidation of a reactive dyebath effluent: comparison of $O_3$, $H_2O_2$/UV-C and $TiO_2$/UV-A processes. *Water Research,* 2002 36, 1143-1154.

[116] Arslan, I; Bacioglu, IA. Effect of common reactive dye auxiliaries on the ozonation of dyehouse effluents contaning vinylsulphone and aminochlorotriazine dyes. *Desalination,* 2000 130, 61-71.
[117] Kušić, H; Lončarić Božić, A; Koprivanac, N; Peternel, I; Vujević, D; Papić, S. Comparative study of several ozone based processes for reactive dye degradation. In: Kruithof J, editor. Proceedings of 17th world congress & exhibition: Ozone & Related Oxidants, Innovative & Current Technologies. *Strasbourg: International Ozone Association;* 2005; VII.3.24.1-10.
[118] Ince, NH; Tezcanli, G. Reactive dyestuff degradation by combined sonolysis and ozonation. *Dyes and Pigments,* 2001 49 (3), 145-153.
[119] Lukes, P. *Water treatment by pulsed streamer corona discharge*. PhD Thesis. Prague: Institute of Plasma Physics AS CR; 2001.
[120] Joshi, AA; Locke, BR; Arce, P; Finney, WC. Formation of hydroxyl radicals, hydrogen peroxide and aqueous electrons by pulsed streamer corona discharge in aqueous solution. *Journal of Hazardous Materials,* 1995 41, 3-30.
[121] Sunka, P; Babicky, V; Clupek, M; Lukes, P; Simek, M; Schmidt, J; Cernak, M. Generation of chemically active species by electrical discharges in water. *Plasma Sources Science and Technology,* 1999 8 (2), 258-265.
[122] Clements, JS; Sato, M; Davis, RH. Preliminary investigation of prebreakdown phenomena and chemical reactions using a pulsed high-voltage discharge in water. IEEE Trans. *Industrial Applications,* 1987 IA-23, 224-235.
[123] Sato, M; Sun, B; Ohshima, T; Sagi, Y. Characteristics of active species and removal of organic compounds by a pulsed corona discharge in water. *Journal of Advanced Oxidation Technologies,* 1999 4, 339-342.
[124] Sun, B; Sato, M; Clements, JS. Optical study of active species produced by a pulsed streamer corona discharge in water. *Journal of Electrostatics,* 1997 39, 189-202.
[125] Grymonpre, DR. *An experimental and theoretical analysis of phenol degradation by pulsed corona discharge*. PhD Thesis. Tallahassee: Florida State University; 2001.
[126] van Craeynest, K; van Langenhove, H; Stuetz, RM. AOPs for VOCs and odour treatment: Non-thermal palsma. In: Parsons SA, editor. *Advanced oxidation processes for water and wastewater treatment.* London: IWA Publishing; 2004; 275-301.
[127] Lukes, P; Appleton, AT; Locke, BR. Hydrogen peroxide and ozone formation in hybrid gas-liquid electrical discharge reactors. *IEEE Trans. Industrial. Applications,* 2004 40 (1), 60–67.
[128] Sugiarto, AT; Ohshima, T; Sato, M. Advanced oxidation processes using pulsed streamer corona discharge in water. *Thin Solid Films*, 2002 407, 174-178.
[129] Sunka, P; Babicky, V; Clupek, M; Fuciman, M; Lukes, P; Simek, M; Benes, J; Locke, BR; Majcherova, Z. Potential applications of pulse electrical discharges in water, *Acta Physica Slovaca,* 2004 54 (2), 135-145.
[130] Loncaric Bozic, A; Koprivanac, N; Sunka, P, Clupek, M; Babicky, V. Organic synthetic dye degradation by modified pinhole discharge. *Czechoslovak Journal of Physics,* 2004 54C, C958-C963.
[131] Koprivanac, N; Kusic, H; Vujevic, D; Peterenel, I; Locke, BR. Influence of iron on degradation of organic dyes in corona. *Journal of Hazardous Materials*, 2005 B117, 113-119.

[132] Kusic, H; Koprivanac, N; Peternel, I; Locke BR. Hybrid gas/liquid electrical discharge reactors with zeolites for colored wastewater degradation. *Journal of Advanced Oxidation Technologies, 2005* 8 (2), 172-181.

[133] Peternel, I; Kusic, H; Koprivanac, N; Locke, BR. The roles of ozone and zeolite on reactive dye degradation in electrical discharge reactors. *Environmental Technology,* 2006 27 (5) 545-558.

[134] Mason, TJ; Petrier, C. Ultrasound processes. In: Parsons SA, editor. *Advanced oxidation processes for water and wastewater treatment.* London: IWA Publishing; 2004; 185-208.

[135] Destaillats, H; Hoffmann, MR; Wallace, HC. Sonochemical degradation of polutants. In: Tarr MA, editor. *Chemical degradation methods for wastes and pollutants - environmental and industrial applications.* New York: Marcel Dekker; 2003; 201-233.

[136] Vajnhandl, S; Majcen Le Marechal, A. Ultrasound in textile dyeing and the decolouration/mineralization of textile dyes. *Dyes and Pigments,* 2005 65, 89-101.

[137] Tezcanli-Guyer, G. Ince, NH. Degradation and toxicity reduction of textile dyestuff by ultrasound. *Ultrasonics Sonochemistry,* 2003 10, 235–240.

[138] Rehorek, A; Tauber M; Gubitz G. Application of power ultrasound for azo dye degradation. *Ultrasonics Sonochemistry,* 2004 11, 177–182.

[139] Cooper, WJ; Gehringer, P; Pikaev, AK; Kurcuz, CN; Mincher, BJ. Radiation processes. In: Parsons SA, editor. *Advanced oxidation processes for water and wastewater treatment.* London: IWA Publishing; 2004; 209-248.

[140] Dajka, K; Takacs, E; Solpan, D; Wojnarovits, L; Guven, O. High-energy irradiation treatment of aqueous solutions of C.I. Reactive Black 5 azo dye: pulse radiolysis experiments. *Radiation Physics and Chemistry,* 2003 67, 535–538.

[141] Zhang, S-J; Yu, H-Q; Li, Q-R. Radiolytic degradation of Acid Orange 7: A mechanistic study. *Chemosphere*, 2005 61, 1003–1011.

[142] Brillas, E; Cabot, P-L; Casado, J. Electrchemical methods for degradation of organic pollutants in aqueous media. In: Tarr MA, editor. *Chemical degradation methods for wastes and pollutants - environmental and industrial applications.* New York: Marcel Dekker; 2003; 235-304.

[143] Catanho, M; Malpass, GRP; Motheo AJ. Photoelectrochemical treatment of the dye reactive red 198 using DSA electrodes. *Applied Catalysis B: Environmental,* 2006 62, 193–200.

[144] Awad, HS; Abo Galwa N. Electrochemical degradation of Acid Blue and Basic Brown dyes on Pb/$PbO_2$ electrode in the presence of different conductive electrolyte and effect of various operating factors. *Chemosphere*, 2005 61, 1327–1335.

[145] Carneiro, PA; Osugi, ME; Fugivara, CS; Boralle, N; Furlan, M; Zanoni, MVB. Evaluation of different electrochemical methods on the oxidation and degradation of Reactive Blue 4 in aqueous solution. *Chemosphere*, 2005 59, 431–439.

[146] Kusic, H; Lonacaric Bozic, A; Koprivanac, N; Papic, S. Fenton type processes for minimization of organic content in coloured wastewaters: Part II: combination with zeolites. Dyes and Pigments, 2006 *in press*

[147] Papić, S; Koprivanac, N; Lončarić Božić, A; Vujević, D; Dragičević Kučar, S; Kušić, H; Peternel, I. AOPs in azo dye wastewater treatment. Water Environment Research, 2006 78 (6), 572-579.

[148] Tomlinson, AAG. *Modern zeolites structure and function in detergents and petrochemicals.* Zurich; Trans Tech Publications; 1998.

[149] www.bza.org/zeolites

[150] www.zeolyst.com

[151] Nicole, I; De Laat, J; Dore, M, Duguet, JP; Bonnel, C. Utilisation du rayonnement ultraviolet dans le traitement des eaux: mesure du flux photonique par actinometrie chimique au peroxyde d'hydrogene: Use of U.V. radiation in water treatment: measurement of photonic flux by hydrogen peroxide actinometry. *Water Research,* 1990 24 (2), 157-168.

[152] APHA Standard methods for the examination of water and wastewater treatment. 20$^{th}$ Edition. Washington DC: American Public Health Association; 1998.

[153] Pretsch, E; Seibl, J; Simon, W. *Tabellen zur strukturaufklarung organischer verbindungen mit spektroskopischen methoden.* Berlin: Springer Verlag; 1981.

[154] Alaton, IA; Balcioglu, IA; Bahnemann, D. Advanced oxidation of reactive dyebath effluent: comparison of $O_3$, $H_2O_2$/UV-A and $TiO_2$/UV-A processes. *Water Research,* 2002 36, 1143-1154.

[155] Gutowska, A; Kaluzna-Czaplinska, J; Jozwiak WK. Degradation mechanism of Reactive Orange 113 dye by $H_2O_2/Fe^{2+}$ and ozone in aqueous solution. *Dyes and Pigments,* 2006 *in press.*

[156] Feng, W; Nansheng, D; Helin H. Degradation mechanism of azo dye C. I. reactive red 2 by iron powder reduction and photooxidation in aqueous solutions. *Chemosphere,* 2000 41, 1233-1238.

[157] Kawai, T; Tsutsumi, K. Adsorption characteristic of surfactants and phenol on modified zeolites from their aqueous solutions. *Colloid Polymeric Science,* 1995 273, 787-792.

[158] Kušić, H; Koprivanac, N; Locke, BR. Decomposition of phenol by hybrid gas/liquid electrical discharge reactors with zeolite catalysts. *Journal of Hazardous Materials,* 2005 125, 190-200.

[159] Daneshvar, N; Salari, D; Khataee AR. Photocatalytic degradation of azo dye acid red 14 in water: investigation of the effect of operational parameters. *Journal of Photochemistry and Photobiology A: Chemistry,* 2003 157, 111-116.

[160] Bilgi, S; Demir, C. Identification of photooxidation degradation products of C.I. Reactive Orange 16 dye by gas chromatography - mass spectrometry. *Dyes and Pigments,* 2005 66, 69-76.

[161] Hoigné, J; Bader, H. Rate constants of reactions of ozone with organic and inorganic compounds in water—II: Dissociating organic compounds. *Water Research,* 1983 17 (2), 185-194.

In: Hazardous Materials and Wastewater
Editor: Allison A. Lewinsky, pp. 201-224
ISBN 1-60021-257-3


*Chapter 6*

# UTILISATION OF FLY ASH BASED ADSORBENTS FOR WASTEWATER TREATMENT

***Shaobin Wang***
Department of Chemical Engineering, Curtin University of Technology, GPO Box U1987, Perth, WA 6845, Australia

## ABSTRACT

Wastewater from various industries and domestic households has been a big environmental problem for human society. Wastewater discharged from industries and homes includes a large quantity of various toxic substances such as inorganic heavy metals and organic compounds. Treatment of wastewater involves several processes and several technologies have been developed such as filtration, chemical precipitation, advanced oxidation, reverse osmosis or adsorption. Among these processes, the adsorption with the selection of a suitable adsorbent can be a simple and effective technique for the removal of inorganic and organic pollutants from wastewater. Activated carbon is usually used as an effective adsorbent. However, due to the high cost and regeneration problem, other low-cost and efficient adsorbents are required. Fly ash as a byproduct generated from thermal power plants is solid waste but available in large quantity. Utilisation of the fly ash will help reduce the environmental burden and enhance economic benefit. In this paper, we report a comprehensive investigation employing various treatment techniques to convert fly ash to various effective adsorbents for removal of organic dyes and heavy metals from wastewater. Fly ash is usually composed of inorganic minerals and unburned carbon. In our investigation, the mineral section of fly ash has been treated by physical and chemical methods to convert it to zeolite or inorganic polymer. Unburned carbon in the fly ash was also used as precursor for production of activated carbon for wastewater treatment. The adsorption kinetics and isotherms of these adsorbents (Fly ash, zeolite, inorganic polymer and unburned carbon) in aqueous solution for uptake of dyes and heavy metals were investigated. The results indicate that modified fly ashes exhibit comparable adsorption capacity as activated carbon.

# 1. INTRODUCTION

Fly-ash is a by-product produced by all coal fired power stations and other biomass combustion. It is estimated that 349 Mt was produced worldwide in 2000. In Australia, annual production of fly ash has been steadily increased to 13 Mt in 2003 (Wang and Wu, 2006). The disposal of fly ash from coal fired power stations causes significant economic and environmental problems. A relatively small percentage of the material finds application as an ingredient in cement and other construction products, but a vast majority of the material generated each year is held in ash dams or similar dumps. This unproductive use of land and the associated long-term financial burden of maintenance have led to realisation that alternative uses for fly ash as a value-added product beyond incorporation in construction materials are needed. Utilisation of fly ash in such areas as novel materials, waste management, recovery of metals and agriculture have been investigated (Iyer and Scott 2001).

Effluents generated in dye houses of the textile industry contain considerable amounts of unexhausted dye (about 20 – 40% of input dye) in addition to other impurities like salt, surfactant, and heavy metals. But the discharge of such coloured dye effluents is aesthetically objectionable. Coloured effluents also interfere with the transmission of sunlight, and upset the biological metabolism processes, which causes the direct destruction of aquatic communities present in ecosystems (Guaratini and Zanoni, 2000). Heavy metals from industrial processes are also among the most important pollutants in source and treated water, and are becoming a severe public health problem (Babel and Kurniawan, 2003).

Nowadays, several technologies have been developed such as filtration, chemical precipitation, advanced oxidation, reverse osmosis or adsorption. Adsorption is the most popular method for wastewater treatment due to its easy and inexpensive operation, but there are certain problems using activated carbon as adsorbents due to the high cost of use and regeneration. The adsorption process will provide an attractive technology if the sorbent is inexpensive and ready for use.

Utilisation of industrial solid wastes for the treatment of wastewater from another industry could be helpful not only to environment in solving the solid waste disposal problem, but also to the economy. Although a lot of investigations have been done in employing fly ash as adsorbent, those researches show that adsorption behaviour of fly ash varies with the source of fly ash and its properties. Moreover, fly ash presents low adsorption capacity (Wang and Wu, 2006). To enhance the adsorption capacity, it is necessary to modify the properties of fly ash. In this paper, we present a systematic investigation on utilisation of fly ash as low cost adsorbent for wastewater treatment including removal of organic dyes and heavy metals. We have investigated physical and chemical treatments of fly ash to convert it into high effective adsorbents including crystalline zeolites and amorphous geopolymer. The adsorption kinetics and isotherm for organic dyes and some heavy metals are also studied.

# 2. COMPARISON OF VARIOUS FLY ASHES FOR DYE ADSORPTION

It is known that fly ash from various sources has different characteristics in chemical composition and physical properties. Table 1 gives the textural properties of five fly ash samples obtained from power stations in Western Australia. It is shown that the surface area and pore volume varies from 4.0 to 21 $m^2/g$ and 0.0049 – 0.020 $cm^3/g$, respectively. The unburned carbon content is also different. Normally, higher unburned carbon in fly ash will not be good for cement production, however, the porous carbon will be beneficial for adsorption.

**Table 1. Textural property of various fly ash samples**

| Fly ash | $S_{BET}$ ($m^2/g$) | V($cm^3/g$) | Unburned char content (%) |
|---|---|---|---|
| WRFL | 10.7 | 0.011 | 12.95 |
| M7LL | 13.2 | 0.013 | 2.20 |
| M7FL | 7.0 | 0.0081 | 2.74 |
| M5LL | 21.0 | 0.020 | 4.84 |
| Unit 5 | 15.6 | 0.016 | 5.20 |

The mineral phases of these fly ashes were determined by X-ray diffraction (XRD), which is presented in Figure 1. One can see that the XRD patterns for all samples are similar. The minerals include quartz, mullite, with minor hematite and magnetite.

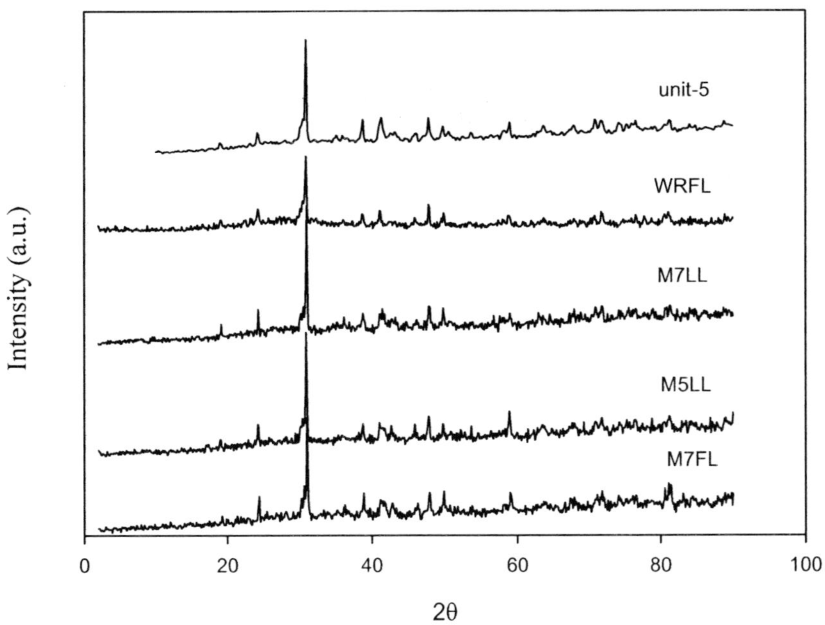


Figure 1. XRD patterns of various fly ash samples.

The dynamic adsorption of a typical dye, methylene blue (MB), on the fly ash samples is conducted at 30 °C and the results are given in Figure 2. It is seen that the fly ash samples present varying adsorption behaviour. The adsorption shows a high rate at the initial 25 h and then approaches to equilibrium. The adsorption equilibrium follows an order of M5LL > Unit 5 > M7LL > WRFL > M7FL. This order is the same as the surface area and pore volume, which suggests that the surface area and pore volume determine the adsorption capacity. The highest adsorption equilibrium can be 2.1 x $10^{-5}$ mol/g for M5LL while the lowest adsorption is only 5 x $10^{-6}$ mol/g for M7FL.

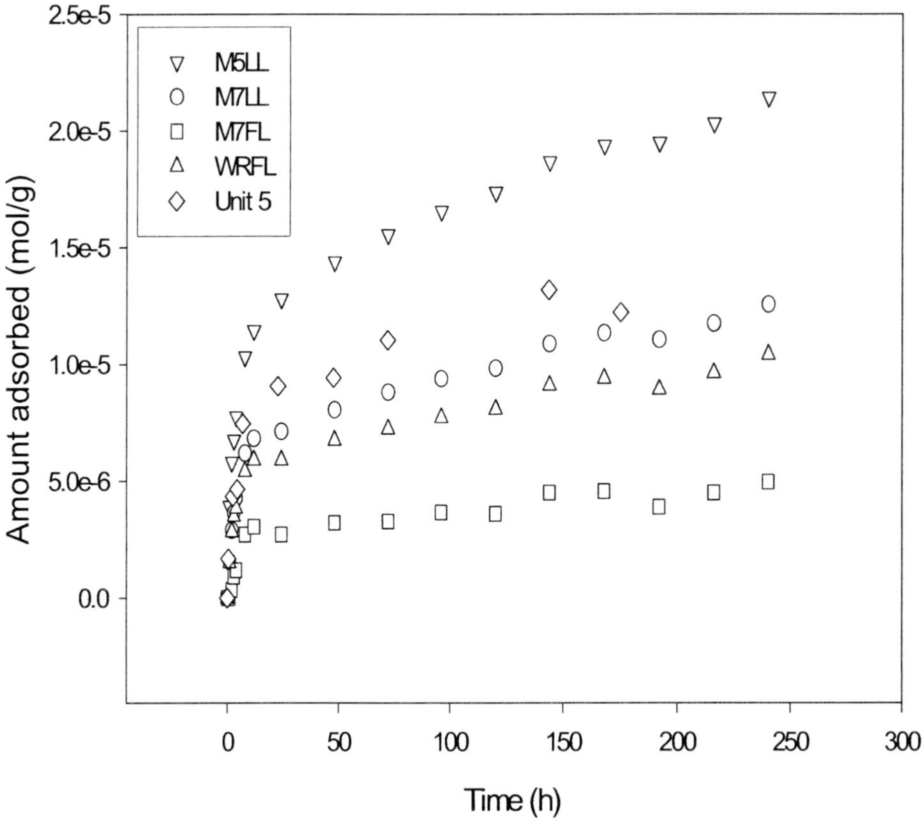


Figure 2. Dynamic adsorption of methylene blue on various fly ashes.

Some investigations using fly ash for dye adsorption have been conducted. Table 2 lists the previous results of adsorption capacity for various dyes. It is seen that dye adsorption depends on source of fly ash and type of dye. The adsorption capacity of dyes varies from 2.3 x $10^{-6}$ – 2.4 x $10^{-4}$ mol/g. Our results are much closer to the previous data.

From the dynamic adsorption curves, the adsorption kinetics can be obtained. There are several kinetic models applied for solid-liquid adsorption. The two important models are the Lagergren first-order model (Ho, 2004) and Ho's pseudo second-order model (Ho and McKay, 1999).

The first-order kinetic equation may be expressed as:

$$\frac{dq_t}{dt} = k_1(q_e - q_t) \quad (1)$$

where $q_e$ and $q_t$ are the sorption capacity (mol/g) of dye at equilibrium and at a time t, respectively, and $k_1$ is the rate constant for pseudo-first order sorption (L/min). After integration and applying boundary conditions, viz. that the initial conditions are $(q_e - q_t) = 0$ at t = 0, equation (1) becomes:

$$\ln(q_e - q_t) = \ln q_e - k_1 t \tag{2}$$

**Table 2. Comparison of dye adsorption on fly ash reported in the literatures**

| Dye | Fly ash type | Adsorption capacity (mol/g) | Reference |
|---|---|---|---|
| Methylene blue | Coal FA | $14.4 \times 10^{-5}$ | (Viraraghavan and Ramakrishna, 1999) |
| Crystal violet | Coal FA | $9.76 \times 10^{-5}$ | (Mohan et al., 2002) |
| Rosaniline hydrochloride | Coal FA | $1.35 \times 10^{-5}$ | (Mohan et al., 2002) |
| Methylene blue | FA-F | $1.89 \times 10^{-5}$ | (Janos et al., 2003) |
| Rhodamine B | FA-F | $1.15 \times 10^{-5}$ | (Janos et al., 2003) |
| Egacid orange II | FA-F | $2.364 \times 10^{-4}$ | (Janos et al., 2003) |
| Egacid Red G | FA-F | $1.405 \times 10^{-4}$ | (Janos et al., 2003) |
| Egacid yellow G | FA-F | $5.2 \times 10^{-5}$ | (Janos et al., 2003) |
| Midlon Black VL | FA-F | $3.3 \times 10^{-5}$ | (Janos et al., 2003) |
| Acid Blue 29 | FA | $3.25 \times 10^{-6}$ | (Ramakrishna and Viraraghavan, 1997) |
| Acid Blue 9 | FA | $5.43 \times 10^{-6}$ | (Ramakrishna and Viraraghavan, 1997) |
| Acid Red 91 | FA | $2.34 \times 10^{-6}$ | (Ramakrishna and Viraraghavan, 1997) |
| Acid Red 1 | FA | $7.26 \times 10^{-6}$ | (Ramakrishna and Viraraghavan, 1997) |
| Congo red | FA-C | $4.47 \times 10^{-5}$ | (Acemioglu, 2004) |
| Methylene blue | FA-F | $3.47 \times 10^{-6}$ | (Woolard et al., 2002) |
| Methylene blue | FA-F | $1.4 \times 10^{-5}$ | (Wang et al., 2005) |
| Methylene blue | FA-$HNO_3$ | $2.2 \times 10^{-5}$ | (Wang et al., 2005) |

Eq.(2) can be transformed to a nonlinear form

$$q_t = q_e(1 - e^{-k_1 t}) \tag{3}$$

The pseudo-second order equation can be written as:

$$\frac{dq_t}{dt} = k_2(q_e - q_t)^2 \tag{4}$$

where $q_e$ and $q_t$ are the sorption capacity (mol/g) of dye at equilibrium and at a time t, respectively, and $k_2$ is the rate constant for pseudo-second order sorption (g/mol h). For the boundary conditions t = 0 to t = t and $q_t = 0$ to $q_t = q_t$, the integrated form of equation (4) becomes:

$$\frac{1}{q_e - q_t} = \frac{1}{q_e} + k_2 t \tag{5}$$

which is the integrated rate law for a pseudo-second order reaction. This has a linear form:

$$\frac{t}{q_t} = \frac{1}{k_2 q_e^2} + \frac{1}{q_e} t \tag{6}$$

and nonlinear form

$$q_t = \frac{q_e^2 k_2 t}{1 + q_e k_2 t} \tag{7}$$

Figure 3 shows a comparison of kinetic models for MB adsorption on M5LL fly ash and the calculated kinetic parameters are presented in Table 3. As shown that the pseudo second-order kinetic model fits the experimental data well while the pseudo first-order does not exhibit good fitting for the whole period.

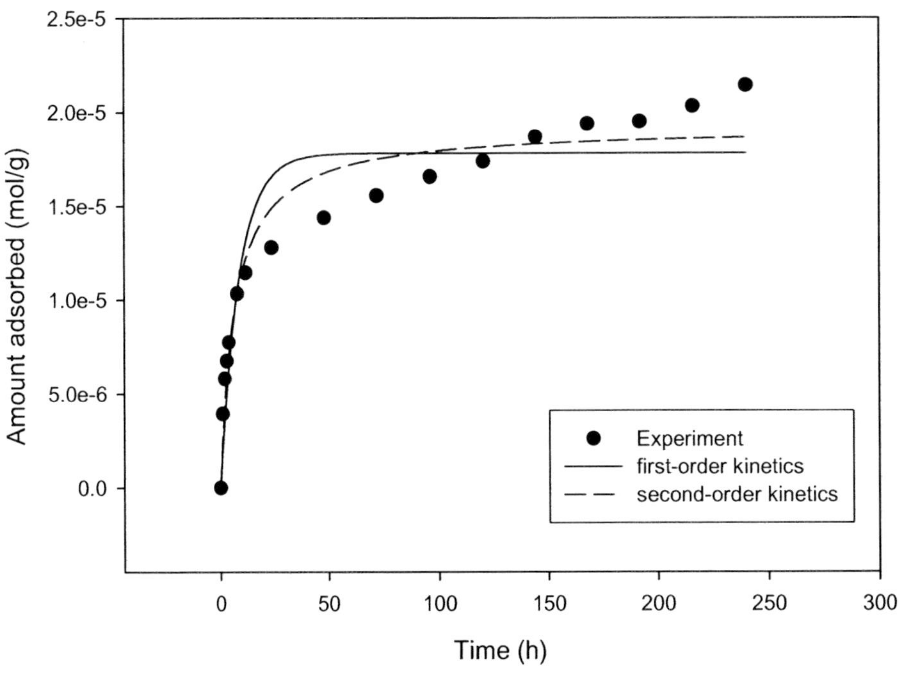


Figure 3. Comparison of kinetic model in dye adsorption

**Table 3. Kinetic parameters of MB adsorption on M5LL fly ash**

| Pseudo first-order kinetics | | | Pseudo second-order kinetics | | |
|---|---|---|---|---|---|
| $k_1$ ($h^{-1}$) | $q_1$(mol/g) | $R^2$ | $k_2$ (g $mol^{-1}$ $h^{-1}$) | $q_2$(mol/g) | $R^2$ |
| 0.110 | $1.78 \times 10^{-5}$ | 0.887 | $7.45 \times 10^{3}$ | $1.92 \times 10^{-5}$ | 0.946 |

For the kinetics of dye adsorption on fly ash, few studies have been carried out. It has been found that the adsorption of dye on fly ash can be described using pseudo first order and pseudo second-order kinetics. Mohan et al.(Mohan et al., 2002) investigated the adsorption of two basic dyes on fly ash and found the adsorption following the first-order rate kinetics. While, Acemioglu reported that the adsorption of Congo red on calcium-rich fly ash obeyed the pseudo-second-order kinetic model (Acemioglu, 2004).

It is known that adsorption is affected by adsorbent loading, solution pH, and temperature. In our investigations, we found that the uptake of the dye increases as the amount of adsorbent material increases and reaches the maximum at the loading of 0.02 – 0.1 g. When more adsorbents are loaded, the adsorption decreases dramatically. This is probably due to the acidity of the fly ash. We found that the pH of the solution with different fly ash loadings shows varying values. The higher the loading of the fly ash in solution, the lower the solution pH. The pH of the solution will influence the adsorption significantly as shown below.

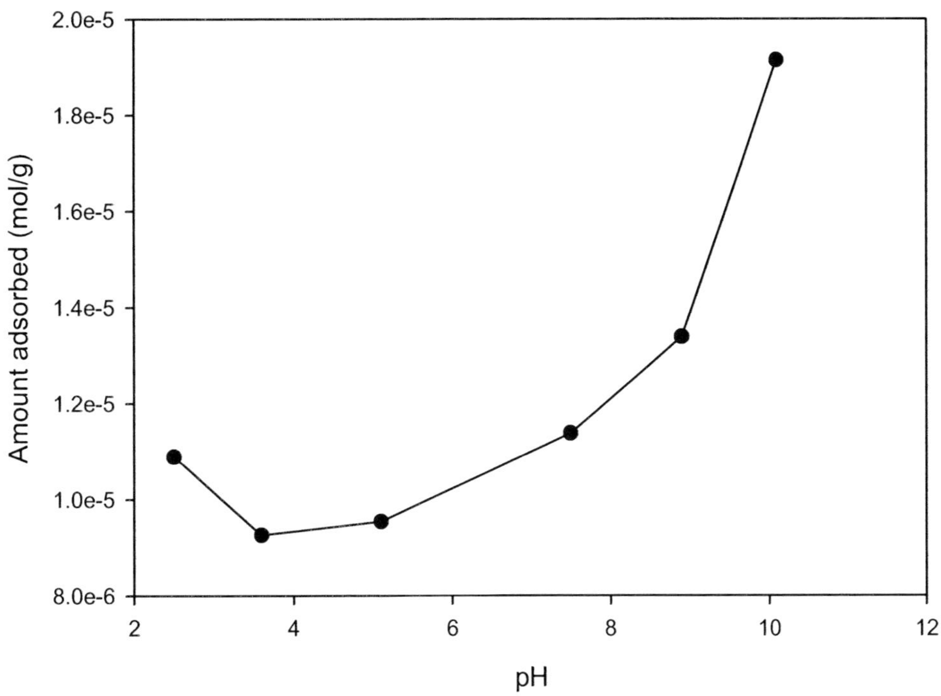


Figure 4. Effect of pH on the adsorption of MB on fly ash.

The variation of MB adsorption on Unit 5 fly ash over a broad range of pH is shown in Figure 4. As seen, the adsorption is lower at pH < 7 and then is increased to higher values at pH > 7. More significant enhancement in the adsorption of dye is reached at pH = 10 than at pH = 8. It is known that upon dissolution ionic dyes release coloured anions/cations into solution. The adsorption of these charged dye groups onto the adsorbent surface is primarily influenced by the surface charge on the adsorbent which is in turn influenced by the solution pH. Methylene blue is a basic dye. In water, it produces cation ($C^+$) and reduced ions ($CH^+$). As the pH of the dye solution becomes higher, the association of dye cations on solid will take place more easily.

## 3. Effect of Physical and Chemical Treatment on Dye Adsorption on Fly Ash

It is seen that raw fly ash exhibits low adsorption compared with activated carbon and zeolite (Wang and Wu, 2006). To improve the adsorption, fly ash can be activated by treatment with physical or chemical methods. Here we attempt to treat the raw fly ash (unit 5) using heat treatment and acid/base treatment. In heat treatment, fly ash was calcined at 600 °C. For chemical activation, fly ash was either treated in acids (HCl and $HNO_3$) or bases (NaOH and KOH) solution at room temperature. Figure 5 shows the adsorption isotherms for raw fly ash and activated fly ash samples at 30 °C.

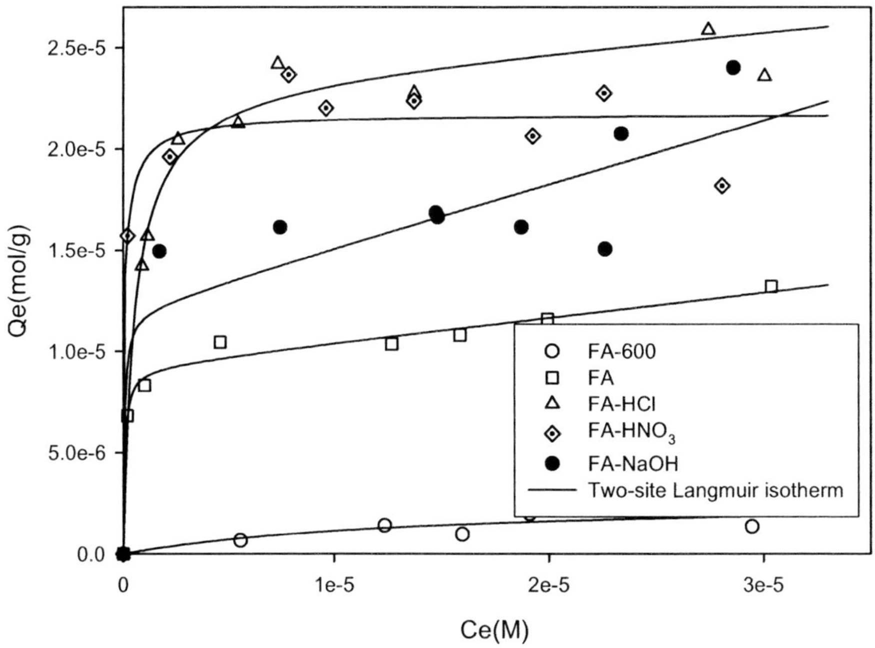


Figure 5. Isotherm of methylene blue adsorption on various fly ashes.

FA-600 exhibits the lowest adsorption while acid treated products have greater adsorption capacity than the base-treated ash. Generally, FA-HCl presents the highest adsorption capacity. The adsorption of FA is around 1.4 x $10^{-5}$ mol/g and FA-$HNO_3$ can reach 2.5 x $10^{-5}$ mol/g. The remarkable reduction of adsorption for FA-600 may be a result of several effects occurring during high temperature treatment: (i) removal of porous unburned carbon, (ii) the decomposition of hydroxyl groups on fly ash, (iii) the removal of most of the micropores due to the calcination. Acid treatment will remove the minerals in unburned carbon, leading to increase in the pore volume and surface area, and also change of the surface groups on carbon. The higher adsorption of methylene blue on FA-HCl than FA-$HNO_3$ is probably due to the adsorption of negative ions (that is $Cl^-$ from HCl) on the positive sites on the surface, so that the surface becomes more negatively charged, making it easy to adsorb positive charged molecules of methylene blue. Treatment by NaOH will induce the formation of porous zeolite from fly ash and also change the negatively charged surface, resulting in an increase of adsorption.

The equilibrium adsorption isotherm is of importance in the design of adsorption systems. Several isotherm equations are available and the two important isotherms are selected in this study, the Langmuir and Freundlich isotherms.

The Langmuir isotherm is based on an assumption that the adsorption occurs at specific homogeneous sites within the adsorbent. The generalised Langmuir isotherm can be written in the form

$$Q_e = \sum_{i=1}^{n} \frac{a_i b_i C_e}{1 + b_i C_e} \tag{8}$$

$Q_e$ is the adsorbed amount of the dye, $C_e$ is the equilibrium concentration of the dye in solution, $a_i$ and $b_i$ are the parameters on the Langmuir isotherm. $A_i$ is the monolayer adsorption capacity and $b_i$ is the constant related to the free energy of adsorption. In most cases, the single- or two-site models give a satisfactory fit to experimental data.

The Freundlich isotherm is an empirical equation employed to describe heterogeneous systems. The Freundlich equation is expressed

$$Q_e = K C_e^{1/n} \tag{9}$$

where K and n are Freundlich adsorption isotherm constants, being indicative of the extent of the adsorption and the degree of non-linearity between solution concentration and adsorption, respectively.

Using above isotherm models to fit the adsorption isotherm, it is found the Freundlich model is better than single-site Langmuir model, and the two-site Langmuir model will be the best for simulation of the adsorption isotherm. This suggests that some heterogeneity in the surface or pores of the fly ash will play a role in dye adsorption.

In recent years, sonochemical and microwave technologies have been applied for materials synthesis recently and they show much higher efficiency than traditional methods. We thus used two techniques to treat fly ash and investigated their effect on adsorption capacity (Wang et al., 2005). Figure 6 shows a comparison of adsorption by various adsorbents for three dyes, methylene blue, crystal violet (CV) and rhodamine B (RB). FA-HCl-S is referred to the sample treated with HCl solution in an ultrasonic bath. Two another samples were obtained by treating with HCl solution under microwave heating at 2 and 10 min, respectively, which are referred to FA-HCl-M1 and FA-HCl-M2. As shown that acid treatment will generally increase the adsorption of dyes and all the adsorbents demonstrate higher adsorption to methylene blue than crystal violet and rhodamine B. The mode of chemical treatment results in different effects on adsorption behaviour. Higher temperature treatment under conventional heating process will definitely enhance the adsorption. Sonochemical treatment seems to have a better effect than the conventional chemical treatment. Microwave chemical treatment will incur the best effect on adsorption. BET surface area measurements indicate that microwave treatment results in high surface area of sample (Wang et al., 2005). For the adsorption of different dyes, it is also seen that the adsorption capacity of dyes follows the order of MB > CV > RB. The difference is probably due to the structure of the dyes. It is known that crystal violet and rhodamine B molecules are much bigger than methylene blue, which will prevent the molecules entering the smaller pores of adsorbents, resulting in lower adsorption.

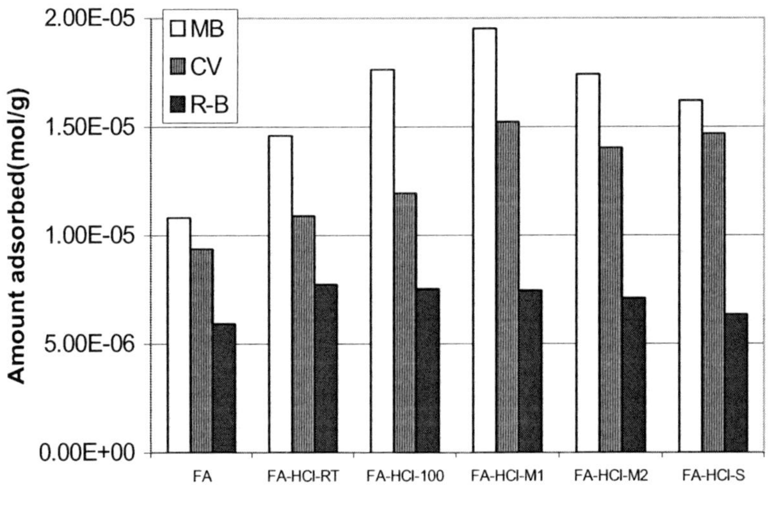


Figure 6. Comparison of dye adsorption on various treated fly ashes

## 4. Hydrothermal Conversion of Fly Ash to Zeolite

In recent years, the production of zeolites using coal fly ash as a resource of aluminosilicate constitutes one important issue of waste management and many techniques have been investigated. Hydrothermal conversion of fly ash to zeolite can produce highly porous materials, enhancing the surface area, pore volume and ion exchange capacity, thus making higher adsorption capacity of the solid. Here, we investigated the conversion of fly ash to zeolite under various conditions and their adsorption capacity.

**Table 4. Properties of fly ash adsorbents**

| Sample | $S_{BET}(m^2/g)$ | pH | Phases |
|---|---|---|---|
| FA | 15.6 | 4.4 | quartz, mullite, with minor hematite and magnetite |
| FA-NaOH-RT | 20.2 | 8.8 | quartz, mullite, with minor hematite and magnetite |
| FA-NaOH-60 | 35.0 | 9.9 | quartz, mullite, with minor hematite and magnetite |
| FA-NaOH-100 | 52.4 | 10.3 | quartz, mullite, P1 zeolite |

Figure 7 shows the XRD patterns of the various FA samples treated by 1M NaOH at different temperatures, room temperature (RT), 60 °C and 100 °C. As seen that XRD profiles do not show significant difference for FA, FA-NaOH-RT and FA-NaOH-60 but the XRD profile of FA-NaOH-100 is quite different, indicating a new phase is formed which is identified as zeolite Na–P1. The surface area measurement shows that the surface area will increase with the increasing temperature when fly ash is treated with NaOH (Table 4). In addition, conversion of fly ash to zeolite increases the basicity of solids.

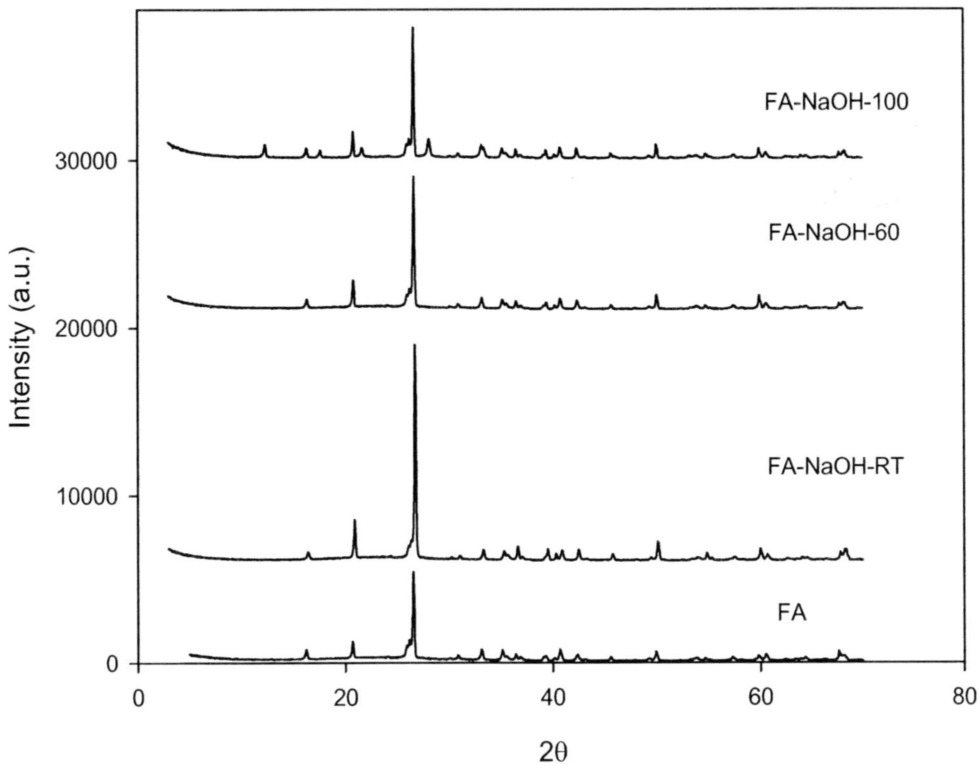


Figure 7. XRD profiles of fly ash and NaOH treated samples at different temperatures.

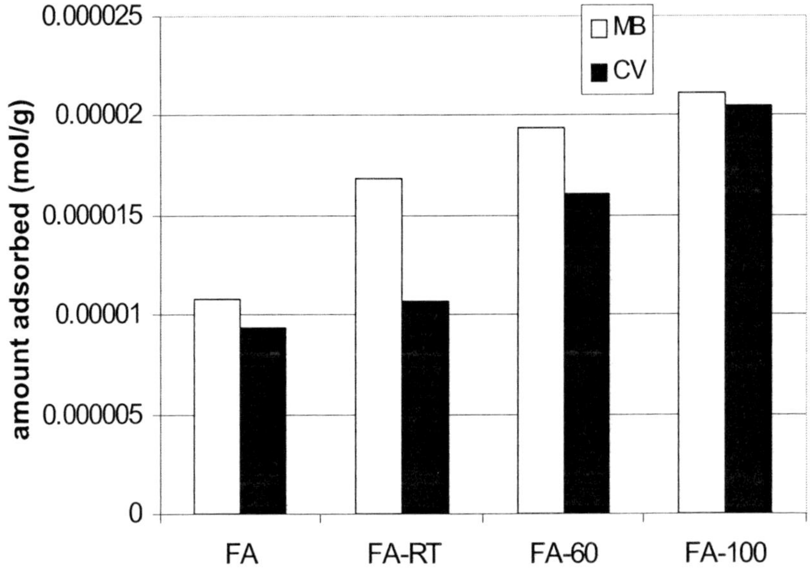


Figure 8. Comparison of MB and CV on various fly ash samples.

Figure 8 presents the adsorption of methylene blue and crystal violet on FA treated with NaOH at different temperatures. It is seen that treatment temperature seems to have positive effect on adsorption. The adsorption of MB and CV can increase to 2.0 x $10^{-5}$ mol/g for FA-NaOH-100. This is because the high surface area of the solids when treated with NaOH at high temperature, thus resulting in higher adsorption for two dyes. Further investigations also indicate that treatment by KOH or NaOH will not produce significant difference, however, NaOH concentration will remarkably influence the adsorption capacity. Higher NaOH concentration makes more fly ash converted to zeolite and thus more surface area will be produced resulting in higher adsorption capacity.

## 5. GEOPOLYMER FROM FLY ASH

Apart from the formation of crystalline zeolite particles, hydrothermal conversion of fly ash can also produce amorphous form aluminosilicate, referred to geopolymers (Davidovits, 1991). Thus, we employed a fusion method to prepare a geopolymer from fly ash. The mineral fly ash was treated with sodium hydroxide at different ratio and various temperatures for 1 h to obtain a fused mass, which was cooled to room temperature. The obtained fused fly ash solid was then mixed with water in a weight ratio of 1:4 and hydrolysed and aged for 1 day at room temperature under agitation. After that, the slurry was filtrated, washed and dried at 110 °C overnight. Figure 9 presents XRD pattern of the geopolymer. It is seen that most of crystalline aluminosilicate in fly ash has been geopolymerised to amorphous phase during the synthesis. Higher temperature fusion of fly ash makes more conversion of crystalline fly ash to geopolymer. Table 5 also suggests that higher temperature treatment results in higher surface area and pore volume. The surface area and pore volume can be as high as 56 $m^2/g$ and 0.14 $cm^3/g$ at treatment temperature of 550 °C.

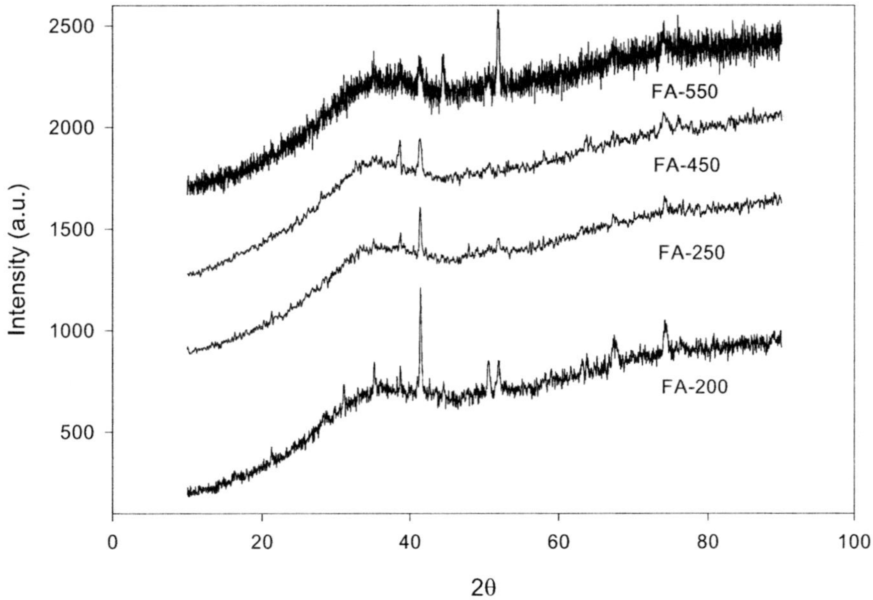


Figure 9. XRD patterns of fusion treated fly ash samples.

**Table 5. Textural properties of fly ash samples**

| Sample | $S_{BET}(m^2/g)$ | $V(cm^3/g)$ | pH |
|---|---|---|---|
| FA | 8.4 | 0.010 | 5.35 |
| FA-100 | 31.8 | 0.054 | 10.05 |
| FA-200 | 36.0 | 0.091 | |
| FA-250 | 44.1 | 0.12 | |
| FA-350 | 38.9 | 0.097 | 11.52 |
| FA-450 | 41.4 | 0.10 | |
| FA-550 | 56.0 | 0.14 | 11.48 |

Figure 10 displays the removal of methylene blue and crystal violet using FA, FA-350, and natural zeolite. As can be seen that three adsorbents exhibit varying adsorption behaviour. The raw fly ash shows much low adsorption capacity for both dyes at around 5 x $10^{-6}$ mol/g. Natural zeolite presents higher adsorption capacity than fly ash around 5 x $10^{-5}$ mol/g and the treated fly ash produced the highest adsorption capacity at 1.2 x $10^{-4}$ and 2.4 x $10^{-4}$ mol/g for methylene blue and crystal violet, respectively, which are much higher than the value of zeolite from fly ash.

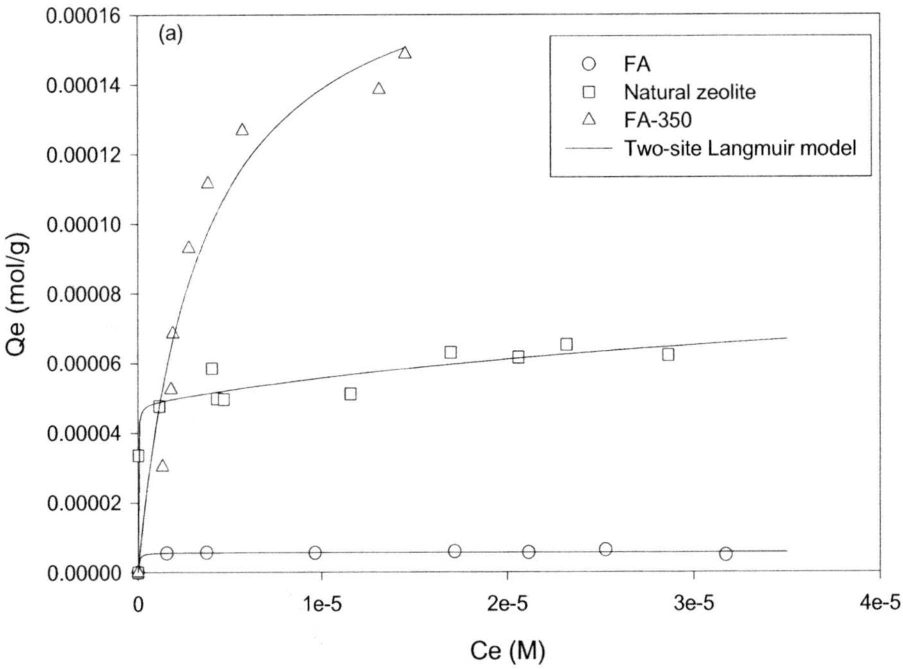


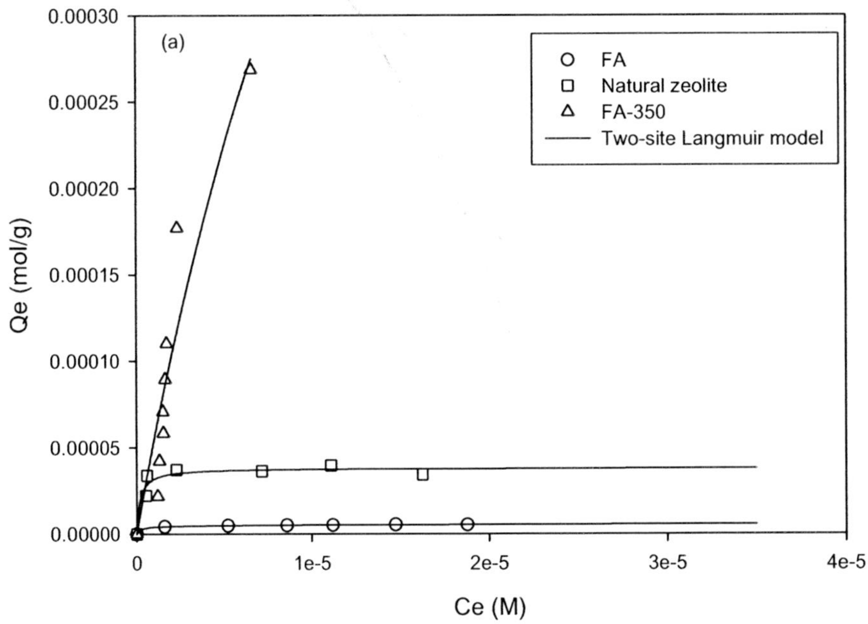


Fig.10. Adsorption isotherms of methylene blue and crystal violet on fly ash and zeolite. (A) methylene blue (B) Crystal violet

## 6. Unburned Carbon from Fly Ash

In fly ash, apart from inorganic minerals, carbon is also an important component and its content varies with the sources of coal and combustion processes. Previous investigations have shown that fly ash with different composition exhibited varying adsorption capacity and it is believed that the unburned carbon will play an important role in adsorption. Here, we prepared several fly ash samples with different content of unburned carbon and investigated their correlation with the adsorption properties for some basic dyes as well as the role of unburned carbon in adsorption. Figure 11 presents the relationship between adsorption capacity of methylene blue (MB) and crystal violet (CV) on FA and the carbon content in FA. As seen that there is a significant linear correlation between carbon content and adsorption capacity, indicating that the adsorption capacity strongly depends on carbon content, not the chemical property (pH) or bulk mineral phase. When the content of unburned carbon is more than 35%, adsorption capacity can be as high as 1.1 $x10^{-4}$ mol/g and 8.0 $x10^{-5}$ mol/g for methylene blue and crystal violet.

Figure 12 shows the effect of temperature on dynamic adsorption of two dyes on unburned carbon. It is seen that higher temperature will result in an increase in dye adsorption on unburned carbon, suggesting that the dye adsorption is an endothermic process. The temperature seems to have a significant effect on acid black adsorption while less effect on basic violet adsorption. For basic violet, the adsorption approaches equilibrium after 100 h and the adsorption capacity will reach 2 $x10^{-4}$ mol/g at 50 °C. However, for acid black, the adsorption shows gradual increase and no equilibrium will be established after 500 h.

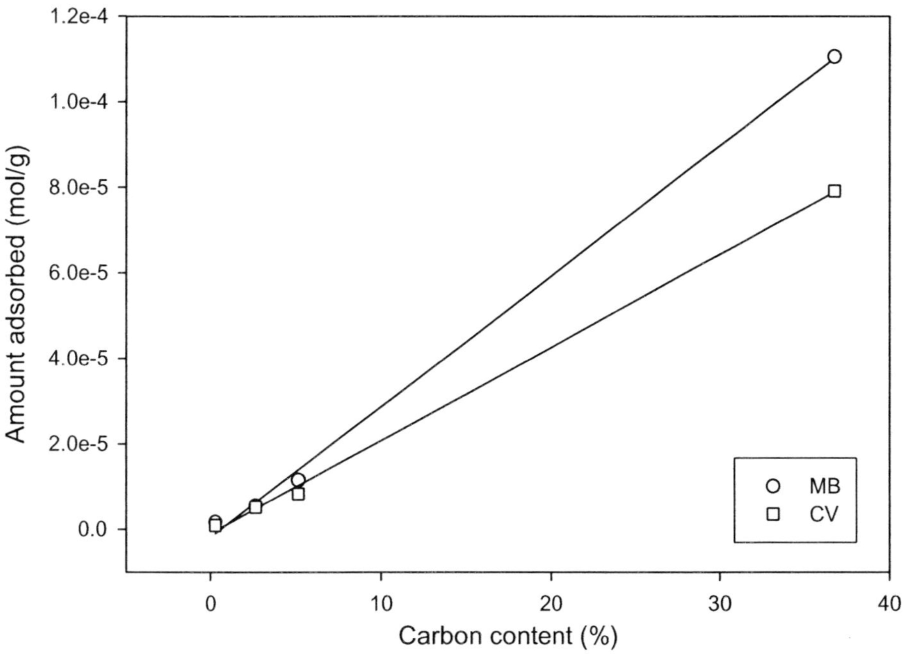


Figure 11. Relationship between adsorption capacity and carbon content in fly ash.

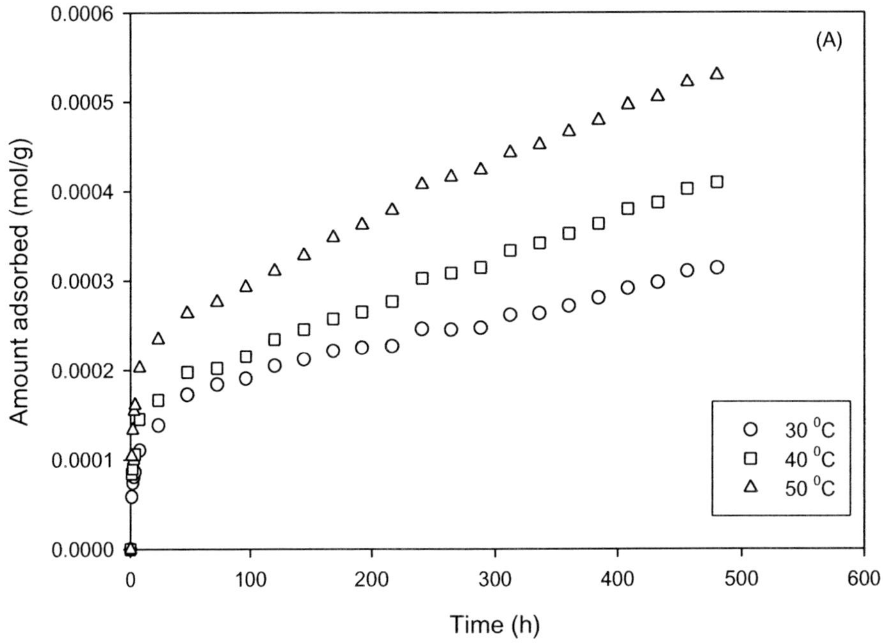


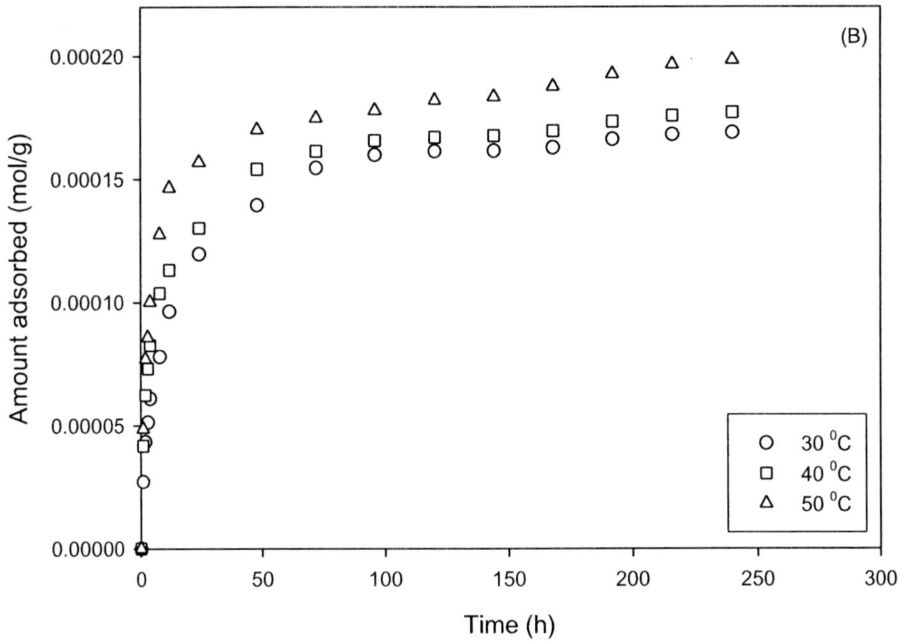


Figure 12. Effect of temperature on dye adsorption on unburned carbon. (A) Acid black 1 (B) Basic violet 3

The kinetics of the adsorption was investigated by the second-order model and Table 6 displays the parameters. As seen that the adsorption equilibrium capacity increases with the increasing temperature for both dyes and that unburned carbon presents higher adsorption capacity for acid black. The pseudo second-order kinetic rate constant, $k_2$, shows different trend variation for the two dyes. For basic violet, $k_2$ increases as the temperature increases from 30 to 40 °C and then it keeps close to the value at 50 °C, indicating that the adsorption mechanism is different at higher temperature from that at lower temperature.

From the kinetic parameters, $k_2$ at varying temperature, the activation energies for the adsorption process can be estimated to be 9.9 and -29.7 kJ/mol for basic violet 3 and acid black 1, respectively. The negative value of the activated energy for acid dye indicates that the adsorption is not a chemical but a physical process while the positive values of activated energies for basic dye indicates that the adsorption process is a combination of chemical and diffusion process.

**Table 6. Kinetic parameters for two dyes obtained at different temperatures**

| Dye | Temp (°C) | $k_2$ (g $mol^{-1}$ $h^{-1}$) | $q_e$ (mol/g) | $R^2$ |
|---|---|---|---|---|
| | 30 | 772.86 | $1.72 \times 10^{-4}$ | 0.964 |
| Basic violet | 40 | 1027.65 | $1.78 \times 10^{-4}$ | 0.999 |
| | 50 | 981.78 | $1.97 \times 10^{-4}$ | 0.998 |
| | 30 | 85.14 | $3.07 \times 10^{-4}$ | 0.973 |
| Acid black | 40 | 47.92 | $4.05 \times 10^{-4}$ | 0.952 |
| | 50 | 41.20 | $5.27 \times 10^{-4}$ | 0.963 |

**Table 7. Comparison of adsorption capacity for various adsorbents**

| Dye | Adsorbent | Capacity (mol/g) | References |
|---|---|---|---|
| Methylene blue | Activated carbon from coconut shell | $0.52 \times 10^{-4}$ | (Singh et al., 2003) |
| Methyl orange | Activated carbon from coconut shell | $0.29 \times 10^{-4}$ | (Singh et al., 2003) |
| Methylene blue | Fly ash | $0.14 \times 10^{-4}$ | (Wang et al., 2005) |
| Methylene blue | Red mud | $0.078 \times 10^{-4}$ | (Wang et al., 2005) |
| Malachite green | Activated carbon from fertiliser waste | $1.25 \times 10^{-4}$ | (Gupta et al., 1997) |
| Malachite green | Blast furnace slag | $1.05 \times 10^{-4}$ | (Gupta et al., 1997) |
| RB | Iron humate | $0.44 \times 10^{-4}$ | (Janos, 2003) |
| RB | rice husk-based porous carbons | $7 \times 10^{-4}$ | (Guo et al., 2005) |
| RB | Bagasse fly ash | $(1.18 – 1.43) \times 10^{-4}$ | (Gupta et al., 2000) |
| RB | Activated carbon from pistachio shells | $(1.6 – 1.9) \times 10^{-4}$ | (Attia et al., 2003) |
| RB | Carbon from fertiliser waste | $(1.7 - 1.9) \times 10^{-4}$ | (Bhatnagar and Jain, 2005) |
| RB | Unburn carbon | $(0.97 – 1.5) \times 10^{-4}$ | (Wang and Li, 2005) |
| Acid black | Unburn carbon | $(3.1 - 5.3) \times 10^{-4}$ | This work |
| CV | Unburn carbon | $(1.7 - 2.0) \times 10^{-4}$ | This work |

Table 7 presents a comparison of the adsorption capacity of the unburned carbon with activated carbon and other adsorbents. It is seen that unburned carbon in this report shows the comparable adsorption capacity to other activated carbons, revealing that unburned carbon can be employed as a promising adsorbent for dye adsorption.

## 7. METAL ADSORPTION BEHAVIOUR ON FLY ASH

Copper and nickel are important heavy metal ions in wastewater systems. Elevated levels of $Cu^{2+}$ and $Ni^{2+}$ in the environment can be attributed to a variety of sources, such as industrial effluents from metal cleaning and plating baths, paper, paperboard mills, wood-pulp production, tire manufacture, and fertilizer industries. Excessive intake of copper results in its accumulation in the liver, leading to copper poisoning and gastrointestinal problems. It is also toxic to aquatic organisms, even at very low concentration

A preliminary test shows that raw fly ash exhibits little adsorption to copper and nickel. In order to increase adsorption of $Cu^{2+}$ and $Ni^{2+}$ on fly ash, a hydrothermal conversion of fly ash to zeolite was conducted and tested for the heavy metal removal. Fig.13 presents a dynamic adsorption of two ions on a treated fly ash sample at different temperatures. The sample was treated with 5 M NaOH at 140 °C. It is seen that adsorption will reach equilibrium around 100 h. Higher temperature results in higher adsorption for $Cu^{2+}$ and $Ni^{2+}$. At 30 °C, equilibrium adsorption for $Cu^{2+}$ and $Ni^{2+}$ is 35 mg/g and 30 mg/g, respective while at 40 °C the equilibrium adsorption for $Cu^{2+}$ and $Ni^{2+}$ is similar.

Figure 13 and Figure 14 show the curve-fitting plots of the Lagergren first-order and pseudo second-order plots, respectively. The parameters obtained for the two models are presented in Table 3. As shown that the pseudo first-order and the pseudo second-order models will be well in fitting the experimental data. The data of equilibrium adsorption from the second-order kinetics will be close to the experimental data and regression coefficients also indicate that the pseudo second-order will show better simulation results.

In addition, a fusion method was also employed for conversion of fly ash to geopolymer. Figure 14 presents the adsorption isotherm of $Cu^{2+}$ adsorption on FA-NaOH-550 at different temperatures. As shown that $Cu^{2+}$u adsorption is increases with the increasing equilibrium concentration. However, the adsorption capacity is much similar at different temperatures, suggesting that temperature is not so important affecting the adsorption capacity. From the results, it is seen also that the adsorption capacity of FA-550 for $Cu^{2+}$ is around 90 mg/g.

**Table 8. Parameters for kinetic models**

| Temp ($^0$C) | Metal | Experimental | First-order kinetics | | | Second-order kinetics | | |
|---|---|---|---|---|---|---|---|---|
| | | $q_e$ (mg/g) | $k_1$(1/h) | $q_e$(mg/g) | $R^2$ | $k_2$(g/mg h) | $q_e$ (mg/g) | $R^2$ |
| 30 | $Cu^{2+}$ | 42.0 | 1.30 | 36.36 | 0.893 | $5.79 \times 10^{-3}$ | 42.99 | 0.998 |
| 40 | | 64.0 | 0.051 | 59.19 | 0.951 | $1.08 \times 10^{-3}$ | 67.38 | 0.996 |
| 30 | $Ni^{2+}$ | 28.0 | 0.38 | 28.18 | 0.961 | 0.4678 | 28.05 | 0.999 |
| 40 | | 65.8 | 0.059 | 61.20 | 0.943 | $1.325 \times 10^{-3}$ | 68.26 | 0.998 |

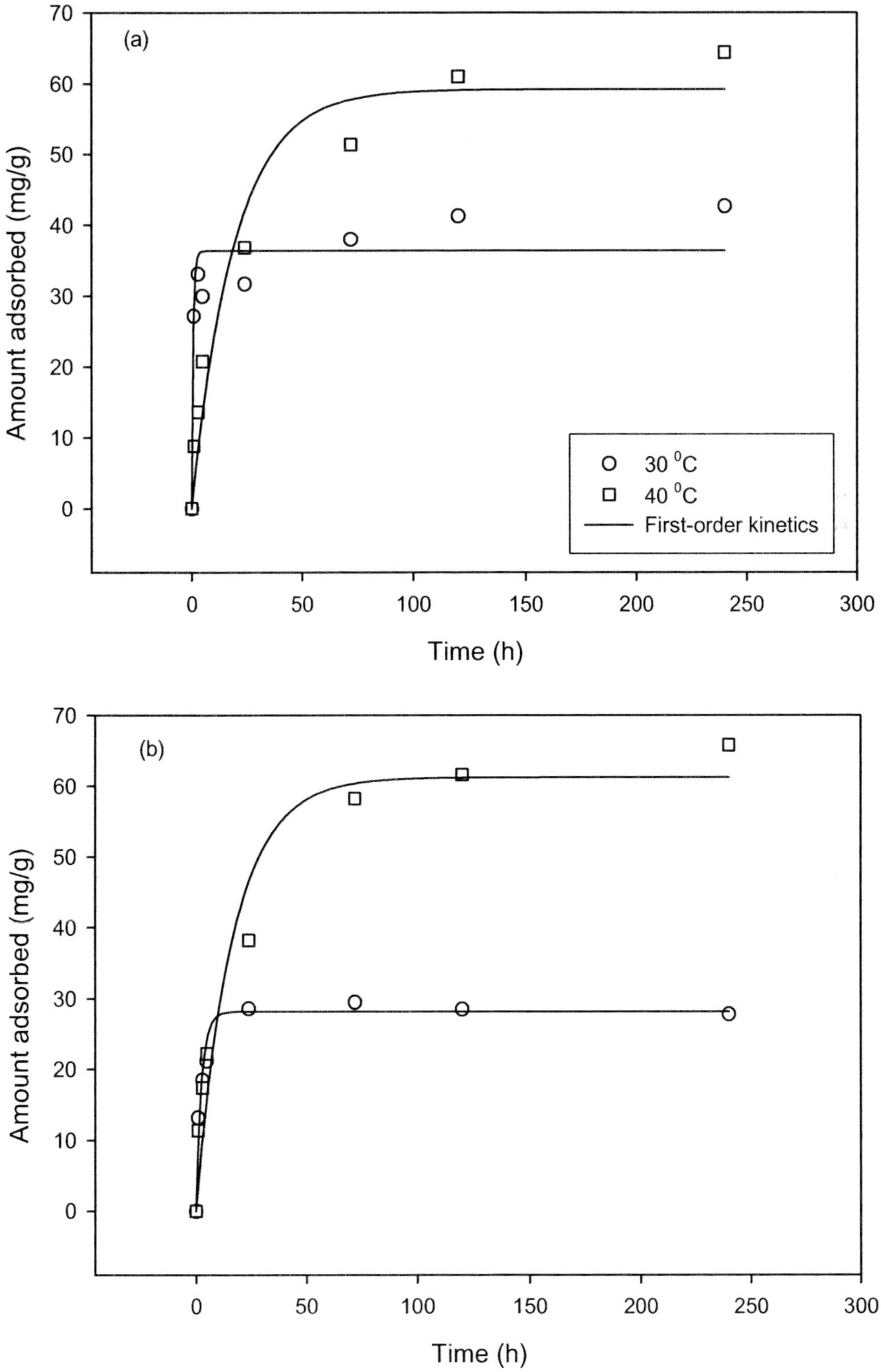


Figure 13. The pseudo first-order kinetics of metal adsorption on treated fly ash. (A) $Cu^{2+}$ (B) $Ni^{2+}$

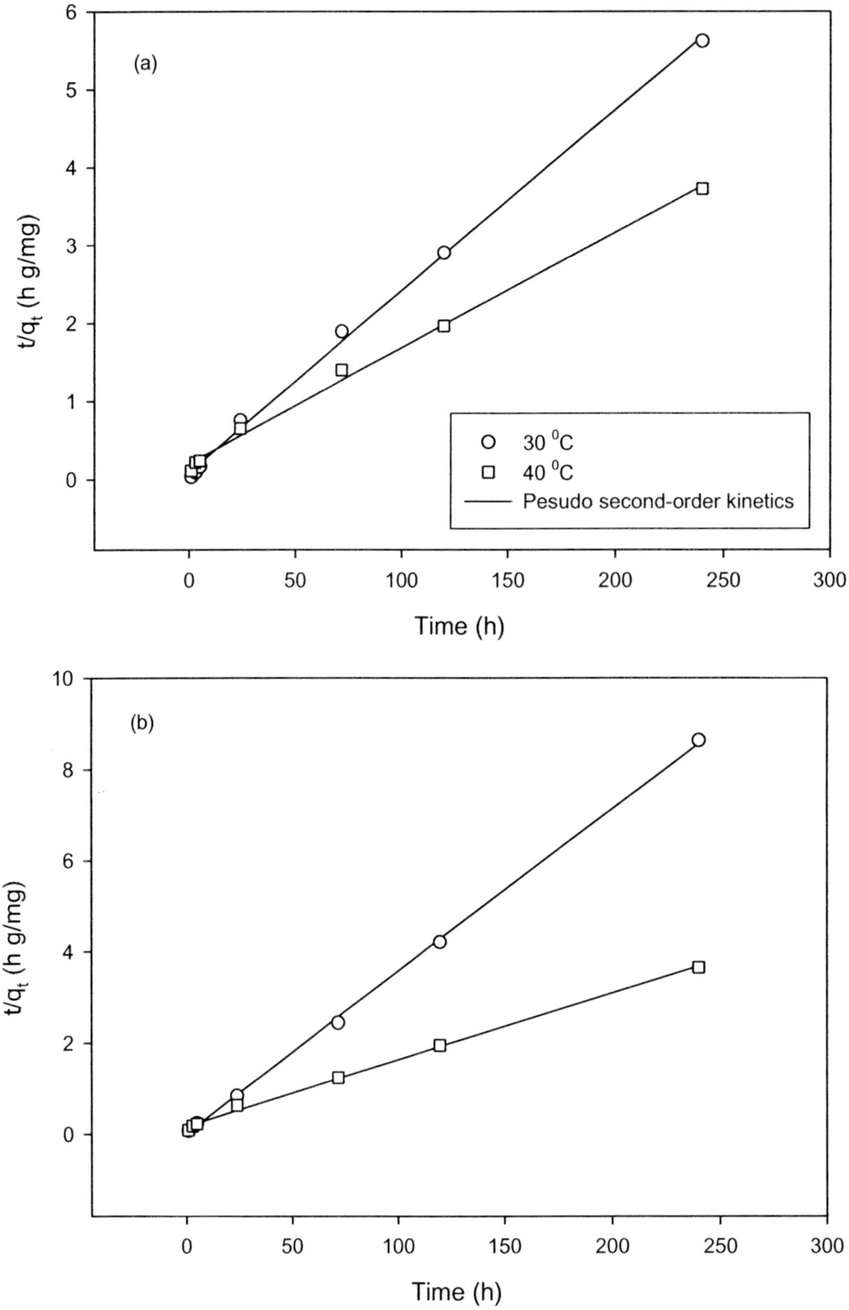


Figure 14. The pseudo second-order kinetics of metal adsorption on treated fly ash. (A) Cu (B) Ni

Two adsorption isotherms, the Langmuir and Freundlich models, have been used for the fitting of the experiment data. The isotherm parameters are given in Table 9. As seen from Figure 15 that both isotherms can fit the data well even though the correlation coefficients are not so high. While it seems from the table that the Langmuir isotherm would be better than the Fredunlich model in data fitting. In addition, the parameters obtained from both isotherms are much close for different temperatures, confirming that the temperature has little effect on adsorption in this case.

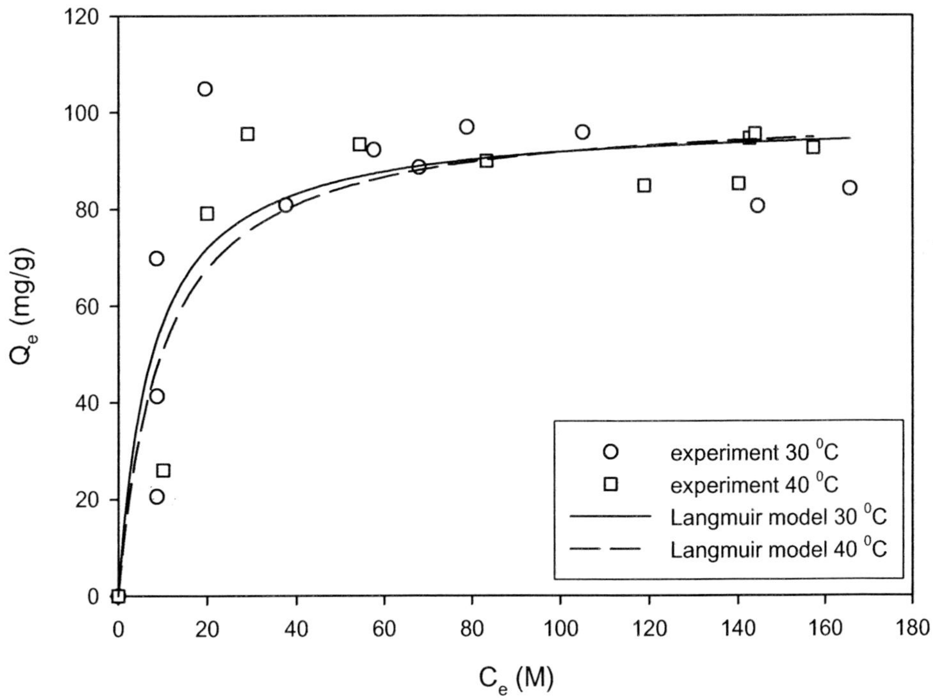


Figure 15. Adsorption isotherms of Cu2[+] on fusion treated fly ash at different temperatures.

**Table 9. Isotherm parameters for $Cu^{2+}$ adsorption on FA-550**

| Temperature ($^0$C) | Langmuir isotherm $K_L$ | $q_{max}$ (mg/g) | $R^2$ | Freundlich isotherm $K_f$ (mg/g) | 1/n | $R^2$ |
|---|---|---|---|---|---|---|
| 30 | 0.1342 | 98.7 | 0.754 | 40.66 | 0.1724 | 0.668 |
| 40 | 0.1020 | 100.8 | 0.866 | 40.10 | 0.1733 | 0.790 |

Table 10 lists the previous results of $Cu^{2+}$ and $Ni^{2+}$ adsorption on various fly ash samples. It is seen that adsorption capacity varies greatly depending on adsorption conditions. For $Cu^{2+}$, the highest adsorption can be around 200 mg/g after fly ash was treated but for $Ni^{2+}$, the highest adsorption is only 15 mg/g. Our results show that raw fly ash exhibits much low adsorption to $Cu^{2+}$ and $Ni^{2+}$. But after treatment, fly ash can reach high adsorption capacity around 100 mg/g for $Cu^{2+}$ and 60 mg/g for $Ni^{2+}$.

**Table 10. Summary of adsorption of $Cu^{2+}$ and Ni ions on fly ash**

| Metals | Adsorbent | Adsorption (mg/g) | References |
|---|---|---|---|
| $Cu^{2+}$ | Fly ash | 1.39 | (Panday et al., 1985) |
| | Fly ash + wollastonite | 1.18 | (Panday et al., 1985) |
| | Fly ash | 1.7 – 8.1 | (Ayala et al., 1998) |
| | Fly ash(I) | 0.34 –1.35 | (Bayat, 2002) |
| | Fly ash (II) | 0.09 –1.25 | (Bayat, 2002) |
| | Fly ash | 207.3 | (Apak et al., 1998) |
| | Fly ash-washed | 205.8 | (Apak et al., 1998) |
| | Fly ash-acid | 198.5 | (Apak et al., 1998) |
| | Fly ash | 0.63 -0.81 | (Lin and Chang, 2001) |
| | bagasse fly ash | 2.26-2.36 | (Gupta and Ali, 2000) |
| | Fly ash | 0.76 | (Rao et al., 2003) |
| $Ni^{2+}$ | Fly ash | 9-14.0 | (Banerjee et al., 2003) |
| | Fe impregnated fly ash | 9.8-14.93 | (Banerjee et al., 2003) |
| | Al impregnated fly ash | 10- 15.75 | (Banerjee et al., 2003) |
| | Fly ash(I) | 0.40-0.98 | (Bayat, 2002) |
| | Fly ash (II) | 0.06 –1.16 | (Bayat, 2002) |
| | bagasse fly ash | 1.12-1.70 | (Gupta et al., 2003) |

## 8. Conclusion

Fly ash is an important solid waste from coal or biomass combustion. It is demonstrated that fly ash can be a good adsorbent for organic dye and heavy metal removal from wastewater. The adsorption capacity depends on the source of fly ash and various treatments will generally increase the adsorption capacity. The mineral section of fly ash can be converted into zeolite and geopolymer which are porous materials and exhibit higher adsorption capacity than raw fly ash. Unburned carbon in fly ash is also playing an important role for adsorption. High content of unburned carbon in fly ash results in higher adsorption. The kinetic and isothermal studies indicate that the adsorption usually follows the pseudo second-order model and two-site Langmuir isotherm will be the better one for adsorption isotherm.

## References

Acemioglu, B. 2004. Adsorption of Congo red from aqueous solution onto calcium-rich fly ash. *Journal of Colloid and Interface Science* 274(2), 371-379.

Apak, R., Tutem, E., Hugul, M. and Hizal, J. 1998. Heavy metal cation retention by unconventional sorbents (red muds and fly ashes). *Water Research* 32(2), 430-440.

Attia, A. A., Girgis, B. S. and Khedr, S. A. 2003. Capacity of activated carbon derived from pistachio shells by H3PO4 in the removal of dyes and phenolics. Journal of Chemical *Technology and Biotechnology* 78(6), 611-619.

Ayala, J., Blanco, F., Garcia, P., Rodriguez, P. and Sancho, J. 1998. Asturian fly ash as a heavy metals removal material. *Fuel* 77(11), 1147-1154.

Babel, S. and Kurniawan, T. A. 2003. Low-cost adsorbents for heavy metals uptake from contaminated water: a review. *Journal of Hazardous Materials* 97(1-3), 219-243.

Banerjee, S. S., Jayaram, R. V. and Joshi, M. V. 2003. Removal of nickel(II) and zinc(II) from wastewater using fly ash and impregnated fly ash. *Separation Science and Technology* 38(5), 1015-1032.

Bayat, B. 2002. Comparative study of adsorption properties of Turkish fly ashes: I. The case of nickel(II), copper(II) and zinc(II). *Journal of Hazardous Materials* 95(3), 251-273.

Bhatnagar, A. and Jain, A. K. 2005. A comparative adsorption study with different industrial wastes as adsorbents for the removal of cationic dyes from water. *Journal of Colloid and Interface Science* 281(1), 49-55.

Davidovits, J. 1991. Geopolymers - Inorganic Polymeric New Materials. *Journal of Thermal Analysis* 37(8), 1633-1656.

Guaratini, C. C. I. and Zanoni, M. V. B. 2000. Textile dyes. *Quimica Nova* 23(1), 71-78.

Guo, Y. P., Zhao, J. Z., Zhang, H., Yang, S. F., Qi, J. R., Wang, Z. C. and Xu, H. D. 2005. Use of rice husk-based porous carbon for adsorption of Rhodamine B from aqueous solutions. *Dyes and Pigments* 66(2), 123-128.

Gupta, V. K. and Ali, I. 2000. Utilisation of bagasse fly ash (a sugar industry waste) for the removal of copper and zinc from wastewater. *Separation and Purification Technology* 18(2), 131-140.

Gupta, V. K., Ali, I., Jain, C. K., Sharma, M. and Saini, V. K. 2003. Removal of cadmium and nickel from wastewater using bagasse fly ash - A sugar industry waste. *Water Research* 37(16), 4038-4044.

Gupta, V. K., Mohan, D., Sharma, S. and Sharma, M. 2000. Removal of basic dyes (rhodamine B and methylene blue) from aqueous solutions using bagasse fly ash. *Separation Science and Technology* 35(13), 2097-2113.

Gupta, V. K., Srivastava, S. K. and Mohan, D. 1997. Equilibrium uptake, sorption dynamics, process optimization, and column operations for the removal and recovery of malachite green from wastewater using activated carbon and activated slag. *Industrial & Engineering Chemistry Research* 36(6), 2207-2218.

Ho, Y. S. 2004. Citation review of Lagergren kinetic rate equation on adsorption reactions. *Scientometrics 59*(1), 171-177.

Ho, Y. S. and McKay, G. 1999. Pseudo-second order model for sorption processes. *Process Biochemistry* 34(5), 451-465.

Iyer, R. S. and Scott, J. A. 2001. Power station fly ash - a review of value-added utilization outside of the construction industry. *Resources Conservation and Recycling* 31(3), 217-228.

Janos, P. 2003. Sorption of basic dyes onto iron humate. *Environmental Science & Technology* 37(24), 5792-5798.

Janos, P., Buchtova, H. and Ryznarova, M. 2003. Sorption of dyes from aqueous solutions onto fly ash. *Water Research* 37(20), 4938-4944.

Lin, C.-J. and Chang, J.-E. 2001. Effect of fly ash characteristics on the removal of Cu(II) from aqueous solution. *Chemosphere* 44(5), 1185-1192.

Mohan, D., Singh, K. P., Singh, G. and Kumar, K. 2002. Removal of dyes from wastewater using flyash, a low-cost adsorbent. Industrial and *Engineering Chemistry Research* 41(15), 3688-3695.

Panday, K. K., Prasad, G. and Singh, V. N. 1985. Copper(II) removal from aqueous solutions by fly ash. *Water Research* 19(7), 869-873.

Ramakrishna, K. R. and Viraraghavan, T. 1997. Dye removal using low cost adsorbents. Water Science and Technology. Proceedings of the 1996 2nd IAWQ International Conference on Pretreatment of Industrial Wastewaters, Oct 16-18 1996 36(2-3), 189-196.

Rao, M., Parwate, A. V., Bhole, A. G. and Kadu, P. A. 2003. Performance of low-cost adsorbents for the removal of copper and lead. *Journal of Water Supply: Research and Technology* - AQUA 52(1), 49-58.

Singh, K. P., Mohan, D., Sinha, S., Tondon, G. S. and Gosh, D. 2003. Color removal from wastewater using low-cost activated carbon derived from agricultural waste material. *Industrial & Engineering Chemistry Research* 42(9), 1965-1976.

Viraraghavan, T. and Ramakrishna, K. R. 1999. Fly ash for colour removal from synthetic dye solutions. *Water Quality Research Journal of Canada* 34(3), 505-517.

Wang, S.B., Boyjoo, Y., Choueib, A. and Zhu, Z. H. 2005. Removal of dyes from aqueous solution using fly ash and red mud. *Water Research* 39(1), 129-138.

Wang, S. B., Boyjoo, Y. and Choueib, A. 2005. A comparative study of dye removal using fly ash treated by different methods. *Chemosphere* 60(10), 1401-1407.

Wang, S. B. and Li, H. T. 2005. Dye adsorption on unburned carbon: Kinetics and equilibrium. *Journal of Hazardous Materials* 126(1-3), 71-77.

Wang, S. B. and Wu, H. W. 2006. Environmental-Benign Utilisation of Fly Ash as Low-Cost Adsorbents. *J. Hazard. Mater.* in press.

Woolard, C. D., Strong, J. and Erasmus, C. R. 2002. Evaluation of the use of modified coal ash as a potential sorbent for organic waste streams. *Applied Geochemistry* 17(8), 1159-1164.

In: Hazardous Materials and Wastewater
Editor: Allison A. Lewinsky, pp. 225-249
ISBN 1-60021-257-3


***Chapter 7***

# ADSORPTION KINETICS AND ISOTHERMS OF ANIONIC DYE OF REACTIVE BLUE 19 FROM AQUEOUS SOLUTIONS ONTO DTMA-SEPIOLITE

***Adnan Özcan*, A. Safa Özcan and Özer Gök***
Department of Chemistry, Faculty of Science, Anadolu University, Eskişehir, Turkey

## ABSTRACT

The objective of this study was to prepare organically modified adsorbent (organoclay) from a low-cost natural clay mineral, which is namely sepiolite by treated with a surfactant for enhancement of the removal of anionic textile dyes due to the fact that these dyes generate a considerable amount of hazardous colored wastewater, which cause for environmental pollution. Reactive Blue 19 (RB19) was selected as adsorbate for evaluating the adsorption performance of sepiolite samples, which was affected significantly by a surfactant [dodecyltrimethyl ammonium (DTMA)-bromide] modification. The surface modification of DTMA-sepiolite was examined using the elemental analysis and FTIR technique. The effects of temperature, pH, contact time and concentration were tested in the experiments. The maximum adsorption capacities, which are obtained from Langmuir isotherm equation, of RB19 onto natural sepiolite and DTMA-sepiolite at pH 1.5 and 20 °C were $2.30 \times 10^{-4}$ and $3.11 \times 10^{-4}$ mol $g^{-1}$, respectively. These results indicate that DTMA-sepiolite has 1.35 times higher adsorption capacity than natural sepiolite. The Langmuir, Freundlich and Dubinin-Radushkevich (D-R) adsorption models were applied to describe the equilibrium isotherms and the isotherm constants were also deduced from experimental data. The results indicated that the Langmuir isotherm model appears to fit better than other adsorption isotherm models for the adsorption of RB19 onto natural sepiolite and DTMA-sepiolite. The first-order, pseudo-second-order kinetic and the intraparticle diffusion models were used to describe

---

* Corresponding author. Dr. Adnan Özcan, Anadolu University, Faculty of Science, Department of Chemistry, Yunusemre Campus, 26470 Eskişehir, Turkey; Tel: +90-222-3350580/Ext.:5815; Fax: +90-222-3204910; E-mail: aozcan@anadolu.edu.tr

the kinetic data and the rate constants were evaluated as well. The experimental data fitted very well by the pseudo-second-order kinetic model with the high correlation coefficients of 0.999 for natural sepiolite and DTMA-sepiolite and also followed by the intraparticle diffusion model up to 60 min, whereas diffusion is not only the rate-controlling step. The rate constants calculated from the pseudo-second-order kinetic model were used to derive the activation energy ($E_a$) of the adsorption. The numerical values of activation energies are 13.58 kJ $mol^{-1}$ for natural sepiolite and 12.49 kJ $mol^{-1}$ for DTMA-sepiolite and these results show that the adsorption process follows a physisorption mechanism. The equilibrium constants were used to calculate thermodynamic parameters including the changes of free energy, enthalpy and entropy. The numerical values of the changes of free energy for natural sepiolite and DTMA-sepiolite were 2.593 and 1.936 kJ $mol^{-1}$ at 20 °C, respectively. The changes of enthalpy values for natural sepiolite and DTMA-sepiolite at a temperature range of 20 – 40 °C were –3.723 and –6.835 kJ $mol^{-1}$, respectively. These results indicated that the adsorption of RB19 onto natural sepiolite and DTMA-sepiolite are not spontaneous according to the change of free energy and are exothermic according to the change of enthalpy. All over these results indicate that effective adsorbents for the removal of RB19 could be prepared by appropriate modification of sepiolite, although natural sepiolite itself can be considered as a potential adsorbent in wastewater treatment systems.

**Keywords:** Clay; Sepiolite; Adsorption; Dye; Hazardous pollutants; Surfactant; DTMA-sepiolite; Organoclay; Kinetics; Isotherms, Thermodynamics

# 1. INTRODUCTION

Synthetic dyes have been increasingly used in the textile, paper-making, rubber, plastics, cosmetics, pharmaceutical, leather and food industries instead of natural dyes due to their ease of use, low-cost of synthesis, stability and variety of color [1,2]. Many of these dyes from wastewater effluents are toxic and even carcinogenic, highly colored and, the disposal of these wastes into receiving waters induces damage to the environment as they may significantly affect photosynthetic activity in aquatic life due to reduced light penetration. Most of synthetic dyes present in these effluents also resist biological oxidation. Color removal from these kinds of textile effluents are a major environmental problem due to the difficulty of treating such stream by conventional physical, chemical and biological treatment methods [3-7].

Among various traditional chemical and physical methods for the treatment of these effluents (e.g. sedimentation, filtration, flocculation, ozonation, oxidation) prove to be insufficient to remove color from wastewater effluents producing by textile dyeing and washing. On the other hand, the adsorption process is one of the effective methods that have been successfully employed for color removal from wastewater due to its low-cost, high efficiency and easy operation [8,9].

Activated carbon is the most powerful and widely used adsorbent, but there are certain problems using activated carbon as an adsorbent due to the high operating costs of use and its regeneration [10]. Therefore, there is a need to search inexpensive, locally available and efficient materials for the removal of dyes from wastewaters. In this way, natural clays for the textile dye removal from wastewater including sepiolite [11-17], kaolinite [18], montmorillonite [19-21], smectite [22-24], zeolite [25-28] and bentonite [29-33] are being considered as alternative inexpensive and efficient adsorbents. A limited number of studies were found to be in the literature using natural clay of sepiolite as an adsorbent such as catalyst support, wastewater treatment, solid wastes, reducing the toxic effect of some heavy metals and pesticides [34], but the equilibrium and kinetics of the adsorption of Reactive Blue 19 (RB19) onto natural sepiolite and dodecyltrimethyl ammonium (DTMA)-bromide modified sepiolite has not been found in the literature up to now.

Sepiolite is a natural clay mineral, with a $(Si_{12})(Mg_8)(O_{30})(OH)_4(OH_2)_4.8H_2O$ [35] chemical formula, as a unit-cell formula is a magnesium hydrosilicate with a micro-fibrous structure. It has a high surface area, high chemical and mechanical stability; therefore it is an attractive adsorbent for organic or inorganic ions. A general structure of sepiolite consists of alternation of blocks and tunnels that grows up in the microfibre direction. Each block has two tetrahedral silica sheets enclosing a central octahedral magnesia sheet but the silica sheets are discontinued and inversion of these sheets that give rise to structural tunnels. Moreover, some isomorphic substitutions in the tetrahedral sheets of the lattice of sepiolite including $Al^{3+}$ instead of $Si^{4+}$ from negatively adsorption sites. Such sites are occupied by exchangeable cations that compensate for the electrical charge [12,35-37].

However, such clays may be ineffective as adsorbents for poorly water-soluble (hydrophobic) organic contaminants such as textile effluents, but replacement of the inorganic cation with an organic cation such as tetraalkylammonium (for instance dodecyltrimethyl ammonium (DTMA)-bromide) by the simple ion-exchange reactions causes considerable change in the clay mineral surfaces. It leads to van der Waals interactions between organic surfactant cations and adsorbate. These kind of adsorbents are known as organoclays because the exchangeable inorganic cations such as $H^+$, $Na^+$, $Ca^{2+}$ are replaced by organic cation (e.g. quaternary ammonium compounds) to cause to transform organophobic to strongly organophilic and therefore the adsorption capacity increases. Thus, organoclays are powerful adsorbents for a wide variety of environmental applications [38-41].

The design and efficient operation of adsorption processes require equilibrium adsorption data for the adsorption kinetics and isotherm models. The adsorption isotherms are also inevitable tool to understand the adsorption mechanism for the theoretical evaluation [2,42].

The purpose of this study was to investigate the possibility of natural sepiolite and DTMA-sepiolite as adsorbents for the removal of Reactive Blue 19 (RB19) dye from aqueous solution by adsorption technique. Sepiolite is an abundant and low-cost natural clay and easily available in large quantity in Turkey. The Langmuir, Freundlich and Dubinin-Radushkevich (D-R) adsorption isotherm equations were used to fit the equilibrium isotherms. The equilibrium kinetic and thermodynamic data for the adsorption mechanism of RB19 onto natural sepiolite and DTMA-sepiolite were deduced from experimental data.

## 1.1. Equilibrium Parameters of Adsorption

Equilibrium data, generally known as adsorption isotherms, are basic requirements to understand the mechanism of the adsorption. Traditional adsorption isotherm models, Langmuir, Freundlich and Dubinin-Radushkevich (D-R), are used to describe the equilibrium between adsorbed RB19 onto the adsorbent ($q_e$) and RB19 in solution ($C_e$) at a constant temperature.

The Langmuir adsorption isotherm model assumes that adsorption occurs at specific homogeneous sites within the adsorbent and has found successful application in much monolayer adsorption process. The linearized form of the Langmuir isotherm equation is [43]:

$$\frac{1}{q_e} = \frac{1}{q_{max}} + \left(\frac{1}{q_{max} K_L}\right)\frac{1}{C_e} \tag{1}$$

where $q_e$ is the equilibrium dye concentration on the adsorbent (mol $g^{-1}$); $C_e$, the equilibrium dye concentration in the solution (mol $dm^{-3}$); $q_{max}$, the monolayer adsorption capacity of the adsorbent (mol $g^{-1}$) and $K_L$ is the Langmuir constant ($dm^3$ $mol^{-1}$) related to the free energy of adsorption. A plot of $1/q_e$ versus $1/C_e$ for the adsorption gives a straight line of slope $1/(q_{max} K_L)$ and intercept ($1/q_{max}$).

The effect of isotherm shape has been discussed [44] with a view to estimate whether an adsorption system is favorable or unfavorable. The essential feature of the Langmuir isotherm can be expressed by means of '$R_L$', a dimensionless constant referred to as separation factor or equilibrium parameter $R_L$ is calculated using the following equation,

$$R_L = \frac{1}{1 + K_L C_o} \tag{2}$$

where $C_o$ is the highest initial dye concentration (mol $dm^{-3}$).

The linearized form of the Freundlich isotherm equation is an empirical equation employed to describe heterogeneous systems and can be written as following [45]:

$$\ln q_e = \ln K_F + \frac{1}{n}\ln C_e \tag{3}$$

where $K_F$ ($dm^3$ $g^{-1}$) and $n$ (dimensionless) are Freundlich isotherm constants, being indicative of the extent of the adsorption and the degree of nonlinearity between solution concentration and adsorption, respectively. The plot of ln $q_e$ against ln $C_e$ for the adsorption was drawn to generate $K_F$ and $n$ from the intercept and the slope values, respectively.

The Dubinin-Radushkevich (D-R) isotherm is more general than the Langmuir isotherm due to the fact that it does not assume a homogeneous surface or constant adsorption potential. It was applied to distinguish between the physical and chemical adsorption [46]. The linear form of D-R isotherm equation [47] is

$$\ln q_e = \ln q_m - \beta\varepsilon^2 \tag{4}$$

where $\beta$ is a constant related to the mean free energy of adsorption per mole of the adsorbate ($mol^2$ $kJ^{-2}$); $q_m$, the theoretical saturation capacity and $\varepsilon$ is the Polanyi potential, which is equal to $RT\ln(1+\frac{1}{C_e})$, where $R$ (J mol$^{-1}$ K$^{-1}$) is the gas constant, and $T$ (K) is the absolute temperature. Hence by plotting ln $q_e$ against $\varepsilon^2$, it is possible to generate the value of $q_m$ (mol g$^{-1}$) from the intercept, and the value of $\beta$ from the slope.

The constant $\beta$ gives an idea about the mean free energy $E$ (kJ mol$^{-1}$) of adsorption per molecule of the adsorbate when it is transferred to the surface of the solid from infinity in the solution and can be calculated using the relationship [9,48,49]:

$$E=\frac{1}{(2\beta)^{1/2}} \tag{5}$$

## 1.2. Kinetic Parameters of Adsorption

Kinetics of adsorption is an inevitable characteristic to be responsible for the efficiency of adsorption. Various kinetic models such as the first-order, pseudo-second-order and intraparticle diffusion have been applied to the experimental data to predict the adsorption kinetics. Among them the first-order rate equation is [50],

$$\frac{1}{q_t}=\frac{1}{q_1}+\frac{k_1}{q_1 t} \tag{6}$$

where $q_1$ and $q_t$ are the amounts of the dye adsorbed at equilibrium and at time $t$, in mol g$^{-1}$, and $k_1$ is the first-order rate constant (min$^{-1}$) of adsorption. Values of $k_1$ can be calculated from the slope of the plot of $1/q_t$ versus $1/t$.

The pseudo-second-order kinetic model [51] is expressed as:

$$\frac{t}{q_t}=\frac{1}{k_2 q_2^2}+\frac{1}{q_2}t \tag{7}$$

where $q_2$ is the maximum adsorption capacity (mol g$^{-1}$) for the pseudo-second-order adsorption; $q_t$, the amount of dye adsorbed at equilibrium at time $t$ (mol g$^{-1}$); $k_2$, the equilibrium rate constant of pseudo-second-order adsorption (g mol$^{-1}$ min$^{-1}$). Values of $k_2$ and $q_2$ can be calculated from the plot of $t/q_t$ against $t$.

The intraparticle diffusion equation [52] can be written as follows:

$$q_t=k_p t^{1/2}+C \tag{8}$$

where $C$ is the intercept, and $k_p$ is the intraparticle diffusion rate constant (mol g$^{-1}$ min$^{-1/2}$).

The first-order and pseudo-second-order kinetic models could not identify the diffusion mechanism and the kinetic results were then analyzed by using the intraparticle diffusion model. According to this model, the plot of uptake, $q_t$, versus the square root of time ($t^{1/2}$)

should be linear if intraparticle diffusion is involved in the adsorption process and if these lines pass through the origin then intraparticle diffusion is the rate-controlling step [53-55]. When the plots do not pass through the origin, this is indicative of some degree of boundary layer control and these further shows that the intraparticle diffusion is not the only rate-limiting step, but also other kinetic models may control the rate of adsorption, all of which may be operating simultaneously. The slope of linear portion from the figure can be used to derive values for the rate parameter, $k_p$, for the intraparticle diffusion.

## 1.3. Activation Energy and Thermodynamic Parameters of Adsorption

Arrhenius relationship was used to evaluate the activation energy of adsorption representing the minimum energy that reactants must have for the adsorption to proceed:

$$\ln k_2 = \ln A - \frac{E_a}{RT} \quad (9)$$

where $E_a$ is the Arrhenius activation energy of adsorption; $A$, the Arrhenius factor; $R$, the gas constant, and is equal to 8.314 J mol$^{-1}$ K$^{-1}$, and $T$ is the solution temperature. When ln $k_2$ is plotted versus $1/T$, a straight line with slope $-E_a/R$ is obtained.

The thermodynamic parameters such as the change in free energy ($\Delta G^o$), enthalpy ($\Delta H^o$) and entropy ($\Delta S^o$) associated to the adsorption process and were determined by using following equations.

$$K_C = \frac{C_A}{C_S} \quad (10)$$

$$\Delta G^o = -RT \ln K_C \quad (11)$$

$$\ln K_C = \frac{\Delta S^o}{R} - \frac{\Delta H^o}{RT} \quad (12)$$

where $K_C$ is the equilibrium constant; $C_A$, the amount of dye adsorbed on the adsorbent at equilibrium (mol dm$^{-3}$) and $C_S$ is the equilibrium dye concentration in the solution (mol dm$^{-3}$). The value of calculated $q_2$ from the pseudo-second-order model was used to obtain $C_A$ and $C_S$. The plot of ln $K_C$ as a function of $1/T$ yields a straight line from which $\Delta H^o$ and $\Delta S^o$ can be calculated from the slope and intercept, respectively.

# 2. MATERIALS AND METHODS

## 2.1. Materials

A commercial textile dye, Reactive Blue 19 (RB19) (Remazol Brilliant Blue R), was obtained from Bursa, Turkey and used without further purification. The chemical structure of dye was depicted in Fig 1. Sepiolite was provided from Eskisehir, Turkey. It was crushed, ground, sieved through a 63-μm sieve, and dried at 110 °C in an oven for 2 h prior to use. The cation exchange capacity (CEC) determined by the methylene blue method [56] is 544.3 mmol $kg^{-1}$. The surface area of sepiolite was obtained by using the Brunauer, Emmett and Teller (BET) method by assuming the section area of nitrogen molecule to be 0.162 $nm^2$. Nitrogen adsorption at 77 K is a standard and widely used method for determining surface area. Nitrogen adsorption isotherm of sepiolite was performed at 77 K using nitrogen adsorption apparatus, Quantachrome Instruments Nova2200e-series. The weight of the sample was around 0.1 g. The sample was degassed under vacuum at 150 °C for 2 h before nitrogen adsorption in a vacuum system. The surface area is 152.1 $m^2\ g^{-1}$. Dodecyltrimethyl ammonium (DTMA) bromide (Merck) was used as a surfactant.

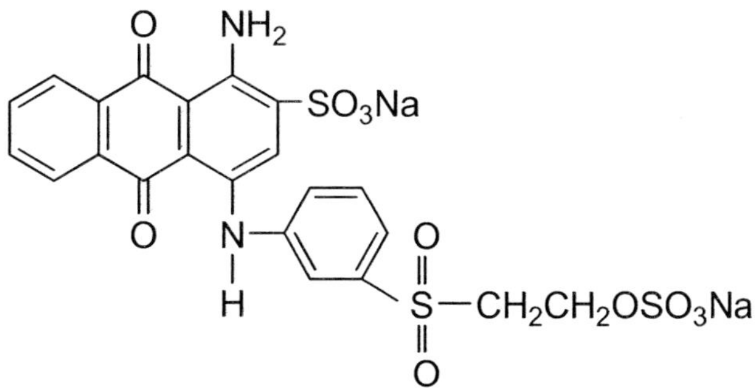


Figure 1. Chemical structure of RB19.

## 2.2. Material Characterization

The chemical analysis of natural sepiolite was determined by using an energy dispersive X-ray spectrometer (EDX-LINK ISIS 300) attached to a scanning electron microscope (SEM-Cam Scan S4). The crystalline phases present in the sepiolite were determined by using X-ray diffractometer (XRD-Rigaku Rint 2000) with Cu $K_\alpha$ radiation.

The elemental analysis (Vario EL III Elemental Analyzer, Hanau, Germany) of DTMA-sepiolite was carried out to determine C/N ratio in DTMA-sepiolite.

FTIR spectra for natural sepiolite and DTMA-sepiolite were recorded (KBr) on a Jasco FT/IR-300E Model Fourier transform infrared spectrometer to confirm the surface modification.

### 2.3. Preparation of DTMA-sepiolite

The Na-exchanged form of clay was prepared by stirring a thirty gram of sample for 24 h with 1 M NaCl. This was followed by several washings with deionized water and filtered to remove the excess NaCl and other exchangeable cations from the clay. The clay was then resuspended and filtered until a negative chloride test was obtained with 0.1 M $AgNO_3$.

Twenty gram of the Na-saturated sepiolite was dispensed in 0.5 $dm^{-3}$ of deionized water. Dodecyltrimethyl ammonium (DTMA) bromide modified sepiolite was prepared by adding quantities of the respective bromide salts equal to twice the cation exchange capacity of the sepiolite and stirring for 24 h. The clay was then washed with deionized water until free of salts and a negative bromide test had been obtained with 0.1 M $AgNO_3$ and was crushed, ground, sieved through a 63-μm sieve and dried. The prepared adsorbent was used for the adsorption studies [57].

### 2.4. Adsorption Experiments

The pH experiments were carried out by 50 ml of $2.00\times10^{-4}$ mol $dm^{-3}$ dye solutions with 0.5 g $dm^{-3}$ of natural sepiolite and DTMA-sepiolite. The pH was carefully adjusted between 1 and 8 with adding a small amount of dilute HCl or NaOH solution using a pH meter (Fisher Accumet AB15). The dye solutions were stirred using a mechanical magnetic stirrer in a 100-ml erlenmeyer flask closed with parafilm to avoid evaporation. The optimum pH was then determined as 1.5 and used throughout all adsorption experiments, which were conducted at various time intervals and temperatures (20, 30 and 40 °C) to determine when adsorption equilibrium was reached and the maximum removal of dye was attained. The solutions were filtered and were then subjected to quantitative analyses. The equilibrium concentrations of each solution were measured by a Shimadzu UV-2101PC spectrophotometer at the $\lambda_{max}$ value of 592 nm for RB19. The amounts of the dye adsorbed onto natural sepiolite and DTMA-sepiolite were determined by the difference between the initial and remaining concentration of dye solutions.

The adsorbed dye in equilibrium was also used to evaluate the adsorption isotherms for the adsorption of RB19 onto natural sepiolite and DTMA-sepiolite.

## 3. Results and Discussion

### 3.1. Chemical Composition of Sepiolite

X-ray diffraction (XRD) analysis together with the chemical analysis of (29.3% $SiO_2$, 12.4% MgO, 9.09% CaO, and 34.99% $CaCO_3$) indicates that sepiolite and dolomite are the major components along with traces of Al, K and Fe oxides in the form of impurities.

### 3.2. Elemental and FTIR Analysis

The ratio of C/N for DTMA-sepiolite from elemental analysis results is 11.31 and the calculated value of C/N ratio for DTMA bromide is 12.86 and the percentage of DTMA onto clay is 13.86. The above results confirm that the intercalation of DTMA molecules between the sepiolite layers occurs, and these results are also consistent with FTIR results.

FTIR spectra of natural sepiolite and DTMA-modified sepiolite were illustrated in Figure 2. The band at 3687 $cm^{-1}$ that corresponds to stretching ($\nu_{OH}$) vibrations of hydroxyl groups (belong to $Mg_3OH$) attached to octahedral Mg ions located in the interior blocks of natural sepiolite and DTMA-modified sepiolite [36, 58]. The band at 3621 $cm^{-1}$ indicates that assigned to H-O-H stretching vibrations of water molecules weakly hydrogen bonded to the Si-O surface in both of samples. The broad band at 3419 $cm^{-1}$, observed each samples, is due to H-O-H vibrations of adsorbed water.

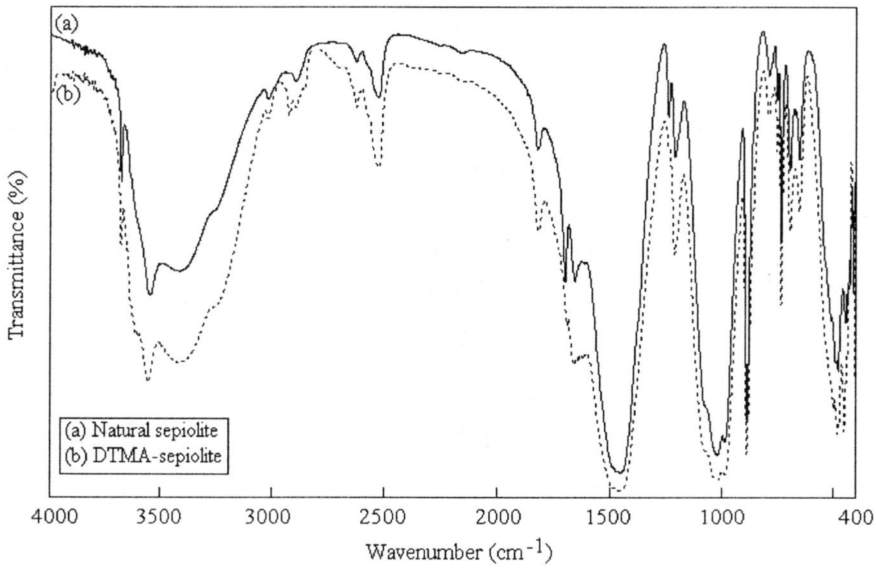


Figure 2. FTIR spectra of (a) natural sepiolite and, (b) DTMA-sepiolite.

A pair of bands at 2854 and 2928 $cm^{-1}$ was observed only with DTMA-modified sepiolite can be assigned to the symmetric and asymmetric stretching vibrations of the methylene group and their bending vibrations between 1380 and 1456 $cm^{-1}$, supporting the intercalation of surfactant (DTMA) molecules between the silica layers, but these stretching bands are not observed in the natural sepiolite [59]. The band at 1660 $cm^{-1}$ corresponds to the OH deformation of water, because the OH stretching band at 3419 $cm^{-1}$ suggests the presence of some interlamellar water [60] and the stretching vibrations obtained at 1630, 1613, and 1456 $cm^{-1}$ could be characteristics of reversibly adsorbed carbonate on the oxide surfaces [61] for both of samples and this result was well agreement with the dolomite content of the XRD results.

The Si-O coordination bands at 1210, 1089, and 981 $cm^{-1}$ are observed as a result of the Si-O vibrations. The deep band at 1018 $cm^{-1}$ represents the stretching of Si-O in the Si-O-Si groups of the tetrahedral sheet [62] and the band at 882 $cm^{-1}$ conducted with bending vibration of carbonate and two peaks at around 690 $cm^{-1}$ is represent the bending vibration of $Mg_3OH$ both of samples. The bands at 503 and 474 $cm^{-1}$ due to Si-O-Al (octahedral) and Si-O-Si bending vibrations respectively, natural sepiolite and DTMA-modified sepiolite [63].

## 3.3. Effect of pH

The effect of pH on the removal of the dye (RB19) onto natural sepiolite and DTMA-sepiolite from aqueous solution was illustrated in Figure 3. As can be seen from Figure 3, the maximum RB19 removal was observed at acidic pH of 1.5. It was observed that the adsorption procedure is highly dependent on the pH of the solution, which affects the surface charge of the adsorbent and the degree of ionization of adsorbate. Generally at low pH values, anionic dye adsorption, which is RB19 here, is increased by the electrostatic attractions between negatively charged dye anions and positively charged adsorption sites and lead to an increase in the dye adsorption. The reason for the high adsorption capacity is the strong electrostatic interactions between the $-N^+(CH_3)_3$ of DTMA-sepiolite and dye anions. When the pH of the solution is increased, the positive charge on the oxide or solution interface decreases and the adsorbent surface seems to be negatively charged. In contrast to, a lower adsorption at higher pH values may be due to the abundance of hydroxide ions and because of ionic repulsion between the negatively charged surface and the anionic dye molecules. Therefore the outer surface of the adsorbent at higher pH values has not exchangeable anions [64].

Figure 3 also shows that DTMA-sepiolite is a more effective adsorbent than that of natural sepiolite for adsorbing RB19. The adsorption capacities of natural sepiolite and DTMA-sepiolite were found to be $9.98 \times 10^{-5}$ mol $g^{-1}$ and $1.23 \times 10^{-4}$ mol $g^{-1}$, respectively, at pH 1.5 and 20 °C.

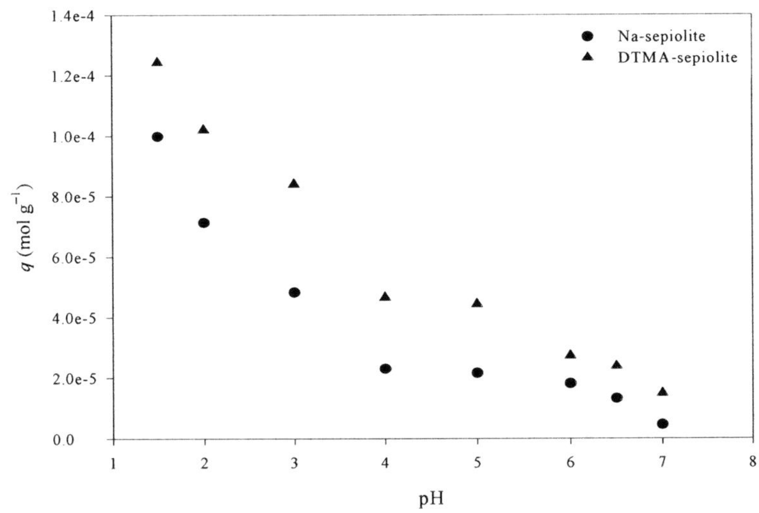


Figure 3. Effect of pH for the adsorption of RB19 onto natural sepiolite and DTMA-sepiolite.

### 3.4. Effect of Initial Dye Concentration, Contact Time and Temperature

The influence of the initial concentration of RB19 in the solutions of the adsorption onto natural sepiolite and DTMA-sepiolite was investigated at the pH value of 1.5 and 20 °C. As shown in Figure 4, when the initial dye concentration was increased from $1.0x10^{-4}$ to $3.5x10^{-4}$ mol dm$^{-3}$, the adsorption capacity of dye increased from $5.95x10^{-5}$ to $1.44x10^{-4}$ mol g$^{-1}$ for natural sepiolite and from $6.89x10^{-5}$ to $1.86x10^{-4}$ mol g$^{-1}$ for DTMA-sepiolite. This indicates that the initial dye concentrations play an important role in the adsorption capacities of RB19 onto natural sepiolite and DTMA-sepiolite.

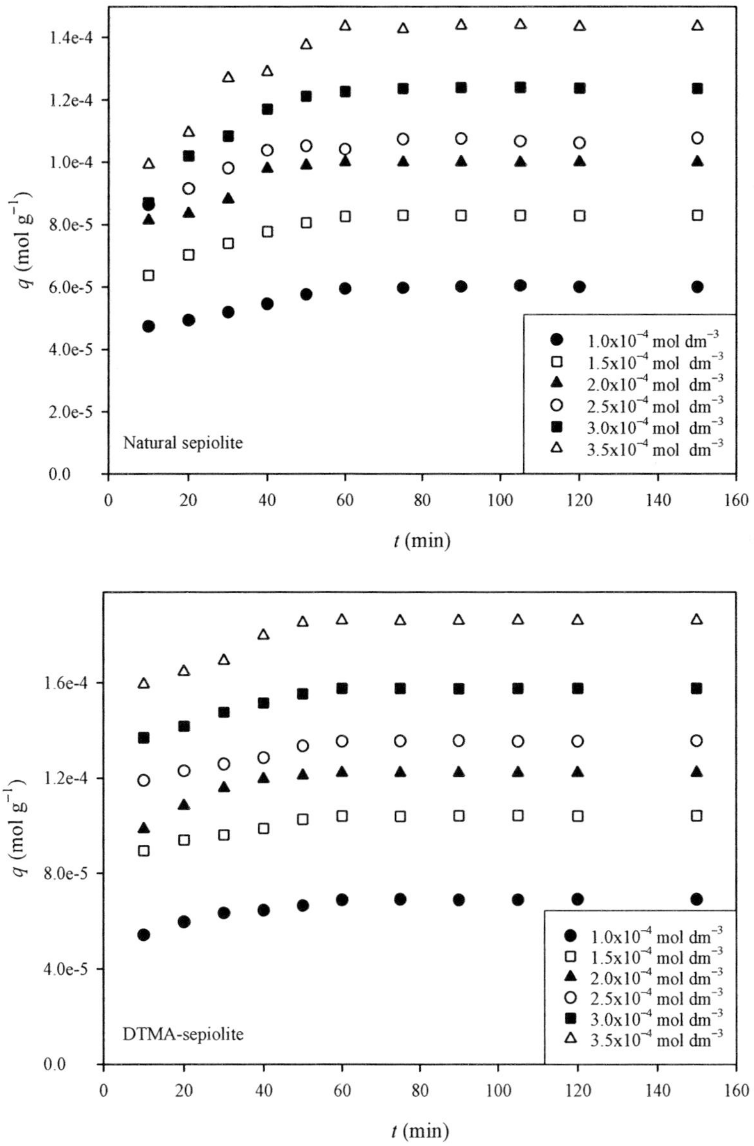


Figure 4. Effect of initial dye concentration and contact time for the adsorption of RB19 onto natural sepiolite and DTMA-sepiolite at 20 °C.

The effect of contact time on the amount of RB19 adsorbed onto natural sepiolite at 20 °C (Figure 4) was also investigated. When the equilibrium time was increased, the amount of adsorbed dye onto natural sepiolite and DTMA-sepiolite was not increased. The removal of RB19 onto natural sepiolite by adsorption was rapidly reached to equilibrium beyond which there was not significant increase in the rate of removal. Maximum adsorption of RB19 onto natural sepiolite and DTMA-sepiolite was observed at 60 min, it can be said that beyond which there is almost no further increase in the adsorption and it is thus fixed as the equilibrium contact time.

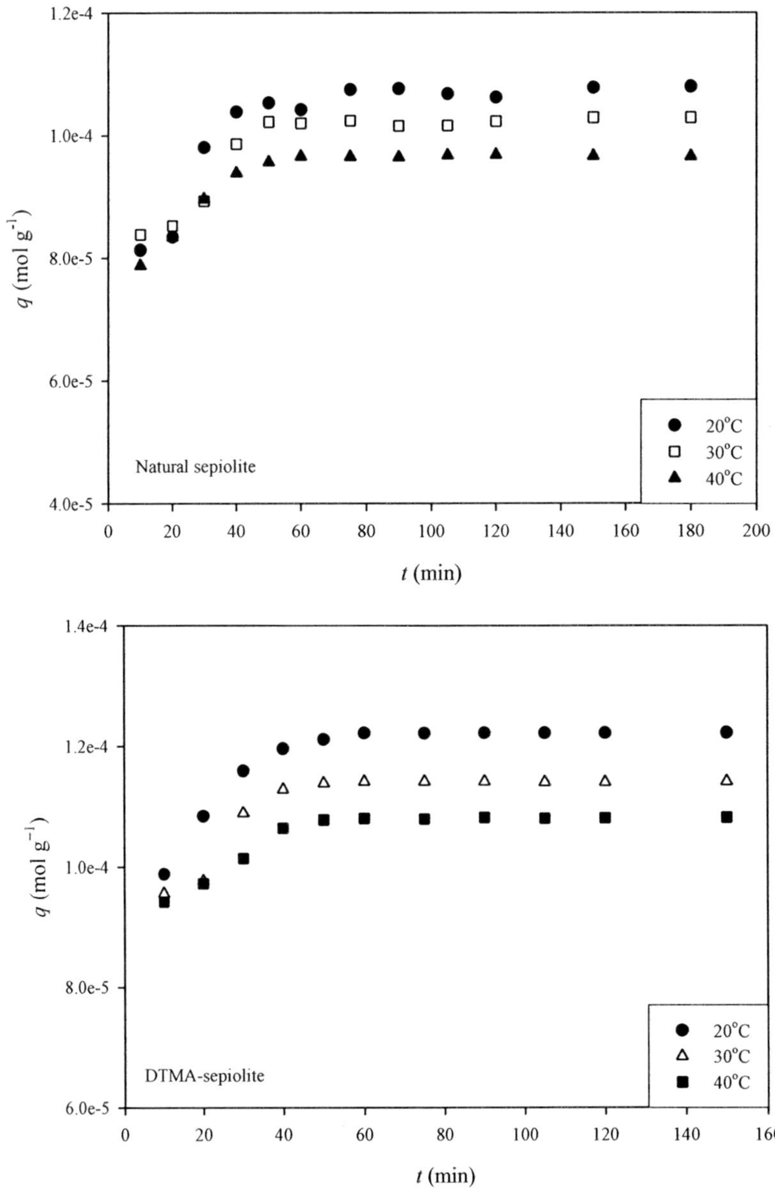


Figure 5. Effect of temperature for the adsorption of RB19 onto natural sepiolite and DTMA-sepiolite at the concentration of $2.00\times10^{-4}$ mol $dm^{-3}$.

The equilibrium adsorption capacity of RB19 onto natural sepiolite and DTMA-sepiolite at a concentration of $2.00 \times 10^{-4}$ mol dm$^{-3}$ was also affected by temperature and decreased from $9.98 \times 10^{-4}$ to $9.38 \times 10^{-5}$ mol g$^{-1}$ for natural sepiolite and from $1.23 \times 10^{-4}$ to $1.08 \times 10^{-4}$ mol g$^{-1}$ for DTMA-sepiolite with increasing temperature from 20 to 40 °C (Figure 5) which indicates that the adsorption of RB19 onto adsorbent surface was favored at lower temperatures and it is controlled by an exothermic process. This is partly due to a weakening of the attractive forces [65,66] between RB19 and natural sepiolite or DTMA-sepiolite. Based on the above results, it implies that physical adsorption mechanism may play a vital role in this system. Before and after equilibrium time, the adsorption capacity indicates different trends at various temperatures. Below the equilibrium time, an increase in the temperature leads to an increase in the dye adsorption, which shows a kinetically controlling process. After the equilibrium attained, the uptake decreases with increasing temperature indicating that the adsorption of RB19 onto natural sepiolite and DTMA-sepiolite from aqueous solution is controlled an exothermic process.

### 3.5. Adsorption Isotherms

The equilibrium adsorption isotherms are one of the most important data to identify the mechanism of the adsorption. Various adsorption isotherm (Langmuir, Freundlich and Dubinin-Radushkevich (D-R)) equations are chosen to analyze the experimental data for this study. The plots of linear form of Langmuir, Freundlich and D-R adsorption isotherms of RB19 obtained at a temperature 20 °C are depicted in Figs. 6-8. The correlation coefficients obtained from Figs. 6-8, for Langmuir, Freundlich and D-R isotherms are 0.990, 0.984 and 0.985, respectively for natural sepiolite and 0.978, 0.971 and 0.973, respectively for DTMA-sepiolite. It is evident from these data that the surface of natural sepiolite and DTMA-sepiolite is made up of homogeneous adsorption patches. In other words, the Langmuir isotherm model fits well with to the experimental data when the $r^2$ values are compared.

The maximum adsorption capacities ($q_{max}$) of natural sepiolite and DTMA-sepiolite calculated from Langmuir isotherm equation are $2.30 \times 10^{-4}$ and $3.11 \times 10^{-4}$ mol g$^{-1}$, respectively and define the total capacity of the adsorbents for RB19. $K_L$ values are $5.06 \times 10^3$ dm$^3$ mol$^{-1}$ for natural sepiolite and $3.71 \times 10^4$ dm$^3$ mol$^{-1}$ for DTMA-sepiolite.

As the value of $R_L$ calculated from eq. (2) lies between 0 and 1, the adsorption process is favorable [44,67]. Further, the $R_L$ values for this study are 0.429 for natural sepiolite and 0.435 for DTMA-sepiolite, so, the adsorption of RB19 onto natural sepiolite and DTMA-sepiolite is favorable.

One of the Freundlich constants $K_F$ indicates the adsorption capacity of the adsorbent and the values of $K_F$ at 20 °C are $1.78 \times 10^{-2}$ dm$^3$ g$^{-1}$ for natural sepiolite and $4.25 \times 10^{-2}$ dm$^3$ g$^{-1}$ for DTMA-sepiolite. The other Freundlich constant $n$ is a measure of the deviation from linearity of the adsorption and the numerical values of $n$ at 20 °C are 1.689 for natural sepiolite and 1.513 for DTMA-sepiolite and are greater than unity, indicating that RB19 is favorably adsorbed by natural sepiolite and DTMA-sepiolite and therefore the adsorption process is physical.

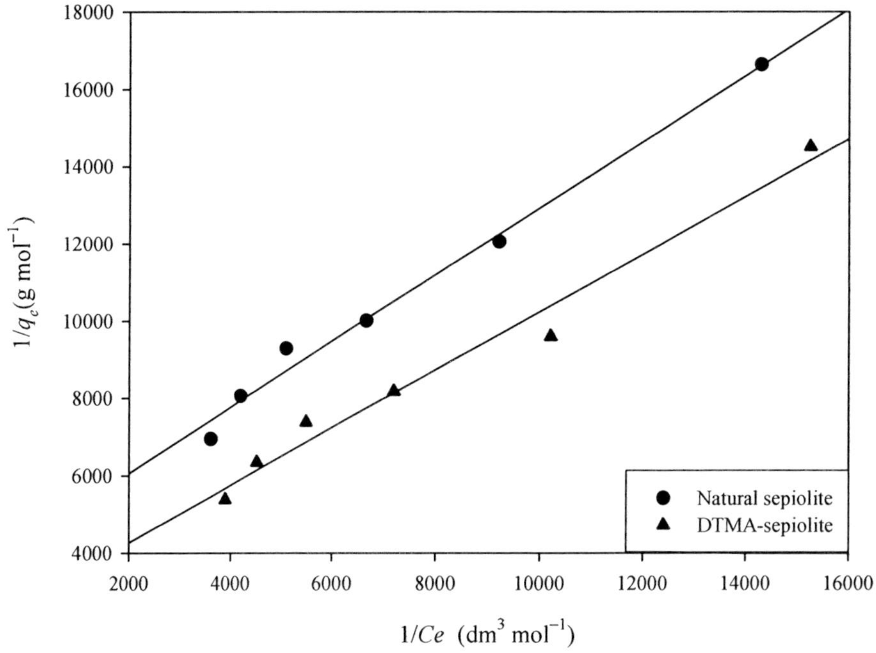


Figure 6. Langmuir adsorption isotherm plots for the adsorption of RB19 onto natural sepiolite and DTMA-sepiolite at 20 °C.

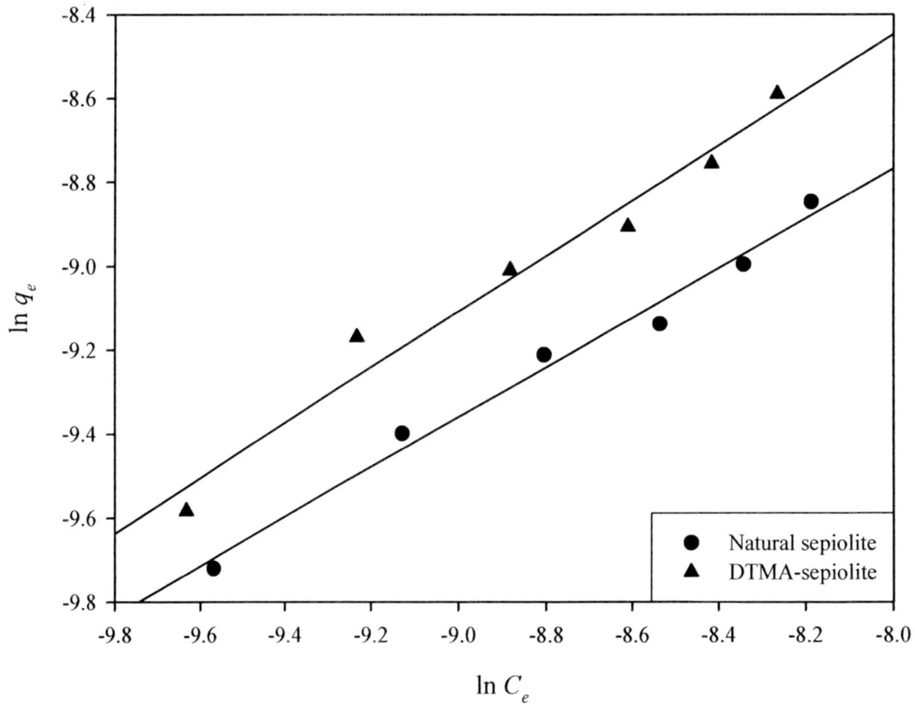


Figure 7. Freundlich adsorption isotherm plots for the adsorption of RB19 onto natural sepiolite and DTMA-sepiolite at 20 °C.

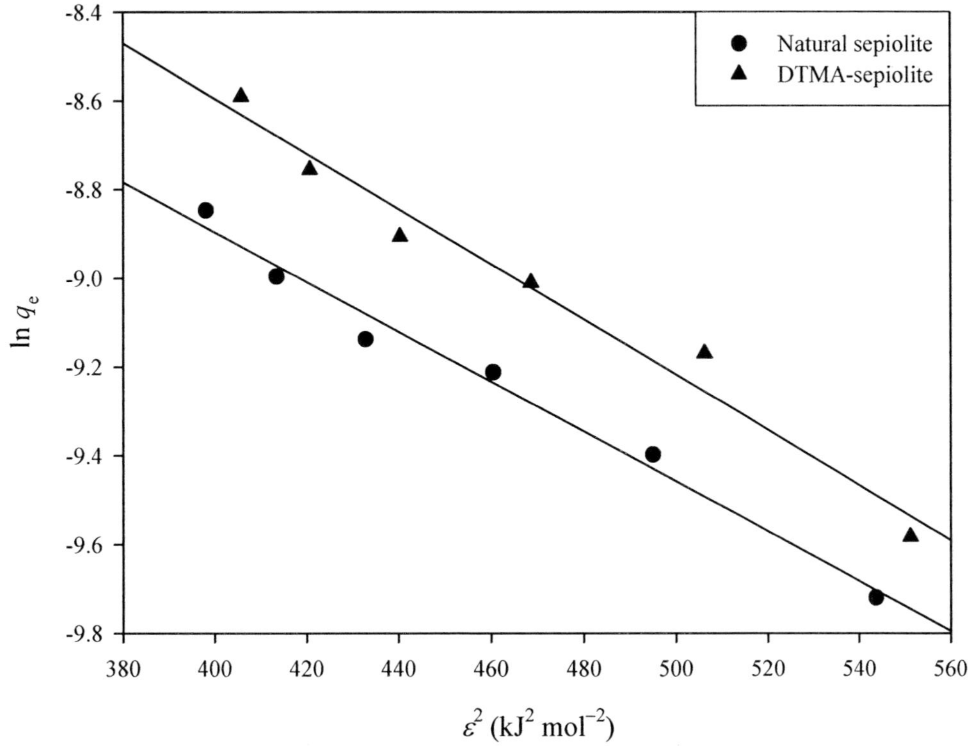


Figure 8. Dubinin-Radushkevich (D-R) adsorption isotherm plots for the adsorption of RB19 onto natural sepiolite and DTMA-sepiolite at 20 °C.

The maximum adsorption capacities of adsorbent from D-R isotherm are 1.30x10$^{-3}$ mol g$^{-1}$ for natural sepiolite and 2.24x10$^{-3}$ mol g$^{-1}$ for DTMA-sepiolite and the values of $\beta$ (D-R constant) are 5.62x10$^{-3}$ mol$^2$ kJ$^{-2}$ for natural sepiolite and 6.23x10$^{-3}$ mol$^2$ kJ$^{-2}$ for DTMA-sepiolite. The mean free energy $E$ (kJ mol$^{-1}$) of adsorption was connected with D-R adsorption isotherm and calculated from eq. (5). This parameter gives information about adsorption mechanism as chemical ion-exchange or physical adsorption. The magnitude of $E$ is between 8 and 16 kJ mol$^{-1}$, the adsorption process follows by chemical ion-exchange [68], while for the values of $E$<8 kJ mol$^{-1}$, the adsorption process is of a physical nature [69]. The numerical values for the adsorption of natural sepiolite and DTMA-sepiolite of the mean free energies are 9.435 and 8.961 kJ mol$^{-1}$, respectively, which may be correspond to a boundary of a physisorption and the predominance of van der Waals forces.

### 3.6. Kinetics of Adsorption

In order to analyze the adsorption kinetics of RB19, three kinetic models, which are the first-order, pseudo-second-order and intraparticle diffusion, have been applied for the experimental data. The plots of linear form of the first-order, pseudo-second-order (Figure 9) and intraparticle diffusion (Figure 10) for the adsorption of RB19 were obtained at various concentrations at 20 °C. The kinetic parameters for the adsorption of RB19 onto natural sepiolite and DTMA-sepiolite are tabulated in Table 1. The plots of $1/q_t$ versus $1/t$ for the first-order equation are not shown as a figure because the correlation coefficients for the first-order model are lower than that of the pseudo-second-order model (Table 1). Therefore, this implies that the adsorption of RB19 onto natural sepiolite and DTMA-sepiolite does not follow the first-order kinetic model.

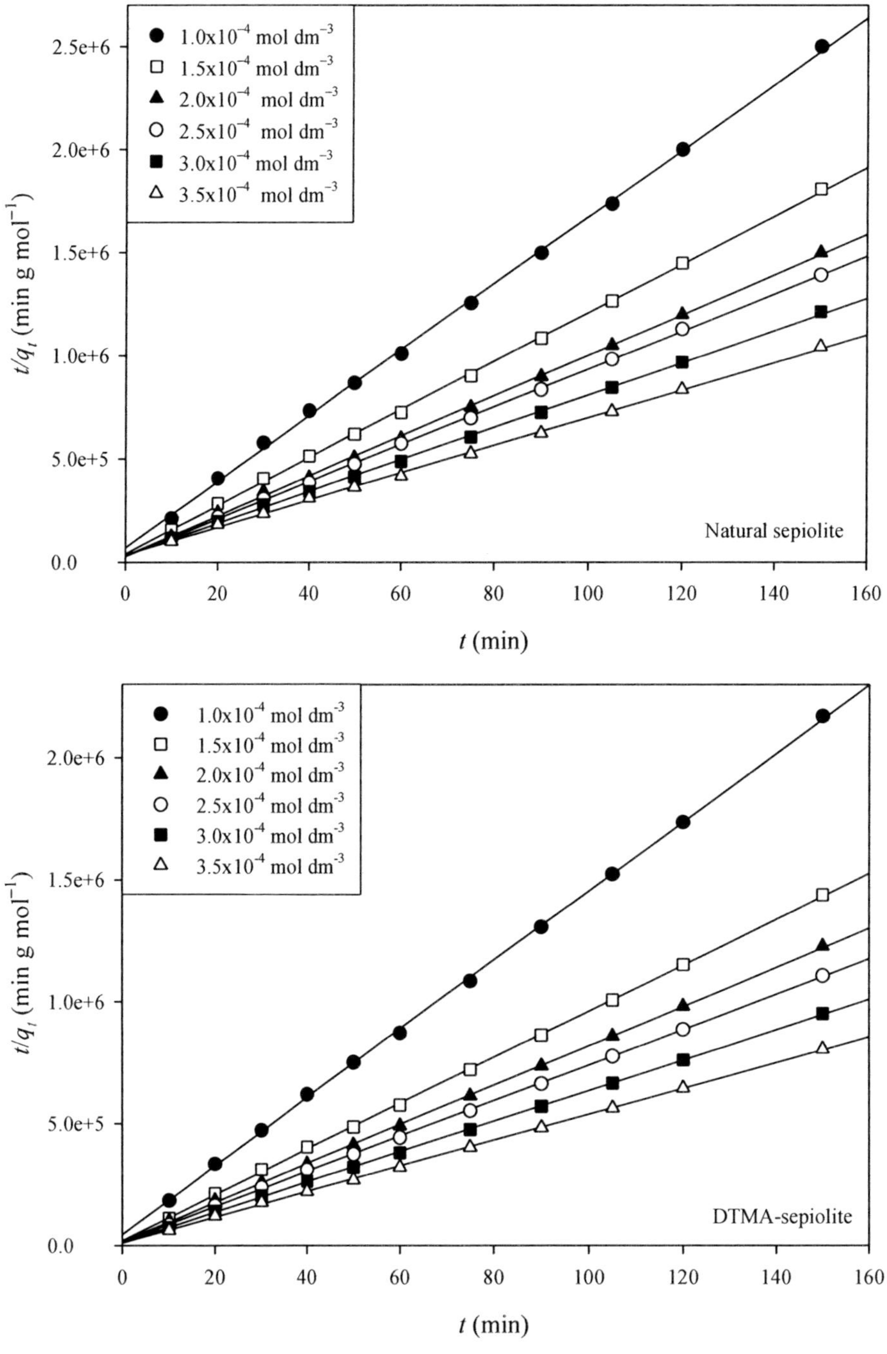


Figure 9. Pseudo-second-order kinetic plots for the adsorption of RB19 onto natural sepiolite and DTMA-sepiolite at 20 °C.

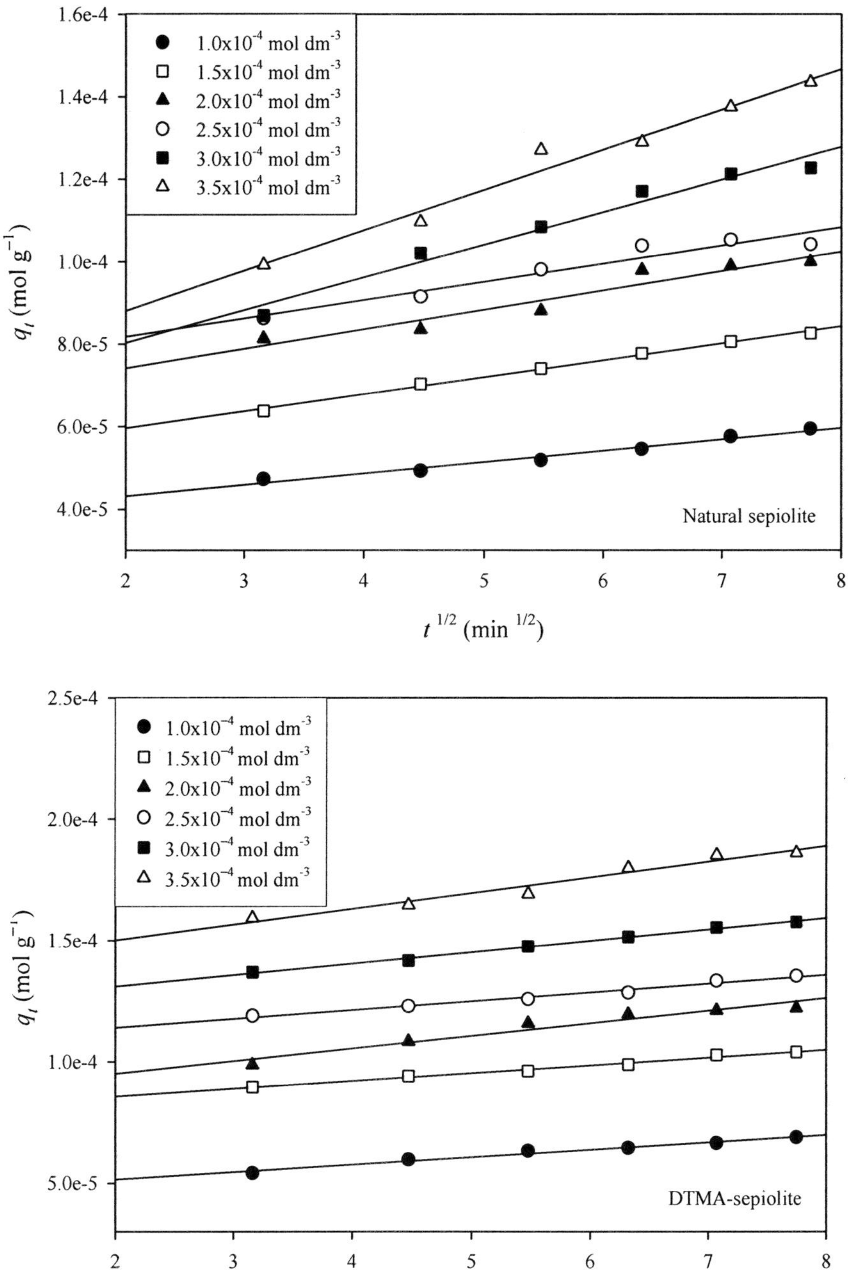


Figure 10. Intraparticle diffusion plots for the adsorption of RB19 onto natural sepiolite and DTMA-sepiolite at 20 °C.

The correlation coefficients obtained are greater than 0.999 for the pseudo-second-order model. These results imply that the adsorption system studied obeys to the pseudo-second-order kinetic model at all time intervals.

The correlation coefficients for the intraparticle diffusion model are lower than that of the pseudo-second-order kinetic model but this model indicates that the adsorption of RB19 onto natural sepiolite and DTMA-sepiolite may be followed by an intraparticle diffusion model up to 60 min.

The pseudo-second-order rate constants indicate a steady increase from $3.33x10^3$ to $4.75x10^3$ g $mol^{-1}$ $min^{-1}$ for natural sepiolite and from $4.02x10^3$ to $5.58x10^3$ g $mol^{-1}$ $min^{-1}$ for DTMA-sepiolite with an increase in the solution temperatures from 20 to 40 °C (Table 1), indicating that the adsorption of RB19 onto natural sepiolite and DTMA-sepiolite are rate-controlled.

## 3.7. Activation Energy and Thermodynamic Parameters of Adsorption

The pseudo-second-order kinetic model rate constants were used to obtained activation energy of adsorption and for this reason, ln $k_2$ is plotted versus $1/T$ (Figure 11) to derive activation energy of adsorption of RB19 onto natural sepiolite and DTMA-sepiolite. The magnitude of activation energy ($E_a$) gives a type of adsorption, which is mainly physical or chemical. The range of 5–40 kJ $mol^{-1}$ of activation energies indicates a physisorption mechanism or the range of 40–800 kJ $mol^{-1}$ suggests a chemisorption mechanism [70]. The results obtained in this study are +13.58 kJ $mol^{-1}$ for natural sepiolite and +12.49 kJ $mol^{-1}$ for DTMA-sepiolite indicating that the adsorption has a low potential barrier and assigned to a physisorption mechanism.

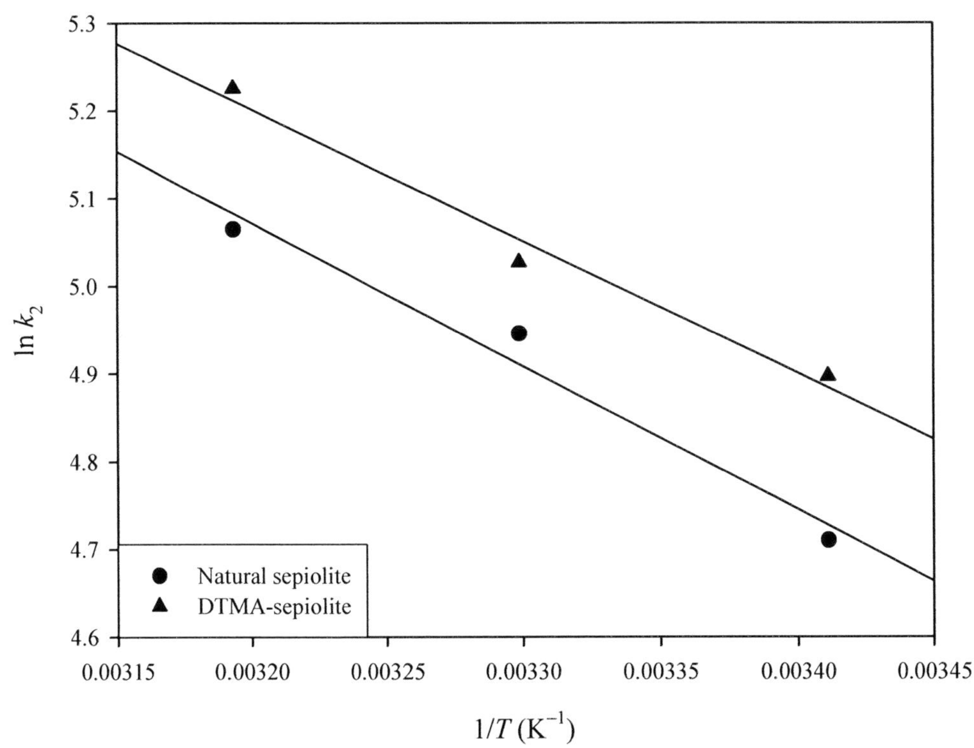


Figure 11. Arrhenius plots for the adsorption of RB19 onto natural sepiolite and DTMA-sepiolite.

**Table 1. Kinetic parameters for the adsorption of RB19 onto natural sepiolite and DTMA-sepiolite at various initial concentrations and temperatures**

| | t (°C) | $C_0$ (mol dm$^{-3}$) | $k_1$ (min$^{-1}$) | $q_1$ (mol g$^{-1}$) | $r_1^2$ | $k_2$ (g mol$^{-1}$ min$^{-1}$) | $q_2$ (mol g$^{-1}$) | $r_2^2$ | $k_p$ (mol g$^{-1}$ min$^{-1/2}$) | C (mol g$^{-1}$) | $r_p^2$ |
|---|---|---|---|---|---|---|---|---|---|---|---|
| Natural sepiolite | 20 | $1.00x10^{-4}$ | 3.437 | $6.08x10^{-5}$ | 0.774 | $3.73x10^{3}$ | $6.23x10^{-5}$ | 0.999 | $2.74x10^{-6}$ | $3.77x10^{-5}$ | 0.974 |
| | | $1.50x10^{-4}$ | 3.627 | $8.55x10^{-4}$ | 0.960 | $3.39x10^{3}$ | $8.55x10^{-5}$ | 0.999 | $4.11x10^{-6}$ | $5.14x10^{-5}$ | 0.995 |
| | | $2.00x10^{-4}$ | 3.000 | $1.02x10^{-4}$ | 0.830 | $3.33x10^{3}$ | $1.03x10^{-4}$ | 0.999 | $4.70x10^{-6}$ | $6.48x10^{-5}$ | 0.920 |
| | | $2.50x10^{-4}$ | 2.955 | $1.10x10^{-4}$ | 0.931 | $3.13x10^{3}$ | $1.10x10^{-4}$ | 0.999 | $4.42x10^{-6}$ | $7.29x10^{-5}$ | 0.929 |
| | | $3.00x10^{-4}$ | 5.101 | $1.30x10^{-4}$ | 0.981 | $1.91x10^{3}$ | $1.28x10^{-4}$ | 0.999 | $7.91x10^{-6}$ | $6.45x10^{-5}$ | 0.970 |
| | | $3.50x10^{-4}$ | 5.593 | $1.51x10^{-4}$ | 0.942 | $1.27x10^{3}$ | $1.50x10^{-4}$ | 0.999 | $9.78x10^{-6}$ | $6.84x10^{-5}$ | 0.976 |
| | 30 | $2.00x10^{-4}$ | 2.812 | $9.94x10^{-5}$ | 0.935 | $4.22x10^{3}$ | $9.88x10^{-5}$ | 0.999 | $4.17x10^{-6}$ | $6.59x10^{-5}$ | 0.971 |
| | 40 | $2.00x10^{-4}$ | 2.275 | $9.53x10^{-5}$ | 0.956 | $4.75x10^{3}$ | $9.54x10^{-5}$ | 0.999 | $3.22x10^{-6}$ | $6.90x10^{-5}$ | 0.986 |
| DTMA-sepiolite | 20 | $1.00x10^{-4}$ | 3.286 | $7.06x10^{-5}$ | 0.911 | $4.47x10^{3}$ | $7.10x10^{-5}$ | 0.999 | $3.05x10^{-6}$ | $4.55x10^{-5}$ | 0.978 |
| | | $1.50x10^{-4}$ | 1.985 | $1.06x10^{-4}$ | 0.909 | $4.33x10^{3}$ | $1.06x10^{-4}$ | 0.999 | $3.20x10^{-6}$ | $7.94x10^{-5}$ | 0.989 |
| | | $2.00x10^{-4}$ | 2.811 | $1.26x10^{-4}$ | 0.975 | $4.02x10^{3}$ | $1.25x10^{-4}$ | 0.999 | $5.19x10^{-6}$ | $8.47x10^{-5}$ | 0.940 |
| | | $2.50x10^{-4}$ | 1.711 | $1.37x10^{-4}$ | 0.870 | $3.61x10^{3}$ | $1.38x10^{-4}$ | 0.999 | $3.62x10^{-6}$ | $1.07x10^{-4}$ | 0.980 |
| | | $3.00x10^{-4}$ | 1.879 | $1.60x10^{-4}$ | 0.911 | $3.21x10^{3}$ | $1.60x10^{-4}$ | 0.999 | $4.71x10^{-6}$ | $1.22x10^{-4}$ | 0.995 |
| | | $3.50x10^{-4}$ | 2.128 | $1.89x10^{-4}$ | 0.858 | $2.43x10^{3}$ | $1.90x10^{-4}$ | 0.999 | $6.47x10^{-6}$ | $1.37x10^{-4}$ | 0.958 |
| | 30 | $2.00x10^{-4}$ | 2.490 | $1.17x10^{-4}$ | 0.860 | $4.58x10^{3}$ | $1.16x10^{-4}$ | 0.999 | $4.66x10^{-6}$ | $8.05x10^{-5}$ | 0.897 |
| | 40 | $2.00x10^{-4}$ | 1.855 | $1.10x10^{-4}$ | 0.898 | $5.58x10^{3}$ | $1.10x10^{-4}$ | 0.999 | $3.38x10^{-6}$ | $8.32x10^{-5}$ | 0.959 |

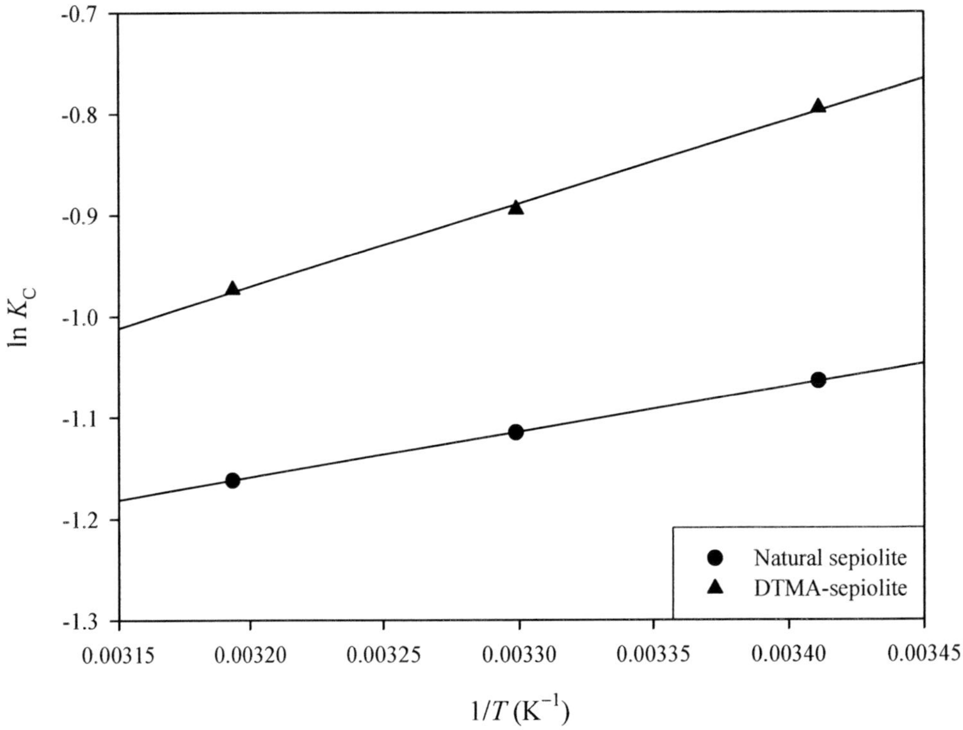


Figure 12. Plots of ln $K_C$ versus $1/T$ for the estimation of thermodynamic parameters for the adsorption of RB19 onto natural sepiolite and DTMA-sepiolite.

The thermodynamic parameters of $\Delta H^o$ and $\Delta S^o$ were obtained from the ln $K_C$ versus $1/T$ plot (Figure 12). The change of free energy for the physisorption is between –20 and 0 kJ $mol^{-1}$, but chemisorption is a range of –80 to –400 kJ $mol^{-1}$ [71]. The overall Gibbs free energy changes ($\Delta G^o$) during the adsorption process in this study, calculated from eq. (11) are 2.593 kJ $mol^{-1}$ at 20 °C, 2.809 kJ $mol^{-1}$ at 30 °C and 3.024 kJ $mol^{-1}$ at 40 °C for natural sepiolite and 1.936 kJ $mol^{-1}$ at 20 °C, 2.254 kJ $mol^{-1}$ at 30 °C and 2.534 kJ $mol^{-1}$ at 40 °C for DTMA-sepiolite. The positive values of $\Delta G^o$ at all studied temperatures are due to the fact that adsorption process is not spontaneous. The negative values of $\Delta H^o$ (–3.723 kJ $mol^{-1}$ for natural sepiolite and –6.835 kJ $mol^{-1}$ for DTMA-sepiolite) suggest the exothermic nature of adsorption. The negative values of $\Delta S^o$ (–21.55 J $K^{-1}$ $mol^{-1}$ for natural sepiolite and –29.94 J $K^{-1}$ $mol^{-1}$ for DTMA-sepiolite) suggest decreased randomness at the solid/solution interface during the adsorption of RB19 onto natural sepiolite and DTMA-sepiolite.

## 4. Conclusion

This study has related that one of the natural clays; sepiolite and surfactant modified sepiolite can be used to remove Reactive Blue 19 (RB19) from aqueous solution. Sepiolite was modified by a surfactant, which is namely dodecyltrimethyl ammonium (DTMA) bromide, to enhance the dye adsorption onto clay surface. The surface modification of natural sepiolite was examined by elemental analysis and FTIR technique. The maximum adsorption capacities of natural sepiolite and DTMA-sepiolite for the removal of dye were obtained at pH 1.5. The surfactant (DTMA)-modified sepiolite is found to be more effective adsorbent than that of natural sepiolite. Experiments were performed as a function of pH, initial dye concentration, contact time and temperature.

The Langmuir, Freundlich and Dubinin-Radushkevich (D-R) adsorption isotherm models were used for the mathematical description of the adsorption of RB19 onto natural sepiolite and DTMA-sepiolite depending on temperature and the isotherm constants were evaluated from these isotherms. Results indicated that the adsorption equilibrium data fitted well to the Langmuir isotherm model in the studied concentration range at 20 °C. The value of mean free energies from D-R isotherm of adsorption were 9.435 kJ $mol^{-1}$ for natural sepiolite and 8.961 kJ $mol^{-1}$ for DTMA-sepiolite, corresponding to a boundary of a physisorption mechanism. The kinetics of RB19 adsorption onto natural sepiolite and DTMA-sepiolite were based on the assumption of the pseudo-second-order mechanism and also followed the intraparticle diffusion model up to 60 min, whereas diffusion is not only the rate controlling step. The activation energies of adsorption were found to be the range of physical adsorption. The negative value of $\Delta H^o$ leads to an exothermic physical nature of adsorption. It may be concluded from above results that sepiolite can be used for elimination of dye pollution from wastewater since it is a low-cost, abundant and locally available adsorbent.

## REFERENCES


[1] Gong, R.; Ding, Y.; Li, M.; Yang, C.; Liu, H. and Sun, Y. (2005). Utilization of powdered peanut hull as biosorbent for removal of anionic dyes from aqueous solutions. *Dyes Pigments,* 64(3), 187-192.

[2] Allen, S.J.; McKay, G. and Porter, J.F. (2004). Adsorption isotherm models for basic dye adsorption by peat in single and binary component systems. *J. Colloid Interface Sci.,* 280(2), 322-333.

[3] Malik, P.K. (2004). Dye removal from wastewater using activated carbon developed from sawdust: Adsorption equilibrium and kinetics. *J. Hazard. Mater.,* 113(1-3), 81-88.

[4] Wong, Y.C.; Szeto, Y.S.; Cheung, W.H. and McKay, G. (2004). Adsorption of acid dyes on chitosan-equilibrium isotherm analyses. *Process Biochem.,* 39(6), 693-702.

[5] Bhattacharyya, K.G. & Sharma, A. (2005). Kinetics and thermodynamics of methylene blue adsorption on Neem (*Azadirachta indica*) leaf powder. *Dyes Pigments,* 65(1), 51-59.

[6] Mon, J.; Flury, M. and Harsh, J.B. (2006). Sorption of four triarylmethane dyes in a sandy soil determined by batch and column experiments. *Geoderma,* 133(3-4), 217-224.

[7] Ketelsen, H. & Meyer-Windel, S. (1999). Adsorption of brilliant blue FCF by soils. *Geoderma*, 90(1-2), 131-145.

[8] Juang, R.S.; Tseng, R.L.; Wu, F.C. and Lee, S.H. (1997). Adsorption behavior of reactive dyes from aqueous solutions on chitosan. *J. Chem. Technol. Biotechnol.,* 70(4), 391-399.

[9] Dubey, S.S. & Gupta, R.K. (2005). Removal behavior of babool bark (*Acacia nilotica*) for submicro concentrations of $Hg^{2+}$ from aqueous solutions: A radiotracer study. *Sep. Purif. Technol.*, 41(1), 21-28.

[10] Wang, S.; Boyjoo, Y.; Choueib, A. and Zhu, Z.H. (2005). Removal of dyes from aqueous solution using fly ash and red mud. *Water Res.*, 39(1), 129-138.

[11] Arbeloa, F.L.; Arbeloa, T.L. and Arbeloa, I.L. (1997). Spectroscopy of Rhodamine 6G adsorbed on sepiolite aqueous suspensions. *J. Colloid Interface Sci.*, 187(1), 105-112.

[12] Rytwo, G.; Tropp, D. and Serban, C. (2002). Adsorption of diquat, paraquat and methyl green on sepiolite: Experimental results and model calculations. *Appl. Clay Sci.*, 20(6), 273-282.

[13] Özcan, A.S.; Tetik, Ş. and Özcan, A. (2004). Adsorption of acid dyes from aqueous solutions onto sepiolite. *Sep. Sci. Technol.*, 39(2), 301-320.

[14] Özcan, A. & Özcan, A.S. (2005). Adsorption of Acid Red 57 from aqueous solutions onto surfactant-modified sepiolite. *J. Hazard. Mater.*, 125(1-3), 252-259.

[15] Özcan, A.; Öncü, E.M. and Özcan, A.S. (2006). Kinetics, isotherm and thermodynamic studies of adsorption of Acid Blue 193 from aqueous solutions onto natural sepiolite. *Colloids Surfaces A: Physicochem. Eng. Aspects,* 277(1-3), 90-97.

[16] Özcan, A.; Öncü, E.M. and Özcan, A.S. (2006). Adsorption of Acid Blue 193 from aqueous solutions onto DEDMA-sepiolite. *J. Hazard. Mater.*, 129(1-3), 244-252.

[17] Ravikumar, K.; Pakshirajan, K.; Swaminathan, T. and Balu, K. (2005). Optimization of batch process parameters using response surface methodology for dye removal by a novel adsorbent. *Chem. Eng. J.*, 105(3), 131-138.

[18] Harris, R.G.; Wells, J.D. and Johnson, B.B. (2001). Selective adsorption of dyes and other organic molecules to kaolinite and oxide surfaces. *Colloids Surfaces A: Physicochem. Eng. Aspects,* 180(1-2), 131-140.

[19] Wang, C.-C.; Juang, L.-C.; Hsu, T.-C.; Lee, C.-K.; Lee, J.-F. and Huang, F.-C. (2004). Adsorption of basic dyes onto montmorillonite. *J. Colloid Interface Sci.*, 273(1), 80-86.

[20] Wibulswas, R. (2004). Batch and fixed bed sorption of methylene blue on precursor and QACs modified montmorillonite. *Sep. Purif. Technol.*, 39(1-2), 3-12.

[21] Polubesova, T.; Epstein, M.; Yariv, S.; Lapides, I. and Nir, S. (2004). Adsorption of alizarinate-micelle complexes on Na-montmorillonite. *Appl. Clay Sci.*, 24(3-4), 177-183.

[22] Ogawa, M.; Kawai, R. and Kuroda, K. (1996). Adsorption and aggregation of a cationic cyanine dye on smectites. *J. Phys. Chem.-US.*, 100(40), 16218-16221.

[23] Arbeloa, F.L.; Herran Martinez, J.M.; Arbeloa, T.L. and Arbeloa, I.L. (1998). Hydrophobic effect on the adsorption of rhodamines in aqueous suspensions of smectites. The rhodamine 3B/Laponite B system. *Langmuir*, 14(16), 4566-4573.

[24] Czímerová, A.; Bujdák, J. and Gáplovský, A. (2004). The aggregation of thionine and methylene blue dye in smectite dispersion. *Colloids Surfaces A: Physicochem. Eng. Aspects,* 243(1-3), 89-96.

[25] Armagan, B.; Ozdemir, O.; Turan, M. and Çelik, M.S. (2003). The removal of reactive azo dyes by natural and modified zeolites. *J. Chem. Technol. Biotechnol.*, 78(7), 725-732.

[26] Armağan, B.; Turan, M. and Çelik, M.S. (2004). Equilibrium studies on the adsorption of reactive azodyes into zeolite. *Desalination*, 170(1), 33-39.

[27] Benkli, Y.E.; Can, M.F.; Turan, M. and Çelik, M.S. (2005). Modification of organo-zeolite surface for the removal of reactive azo dyes in fixed-bed reactors. *Water Res.*, 39(2-3), 487-493.

[28] Wang, S.; Li, H. and Xu, L. (2006). Application of zeolite MCM-22 for basic dye removal from wastewater. *J. Colloid Interface Sci.*, 295(1), 71-78.

[29] Özcan, A.S. & Özcan, A. (2004). Adsorption of acid dyes from aqueous solutions onto acid-activated bentonite. *J. Colloid Interface Sci.*, 276(1), 39-46.

[30] Özcan, A.S.; Erdem, B. and Özcan, A. (2004). Adsorption of Acid Blue 193 from aqueous solutions onto Na-bentonite and DTMA-bentonite. *J. Colloid Interface Sci.*, 280(1), 44-54.

[31] Özcan, A.S.; Erdem, B. and Özcan, A. (2005). Adsorption of Acid Blue 193 from aqueous solutions onto BTMA-bentonite. *Colloids Surfaces A: Physicochem. Eng. Aspects,* 266(1-3), 73-81.

[32] Hu, Q.H.; Qiao, S.Z.; Haghseresht, F.; Wilson, M.A. and Lu, G.Q. (2006). Adsorption study for removal of basic red dye using bentonite. *Ind. Eng. Chem. Res.*, 45(2), 733-738.

[33] Al-Asheh, S.; Banat, F. and Abu-Aitah, L. (2003). The removal of methylene blue dye from aqueous solutions using activated and non-activated bentonites. *Adsorpt. Sci. Technol.*, 21(5), 451-462.

[34] Gonzales-Pradas, E.; Villafranca-Sanchez, M.; Socias-Viciana, M.; Fernandez-Perez, M. and Urena- Amate, M.D. (1999). Preliminary studies in removing atrazine, isoproturon and imidacloprid from water by natural sepiolite. *J. Chem. Technol. Biotechnol.*, 74(5), 417-422.

[35] Grim, R.E. (1968). *Clay Mineralogy,* New York, McGraw Hill.

[36] Rytwo, G.; Nir, S.; Serban, C.; Margulies, L.; Casal, B.; Merino, J.; Ruiz-Hitzky, E. and Serratosa, J.M. (1998). Adsorption of monovalent organic cations on sepiolite: Experimental results and model calculations. *Clays Clay Miner.*, 46(3), 340-348.

[37] Sabah, E.; Turan, M. and Celik, M.S. (2002). Adsorption mechanism of cationic surfactants onto acid- and heat-activated sepiolites. *Water Res.*, 36(16), 3957-3964.

[38] Sheng, G.; Xu, S. and Boyd, S.A. (1996). Cosorption of organic contaminants from water by hexadecyltrimethylammonium-exchanged clays. *Water Res.*, 30(6), 1483-1489.

[39] Stucki, J.W.; Wu, J.; Gan, H.M.; Komadel, P. and Banin, A. (2000). Effects of iron oxidation state and organic cations on dioctahedral smectite hydration. *Clays Clay Miner.*, 48(2), 290-298.

[40] Dentel, S.K.; Jamrah, A.I. and Sparks, D.L. (1998). Sorption and cosorption of 1,2,4-trichlorobenzene and tannic acid by organo-clays. *Water Res.*, 32(12), 3689-3697.

[41] Shen, Y.H. (2001). Preparation of organobentonite using non-ionic surfactants. *Chemosphere*, 44(5), 989-995.

[42] Sohn, S. & Kim, D. (2005). Modification of Langmuir isotherm in solution systems-definition and utilization of concentration dependent factor. *Chemosphere*, 58(1), 115-123.

[43] Langmuir, I. (1918). The adsorption of gases on plane surfaces of glass, mica and platinum. *J. Am. Chem. Soc.*, 40(9), 1361-1403.

[44] Hall, K.R.; Eagleton, L.C.; Acrivos, A. and Vermeulen, T. (1966). Pore- and solid-diffusion kinetics in fixed-bed adsorption under constant-pattern conditions. *Ind. Eng. Chem. Fundam.*, 5(2), 212-223.

[45] Freundlich, H.M.F. (1906). Über die adsorption in lösungen. *Z. Phys. Chem.*, 57, 385-470.

[46] Benhammou, A.; Yaacoubi, A.; Nibou, L. and Tanouti, B. (2005). Adsorption of metal ions onto Moroccan Stevensite: Kinetic and isotherm studies. *J. Colloid Interface Sci.*, 282(2), 320-326.

[47] Dubinin, M.M. & Radushkevich, L.V. (1947). Proc. Acad. Sci. U.S.S.R. Phys. Chem. Sect., 55, 331.

[48] Hobson, J.P. (1969). Physical adsorption isotherms extending from ultrahigh vacuum to vapor pressure. *J. Phys. Chem.*, 73(8), 2720-2727.

[49] Hasany, S.M. & Chaudhary, M.H. (1996). Sorption potential of Hare River sand for the removal of antimony from acidic aqueous solution. *Appl. Radiat. Isot.*, 47(4), 467-471.

[50] Kannan N. & Sundaram, M.M. (2001). Kinetics and mechanism of removal of methylene blue by adsorption on various carbons- a comparative study. *Dyes Pigments,* 51(1), 25-40.

[51] Ho, Y.S. & McKay, G. (1998). Kinetic models for the sorption of dye from aqueous solution by wood. *Process Safety Environ. Protect.*, 76(B2), 183-191.

[52] Weber Jr., W.J. and Morriss, J.C. (1963). Kinetics of adsorption on carbon from solution. *J. Sanitary Eng. Div. Am. Soc. Civ. Eng.*, 89, 31-60.

[53] Jain, C.K.; Singhal, D.C. and Sharma, M.K. (2004). Adsorption of zinc on bed sediment of River Hindon: Adsorption models and kinetics. *J. Hazard. Mater.*, 114(1), 231-239.

[54] Sarkar, M.; Acharya, P.K. and Bhattacharya, B. (2003). Modeling the adsorption kinetics of some priority organic pollutants in water from diffusion and activation energy parameters. *J. Colloid Interface Sci.*, 266(1), 28-32.

[55] Choy, K.K.H.; Porter, J.F. and McKay, G. (2004). Intraparticle diffusion in single and multicomponent acid dye adsorption from wastewater onto carbon. *Chem. Eng. J.*, 103(1-3), 133-145.

[56] Taylor, R.K. (1985). Cation exchange in clays and mudrocks by methylene blue. *J. Chem. Technol. Biotechnol.*, 35A, 195-207.

[57] Li, Z. & Bowman, R.S. (2001). Retention of inorganic oxyanions by organo-kaolinite. *Water Res.*, 35(16), 3771-3776.

[58] Shariatmadari, H.; Mermut, A.R. and Benke, M.B. (1999). Sorption of selected cationic and neutral organic molecules on palygorskite and sepiolite. *Clays Clay Miner.*, 47(1), 44-53.

[59] Madejová, J. (2003). FTIR techniques in clay mineral studies. *Vib. Spectrosc.*, 31(1), 1-10.

[60] de Haro, M.C.J.; Ruiz-Conde, A. and Perez-Rodriguez, J.L. (1998). Stability of n-butylammonium vermiculite in powder and flake forms. *Clays Clay Miner.*, 46(6), 687-693.

[61] Das, D.P.; Das, J. and Parida, K. (2003). Physicochemical characterization and adsorption behavior of calcined Zn/Al hydrotalcite-like compound (HTlc) towards removal of fluoride from aqueous solution. *J. Colloid Interface Sci.*, 261(2), 213-220.

[62] Sabah, E. & Celik, M.S. (2002). Interaction of pyridine derivatives with sepiolite. *J. Colloid Interface Sci.,* 251(1), 33-38.

[63] Temuujin, J.; Jadambaa, Ts.; Burmaa, G.; Erdenechimeg, Sh.; Amarsanaa, J. and MacKenzie, K.J.D. (2004). Characterisation of acid activated montmorillonite clay from Tuulant (Mongolia). *Ceramics Int.,* 30(2), 251-255.

[64] Wu, Z.; Ahn, I.-S.; Lee, C.-H.; Kim, J.-H.; Shul, Y.G. and Lee, K. (2004). Enhancing the organic dye adsorption on porous xerogels. *Colloids Surfaces A: Physicochem. Eng. Aspects,* 240(1-3), 157-164.

[65] Singh, D. (1998). Effect of different factors on the adsorption of phosphamidon on two different types of Indian soil. *Adsorpt. Sci. Technol.,* 16(8), 583-594.

[66] Tsai, W.T.; Lai, C.W. and Hsien, K.J. (2004). Adsorption kinetics of herbicide paraquat from aqueous solution onto activated bleaching earth. *Chemosphere,* 55(6), 829-837.

[67] Weber, T.W. & Chakravorti, R.K. (1974). Pore and solid diffusion models for fixed-bed adsorbers. *J. Am. Inst. Chem. Eng.,* 20(2), 228-238.

[68] Helfferich, F. (1962). *Ion Exchange,* New York, McGraw-Hill.

[69] Onyango, M.S.; Kojima, Y.; Aoyi, O.; Bernardo, E.C. and Matsuda, H. (2004). Adsorption equilibrium modelling and solution chemistry dependence of fluoride removal from water by trivalent-cation-exchanged zeolite F-9. *J. Colloid Interface Sci.,* 279(2), 341-350.

[70] Nollet, H.; Roels, M.; Lutgen, P.; Meeren, P.V. der and Verstraete W. (2003). Removal of PCBs from wastewater using fly ash. *Chemosphere,* 53(6), 655-665.

[71] Jaycock, M.J. and Parfitt, G.D. (1981). *Chemistry of Interfaces,* Onichester, Ellis Horwood Ltd.

In: Hazardous Materials and Wastewater
Editor: Allison A. Lewinsky, pp. 251-273
ISBN 1-60021-257-3


*Chapter 8*

# REMOVAL OF AROMATIC AMINES FROM AQUEOUS SOLUTIONS BY LIGAND EXCHANGER COBALT(II)-CDAE-SPOROPOLLENIN

***Ahmet Ayar***
Department of Chemistry, Faculty of Sciences and Arts,
University of Niğde, Niğde, Turkey

## ABSTRACT

Aromatic amines, such as aniline and its substituted derivatives, are widely used as raw materials in many industrial processes, and very harmful chemical species constituting the focus of many environmental concerns. Although it is well known that they can easily seep through soil and contaminate groundwater due to their high solubility in water, unfortunately, these compounds are still being released to the environment directly through some industrial processes and/or indirectly through decomposition of herbicides and pesticides. Therefore, removal of such types of compounds from industrial wastewaters has become an of prime importance environmental concern in many research areas. In the past decades, adsorption on activated carbon had been a primary way to overcome mentioned environmental troubles; however, nowadays this is not the case due to the irreversible adsorptive characteristics of activated carbon, and accordingly high regeneration costs coming out. Nowadays, attempts have, therefore, intensified towards developing alternative adsorbents, ligand-exchange based adsorbents, which can be easily regenerated even at relatively low temperatures. Nowadays, the ligand-exchange processes have been increasingly employed to overcome inherent drawbacks of some traditional processes. Ligand-exchange behaviors of aromatic amines have been investigated using ligand-exchange sporopollenin in several column studies, which deal with chromatographic separation, adsorption and kinetics. Adsorption and kinetics studies are essentially based on the adsorption data obtained from breakthrough curves, and the adsorption behaviors of ligands have been analyzed by Langmuir and Freundlich isotherm models. The characteristics of particular adsorption process have also been investigated using

Scatchard plot analysis. In this review, adsorption of aniline and its some substituted derivatives on a modified sporopollenin exhibiting ligand exchanger properties is discussed in detail. The scope of the present article is constructed on the basis of experimental results obtained through some previously conducted systematic column studies, and the results of these studies are evaluated in a collective manner. Thus, some proposed modification methods of sporopollenin, the theory of ligand-exchange, breakthrough technique, related adsorption isotherms and Scatchard plot analysis are described, and the adsorption phenomena of the studied resin, including column operation and rate-controlling step of processes are also discussed in detail.

## INTRODUCTION

Aromatic amines, such as aniline and its substituted derivatives, are widely used in the manufacturing of azo dyes, rubbers, plastics cosmetics, pharmaceuticals, carbamate and pesticides [1, 2, 3, 4, 5]. However, these compounds which find many useful applications in our life, are very harmful if they are discharged into natural water resources [6, 7]. They are known to be toxic water pollutants and their presence in wastewater, even in very low concentrations, may create serious public health and environmental problems [3, 8, 9]. Although they can easily seep through soil and contaminate groundwater due to their high solubility in water, unfortunately, these compounds are still being released to the environment directly as industrial residues or indirectly as breakdown products of herbicides and pesticides [10]. Hence, aromatic amines constitute a group of priority water pollutants which are present at high concentrations in groundwater and soils nearby many industrial contaminated zones, and concentration of these chemical species have to be continuously controlled. Therefore, effective removal of such type of organic pollutants from wastewaters and industrial wastes still remains a major topic of recent researches. Being important water pollutants these compounds have to be under continuous controlling in industrial wastewater, even in very low concentrations.

Several attempts have been practiced to treat various effluents for the removal of organic pollutants from wastewaters. Owing to their ligand-exchange or specific surface properties, chelating resins [11, 12] and activated carbon [13, 14] are the most commonly used adsorbents in adsorption process. Adsorption on activated carbon has been a primary choice to overcome these environmental troubles; however, from the point of operational cost view it is relatively expensive to use activated carbon because of regeneration troubles. In order to decrease the cost of treatment, attempts have intensified towards developing chelating resins (coordinating resins), which can easily be regenerated even at relatively low temperatures. Hence, nowadays, the ligand-exchange processes have been increasingly employed to overcome inherent drawbacks of some traditional processes. In these processes, ligands in liquid phase are exchanged with the ligands in solid phase. The ligands taken up from the external solutions replace solvent molecules in the solvation shell of the metal ion or replace other ligands in previously formed complexes [15]. Although ligand exchange is being increasingly employed for the chromatographic separation and purification of organic compounds, the possibility of its application to wastewater treatment and pollution control has not received any significant attention. Therefore, there is a need for the development of different adsorbents on which organic compounds can be removed satisfactorily.

In general, a solid phase support includes two aspects: the base support and the stationary phase that is chemically or physically immobilized onto a solid support and carries out the necessary functions. The base support plays a dominant role in the mechanical, chemical and thermal stability of packing materials. One of such materials is sporopollenin which possesses a high content of functional groups available for modification. This material may be successfully used at high temperatures, since it has a constant chemical structure that is stable under high temperature and in the presence of concentrated acids [16, 17, 18, 19, 20, 21, 22].

In this article, the technical feasibility of sporopollenin, chemically modified with a cobalt(II) loaded carboxylated-ethylenediamine, has been reviewed for removal of aromatic amines from aqueous solution. Ligand exchange reactions of aromatic amines on modified sporopollenin resin have been investigated in several studies, which deal with the exchange equilibrium, separation and some other adsorption parameters of these reactions and the kinetics of the ligand exchange reactions [23, 24, 25, 26, 27].The purpose of this review is to provide a summary of recent information concerning the use of sporopollenin as a solid support, and to represent some aspects of techniques regarding adsorption isotherms, kinetics, and column operation.

# THEORY

## General Principles of Ligand Adsorption

Adsorption processes have proven successful in removal of neutral compounds with chelating resins; being the most widely used type of adsorbents due to their high selectivity for the ligand adsorption. So, ligand adsorption is one of the most widely used techniques for removal of neutral compounds because of its simplicity, reversibility and widespread applicability. This technique has recently become more useful in recent years due to the development of effective chelating resins.

Ligand adsorption depends upon types of interactions between solute molecules and metal immobilized on a solid phase. Adsorption of ligands onto ligand-exchange resin takes place as a result of versatile inter-molecular forces that involve several kinds of interactions, which attract the ligand to the stationary phase from solution. However, in ligand adsorption, the primary effect is the metal–ligand coordination, and the ligand-exchange is the most fundamental reaction. The ligands, taken up from the external solutions, replace solvent molecules in the solvation shell of the metal ion or replace other ligands in previously formed complexes [15]. Briefly, ligand-exchange is based on the formation of metal complexes between the central metal and the analytes.

Ligand-exchange was conceptualized and formally introduced by Helfferich in 1961[28]. Cu(II)-or Ni(II)-loaded weak acid cation-exchange resins were used by Helfferich to enhance the sorption of various ligands through relatively strong Lewis acid-base interactions. In such ligand-exchange processes, the water molecules (weak ligands) present at the coordination spheres of metal ion immobilized on the cation-exchange resin are replaced by relatively strong ligands. By using ligand-exchange technique, ligands such as aromatic amines that possess pairs of free electrons can be removed from their medium by the formation of complexes with the metal attached to the solid support, and water or liquid coordinates of the metal will be displaced. The selectivity for the ligand is high when the complexes formed with the metal ion are strong. There is a strong 'driving force' for sorption because the ligands are efficiently sequestered by the metal ion in the resin. The coordinatively saturated complexes are formed in the resin even when the ligand concentration in the external solution is very low [29]. Consequently, ligand-exchange resins (i.e. ion exchangers containing complexing metal ions) have the particular advantage of combining high sorption capacity with high selectivity for the ligand.

Another application for this technique is the ligand-exchange chromatography that has been reviewed by Davankov and Semechkin [30]. Separation in ligand-exchange chromatography depends upon the reversible adsorption of neutral solute molecules to immobilized matrix of resin. In this technique, compounds are separated through the ligands formation and breaking of labile coordinate bonds to a central metal atom on the stationary phase as a partition between the mobile and the stationary phase [31]. A cation-exchange resin can also be used to separate compounds that can form coordination complexes with the metal ions attached to the resin. For example, a CDAE-sporopollenin resin in the cobalt(II) form can be used to separate amino acids which are able to form coordination compounds with metal ions using ammonia as a competing ligand [32]. Ligand-exchange sporopollenin resin has been shown to have great potentialities for the separation and removal of such ligands as amino acids, nucleic acid bases, nucleosides and others from liquid media by the formation of complexes with a transition metal attached to a solid support [33, 34].

## Ligand Binding and Scatchard Plot Analysis

The optimization of a ligand adsorption process requires an understanding of the factors that govern the interaction between ligands and immobilized metal ions on the matrix of the solid phase. In a heterogeneous system, the interactions between immobilized metal ions and ligands are extremely complex since ligand retention on a ligand-exchange phase is a combined effect of donor–acceptor complex formation, electrostatic and/or hydrophobic interaction. These interactions often involve multiple-site interactions between a single ligand and a number of binding sites distributed on the adsorbent surface. Hence, it is believed that the mechanism of ligand adsorption is not only determined by the donor–acceptor complex formation.

Ligand adsorption is considerably different from conventional adsorption processes because of the non-specific ligand binding including multivalent interactions. Although, the semi-reciprocal and double-reciprocal transformations of the classical Langmuir equation are widely used to describe the adsorption behavior of the solutes in adsorption processes, types of interactions usually can not be satisfactorily elucidated. However, previous studies have shown that Scatchard transformation [35] of the Langmuir equation offers more compact information about the characteristics of the isotherm in a simple manner [36, 37, 38, 39, 40].

Analysis of ligand binding is based on the prudence that the interaction between the ligand and immobilized metal ions on the matrix of the solid phase may be represented by the following equilibrium:

$$\overline{ML_{n-1}} + L \underset{k_2}{\overset{k_1}{\rightleftharpoons}} \overline{ML_n} \qquad n = 1,2,3,\ldots \qquad (1)$$

where M is metal ion; L is the ligand; n is the coordinative valence of metal ion for the ligand; $k_1$ and $k_2$ are the respective association and dissociation rate constants take into account the mass transfer resistances, which include the ligand movement from the mobile phase to the immobilized metal complex. The bar denotes the solid phase. The mass-transfer rate equation that is consistent with the Langmuir isotherm can be described by

$$\frac{dq}{dt} = k_1 C_e (q_m - q_e) - k_{-2} q_e \quad (2)$$

where $C_e$ is the concentration of the ligand in the mobile phase; $q_e$ is the amount of ligand adsorbed onto the matrix, and $q_m$ is the maximum capacity of the matrix. The definition of the $(q_m - q_e)$ concentration is empirical and has no actual physical meaning. At equilibrium, the rate of forward interaction becomes equal to the rate of reverse interaction, dq/dt becomes zero and Eq. (3) can be converted to several linear transformations of the Langmuir equation, including the semi-reciprocal, double reciprocal, and Scatchard equations, respectively.

$$\frac{C_e}{q_e} = \frac{1}{q_m K_b} + \frac{C}{q_m} \quad (3)$$

$$\frac{1}{q_e} = \frac{1}{q_m} + \frac{1}{q_m K_b} \cdot \quad \frac{1}{C_e} \quad (4)$$

$$\frac{q_e}{C_e} = K_b (q_m - q_e) \quad (5)$$

Eq. (6) represents a straight line whose slope will be the association binding constant of the ligand ($K_b$) and the intercepts on ordinate will be $K_b q_m$. The graphical representation of such an equation is called the Scatchard plot analysis.

The shape of the Scatchard plot is related to the type of interaction of the ligand with the resin matrix. Scatchard plot should be a straight line for a simple binding site, which follows the Langmuir model. The presence of a deviation from linearity on a plot based on Scatchard analysis usually points out that the adsorption process is governed by more than one type of the binding sites for studied concentration range, whilst the linearity of the Scatchard plot indicates that the binding sites are identical and independent. For instance, the presence of two slopes in a curvilinear Scatchard plot may reveal the existence of two type of binding sites [41]. If the Scatchard plot is linear with a negative slope, it is related to interaction between the ligand and the binding sites that follows the Langmuir model. If it is convex, negative cooperativity is observed. Moreover, if $q_e/C_e$ versus $q_e$ plot shows a concave curve, the interaction of the ligand is governed by a positive cooperativity phenomenon between strong and weak binding sites [42, 43, 44]. The Scatchard transformation is more robust than the semi-reciprocal or double reciprocal transformation of the Langmuir equation, since it is more sensitive to changes in slopes in a plot analysis [45].

In ligand adsorption systems, the experimental data though occasionally reveal that the Scatchard plots display serious deviations from straight lines because of the participation of nonindependent binding sites, and Langmuir isotherm model cannot describe the characteristics of the experimental data satisfactorily. Then, the Freundlich isotherm which is concerned with adsorption kind that occurs as the power function of solute concentration can be used to describe such adsorption behavior. However, as pointed out by Sharma [43], that the amount of adsorbed solute increases indefinitely with the solute concentration in solution is a shortcoming of the Freundlich isotherm, and sometimes unrealistic adsorption parameters may come out.

# METHOD

## Column Technique and Breakthrough Data

Adsorption processes can be performed in two distinct modes; the batch and column methods. Admittedly, the adsorption data needed can be acquired by using a classical batch method, however, in most ligand-exchange processes, the ligand exchanger adsorbents are usually used under the column conditions due to the ease of operation. Furthermore, the column method allows the usage of breakthrough data to acquire adsorption isotherms, and the experimental data can be acquired in the frontal analysis mode at equilibrium because there is no separation between solid and liquid phase. To get highly accurate adsorption data, the most convenient manner is to constitute the breakthrough curve for each concentration of the solute by frontal analysis. This analysis technique had been called "frontal analysis" by James and Phillips [46]. Although the frontal analysis needs more solute than zonal elution studies, that frontal analysis can provide extensive information on the adsorption behavior of the solute make it preferable for adsorption experiments. Frontal analysis is also considered to be the most accurate and reliable method of measurement of the isotherm data [47]. As a result, in comparison to the batch method, a simple frontal analysis provides more compact information about adsorption, adsorption kinetics and chromatographic behaviors of solute on the same adsorbent. Despite the simplicity and ease of column operations, the required equipments are, generally, more complicated and cost-effective comparing to the batch method, being unique drawback of the column methods.

In a column method, the ligand-exchange solid phase is packed into a column by using dry and conventional slurry packing methods. However, slurry packing method is more general than the dry packing because of its effectiveness. In the slurry packing method, as in common liquid chromatography, the solid phase is loaded as its suspension into the column that previously filled with diluents, and by means of gravity the solid phase is packed to the column entirely. It should be noted that in column methods the solid phase is never allowed to dry, and so it is left in contact with a suitable conditioning agent.

After the column has been conditioned, the sample is continuously fed into the column until the solute concentration in the effluent equals to the influent concentration. The solute concentration in the effluent can easily be evaluated on the basis of changing in the absorbance intensity by using a detector adapted to the outlet of the column, and the data is recorded by the data logger as the feed solution is loaded. When the absorbance intensity of the effluent portion passing through the flow-cell of the detector is equal to that of influent, the process is terminated. In this period, as the ligand-exchange resin becomes saturated, the absorbance intensity of the effluent regularly increases and reaches a stable value that equals to the absorbance intensity of the influent. If $C_0$ is the known concentration of solute entering the exchanger bed and C is the concentration of the ligand that emerges, then a plot of $C/C_0$ versus the volume of effluent gives a typical breakthrough curve [48]. The breakthrough and over-all capacities of a column can be readily determined from the effluent volume. Total adsorption capacities of the resin for each concentration of the solute are determined from the breakthrough curves obtained under the determined optimum conditions, using the following equation:

$$q_e = \frac{C_0\,V}{W} \cdot \frac{A_U}{A_U + A_D} \tag{6}$$

where $C_0$ denotes the initial concentration of the solute in the aqueous phase, V is the volume of the aqueous phase, and W is the dry weight of the resin. $A_U$ and $A_D$ denote that adsorbed and non-retained solute percentages of solute, respectively, for breakthrough curves. The areas ($A_U$ and $A_D$) on the left and right side of the breakthrough curves are proportional to the amount of adsorbed and non-retained solute, respectively. In order to determine the dynamic adsorption capacity ($q_e$), it is necessary to solve the areas of $A_U$ and $A_D$ (Figure 1). This can be easily performed by a computer program.

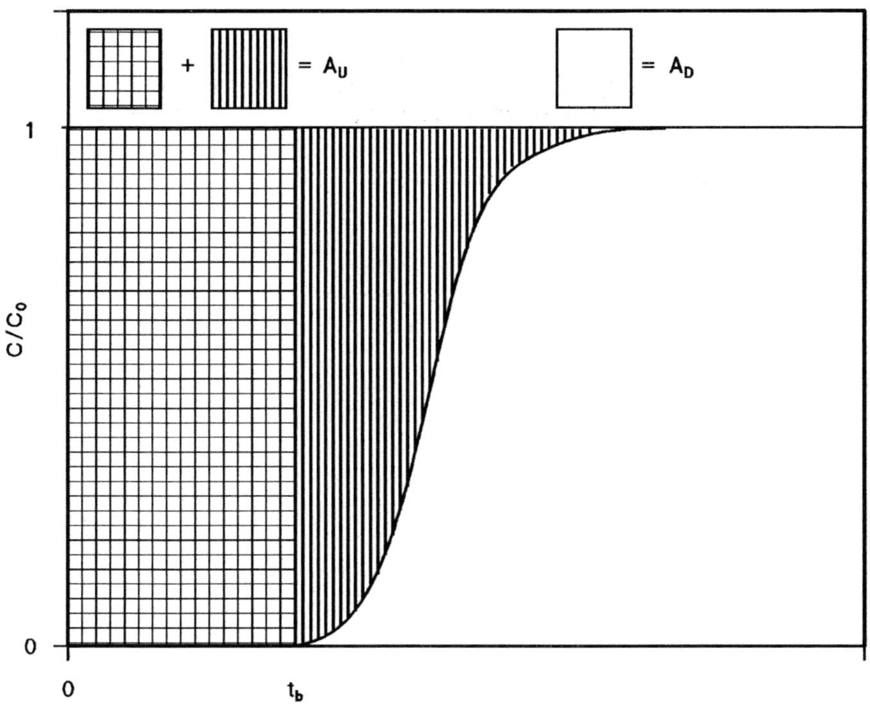


Fig 1. A schematic view of the technique used in the determination of the breakthrough and total adsorption data in a column operation. $A_U$ and $A_D$ denote that adsorbed and non-retained solute percentages of solute, respectively.

## Determination of Rate-controlling Step

A common ligand-exchange process on solid phase differs from an actual ligand-exchange reaction in a number of important aspects. It is accepted that the adsorption processes are normally diffusion controlled and not reaction controlled [15]. Ligand-exchange process, as in other heterogeneous reactions between solids and liquids, are mainly controlled by various factors including the following steps: (i) diffusion of ligands through the liquid film surrounding the particle; film diffusion (ii) diffusion of ligands through the resin particles; particle diffusion and (iii) a chemical reaction between ligands and functional groups present in the matrix; ligand-exchange. One of these steps usually offers much greater resistance than the others, so it can be considered as the rate-controlling step of the process. Actually, the ligand-exchange processes, in which the rate is controlled by a chemical reaction, are very few. Hence, film diffusion or particle diffusion is usually the rate-controlling step. Film diffusion depends mainly on solution concentration at constant particle size. The conditions for film-diffusion control of the overall ligand-exchange adsorption rate are well known and comprise low solution concentration. Particle diffusion relates to the ligand diffusion rate within the exchange matrix.

In a ligand-exchange system, depending on the concentration of the ligand, there are three alternative rate- controlling steps when ligands bind to the metal ions immobilized on the matrix: film diffusion, particle diffusion, and chemical reaction on the fixed ionic groups [15]. In the case of infinite solution volume, the diffusion controlled mass transport based on a homogeneous diffusion model can be described by the following equations for film and particle diffusion control, respectively:

$$U = 1 - \exp\left(-\frac{3\,D_f C t}{r_0\,\delta\,C_r}\right) \tag{7}$$

$$U^2 = 1 - \exp\left(-\frac{D_s t\,\pi^2}{r_0^2}\right) \tag{8}$$

Taking the natural logarithms, the equations can then be rearranged into linear forms as [49, 50]:

$$-\ln[1-U] = k_f t \qquad \text{where } k = \frac{3\,D_f C}{r_0\,\delta\,C_r} \tag{9}$$

$$-\ln[1-U^2] = k_p t \qquad \text{where } k = \frac{D_s \pi^2}{r_0^2} \tag{10}$$

where U is the fractional attainment of equilibrium adsorption at time t; $D_f$ is the diffusion coefficient in solution phase; C is the total concentration of both exchanging species; $\delta$is the liquid film thickness; $r_0$ is radius of spherical resin bead; $C_r$ is the total concentration of both exchanging species in the ligand-exchanger and $D_s$ is diffusion coefficient in solid-phase.

In a batch adsorption system, to determine the rate-controlling step, the homogeneous diffusion model has been extensively used by numerous researchers [51, 52, 53, 54, 55]. In order to examine the adsorption kinetics of the ligand-exchange resin in a column system, appropriate breakthrough data should be acquired for single solute adsorption in a fixed bed. Various methods have been recently presented by several researchers in order to determine the breakthrough data [42, 50, 56, 57, 58, 59, 60, 61, 62], however, these methods are rather complicated. In this sense, our research group have recently proposed a quite simple model to determine the breakthrough data for the single component systems, and the model has been named as "equivalent points method" [48].

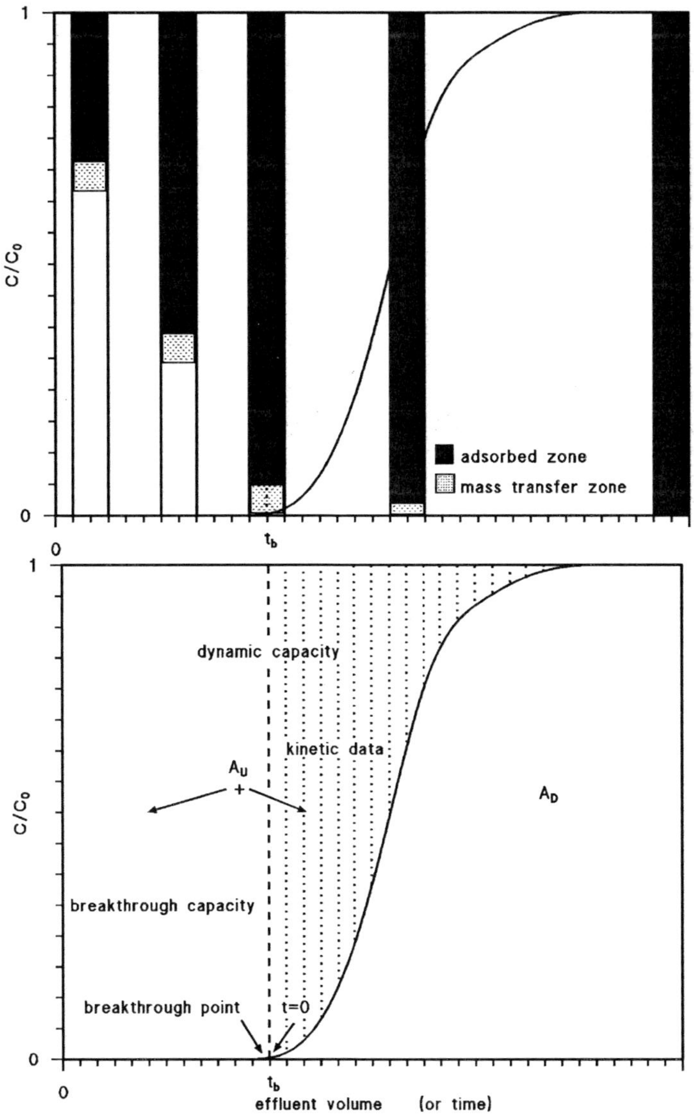


Figure 2. A schematic view of the technique used in the determination of the adsorption and kinetics data in a column operation. Reproduced from (*Gürten, A. A., Uçan, S., Özler, M. A., and Ayar, A.,* Removal of aniline from aqueous solution by PVC-CDAE ligand-exchanger, Journal of Hazardous Materials B120, 81, 2005).

By using the equivalent points method, homogeneous diffusion model has been adapted to a column system for describing the changing in the ligand concentration of effluent beginning from breakthrough point as a function of time. As the ligand first appears in the effluent, the particular breakthrough curve begins to go upward until it reaches a stable plateau. In a frontal analysis, this time period, which the time required to reach the plateau concentration in the effluent, corresponded to the mass transfer zone in the bed (as shown in Figure 2) [48].

The basic idea of this attempt is practically similar to that originally employed by Helfferich for describing the kinetics of batch adsorption systems [63]. In a column system, the batch adsorption conditions can obtain only at the zones of mass transfer beginning from breakthrough point. In these points, a part of ligand in the aqueous solution adsorbs onto the resin, while the others passes through the column and the ligand in solution is infinite for the adsorbent. Thus, the conditions of *zero length column technique* [64], which is used to determine the controlling mechanism of diffusion can be obtained. Note that the amount of ligand being adsorbed at each $t$ time is proportional to the total of each area denoted by punctuated line. In using the homogeneous diffusion model for column adsorption kinetics, following assumption has been made that were examined in detail elsewhere [27, 48, 65]:

> "The kinetic data following after the breakthrough point in a column operation are equivalent to the kinetic data that can be obtained from the batch process".

With this assumption, the adsorption kinetics of solutes in a column system may be analyzed using the homogeneous diffusion model which is constructed on the basis of the fact "The solute concentration in the mobile phase is in equilibrium with the concentration of the adsorbed ligand on the resin surface at each $t$ time after the breakthrough point" [48].

## SPOROPOLLENIN

Ligand exchangers have the particular advantages of combining high adsorption capacity with high selectivity for the ligand [15]. Their adsorption mechanism is supposed to be through complex formation instead of simple ion-exchange, and therefore they can exhibit much more selectivity than most of the ion-exchange resins. The selectivity for the ligands is expected to be high when the complexes with the metal ion are stable. Briefly, in the case of a ligand-exchange process, a strong "driving force" for ligand adsorption is expected because the ligands are efficiently sequestered by the metal ion in the resin. Nevertheless, another important aspect of adsorption is the adsorbent support. Packing materials for adsorption of ligands are continuously improved in order to increase adsorption efficiency. In general, an ideal solid support must have the following attributes:

1. Resistance to chemicals in the context of the mobile phase chemicals,
2. Structural stability and high mechanical strength,
3. Convenience for good solution diffusion in the solid phase,
4. Convenience for further modifications,
5. Uniform particle size,

6. Low or no swelling and coiling,
7. No or very limited secondary interactions with analyte.

Sporopollenin is a natural biopolymer obtained from *Lycopodium clavatum*, which is highly resistant to chemicals, has a high-capacity, stable, constant chemical structure, occurs naturally as a component of spore walls, and exhibits very good stability after even prolonged exposure to mineral acids and alkalines [16, 17, 66, 67]. Consolidation of the sporopollenin are not yet understood [22]; infra-red and $^{13}C$ NMR spectroscopic studies on sporopollenin have shown that sporopollenin has aliphatic, aromatic, hydroxyl, carbonyl/carboxyl and ether functions in various portions in its polymeric structure [20]. Sporopollenin shows good permeability functionality [68, 69], and easily available chelating functional groups on it, is suitable to evaluate it as stationary/solid phase adding some functional groups [70].

Using sporopollenin, several possibilities have been evaluated for preparation of solid phases based on immobilized metal ions suitable for the adsorption or chromatographic processes. In the past two decades, several modified forms of sporopollenin have been utilized as ion exchanger, and ligand exchanger for removal and separation of heavy metal ions and biological and organic compounds from aqueous solutions [24, 25, 26, 27, 32, 65, 71, 72, 73, 74, 75, 76, 77, 78]. Through a literature search, the studies intensified on unmodified sporopollenin were found to be very few [70, 79].

## Modification of Sporopollenin into a Ligand Exchanger Form

The preparation of cobalt(II) immobilized carboxylated-1,2-diaminoethane derivative of sporopollenin comprises of three main steps.

1. Introducing the ethylenediamine functionality through a substitution reaction between 1,2-diaminoethane and sporopollenin, which results in 1,2-diaminoethane sporopollenin (DAE-S),
2. Introducing the $-CH_2COOH$ functionality through a reaction between DAE-S and bromoacetic acid,
3. Immobilization of Co(II) in/on CDAE-S

## Preparation of Diaminoethyl-Sporopollenin

A suspansion of sporopollenin from *Lycopodium Clavatum* in dry toluene (150 mL) containing 1,2-Diaminoethane (50 mL) is aminated by refluxing for 9 hours [34, 72]. The reaction is as follows:

$$Ⓢ + NH_2(CH_2)_2NH_2 \longrightarrow Ⓢ-NH(CH_2)_2NH_2$$

DAE-sporopollenin

where Ⓢ indicates sporopollenin.

## Preparation of Carboxylated-diaminoethyl-Sporopollenin

A suspension of DAE-sporopollenin (diaminoethyl sporopollenin) is placed in a reaction vessel, and bromoacetic acid neutralized by 2M NaOH and 1M $NaHCO_3$ is added to the resin. By stirring the suspension for 24 hours and washing it with water and dilute acetic acid, CDAE-sporopollenin (carboxylated diaminoethyl sporopollenin) is obtained [34, 72]. CDAE-sporopollenin formation reaction is as follows:

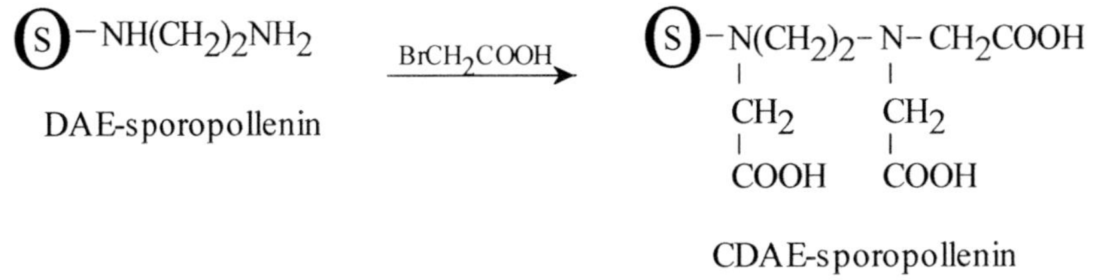


## Preparation of Cobalt(II) Immobilized Carboxylated-diaminoethyl Sporopollenin

A suspension of CDAE-sporopollenin is placed in a reaction vessel, and a 1M $CoCl_2$ solution is added to this suspension. By adding 1M $CoCl_2$ solution to the CDAE-sporopollenin, cobalt(II) are fixed to the resin matrix [25, 32, 72, 75]. Complex formation believed to be is as follows:

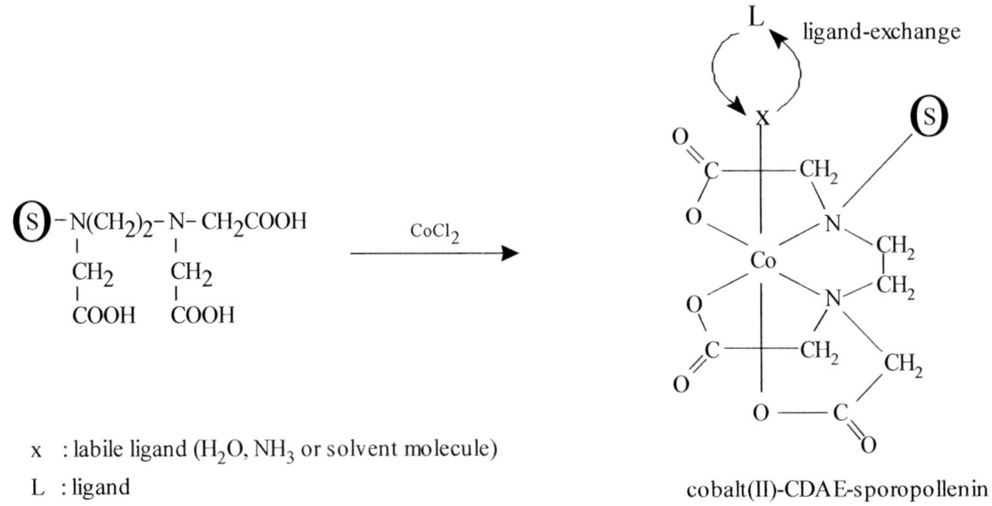


In the ligand-exchange process, the solution pH during loading of the resin with metal ions is an important parameter because of the pH dependencies of the complexation reactions. The experimental studies has been carried out in this manner with various metal ions, which point out that the pH affects the formation of metal-ligand complexes, and optimum metal binding takes place at higher pH (6–8) ranges [72, 73, 80]. Similar results have been achieved also for cobalt ion on the same resin by Ayar and Yıldız [75]. Henceforth the complex formation of the CDAE-sporopollenin with the cobalt ion has been carried out at pH 6. [25, 26, 27, 65, 76, 77, 78].

CDAE-sporopollenin resin bears two important functional groups, having a high affinity for the metal ions: Ethylenediamine and carboxyl groups. Whilst ethylenediamine groups possess a very stable structure with a very minor dissociation tendency, carboxyl groups bind the metal ions strongly [31]. Sporopollenin is an extraordinary resistant biopolymer whose the outer wall is composed of spores and pollen [19]; consequently, ligand-exchange activity of the Co(II)-CDAE-S is only attributed to the carboxylated ethylenediamine functionalities. Carboxylated ethylenediamine groups have a very stable structure with little possibility of dissociation; they are very suitable functional groups for a ligand-exchanger matrix [31]. For such ligand-exchange resin, it is often expressed that the metal leakage is not observed from the ligand-exchanger column during the elution operation of the ligands.

## Adsorption of Aromatic Amines on Ligand-Exchange Sporopollenin

Recently, various researches on the interaction of ligand-exchange sporopollenin with aromatic amines, which deal with adsorption, kinetics, and chromatographic separation have been done. The first paper dealing with the adsorption of aromatic amines on the ligand-exchange sporopollenin has been published by Uçan and Ayar [25]. In that study, the adsorption of the chlorinated anilines such as 2-chloroaniline, 3-chloroaniline, 4-chloroaniline and 2,5-dichloroaniline on the Co(II)-CDAE-sporopollenin has been investigated as a function of the solute concentration at a constant temperature of 25 °C. The adsorption behavior of chlorinated anilines could be expressed by the Langmuir and Freundlich isotherms, and the characteristics of the adsorption process were also investigated using Scatchard plot analysis. The characteristics of the adsorption process were also investigated using Scatchard plot analysis. Another similar research has been performed also to determine the binding constants of aniline and o-anisdine on the same resin by using breakthrough technique [26]. In a further investigation the rate-controlling step of chlorinated anilines adsorption on the Co(II)-CDAE-sporopollenin has been realized, and type of the ligand diffusion has been determined using a novel measurement method, "equivalent points methods" [27]. In this study, the "equivalent points methods" has been applied to the breakthrough and tested for the first time. However, the method was named as "equivalent points methods" by Ayar et al. elsewhere [48].

A single attempt to accomplish the chromatographic separation of the aromatic amines by using the Co(II)-CDAE-sporopollenin as a stationary phase is the study that conducted by the author and his research group for nitroanilines. Using a solution of 0.05 M $NH_4OH$ in ethanol–water as mobile phase, the retention behavior of 3- and 4-nitroanilines was investigated at various column temperatures (15–60 °C) [77]. As mentioned in the "*General Principles of Ligand Adsorption*" section, in all of those studies, mechanism of the ligand binding is based on the prudence the metal–ligand coordination between the ligand and immobilized metal ions on the matrix of the solid phase. The cited research papers cited regarding the adsorption and kinetics, are briefly reviewed in the following sections.

## Adsorption of Aromatic Amines

### *Chlorinated Anilines*

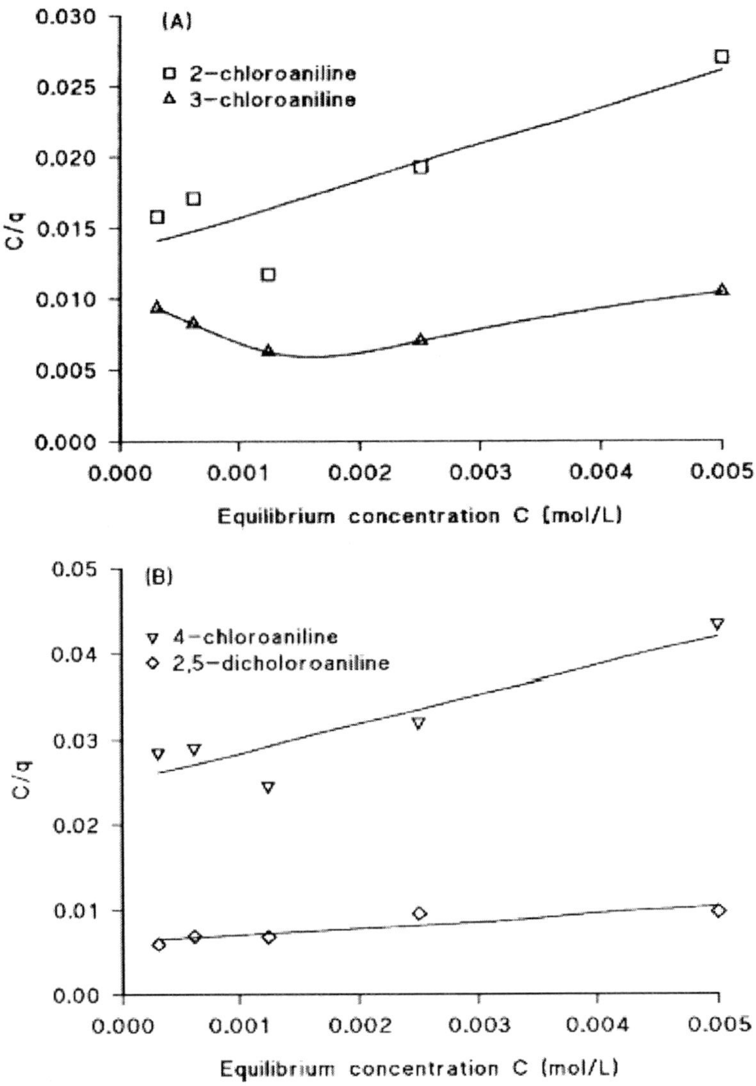


Figure 3. Semi-reciprocal plot of the experimental data for the adsorption of chlorinated anilines. Reproduced from (*Uçan, M., and Ayar, A.,* Sorption equilibria of chlorinated anilines in aqueous solution on resin-bound cobalt ion, Colloids and Surfaces A: Physicochemical and Engineering Aspects, 207, 41, 2002).

In the study that reported by Uçan and Ayar [25], the interactions of chlorinated anilines were shown to involved different types interactions depending upon the positions of the substituted groups on the ligand-exchange resin. Figure 3. shows semi-reciprocal plot of the Langmuir isotherm of the experimental data, whilst Figure 4 presents the Scatchard plot of the experimental data.

In the Fig(s). 3 and 4., the equilibrium binding data for 2-chloroaniline, 4-chloroaniline and 2,5-dichloroaniline seems to be closely clustered around the predicted line indicating a linear dependency of plots, revealing that only one type of the binding sites may be active for these ligands. Probably, such an interaction is thought to be a partially specific one, and occurs through the participation of free N-electrons of the ligands. However, the equilibrium binding data obtained for 3-chloroaniline, which exhibit a divergence from linearity, indicating the presence of nonspecific interaction consistent with the participation of different interactions. This divergence may partly be explained by the participation of secondary equilibrium effects formed between *meta*-Cl atoms and immobilized ligands on the matrix. It should be noted, that the Scatchard plot supports the information obtained from the isotherms, and gives further compact information about the binding characteristics.

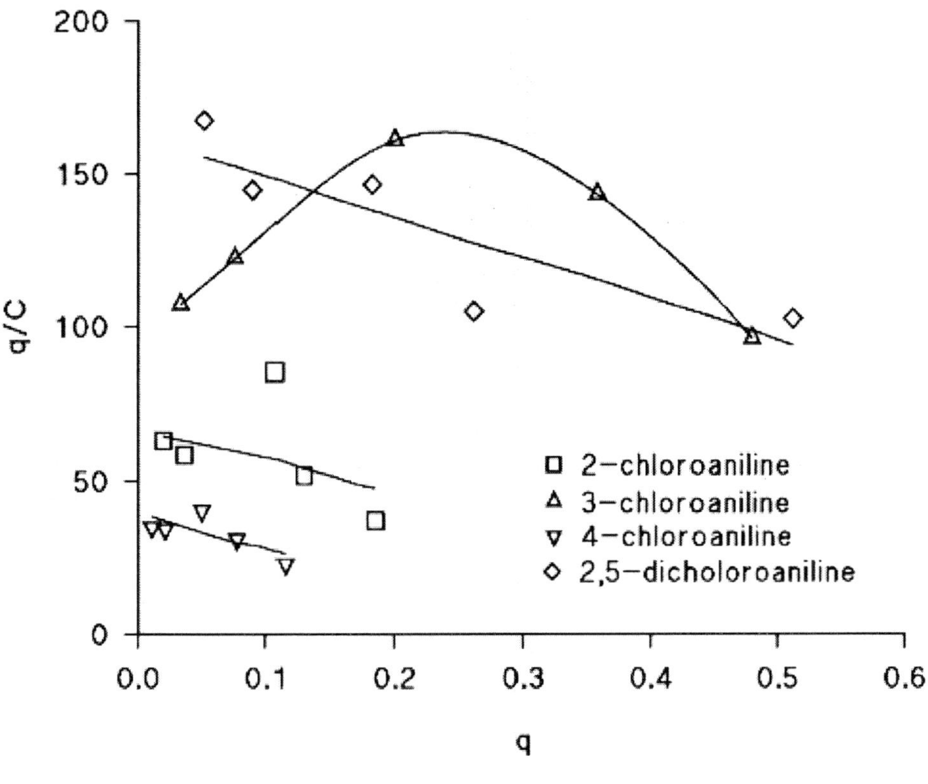


Figure 4. Scatchard plot of the experimental data for the adsorption of chlorinated anilines. Reproduced from (*Uçan, M., and Ayar, A.,* Sorption equilibria of chlorinated anilines in aqueous solution on resin-bound cobalt ion, Colloids and Surfaces A: Physicochemical and Engineering Aspects, 207, 41, 2002).

### *Aniline and o-anisidine*

In order to confirm the possibilities of determining binding parameters of labile ligands in metal-ligand complexes immobilized on sporopollenin, aniline and o-anisidine were selected as labile ligands, and studied in a column design by Ayar et al. [26]. In that study, firstly, the equilibrium adsorption of aniline and o-anisidine was determined in order to obtain the adsorption capacity of cobalt-(II)-CDAE-sporopollenin for each ligand on the basis of particular by breakthrough curves. The characteristics of the adsorption process were investigated using the double reciprocal transformation of the Langmuir plot and Scatchard plot analysis (Figs. 5 and 6).

As it can be seen from Fig.6, in the adsorption of o-anisidine, the equilibrium binding data exhibits a deviation from linearity, indicating the presence of a nonspecific interaction consistent with the participation of different interactions. However, in double reciprocal transformations of the Langmuir plot, the equilibrium binding data were closely clustered around the predicted line for a linear dependency of plot (Figure 5). The results clearly show that the Scatchard plot analysis can be reliably used to reveal the adsorption behaviors of the ligands. Consequently, the Scatchard transformation gives more comprehensive information than the semi-reciprocal or double reciprocal transformation of the Langmuir equation.

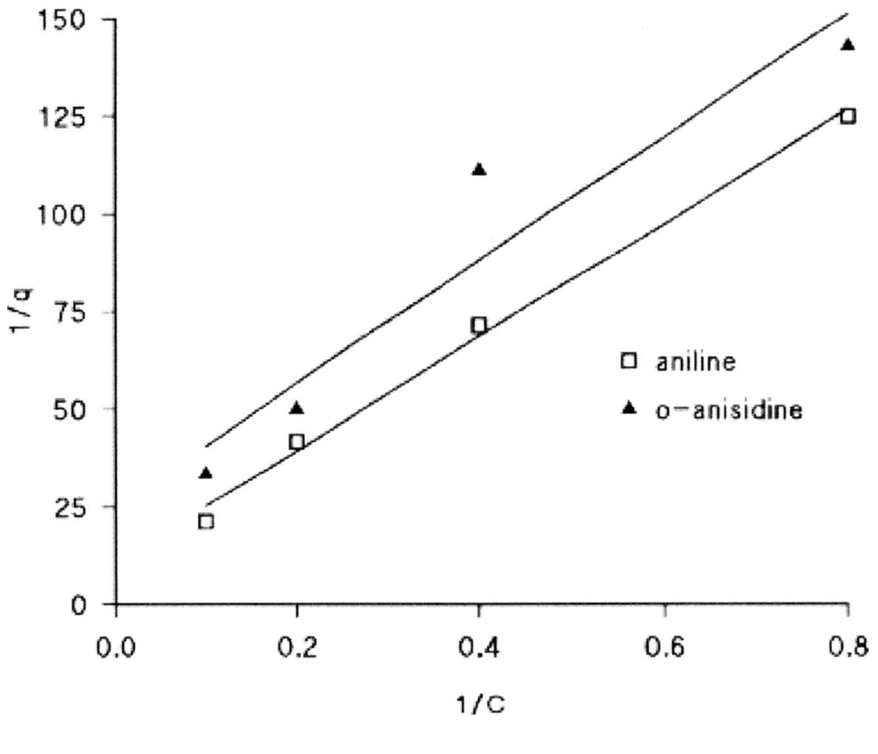


Figure 5. Double reciprocal plot of the experimental data for the adsorption of aniline and o-anisidine. Reproduced from (*Ayar, A., Pekacar, A.I., and Mercimek, B.,* Determination of binding constants of labile ligands in metal-ligand complexes containing carboxylic groups bound polymer by using breakthrough technique, Colloids and Surfaces A: Physicochem. Eng. Aspects 212, 51, 2003).

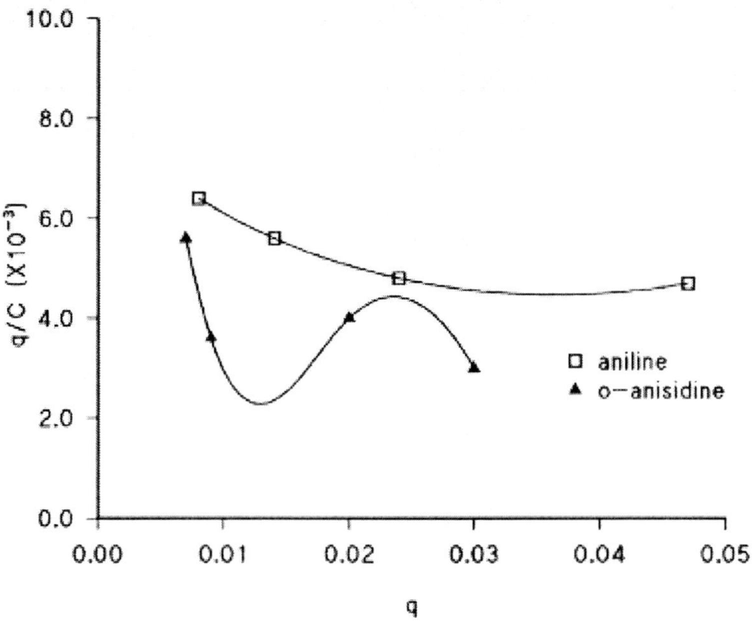


Figure 6. Scatchard plot of the experimental data for the adsorption of aniline and o-anisidine. Reproduced from (*Ayar, A., Pekacar, A.I., and Mercimek, B.,* Determination of binding constants of labile ligands in metal_ ligand complexes containing carboxylic groups bound polymer by using breakthrough technique, Colloids and Surfaces A: Physicochem. Eng. Aspects 212, 51, 2003).

## Determination of Rate-conrolling step of Aromatic Amines

The homogeneous diffusion model has been extensively used by numerous researchers to describe the uptake of adsorbate by an adsorbent, however, only for the batch adsorption system. Ayar and coworkers [27] studied breakthrough profiles of chlorinated anilines on a Co(II)-CDAE-sporopollenin-mediated column system and adapted the homogeneous diffusion model to the data obtained from a column system for determining the rate-controlling step of chlorinated anilines. In that study, the adsorption of chlorinated anilines from solutions of different initial concentrations onto resin was determined as fractional attainment of equilibrium adsorption as a function of time using breakthrough technique. Thus, -ln (1-U) was plotted as a function of time for film diffusion as shown in Figure 7.

As it can be seen in the Figure 7, for 4- and 2,5-dichloroaniline, the graphs obtained at different concentrations fit to a straight line, which suggests that fractional attainment of equilibrium adsorption under film diffusion control is reasonable. On the contrary, for 2-and 3-chloroaniline, the graphs do not fit to a typical straight line; revealing the acceptability of an adsorption under particle diffusion control. When the adsorption of ligand onto the resin is under film diffusion control, a graph of –ln [1-U] versus t gives a straight line. Since this relation is thought to be the case, the adsorption may be considered to be under film diffusion control. In the case of 4- and 2,5-dichloroaniline, in all of the concentration range, the graphs are a straight line, which suggests that fractional attainment of equilibrium adsorption under film diffusion control is observed. On the contrary, for 2- and 3-chloroaniline, the graphs were not a straight line and have undergone adsorption under particle diffusion control.

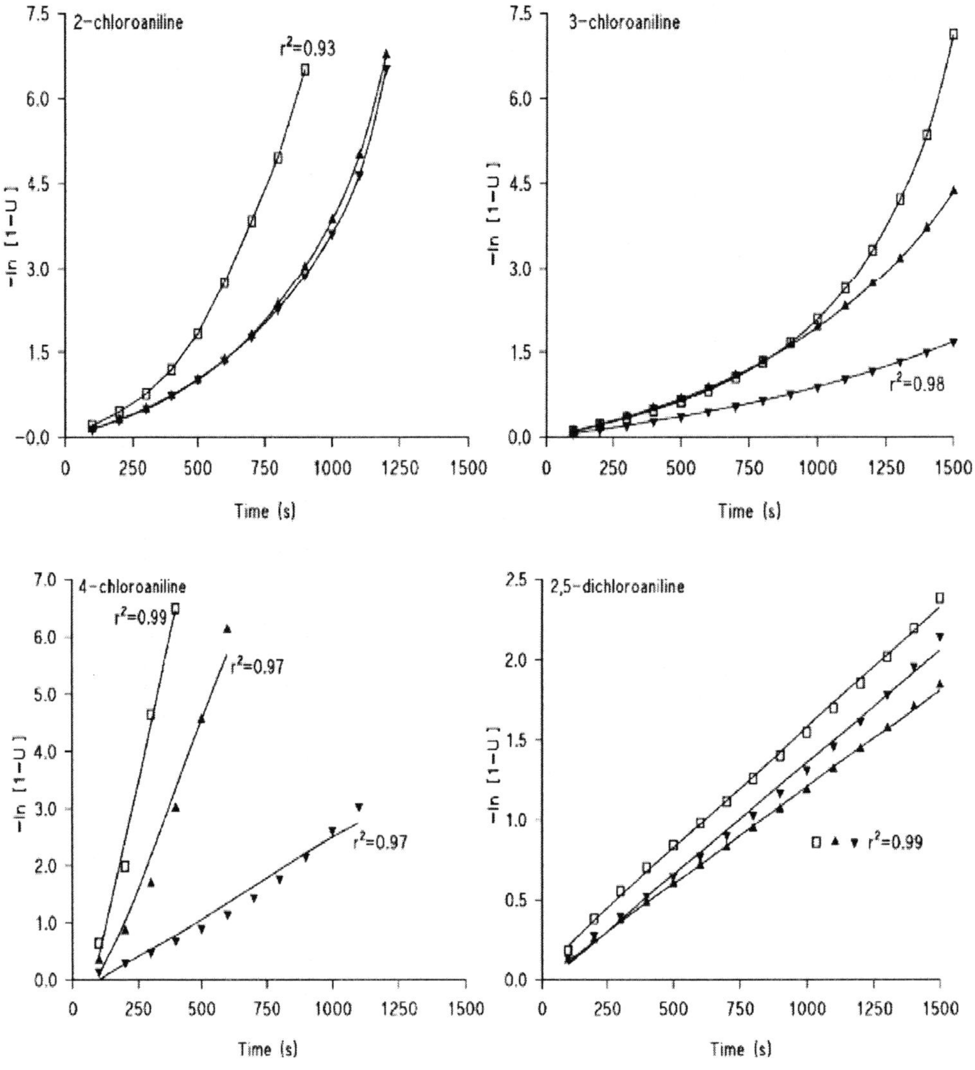


Figure 7. Test of – ln[1-U] vs t for adsorption of ligands at different initial concentrations. Initial concentrations, (▼) 6.0 x $10^{-4}$ M; (▲) 1.3 x $10^{-3}$ M; (□) 2.5 x $10^{-3}$ M. Reproduced from (*Uçan, M., Gürten, A., and Ayar, A.,* Determination of the rate control step of chlorinated anilines in ligand exchange reactions on solid phase by using breakthrough technique, Colloids and Surfaces A: Physicochem. Eng. Aspects 219, 193, 2003).

## ACKNOWLEDGEMENTS

The author thanks Orhan Gezici, Niğde University/Turkey, for helpful discussions and guidance.

## REFERENCES

[1] Gerhartz, W. (Ed.), Ullmann's Encyclopedia of Industrial Chemistry, Vol. 2A, fifth ed., VCH, Weinheim, 1985.

[2] Kijima, K., Kataoka, H., and Makita, M., Determination of aromatic amines as their N-dimethylthiophosphoryl derivatives by gas chromatography with flame photometric detection, *Journal of Chromatography A,* 738, 83, 1996.

[3] Fishbein, L., Aromatic amines, p. 1–40. In Hutzinger, O. (Ed.), *The Handbook of environmental chemistry,* Vol. 3, Part C. Springer-Verlag, 1994, New York.

[4] Hartter, D. R., The use and importance of nitroaromatic chemicals in the chemical industry, p. 1–14., In D. E. Rickert (ed.), Toxicity of nitroaromatic compounds. Chemical Industry Institute of Toxicology Series, Chemisphere, 1985, Washington, D.C.

[5] Chung, K.-T., Kirkovsky, L., Kirkovsky, A., and Purcell, W.P., Review of mutagenicity of monocyclic aromatic amines: quantitative structure–activity relationships, *Mutation Research 387*, 1, 1997.

[6] Sarasa, J., Roche, M.P., Ormad, M.P., Gimeno, E., Puig, A., and Ovelleiro, J.L., Treatment of a wastewater resulting from dyes manufacturing with ozone and chemical coagulation, *Water Research* 32, 2721, 1998.

[7] Saupe, A., High-rate biodegradation of 3- and 4-nitroaniline, Chemosphere 39 (13), 2325 1999.

[8] Kataoka, H., Derivatization reactions for the determination of amines by gas chromatography and their applications in environmental analysis, *Journal of Chromatography A* 733, 19, 1996.

[9] White, P.A., and Claxton, LD., Mutagens in contaminated soil: a review, *Mutation Research,* 567, 227, 2004.

[10] Voyksner, R.D., Straub, R., Keever, J.T., Freeman, H.S., and Hsu, W.N., *Environ. Sci. Technol.* 27, 1665, 1993.

[11] Chanda, M., O'Driscoll, K.F., and Rempel, G.L., Ligand sorption of aromatic amines on resin-bound ferrous ion, *Reactive Polymers* 2, 279, 1984.

[12] Zaki, A.B., El-Sheikh, M.Y., Evans, J., and El-Safty, S.A., Kinetics and Mechanism of the Sorption of Some Aromatic Amines onto Amberlite IRA-904 Anion-Exchange Resin, *Journal of Colloid and Interface Science* 221, 58, 2000.

[13] Aarden, F.B., Adsorption onto heterogeneous porous materials: equilibria and kinetics, Ph.D. thesis, *Technical University of Eindhoven,* 2001.

[14] Arafat, H.A., Ahnert, F., and Pinto, N.G., On the adsorption of aromatics on oxygenated activated carbon in Nonaqueous Adsorption Media, *Separation Science and Technology,* 39 (1), 43, 2004.

[15] Helfferich, F., *Ion Exchange,* Chapter 5, McGraw-Hill Book Company, Inc., USA, 1962 (reprint Dover, Mineola 1995).

[16] Brooks J, Shaw G., Identity of sporopollenin with older kerogen and new evidence for the possible biological source of chemicals in sedimentary rocks, *Nature*, 220, 678, 1968.

[17] Shaw G., In: Harbone J.B., (Ed.), *Sporopollenin in phyto chemical phylogency,* Chapter 3, London, New York, Academic Press, 1970.

[18] Shaw, G., The chemistry of sporopollenin, in: J. Brooks, P.R. Grant, M.D. Muir, P. van Gijzel, G. Shaw (Eds.), *Sporopollenin*, Academic Press, London, pp. 305–351, 1971.

[19] Ahlers, F., Thom, I., Lambert, J., Kuckuk, R., and Wiermann, R., 1H NMR analysis of sporopollenin from Typha Angustifolia, *Phytochemistry* 50, 1095, 1999.

[20] Wilmesmeier, S., Steuernagel, S., and Wiermann, R., Comparative FTIR and 13C CP/MAS NMR spectroscopic investigations on sporopollenin for different systematic origins, Z. Naturforsch, 48, 697, 1993.

[21] Bubert H., Lambert, J., Steuernagel, S., and Ahlers, F., Wiermann R. Continuous decomposition of sporopollenin from pollen of Typha angustifolia L. By acidic methanolysis. Z Naturforsch, 57c, 1035, 2002.

[22] Bohne, G., Richter, E., and Woehlecke, E.R., Exchange of submicrometer particles, polymers and low- molecular-weight solutes with tripartite sporopollenin microcapsules prepared from pine pollen, *Ann. Bot*. 92, 289, 2003.

[23] Vural, U.S., Ersöz, M. and Pehlivan, M., Ligand sorption kinetics of aromatic amines on new ligand-exchanger sporopollenin in cobalt form, *J. Appl. Polym. Sci.* 58, 2423, 1995.

[24] Pehlivan, E., Vural, U.S., Ayar, A., and Yıldız, S., Ligand-Exchange Chromatography of Aromatic Amines on Resin-Bound Cobalt Ion, *Sep. Sci. Technol.*,31(11), 1643, 1996.

[25] Uçan, M., and Ayar, A., Sorption equilibria of chlorinated anilines in aqueous solution on resin-bound cobalt ion, *Colloids and Surfaces A: Physicochemical and Engineering Aspects,* 207, 41, 2002.

[26] Ayar, A., Pekacar, A.I., and Mercimek, B., Determination of binding constants of labile ligands in metal_ ligand complexes containing carboxylic groups bound polymer by using breakthrough technique, *Colloids and Surfaces A: Physicochem. Eng. Aspects* 212, 51, 2003.

[27] Uçan, M., Gürten, A., and Ayar, A., Determination of the rate control step of chlorinated anilines in ligand exchange reactions on solid phase by using breakthrough technique, *Colloids and Surfaces A: Physicochem. Eng. Aspects* 219, 193, 2003.

[28] Helfferich, F., Ligand exchange: A novel separation technique, Nature 189, 1001, 1961.

[29] Walton, H.F., and Stokes, H., Metal amine complexes in ion exchange, *J. Am. Chem. Soc.* 76, 3327, 1954.

[30] Davankov, V.A., and Semechkin, A.V., Ligand-exchange chromatography, *J Chromatogr.* 141, 313, 1977.

[31] Davankov, V.A, Navratil, J.D., and Walton H.F., *Ligand exchange chromatography,* Boca Raton, FL: CRC Press, 1988.

[32] Ayar, A., Yıldız, S., and Pehlivan, E., Ligand exchange chromatography of some amino acids on Co(II)-loaded CMDAE-sporopollenin resin, *Sep. Sci. Technol.* 30 (15), 3081, 1995.

[33] Yıldız, S., Pehlivan, E., Ersöz, M., and Pehlivan, M., Ligand exchange chromatography of nucleosides and nucleic acid bases on a novel modified Co(II)-loaded Lycopodium Clavatum and kinetics in the resin, *J. Chromatogr.* Sci. 31, 150, 1993.

[34] Pehlivan, E., and Yıldız, S., Modified sporopollenin as a novel anion, cation and ligand exchange medium, *Anal. Lett.* 21, 297, 1988.

[35] Scatchard G., The attractions of protein for small molecules and ions. *Ann NY Acad Sci* 51, 660, 1949.

[36] Zhou, J., He, X., and Li, Y., Binding study on 5,5-diphenylhydantoin imprinted polymer constructed by utilizing an amide functional group, *Anal Chim Acta* 394, 353, 1999.
[37] Chaouk, H., Hearn, M.T.W., New ligand, N-(2-pyridylmethyl) aminoacetate, for use in the immobilised metal ion affinity chromatographic separation of proteins, *J Chromatogr* A 852, 105, 1999.
[38] Chaouk, H., and Hearn, M.T.W., Examination of the protein binding behaviour of immobilised copper (II)-2,6-diaminomethylpyridine and its application in the immobilised metal ion affinity chromatographic separation of several human serum proteins. *J Biochem Biophys Methods* 39, 161, 1999.
[39] Noto, V.D., Via, L.D., and Zatta, P., Review of binding methods and detection of Al(III) binding events in trypsin and DL-DPPC liposomes by a general thermodynamic model. *Coord. Chem Rev.,* 228, 343, 2002.
[40] Hidayat, C., Nakajima, M., and Takagi, M., Multivalent Binding Interaction of Alcohol Dehydrogenase on Dye-Metal Affinity Matrix. *Journal of Bioscience and Bioengineering,* 96(2), 168, 2003.
[41] Cerf, A.M.C., Dehaye, J.-P., and Devleeschouwer, M.J., Analysis and modelling of *Pseudomonas aeruginosa* adherence to human buccal epithelial cells, *Colloids and Surfaces B: Biointerfaces* 8,267, 1997.
[42] Lin, S.H., and Huang, C.P., Adsorption of hydrazoic acid from aqueous solution by macroreticular resin. *Journal of Hazardous Materials* B84, 217, 2001.
[43] Sharma, S., and Agarwal, G.P., Interactions of Proteins with Immobilized Metal Ions (Role of Ionic Strength and pH), *Journal of Colloid and Interface Science* 243, 61, 2001.
[44] Yang, L., and Chen, P., Chitosan/coarse filter paper composite membrane for fast purification of IgG from human serum. *Journal of Membrane Science* 205, 141, 2002.
[45] Doyle, R.J. in N. Mozes, P.S. Handley, H.J. Busscher and P.G. Rouxhet (Eds.), *Microbial Cell Surface Analysis: Structural and Physicochemical Methods,* pp. 291 316, VCH, New York, 1991.
[46] James, D.H., and Phillips, C.S.G., The chromatography of gases and vapours. Part III, The determination of adsorption isotherms, *J. Chem. Soc.,* pp. 1066–1070, 1954.
[47] Hage, D.S., High-performance affinity chromatography: a powerful tool for studying serum protein binding, *Journal of Chromatography* B, 768, 3, 2002.
[48] Gürten, A. A., Uçan, S., Özler, M. A. and Ayar, A., Removal of aniline from aqueous solution by PVC-CDAE ligand-exchanger, *Journal of Hazardous Materials* B120, 81, 2005.
[49] Liberti, L., Passino, R., in: J.A. Marinsky, Y. Marcus (Eds.), *Ion-exchange and Solvent Extraction,* Vol 7, Chapter 3, Marcel Dekker, Inc., New York, 1977.
[50] Cortina, J.L., Arad-Yellin, R., Miralles, N., Sastre, A.M., and Warshawsky, A., Kinetics studies on heavy metal ions extraction by Amberlite XAD2 impregnated resins containing a bifunctional organophosphorous extractant, *Reactive & Functional Polymers* 38, 269, 1998.
[51] Juang, R.-S., Preparation, Properties and Sorption Behavior of Impregnated Resins Containing Acidic Organophosphorus Extractants, *Proc. Natl. Sci. Counc.* ROC(A), 23 (3), 353, 1999.

[52] Cortina, J.L., Warshawsky, A., Kahana, N., Kampel, V., Sampaio, C.H., and Kautzmann, R.M., Kinetics of goldcyanide extraction using ion-exchange resins containing piperazine functionality, *Reactive & Functional Polymers* 54, 25, 2003.

[53] Saha, B., Gill, R.J., Bailey, D.G., Kabay, N., and Arda, M., Sorption of Cr(VI) from aqueous solution by Amberlite XAD-7 resin impregnated with Aliquat 336. *Reactive & Functional Polymers* 60, 223, 2004.

[54] McKay, G., Solution to the homogeneous surface diffusion model for batch adsorption systems using orthogonal collocation, *Chemical Engineering Journal* 81, 213, 2001.

[55] Chen, W.-D., Dong, X.-Y., and Sun, Y., Analysis of diffusion models for protein adsorption to porous anion-exchange adsorbent, *Journal of Chromatography A,* 962, 29, 2002.

[56] Tsai, W.-T., Chang, C.-Y., Ho, C.-Y., and Chen, L.-Y., Adsorption properties and breakthrough model of 1,1-dichloro-1-fluoroethane on activated carbons, *Journal of Hazardous Materials* B69, 53, 1999.

[57] Arévalo, E., Rendueles, M., Fernández, A., and Díaz, M., Equilibrium and simulation of the operation for the adsorption of albumin proteins in an iminodiacetic-Cu bounded ion exchange resin (IMAC), *Separation and Purification Technology* 18, 217, 2000.

[58] Kamio, E., Matsumoto, M., and Kondo, Kazuo., Uptakes of rare metal ionic species by a column packed with microcapsules containing an extractant, *Separation and Purification Technology* 29, 121, 2002.

[59] Wood, G.O., A review of the effects of covapors on adsorption rate coefficients of organic vapors adsorbed onto activated carbon from flowing gases, *Carbon* 40, 685, 2002.

[60] Wood, G.O., Quantification and application of skew of breakthrough curves for gases and vapors eluting from activated carbon beds, *Carbon* 40, 1883, 2002.

[61] Tranter, T.J., Herbst, R.S., Todd, T.A., Olson, A.L., and Eldredge, H.B., Application of a second order kinetic model for the preliminary design of an adsorption bed system using ammonium molybdophosphate–polyacrylonitrile for the removal of 137Cs from acidic nuclear waste solutions, *Advances in Environmental Research* 7, 913, 2003.

[62] Tsysin, G.I., Kovalev, I.A., Nesterenko, P.N., Penner, N.A., and Filippov, O.A., Application of linear model of sorption dynamics to the comparison of solid phase extraction systems of phenol, *Separation and Purification Technology* 33, 11, 2003.

[63] Helfferich, F., *Ion Exchange,* Chapter 6, McGraw-Hill Book Company, Inc., USA, 1962.

[64] Grande, C.A., Silva, V. M.T.M., Gigola, C., and Rodrigues, A.E., Adsorption of propane and propylene onto carbon molecular sieve, *Carbon* 41, 2533, 2003.

[65] Ayar, A., and Gürten, A.A., Determination of the rate control step of purin and pyrimidine bases adsorption on cobalt(II)–CDAE-sporopollenin, *Colloids and Surfaces A: Physicochem. Eng. Aspects* 229, 149, 2003.

[66] Zetzche, F., and Vicari, H., Untersuchungen uber die Membrane der Sporen und Pollen, *Helv. Chim. Acta* 14, 58, 1931.

[67] Martin, J.T., Juniper, B.E., *The Cuticles of Plants*, R. and R. Clark Ltd., Edinburgh, Great Britain, 1970.

[68] Woehlecke, H., Lerche, D., and Ehwald, R., Sporopollenin microcapsules: application for chromatography and encapsulation, in: Proceedings of the X International BRG

Workshop on Bioencapsulation "*Cell Physiology and Interactions of Biomaterials and Matrices*", Prague, Czech Republic Poster P-19, 26–28, pp. 183–186, April, 2002.

[69] Ehwald, R., Bohne, G., Richter, E., Woehlecke, H., and Lerche, D., Permeability of sporopollenin capsules prepared from birch and pine pollen, in: Proceedings of the X International BRG Workshop on Bioencapsulation *"Cell Physiology and Interactions of Biomaterials and Matrices*", Prague, Czech Republic, 26–28, pp. 123–126, April, 2002.

[70] Ünlü, N., and Ersoz, M., Adsorption characteristics of heavy metal ions onto a low cost biopolymeric sorbent from aqueous solutions, Journal of Hazardous Materials xxx, xxx, 2006 (in press).

[71] Ersöz, M., Yıldız, S., and Pehlivan, E., Separation of nucleosides and nucleicacid bases by ligand-exchange chromatography using Cu2+ and Ni2+-loaded glyoximated diaminosporopollenin derivatives, *J. Chromatogr. Sci.* 31,61, 1993.

[72] Pehlivan, E., Ersöz, M., Yıldız, S., Duncan, H., Sorption of heavy metal ions on new metal–ligand complexes chemically derived from *Lycopodium clavatum*, *Sep. Sci. Technol.* 29 (13), 1757, 1994.

[73] Pehlivan, E., Ersöz, M., Pehlivan, M, and Yıldız, S., Duncan, H., The effect of pH and temperature on the sorption of zinc(II), cadmium(II) and aluminum(III) onto new metal–ligand complexes of sporopollenin, *J. Colloid Interf. Sci.* 170, 320, 1995.

[74] Ersöz, M., Yiğitoğlu, M., and Ayar, A., The study of kinetic models for ligand exchange of anions as ligands using ligand-exchange resin, *J. Appl. Polym. Sci.* 64, 1225, 1997.

[75] Ayar, A., and Yıldız, S., Sorption of purin and pyrimidine bases onto chelating resin, *Colloids and SurfacesA: Physicochemical and Engineering Aspects,* 201, 291, 2002.

[76] Erciyes, N., Gürten, A.A., Abdullah, M.I., and Ayar, A., Adsorption of indole and 2-methylindole on ligand-exchange matrix, *Journal of Colloid and Interface Science* 278, 91, 2004.

[77] Gürten, A.A., Uçan, M., Abdullah, M.I., and Ayar, A., Effect of the temperature and mobile phase composition on the retention behavior of nitroanilines on ligand-exchange stationary phase, *Journal of Hazardous Materials* xxx (2005) xxx–xxx

[78] Ayar, A., and Mercimek, B., Interaction of nucleic acid bases and nucleosides with cobalt ions immobilized in a column system, Process Biochemistry xxx (2006) xxx–xxx

[79] Arslan, M, Temoçin, Z., and Yiğitoğlu, M., Removal of cadmium(II) from aqueous solutions using sporopollenin, *Fresenius Environmental Bulletin,* 13(7), 616, 2004.

[80] Ersoz, M., Pehlivan, E., Duncan, H.J., Yildiz, S., and Pehlivan, M., Ion exchange equilibria of heavy metals in aqueous solution on new chelating resins of sporopollenin, *Reactive Polymers,* 24, 195, 1995.

In: Hazardous Materials and Wastewater
Editor: Allison A. Lewinsky, pp. 275-308

ISBN 1-60021-257-3


*Chapter 9*

# REMOVAL OF CHROMIUM IONS FROM AQUEOUS SOLUTIONS BY ADSORPTION METHOD

***Fethiye Gode****

Department of Chemistry, Suleyman Demirel University, Isparta, Turkey

## ABSTRACT

Undesirable ions such as Cr(III), and Cr(VI) are removed from water or wastewater by adsorption. The method is technologically simple and enables efficient removal of even traces of impurities from aqueous solutions. Mechanisms including ion exchange, complexation, and adsorption to the surface are possible in the sorption process. Examples of removal of chromium ions by adsorption are presented in here. Suitability of synthetic ion exchangers such as Chelex-100 (chelating resin), Lewatit TP 207 (chelating resin), Lewatit S 100 (cation exchanger resin), Lewatit MP 62 (anion exchanger resin), Lewatit M 610 (anion exchanger resin), a novel chelating resin b-DAEG-sporopollenin (bis-diaminoethyl glyoxim-sporopollenin, CEP-sporopollenin (carboxylated epichlorohydrine-sporopollenin), and natural adsorbents (brown coals, montmorillonite) as highly efficient adsorbent for the removal of chromium was studied and compared with each other. Batch sorption experiments were carried out with chromium to quantify the sorption kinetics. Sorption was found as a function of initial concentration, pH, temperature, shaking time, and adsorbent dosage. The Langmuir and Freundlich adsorption isotherms were used to describe the observed sorption phenomena. Sorption equilibria, calculated from adsorption data, explained favorable performance of adsorption system.

* To whom correspondence should be addressed. E-mail: fgode@fef.sdu.edu.tr(F. Gode). Tel.: +90 246 2114078 fax: +90 246 2371106

# 1. INTRODUCTION

The increasing contamination of urban and industrial wastewaters by toxic metal ions is a worrying environmental problem [1-8]. Removal of heavy metals from aqueous solutions is a basic in the treatment of wastewater from many industries, including metal plating, mining and battery manufacturing. Sequestration of heavy metals from aqueous solutions especially wastewater has received a considerable amount of attention in recent years due to the concern that heavy metals can be readily absorbed by organisms. These inorganic micro-pollutants are of considerable concern because they are non-biodegradable, highly toxic and have a probable carcinogenic effect [3, 9, 10]. These absorbed heavy metals consequently enter human food chains, and thus exert a high health risk to the community and biosphere in general [11, 12]. According to the World Health Organization, the metals of most immediate concern are aluminum, chromium, manganese, iron, cobalt, nickel, copper, zinc, cadmium, mercury, and lead [8]. The concentrations of these pollutants must be reduced by means of treatment to meet legislative standards [8]. In recent years, there has been a growing concern with environmental protection [1, 2, 13, 14]. This can be achieved either by decreasing the influx of pollutants to the environment or by their removal from contaminated media. The former is a feasible choice only for pollutants of anthropogenic origin, whereas the latter is unavailable for those of natural origin [1]. It is necessary to study the migration of trace elements that participate in many geochemical processes, and to monitor the highly toxic micro component content at all stages of the ecological chain (soils, waters, biological systems), since there are the immediate sources of heavy metals that reach the human organism [14, 15].

The demand for rapid and sensitive analytical methods for the determination of chemical forms of toxic elements in environmental samples is increasing [2, 14, 16]. Among various methods for the removal of heavy metals from aqueous solutions including chemical precipitation, reverse osmosis, ion-flotation, evaporation, membrane filtration, ion exchange and adsorption, the ion exchange and adsorption processes is considred being a more useful method for metal removal than the other processes [1, 7, 10, 12, 15, 17-23]. In industrial applicants, precipitation has been the most common technology for the removal of metals. The motivation for change away from precipitation technologies in metals control applications derives from the fact that precipitation can only reduce the dissolved metal concentration to the solubility product level, which is often out of compliance with stringent discharge permit standards and necessitates further polishing stages. Furthermore the increasing costs and liability issues associated with metals contaminated sludge management is a burgeoning deterrent. Another difficulty is treatment of wastewaters containing low concentrations of heavy metals. In this case the common methods for removing metals are either economically unfavorable or technically complicated. Such problems make the arena ripe for new and innovative methods for treatment of metals contaminated wastewater [8, 13].

The main advantages of ion-exchange over chemical precipitation are recovery of metal value, selectivity, less sludge volume produced and the meeting of strict discharge specifications. In ion exchange systems, polymeric resins are usually employed [1, 5, 17, 22].

A conventional ion exchange technique that employs a strong acid cation-exchanger can efficiently remove the heavy metal ions; however, it may also remove innocuous cations such as sodium, calcium, and magnesium. In such cases, selectivity can be enhanced by the use of weak acid cation-exchange resins, wherein carboxlyate groups are linked to either a styrenic or ethylenic backbone [11, 22]. Ion exchange of a specific cation is strongly influenced by the presence of competitive cations and complexing reagents, such as anions. It is expected that in municipal and industrial waste water, where the complexity of the system is high, several cations and anions could influence the removal of specific target metals [22]. It is well known that ion exchange is strongly affected by complex formation of the various mobile species with one another. Furthermore, transition metals can form stable complexes due to their electronic structure (availability of d orbitals, as well as s and p orbitals) [22]. Adsorption and/or ion-exchange onto ion-exchange resins and low cost particulate media such as coal or clay now offer an attractive, inexpensive option for the removal of colloidal and dissolved metals [10, 13].

A strong-base anion-exchange resin will absorb chromates from water according to the following reactions [24]:

$$RCl + HCrO_4^- \leftrightarrow R\text{-}HCrO_4 + Cl^- \quad (1)$$

$$2\,RCl + CrO_4^{2-} \leftrightarrow R_2\text{-}CrO_4 + 2Cl^- \quad (2)$$

Adsorption processes have been widely used in the treatment of industrial wastewater due to their many advantages [7], such as high treatment capacity, high removal efficiency, and fast kinetics. Adsorption has been used as one of the methods to remove chromium species from aqueous solutions with relatively low chromium concentrations. There are many types of adsorbents, including activated carbon [25-28], activated sludge [29], bentonite [30], carbon nanotubes [31], and biomaterials [32-40], that have been studied for the adsorption of chromium from aqueous solutions. However some of these adsorbents do not have high adsorption capacities or need long adsorption equilibrium times while others may have been difficult for regeneration and reuse even though high adsorption capacity and fast adsorption equilibrium were reported [41]. Additionally, the effects of other factors on adsorption behavior have been investigated, such as pH, anions, and temperature. In addition, several recent publications concerning the separation of Cr(III) in various aqueous solutions generally showed that carboxylic resin also offers a higher apparent capacity and slow sorption kinetics [42].

Up to now, XAD resin [43], ion-exchange resins [44-51], chelating resins [52-58] and various polymers [59] have been used as adsorbents. Adsorption on activated carbon or synthetic ion exchanger have been found to be an effective process for Cr(VI) removal, but it is too expensive. Consequently numerous low cost alternatives have been studied including treated sawdust [60], coal [61-64], humic acids [65-67], Moroccan stevensite [68], peanut shells [69], clay [70-72], clinoptilolite [73-75], montmorillonite [76-79], kaolinite [80], modified kaolinite [81], bentonite [82], sepiolite [83], grafted silica [84], siderite [85], coir [86], modified coir fibres [87], tree fern [88], wheat bran [89], juniper fiber [90], rice bran [91], chitosan [92, 93], by products as anion exchange resins [94], alginate and gelatin [95], lignin [96], and zeolite [27, 97-99]. Natural zeolites are capable of removing several cations from aqueous solutions by utilizing ion exchange. Montmorillonite is one of the zeolites that have received extensive attention due to its attractive selectivity for certain heavy metal cations such as lead, cadmium, and nickel. Overall rates of adsorption and ion exchange in porous materials are generally controlled by mass transport within the pore network, rather than by the kinetics of sorption or ion exchange itself. The most important kinetic parameter is then the pore or solid diffusion coefficient. Actually, the effective diffusion coefficient is a macroscopic average over a large number of ions in pores of all different sizes, shapes, and directions and at larger and smaller distances from the pore walls, expressing the average ability of the species to make headway in every given direction [5].

Since the surface properties of various metal ion species and the adsorbents used to remove these metal species can vary with the solution conditions, the mechanisms of metal ion adsorption on the adsorbents can also vary significantly. Commonly, the electrostatic interaction, surface complexation, and ion exchange mechanisms may be identified as the major adsorption mechanisms that work individually or in combinations for the adsorption of metal ions on the adsorbents [41]. In particular, the carboxyl and phenolic groups on the surface of an adsorbent have been reported to be effective in the adsorption of chromium ions, though the specific adsorption mechanisms for various chromium species on the adsorbent have not been fully explored [41, 63].

Chromium is a relatively common element and occupies the 21$^{st}$ position on the index of elements occurring most commonly in the earth crust [100, 101] and is a common pollutant introduced into natural waters from a variety of industrial wastewaters including those from the textile, leather tanning, electroplating, steel works, cooling water towers, oxidative dyeing and metal finishing industries. As a result of unregulated applications and inappropriate waste-disposal practices, chromium contamination of surface and ground water has become a significant environmental problem [85]. In natural waters the two main oxidation states for chromium are (III) and (VI), characterized by a markedly different chemical behavior [16, 17, 18, 20, 41, 42, 85, 100-110]. Chromium speciation depends mainly on redox potential, pH of the media: above pH 7 Cr(III) predominates, below pH 6 Cr(VI) predominates [14], the presence of oxidizing or reducing compounds, the kinetics of the redox reactions, the formation of Cr(III) complexes or insoluble Cr(III) salts, and total Cr concentration [41, 101, 111]. The divalent state is unstable with respect to evolution of oxygen, the trivalent state has broad stability, and hexavalent Cr occurs under strongly oxidizing conditions. The oxidation state of an element can have an important effect on bioavailability and the toxicity [42]. Therefore, an accurate determination of individual Cr species rather than total Cr concentration is required to evaluate the physiological and toxicological effects [107, 111, 112].

In the 1980s, the U.S. Environmental Protection Agency (USEPA) developed and implemented a program to monitor and control the aquatic toxicity of municipal and industrial effluent discharges to surface waters [113]. The office of Environmental Health Hazard Assessment of the California EPA established a public health goal as 2.5 μg/L of Cr in drinking water, using the assumption that hexavalent Cr does not exceed 7 % of total Cr. In UK the allowable concentration of Cr is limited to 15 μg/L for surface waters and in the EU states its maximum allowable concentration in drinking water is 50 μg/L. The concentration for total Cr is limited to 0.5 mg/m$^3$ and for Cr(VI) to 0.05 mg/m$^3$ in indoor air [100]. In waste waters, a maximum permissible concentration for total Cr of 0.5 mg/L and for Cr(VI) of 0.05 mg/L in Japan is legally prescribed [114]. In the USA, the concentration of Cr in drinking water has been regulated with a maximum allowable level of 0.1 mg/L for total Cr [41]. Many countries have developed laws along this line, but since the legislation permits a larger content of Cr(III) than Cr(VI), speciation of Cr in environmental samples is very important [103]. The daily Cr requirement for adults is estimated to be 0.5-2 μg of absorbable Cr(III). If a fractional absorption value of 25 % for "biologically incorporated" Cr(III) in food is assumed, this is provided by a daily dietary intake of 2-8 μg of Cr(III), equivalent to 0.03-0.13 μg of Cr(III) per kg of body weight per day for a 60-kg adult [101, 111]. The recommended limits of Cr in potable water are 0.05 mg/L. So the elimination of Cr(VI) from drinking water is of great concern [17].

Cr, in the hexavalent form, is one of the undesirable heavy metals because it affects human physiology, accumulates in the food chain and cause several ailments [60]. Cr(VI) species, having mobile and strongly oxidant characters, are known as mutagen and potential carcinogen. Cr(VI) as a strong oxidizer is highly toxic [17, 20, 105, 107, 112] and can affect lungs, liver and kidneys, and it also produces gastrointestinal disorders, dermatitis and ulceration of skin in man [16, 100-102, 104, 106]. In contrast, Cr(III), having a limited hydroxide solubility and low toxicity, is generally regarded as a nondangerous pollutant. Cr(III) is considered as indispensable for the metabolism of glucose, lipids and proteins in living organisms [106, 112] and interferes the enzymatic sulfur uptake of the cell [100, 102-105]. The high presence of Cr(III) in the environment occurs as a result of effluent discharge from tanning industries, metal processing activities such as electroplating, wood preservatives, paint and pigments, textile, dyeing, steel fabrication, canning industry, and other metal-related industries [12, 18, 22, 41, 42, 60, 85, 101, 104, 105, 112, 114].

Cr(III) is also mutagenic, carcinogenic and teratogenic [100]. Because of these dramatic differences in physical and chemical properties of two Cr types and benign character of Cr(III), detoxification and immobilization processing of Cr(VI) are based on its reduction to Cr(III) [2, 85]. However, some organisms, namely Bacillus QC 1-2, Escherichia coli ATCC 33456 and sulphate reducing bacteria can tolerate and reduce hexavalent Cr, providing a biological route to eliminate Cr(VI) [20].

In contrast to Cr(VI), Cr(III) is considered to be less toxic [24, 41, 114]. Since Cr(III) is insoluble at neutral and high pH, chemical reduction of Cr(VI) to Cr(III) followed by precipitation of $Cr(OH)_3$, has long been applied as a means of ridding contaminated water of chromate. There is, however, evidence that $Cr(OH)_3$ may itself be oxidized to Cr(VI), leading to contamination of ground water [24, 41].

In aqueous solutions, Cr(VI) exists in the form of chromic acid ($H_2CrO_4$) and its salts ($pK_1$= 0.8, $pK_2$= 6.5) and in the form of dichromate ($Cr_2O_7^{2-}$). In neutral solutions at low concentrations, Cr(VI) will be present in the form of $HCrO_4^-$ and $CrO_4^{2-}$ and Cr(III) will be present in the most probable species being $Cr(OH)^{2+}.4H_2O$ [16, 24, 41, 101, 102, 104]. This species is found to be a component of insulin cofactor and also shows lower toxicity in biological systems [16]. Due to the repulsive electrostatic interactions, these Cr(VI) anion species are generally poorly adsorbed by the negatively charged soil particles in the environment and, hence, they can transfer freely in the aqueous environments. The Cr(III) species in aqueous solutions however may take the form of $Cr^{3+}$, $Cr(OH)^{2+}$, or $Cr(OH)_2^+$, depending on the solution pH values. As these species normally carry positive electric charges, they are relatively easy to be adsorbed on the negatively charged soil particles and thus are less mobile than the Cr(VI) species in the environment [41].

In water, Cr(III) is a positive ion that forms hydroxides and complexes, and is adsorbed at relatively high pH values. In surface waters, the ratio of Cr(III) to Cr(VI) varies widely, and relatively high concentrations of the latter can be found locally. In general, Cr(VI) salts are more soluble than those of Cr(III), making Cr(VI) relatively mobile. In air, Cr is present in the form of aerosols. It can be removed from the atmosphere by wet and dry deposition. Both trivalent and hexavalent Cr are released into the air [101]. Moreover, in air Cr particulates oxidize sulfur dioxide forming acidic aerosol which causes global acid rain [104]. Because of analytical difficulties, data on Cr speciation in ambient air are rarely available, but the proportion present as Cr(VI) has been estimated as 0.01-30 %, based on one study [101].

Many techniques can be used for the determination of total Cr, including atomic absorption spectroscopy [21, 116], inductively coupled plasma-atomic emission spectrometry (ICP) [117, 118], GFAAS [100], solid phase extraction [66, 100, 119, 120], ion chromatography [121], emission spectroscopy, X-ray fluorescence, and neutron activation analysis. Detection limits for AAS are in the range 0.05-0.2 g/L [101]. For determining chelated Cr or the hexavalent or trivalent form only, such methods as gas chromatography, polarography, and spectrophotometry can be used. The determination of Cr species is currently a very sophisticated procedure, and few analytical data are available [101, 112].

Obviously, the ion exchange process has been a strong alternative to precipitation because of its ability to reduce dissolved chromate concentration to no detectable levels in the water treatment plants. The concentrations of the substances that have to be removed are normally in the lower ppm range. Lewatit ion exchange resins are particularly suited for this type of applications. There are basically two types of strong base anions, one uses trimethyl amine (TMA) and the other uses dimethylethanolamine (DMEA) to functionalize the resin. Lewatit M 610 resin which has DMEA functional group has a greater capacity than Lewatit MP 62 resin which is functionalized with a tertiary amine.

Carboxylated resins are a special category of weak cation exchange resins that are useful in Cr metal recovery applications. Their pH range is lower for many metals and the advantage of low pH range is that the metals do not precipitate as oxides or hydroxides. Investigations on the metal sorption properties of iminodiacetic acid (IDA) resins at different temperatures and pH values are recent [37, 122, 123]. Utilization of Chelex-100 and Lewatit TP 207 is gaining attention as a simple, effective and economic means of treating wastewater containing heavy metals. Chelex-100 and Lewatit TP 207 are abundant, inexpensive and readily available.

Lewatit S 100 is a gel type strongly acidic resin of sulphonated polystyrene type. Its principal characteristics are excellent physical, chemical and thermal stability, good ion exchange kinetics and high exchange capacity [124].

Sporopollenin is a natural polymer obtained from *Lycopodium clavatum* which is highly resistant to chemical attack, has a high capacity, is stable, has a constant chemical structure and occurs naturally as a component of spore walls, and exhibits very good stability even after prolonged exposure to mineral acids and alkalies. Sporopollenin is produced by oxidative polymerization of carotenoids and carotenoid esters, which lead to the proposed monomer structures of the macromolecular sporopollenin. At present sporopollenin is generally considered to be a biopolymer and the detailed chemical structure of sporopollenin is as yet unknown. Its empirical formula has been found to be $C_{90}H_{144}O_{27}$ [125, 126]. Therefore, it is suitable to use it as adsorbent.

Although the adsorption capacities of low-rank coals are lower in comparison to synthetic ion exchange materials, the substantially lower cost of the bulk material shows great potential for the usage of low-rank coals as a mode to displace heavy metal ions from wastewater [127]. Brown coals are amorphous, aromatic biopolymers which offer a valuable resource that is inexpensive, renewable, and available in abundant quantities. Brown coals are characterized by their low calorific value, as well as high oxygen and moisture contents. The high oxygen content of low-rank coals allows, however, the unique capability for brown coal to remove metals from solution via ion exchange with carboxylic acid and phenolic hydroxyl functional groups on the surface of coals. Recent studies have shown that brown coals and other naturally occurring biomass can serve as effective adsorbents for metal removal from wastewater [127, 128]. Brown coal contains several functional groups such as methoxyl, hydroxyl groups, aliphatic and phenolic, and carbonyl groups. It is hypothesized that these functional groups serve as sites of adsorption of metal ions [13, 17, 129, 131]. These groups are active centers of the natural polymer for ion-exchange. The brown coals can also be considered as ion-exchange materials [132-140]. Carboxyl or hydroxyl groups are able to take part in the ion exchange according to Eqs. (3) and (4).

$$R\text{-}COOH + M^+ \leftrightarrow R\text{-}COOM + H^+ \quad (3)$$

$$R\text{-}OH + M^+ \leftrightarrow R\text{-}OM + H^+ \quad (4)$$

where M is metal. Because of the polar character of brown coal the specific adsorption potential for dissolved metals is quite high.

Use of clays as sorbents to remove contaminants has been increasingly obtained attention of their low price compared to other materials such as activated carbon, synthetic resins, and synthetic zeolite [76]. Clay linings have been used as barriers in landfills to prevent contamination of groundwater and subsoil by leachates containing metals. These clays are chosen to avoid pollutant release into the environment owing to their high specific surface areas, high chemical and mechanical stability, a variety of surface and structural properties, the degree of surface heterogeneity due to the existence of crystal edges, broken bonds, and imperfection on the surfaces, soil organic matter, low cost, and ubiquitous presence in most soils [71, 76, 78]. In other words, the chemical nature and pore structure generally determine the sorption ability of clay.

Montmorillonite can adsorb heavy metals via two different mechanisms: (1) cation exchange in the interlayers resulting from the interactions between ions and negative permanent charge and (2) formation of inner-sphere complexes through $Si\text{-}O^-$ and $Al\text{-}O^-$ groups at the clay particle edges. Both mechanisms are pH dependent, because in acid conditions ($pH < 4$) most silanol and aluminol groups are protonated; therefore, in particular for the latter, an acidification can lead to an increase in mobility of metals bound to soil. For this reason, it is necessary to improve the knowledge of the effect of pH on the sorption capacity of montmorillonite in soil-solution system [78].

Depending on the type of the clay, clay minerals show a variety of surfaces which exhibit electrical charge properties having a strong influence on the sorption of ionic and polar species. The charge on the surface of the soils must be counterbalanced in the aqueous phase to maintain electro neutrality. As a result, electrical double layer exists at the clay/water interfaces.

$MOH^+$ is formed and adsorption process is rapidly completed by interaction with negative charge on clay, but also clay absorbs metal ions by their outer sheet of hydroxyl groups:

$$M^{2+} + H_2O \Leftrightarrow MOH^+ + H^+, \quad (5)$$

$$SiO^- + MOH^+ \Leftrightarrow SiOMOH, \tag{6}$$

$$nSi\text{-}OH + M^{n+} \Leftrightarrow (Si\text{-}O)_n - M + nH^+ \tag{7}$$

The clay surface may also be negatively charged, providing adsorption sites for metal ions [141]:

$$n\ SiO^- + M^{n+} \Leftrightarrow (Si\text{-}O)_n - M \tag{8}$$

## 2. EXPERIMENTAL

### 2.1. Materials

As chelating resins Chelex-100 (from Merck) and Lewatit TP 207 (from Bayer) were used. Chelex-100 is a chelating ion-exchange resin having functional IDA groups in a styrene-divinylbenzene matrix. Lewatit TP 207 is a chelating ion-exchange resin having functional iminodiacetic acid groups in a styrene matrix. Their physical properties and specifications as reported by the suppliers are shown in Table 1.

Another resin used was Lewatit S 100 from Bayer. The physical and chemical properties of the resin were shown in Table 1. Lewatit MP 62 and Lewatit M 610 resins were received from Bayer AG. Their physical properties are shown in Table 1.

Another chelating resin used was *Lycopodium clavatum* with 20 μm particle-size mesh from BDH Chemicals. The modification reactions of the metal-ligand complex [carboxylated epichlorohydrine-sporopollenin(CEP-sporopollenin) and bis-diaminoethyl glyoxim-sporopollenin(b-DAEG-sporopollenin)] of sporopollenin for Cr(III) is shown in previous works [142].

The brown coal samples collected from Isparta-Yalvaç-Yarıkkaya(YK) and Kaşıkara(KK) coal reserves placed in the north west part of Anatolia. The material was dried in an oven at 105 °C for 24 h, ground and sieved in the size range of 80-100 mesh ASTM and the material was placed in an airtight container for further use. All adsorbents were dried at 105 °C overnight before the adsorption experiments.

The clay used was collected from a deposit in the western part of Turkey (Yalvaç-Isparta), which had an idealized formula of $(Na,Ca)_{0.3}\ (Al,Mg)_2\ Si_4O_{10}\ (OH)_2\ .xH_2O$ (with X-ray diffraction patterns (Panalytical X'PertPro)). It was ground and sieved to a grain size of -65 meshes ASTM before use in the sorption experiments. The samples were washed prior to the adsorption experiments in order to remove the surface dust, and finally kept in bottles for future usage. Samples were dried in a vacuum oven at 60 °C for an hour before use.

All chemicals used were of analytical grade and obtained from Merck. A stock solution of $1x10^{-3}$ mol/L of Cr was prepared by dissolving $CrCl_3.6H_2O$ 1000 mL distilled water. Cr(VI) stock solution (from Merck) was prepared from potassium salts of dichromate ($K_2Cr_2O_7$) of analytical grade. A series of standard metal solution by appropriate dilution of the stock metal solution were prepared. Solutions of 0.1 M NaOH and HCl were used for pH adjustment. Dissolved metals in the samples at least three concentrations of each were determined from linear graph of calibration curve.

**Table 1. Some properties of adsorbents**

**a)** Elemental analysis of the brown coals and montmorillonite

| | **C,** (%) | **H,** (%) | **N,** (%) | **S,** (%) |
|---|---|---|---|---|
| ***YK brown coal*** | 51.73 | 4.95 | 1.03 | 2.98 |
| ***KK brown coal*** | 47.21 | 4.93 | 0.25 | 0.30 |
| ***Montmorillonite*** | 0.56 | 1.90 | 0.02 | 0.05 |

**b)** Chemical properties of YK and KK brown coals

| **Coals** | **Calorific values of dry coal (cal/g)** | **Moisture content (%)** | **Ash (%)** | **Total Sulfur (%)** | **Volatile matter (%)** | **Fixed carbon (%)** |
|---|---|---|---|---|---|---|
| **YK** | 5547 | 27.83 | 12.62 | 1.88 | 49.71 | 43.73 |
| **KK** | 4621 | 29.57 | 24.31 | 2.50 | 46.45 | 23.70 |

**c)** Properties and speciations of Chelex-100, Lewatit TP 207 and Lewatit S 100 resins

| **Resin** | **Lewatit TP 207** | **Chelex-100** | **Lewatit S 100** |
|---|---|---|---|
| Ionic form | $Na^+$ | $Na^+$ | $Na^+$ |
| Functional group | Iminodiacetic acid | Iminodiacetic acid | Sulfonic acid |
| Matrix | Cross linked polystyrene | Polystyrene-divinylbenzene | Crosslinked polystyrene |
| Structure | Macro porous | Macro porous | Gel type beads |
| pH range | 0-14 | 0-14 | 0-14 |
| Bead size (mm) | 0.4-1.25 | 0.3-1.0 | 0.315-1.25 |
| Capacity | 2.4 eq/l | 0.4 mmol/ml | 2.0 (eq/l) |

**d)** Properties and speciations of Lewatit MP 62 and Lewatit M 610 Anion Exchange Resins

| **Resin** | **Lewatit MP 62** | **Lewatit M 610** |
|---|---|---|
| Type | Anion Exchange Resins, weakly basic | Anion Exchange Resins, strongly basic |
| Functional Group | Tertiary amine | Tertiary amine (Dimethylamino ethanol) |
| Matrix | Macroporous crosslinked polystyrene | Macroporous crosslinked polystyrene |
| Bead Size (mm) | 0.31-1.25 | 0.30-1.20 |
| Operating pH Range | 0-8 | 1-14 |
| Total Exchange Capacity min. | 1.7 eq/l | 3.7 mmol H+g-1 dry |
| Thermal Stability | -20 to 100 | -20 to 70 |

## 2.2. Sorption Experiments

Inductively Coupled Plasma Spectrometry (Varian Vista ICP-AES) and AAS (Perkin Elmer AA800 Model ) were used for Cr(III) and Cr(VI) analysis. pH was measured by using a glass electrode (Jenway 3010 model pH meter and Crison Basic20 Model pH meter). A shaker with thermostat (Gallenkamp Incubator) of Orbital model and a Memmert WB29 model thermostated water bath shaker were used for adsorption experiments and temperature studies.

Adsorption isotherm was carried out with different initial concentrations of Cr(III) and Cr(VI) while maintaining the adsorbent dosage at constant level. Known amounts of the adsorbents and 30 mL ($1x10^{-4}$-$1x10^{-3}$ mol/liter) of aqueous solution containing Cr ions were shaken in an incubator to achieve equilibration. The samples were withdrawn from the shaker at the predetermined time intervals, and the adsorbents were separated by filtration and initial and equilibrated metal concentrations in the aqueous solution were determined by ICP-AES or AAS. The ionic strength of the solution was adjusted to certain concentration by adding appropriate amounts of sodium nitrate.

Sorption isotherms are plots of the Cr(III) and Cr(VI) uptake and the final equilibrium concentration of the residual sorbate remaining in the solution. The classical models of Langmuir and Freundlich were used to compare the single metal sorption performance and to determine the capacity of all adsorbents for Cr(III) or Cr(VI) ions.

The effect of initial pH studies on the removal of Cr, about 30 mL of Cr solution of desired concentration were taken in bottles and known amount of CEP-sporopollenin, b-DAEG-sporopollenin chelating resins, Chelex-100 chelate resin, Lewatit TP 207 chelate resin, brown coals, montmorillonite, Lewatit S 100 strongly acidic cation exchange resin, Lewatit MP 62 and Lewatit M 610 anion exchangers were added separately. The solution is adjusted to a pH, which is optimum for the adsorbent but low enough to prevent hydroxides from forming and high enough to ensure the capture of the metals.

For temperature effects on the removal of Cr(III) and Cr(VI) at 20, 35, 50, and 65 °C were carried out on maximum adsorption and each treatment was replicated three times before initiating the experiment.

Effect of adsorbent dosage was studied with a 30 mL solution of $1x10^{-3}$ mol/L Cr solution and varying amounts of adsorbents.

The amount of sorbed Cr ion was calculated from the change in the Cr concentration in the aqueous solution before and after equilibrium and the weight of the dry resin used. In all cases, mass balance was confirmed.

The adsorbed quantities were then determined using the mass balance equation:

$$q = [V ( Co - C )] / m \qquad (9)$$

where q is adsorbed metal (mmol/g adsorbent) on the samples, m is the weight of adsorbent (g), V is volume of metal solution (L), Co is initial metal concentration (mmol/L), and C is metal equilibrium concentration (mmol/L), respectively.

## 3. Results and Discussion

### 3.1. Effect of pH on Cr Removal

The variable that has the most significant effect on the affinity of a resin for a particular ion is pH. The interactions of Cr with all adsorbents are complex, probably simultaneously dominated by adsorption, ion exchange. The pH dependence of adsorption may suggest that metal ions are adsorbed according to the ion-exchange mechanism [1]. In order to optimize the pH for maximum removal efficiency, experiments were conducted with 30 mL of $1x10^{-3}$ mol/L of Cr solution containing about 0.1 g of Chelex-100, Lewatit TP 207, b-DAEG-sporopollenin and CEP-sporopollenin chelating exchangers, Lewatit S 100 strongly acidic ion exchanger, Lewatit MP 62 and Lewatit M 610 anion exchangers, YK and KK coals, and montmorillonite in the different pH range.

Percentage removal of Cr(III) was maximum at the initial pH of about 6 and decreased at lower and higher initial pH values. For b-DAEG-sporopollenin and CEP-sporopollenin chelating resins, the observed change in pH during each isotherm experiment was likely due to ion exchange between protons on b-DAEG-sporopollenin and CEP-sporopollenin and metal ions. The pH was decreased to 2.6-2.9 during adsorption because the adsorption reactions were consistent with the following ion exchange process. Apparently, the amount of sorption $q_e$ increases with increasing pH for Cr ions. The sorption of metal ions is primarily affected by the surface charge on the sorbent, which in term is influenced by solution pH. The low $q_e$ at low pH may be explained by the competitive sorption between proton and metal ions. As the solution pH increases, the number of negatively charged sites increases which favors the sorption of metal cations [60, 76, 78, 130, 142]. The nature of the adsorption depends on several parameters related to the adsorbent (ionic charge), to the solution (pH) and the chemistry of the metal ion: ionic charge, ability to be hydrolyzed and to form polynuclear species. Before batch experiments started, the effect of the pH on metal ion adsorption was investigated. The percentage of bound metal increases with pH from 3 to 5 with maximum binding occurring between pH 5 and 6. At pH 6, 90.25 % for b-DAEG-sporopollenin, 88.94 % for CEP-sporopollenin of Cr in solution were removed. But after pH 6, for avoiding precipitation of $Cr(OH)_3$, pH 5 was chosen as the optimum pH [42, 115, 142].

It can be seen that pH of the solution plays a very important role in the adsorption of Cr on b-DAEG-sporopollenin and CEP-sporopollenin chelating resin. The percentage of adsorbed is at a maximum in the pH range 3-6, for a solution containing $1x10^{-3}$ mol/L of Cr.

Chelex-100 and Lewatit TP 207 chelating ion exchangers with iminodiacetic acid groups have strong complexing properties forming heterocyclic chelates with Cr(III) ions depending on the pH of the mobile phase [143, 144, 164]. The pH value of the solution is an important factor that controls the sorption of Cr(III). The sorption of Cr(III) ion on the resins is clearly very favorable at high pH values at pH 4 and 5. At lower pH, the isotherms become more linear. There is an abrupt increase in Cr(III) sorption when pH is raised from 3 to 4.5. The pH of the aqueous solution of Cr(III) affects its uptake on the resins and in general the uptake decreases at high pH values [129, 164]. It is evident that the removal efficiency of Cr by both resins is reasonably high in the pH range 4.5–5.

Cr(III) metal showed strong affinity towards the resin in the narrow pH range and it can be seen that the percentage adsorbed is at a maximum pH 4.5. The retention of Cr(III) by both resins decreased markedly from pH 5 to 6. Solution pH had a strong effect on the equilibrium constant of Cr(III). The equilibrium constant increased from a value of 0.1 (at approximately pH 2) to 320 (at pH 4.5) for Lewatit TP 207 and 0.1–7 for Chelex-100. This increase is rational; the exchanger being weakly acidic, is practically in the hydrogen form at pH 2 and is gradually converted into metal form as the pH increased. Hence, the following experiments were performed in the solution pH of 4.5. This sorption trend can likely be ascribed to the effect of competitive binding between Cr(III) and hydrogen ions for the binding sides on the surface of the resins. At low pH, an excess of hydrogen ions can compete effectively with

Cr(III) for binding sites, resulting in a lower level of Cr(III) uptake. At pH values below two, hydrogen ions are likely to compete with Cr(III) ions and at pH values above six, Cr(III) ions might precipitate. At higher pH values greater than six, the presence of $OH^-$ ions forms the hydroxyl complexes of Cr, $Cr(OH)_3$ [73, 76, 92, 129, 143-146, 164]. In chelating resins and other adsorbents, in pH range of 1-2, there is little or no adsorption. The percent adsorption increases in the pH range of 3-6, showing the maximum adsorption at pH 5-6 [147, 148, 164].

For Lewatit S 100 resin, the effect of pH on adsorption of Cr(III) was studied at room temperature varying the pH of metal solution-ion exchanger from 2 to 6. In pH range of 2-3, there is little or no sorption, as I mention before. The percent adsorption increases in the pH range of 3-4, showing the maximum adsorption at pH 4 [144, 145, 150]. The sites responsible for the ion exchange process are due to the $-SO_3H$ groups in the resin matrix [147].

The sorption of Cr ions on montmorillonite is primarily affected by the surface charge on the sorbent, which in term is influenced by solution pH [73, 76, 89, 92, 141], and the aluminol and silanol groups are more protonated and, hence, they are less available to retain the investigated metals. This effect is strongly evident for Cu, Pb, and Cd, and less pronounced for the others. The reason of this behavior is that the surface complexation reactions are influenced also by the electrostatic attraction between the surface charge and the dissolved ions [78]. Cd ions are adsorbed less on clay even if they have same valence as Cu and Zn because they are characterized by larger ionic radii and, then, their introduction in interlayer and complexation with surface sites is limited by steric hindrance and lower electrostatic attraction. Cr(III) ion, instead, has a higher charge density and, then, the columbic attraction towards the superficial sites of the clay is greater. For this reason the total capacity of montmorillonite towards the trivalent ion considered is the highest. In general, however, montmorillonite was found to have a good total capacity towards all considered metals [78, 141].

The interaction of Cr(VI) with Lewatit MP 62 and Lewatit M 610 anion exchange resins were also investigated at the different pH values. There are two principal means by which anion-exchanger and ions can interact with each other in aqueous solutions. The most common way involves ion exchange or chelation of metal ions. These interactions are characterized by the resin structure in terms of the functional groups present. Amine groups bound to carbon is considered to be the reaction partner of Cr(VI). In order to find out the optimum pH for maximum removal efficiency, experiments were conducted with changing the pH of Cr(VI) solution from 2 to 6. The maximum recovery (> 95 %) was obtained in the range of 4-5. As I mentioned before, in the experiments removal of Cr(VI) from aqueous solution, no tests were also performed pH $\geq$ 6 because of the possibility of $Cr(OH)_3$ precipitation.

In aqueous solutions, Cr(VI) exists in the form of chromic acid ($H_2CrO_4$) and in the form of dichromate ($Cr_2O_7^-$). In neutral solutions at low concentrations, Cr(VI) will be present in the form of $HCrO_4^-$ and $CrO_4^{-2}$. A strong-base anion-exchange resin will absorb chromates from aqueous solution according to reactions shown in the introduction section. The explanation for this finding lies in the influence of the functional group of the quaternary ammonium moiety of the resin, which has a significant influence on the uptake of Cr(VI). The pH value of the solution is an important factor that controls the sorption of Cr(VI). The sorption of Cr(VI) ion on the resins is clearly very favorable at pH values between 2.5 and 5.0. Cr(VI) is very soluble in aqueous solutions, and their solubilities increase with pH; therefore, it was practical to employ resins at solution pH values of 5.0 and below. For Lewatit MP 62 and Lewatit M 610 anion exchange resins, the removal efficiency of Cr(VI)

clay is able to achieve. Therefore, equilibrium time for metal sorption on montmorillonite appears to be dependent on initial metal concentrations as occurs for other sorbents [84, 99, 141].

High adsorption rates of Cr for Lewatit MP 62 and Lewatit M 610 resin are observed at the onset; and then plateau values are gradually reached within 10 min. The increasing contact time increased the Cr(VI) adsorption and it remains constant after equilibrium reached in 20 minutes for initial concentration of 100 ppm. In each case, the decreasing concentration of Cr(VI) remaining in the solution indicates that Cr was adsorbed strongly by both resins. The Cr(VI) adsorption respect to time curves are smooth and continuous leading to saturation showing monolayer coverage of metal ion on the surface of resin [146].

## 3.4. Effect of Temperature

The adsorption mechanism, (i.e., chemical or physical) is often an important indicator to describe the type and level of interactions between the adsorbate and adsorbent. If adsorption decreases with increasing temperature it may be indicative of physical adsorption and the reverse is generally true for chemisorption. However, there are a number of contradictory cases in the literature [150, 157].

The amounts of sorption of Cr(III) ions by b-DAEG-sporopollenin and CEP-sporopollenin are measured in a temperature range 20-65±1 °C. This is due to the exothermic adsorption reactions of Cr(III) ion with resins. The equilibrium constants for such reactions slightly increase with temperature. The carboxylic groups of iminodiacedic acid resin are partially protonated at all temperatures but their deprotonation degree increases at all temperature resulting in slight increase metal capacities at high temperature. Where only ion exchange is involved, the temperature effect on retention time is relatively small. Few studies of temperature changes have been carried out on chelating ion exchangers [142]. Empirical studies show that temperature has a significant effect on retention in chelation. The "chelate effect" concerns a large positive change in entropy on formation of a chelate complex, which means that the change of Gibbs free energy with temperature will be negative. Considering other thermodynamic factors, it is possible that the entropy change in chelation reactions may have a more important bearing on temperature effects than the enthalpy of sorption. Thus, as the temperature increases, the resulting $\Delta G$ will become more negative and so the equilibrium constant will become larger as temperature increases [17, 158-161].

The uptake of Cr(III) increases with increasing temperature, indicating that higher temperature favor the adsorption for CEP-sporopollenin, whereas the uptake of Cr(III) decreases with increasing temperature for b-DAEG-sporopollenin. Equilibrium constants for Cr(III) were high and adsorption increased slightly with temperature for Chelex-100, Lewatit TP 207 and Lewatit S 100 resins.

Assuming that the activity coefficients are unity at low concentrations, (the Henry's law sense), thermodynamic parameters $\Delta G$, $\Delta H$, and $\Delta S$ were calculated using the following relations [144].

$$K_c = \frac{C_{Ae}}{C_e} \tag{10}$$

$$\Delta G^o = -RT \ln K_c \tag{11}$$

$$\log K_c = \frac{\Delta S^o}{2.303R} - \frac{\Delta H^o}{2.303RT} \tag{12}$$

where $K_c$ is the equilibrium constant at different initial concentrations, $C_e$ the equilibrium concentration in solution (mg/L) and $C_{Ae}$ the solid-phase concentration at equilibrium (mg/L). Where R is the ideal gas constant (8,314 J mol$^{-1}$ K$^{-1}$) and T is temperature in K. $\Delta G$, $\Delta H$, and $\Delta S$ are changes in free energy, enthalpy, and entropy, respectively. $\Delta H$ and $\Delta S$ were obtained from the linear Van't Hoff plot of log$K_c$ versus $1/T$ and presented in Table 2 for all adsorbents. Negative $\Delta G$ values confirm the feasibility of the process and spontaneous nature of the adsorption with high preference of Cr(III) for the resins. The negative values of $\Delta H$ indicate the exothermic nature of the process while the negative $\Delta S$ corresponds to a decrease in the degree of freedom of the adsorbed species [17, 147, 155-157, 162-164].

The change in free energy ($\Delta$G) for Cr(III) is calculated to be –9576.38, -1286.45 J.mol$^{-1}$, respectively for CEP-sporopollenin and b-DAEG-sporopollenin from Eq.(11). The negative values of $\Delta$G mean that the sorption of Cr ions on b-DAEG-sporopollenin and CEP-sporopollenin is spontaneous, perhaps because the surfactant molecules have more affinity toward the metal ions. As clearly shown in Eq.(12), a plot of log$K_c$ versus 1/T would give $\Delta$H and $\Delta$S. As shown in Table 2, the values of $\Delta$H and $\Delta$S (20 °C) for Cr(III), are -7.9346 J.mol$^{-1}$ and 32.7110 J.K$^{-1}$.mol$^{-1}$ for CEP-sporopollenin and –28.2248 J.mol$^{-1}$ and 118.7161 J.K$^{-1}$.mol$^{-1}$ for b-DAEG-sporopollenin, respectively. A slightly positive entropy change for sorption of Cr ions on b-DAEG-sporopollenin and CEP-sporopollenin is due to the fixation of ions on the exchange sites of the randomly distributed surfactant species [76, 78, 80, 142].

In the effect of temperature on Cr(VI) sorption by Lewatit MP 62 and Lewatit M 610 resins, the adsorption rate and equilibrium adsorption level decreases with increasing temperature [146]. The removal of Cr(VI) from the solution of initial concentration decreases as the temperature increase, and agitation rate of 180 rpm with resin indicates the exothermic nature of the process. This may be attributed to a relative increase in the escaping tendency of the solute from the resin phase to the bulk phase with the rise in the temperature of the solution [163]. Therefore, a general temperature range is suggested to be 25-35 °C [146].

The thermodynamic equilibrium constant obtained is used to calculate all other thermodynamic parameters (Table 2). The Gibbs free energy for the adsorption process was obtained at 20 °C using Eq. 11. The Gibbs free energy indicates the degree of spontaneity of the adsorption process, where more negative values reflect a more energetically favorable adsorption process [156]. The negative $\Delta$G values obtained in our study for YK and KK lignite coals confirm the feasibility of these adsorbents and spontaneity of the adsorption [63, 64].

**Table 2. Freundlich and Langmuir adsorption isotherm parameters of adsorbents**

**a)** Freundlich and Langmuir adsorption isotherm parameters of adsorbents for Cr(III) removal

| Adsorbents | Freundlich Isotherm | | | Langmuir Isotherm | | |
|---|---|---|---|---|---|---|
| | $K_f$ | $n$ | $R^2$ | $Q_0$ | $b$ | $R^2$ |
| CEP-spr | 10.39 | 0.99 | 0.99 | 133.33 | 0.10 | 0.93 |
| b-DAEG-spr | 84.84 | 0.80 | 0.88 | 1.23 | 29.51 | 0.93 |
| Chelex-100 | 0.03 | 1.48 | 0.89 | 0.29 | 0.09 | 0.90 |
| Lewatit TP 207 | 0.04 | 1.25 | 0.87 | 0.34 | 0.22 | 0.95 |
| Lewatit S 100 | 45081.67 | 0.71 | 0.89 | 0.39 | 6445.75 | 0.99 |
| YK brown coal | 1.95 | 2.37 | 0.78 | 0.26 | 964.13 | 0.99 |
| KK brown coal | 209.65 | 0.93 | 0.83 | 0.05 | 832.11 | 0.98 |
| Montmorillonite | 0.18 | 3.92 | 0.52 | 0.05 | 5201.02 | 0.99 |

**b)** Langmuir adsorption isotherm parameters of adsorbents for Cr(VI) removal

| Adsorbents | Langmuir Isotherm | | |
|---|---|---|---|
| | $Q_0$ | $B$ | $R^2$ |
| Lewatit MP 62 | 0.40 | 25234 | 0.94 |
| Lewatit M 610 | 0.41 | 35090 | 0.99 |
| YK brown coal | 0.92 | 15.58 | 0.90 |
| KK brown coal | 0.98 | 15.01 | 0.98 |

$K_f$ and $Q_0$: mmol/g adsorbent; $R^2$:Correlation coefficient

## 3.5. Adsorption Isotherms

To have a quantitative means of comparing metal adsorption strength and to design adsorption process effectively, it is useful to employ mathematical models predicting the metal adsorption [59]. It was found that the isotherms of Cr(III) fit well both the Freundlich model (indicating that the adsorption amount increases significantly with increasing Cr(III) equilibrium concentration) and the Langmuir model (suggesting that the adsorption almost reach saturation). The adsorption capacity depends on the surface chemical and physical properties of the adsorbent, in which the porosity is one of the important factors [60, 148].

The Freundlich and Langmuir adsorption isotherms were used to analyze the results and these isotherms have been proved to be useful in describing adsorption behavior of the adsorbents. The Freundlich model is defined by [1, 37, 75, 115, 147, 165]

$$q = K_f\, C^{1/n} \tag{13}$$

where C and q are the concentration of Cr(III) in solution and adsorbent phase, respectively ($K_f$ and n are constants). A linear relationship was observed among the plotted parameters at 20±1 °C temperature which indicates the applicability of Freundlich equation. The linearised form of the Freundlich adsorption isotherm equation can be as follows:

$$\log q = 1/n \log C + \log K_f \tag{14}$$

where n is the Freundlich constant, and $K_f$ is adsorption coefficient.

The Langmuir adsorption isotherm equation can be seen as follows:

$$\frac{C_e}{q_e} = \frac{C_e}{Q_0} + \frac{1}{Q_0 b} \tag{15}$$

where $C_e$ (mg/L) is the equilibrium concentration, $q_e$ (mg/L) is the amount of Cr(III) sorbed at equilibrium, $Q_0$, and $b$ are Langmuir constants related to adsorption capacity and energy of adsorption, respectively.

A linear plot of $C_e/q_e$ vs. $C_e$ shows that the sorption follows the Langmuir isotherm model. $Q_0$ and $b$ were determined from the slope and intercept of the plots [103, 129, 164, 166].

Adsorption data for a wide range of adsorbate concentrations are most accurately described by adsorption isotherms, such as those of Langmuir or Freundlich, which relate adsorption amount to the equilibrium adsorbate concentration in the bulk fluid phase [11, 167, 168].

The adsorption isotherms revealed that metal adsorption increased with increasing adsorbed metal concentration. At the equilibrium point of the isotherm, however, the amount of adsorbed metal remained constant despite increases in metal concentration. This phenomenon can be compared to the Langmuir sorption isotherm model, and the agreement of the experimental data with the Langmuir isotherm suggests that the adsorbed metals form a monolayer coverage on the adsorbent surface, that the energy of adsorption is constant, and that there is no transmigration of adsorbate molecules in the surface plane [64, 169-171]. The Langmuir model assumes a uniform distribution of energetic adsorption sites [75]. The adsorption data of Chelex-100 and Lewatit TP 207 resins fit the Langmuir isotherm equation. The Langmuir isotherm applies to adsorption on a completely homogenous surface with negligible interaction between adsorbed molecules [159, 161, 164].

As in Table 2, evaluation of the equilibrium adsorption constant, *Qo*, and of the maximum adsorption capacity, *b*, was readily obtained from experimental data by plotting the reciprocal amount of adsorbed Cr ions ($C_e/q_e$) versus the reciprocal equilibrium concentration of Cr ions ($q_e$) [1, 11, 37, 76, 80, 115, 165].

One of the essential characteristics of the Langmuir equation could be expressed by dimensionless constant called equilibrium parameter, $R_L$ [162, 164]

$$R_L = 1 / (1 + K_{ads} C_o) \tag{16}$$

where $C_o$ is the highest initial solute concentration (mM). The value of $R_L$ indicates the type of isotherm to be irreversible ($R_L$=0), favorable (0< $R_L$<1), linear ($R_L$=1) or unfavorable ($R_L$>1) [76]. It is shown that the sorption of Cr(III) ions on all adsorbents is favorable.

The Langmuir isotherm equation provided an excellent fit to the equilibrium sorption data, giving correlation coefficients of 0.8955 and 0.9508 on the Chelex-100 and Lewatit TP 207 sorbents, respectively. In our work, the values of $R_L$ obtained are 0.9999. All ion-exchange reactions realized in the sorptions produced favorable isotherms [164].

The ion exchange isotherm characterizes the equilibrium of an ion in solid phase with the concentration of the ion in solution. Its graphical representation covers usually all possible experimental conditions at a given temperature.

Table 2 shows the $K_f$ and n values for Cr(III). If n<1, this implies that increased adsorption modifies the sorbent in a manner that increases the sorption capacity, such as forming new adsorption sites. When n>1 or becomes larger (n>>1), however, the adsorption bond becomes weak, and the value of q changes significantly with small changes in C [147, 172]. The values of $0.1 < 1/n < 1.0$ shows that adsorption of Cr(III) on the CEP-sporopollenin and b-DAEG-sporopollenin resins is unfavorable according to Freundlich isotherms.

As we can see, the Freundlich type isotherm of CEP-sporopollenin is more favorable than that of b-DAEG-sporopollenin. Significantly better are the results for Langmuir-type equation thus confirming that this simplified expression is valid not only in the cases when cations in liquid and resin phase are singly charged, but also when they have unequal charges and compete with each other [75, 142].

The experimental results obtained for the adsorption isotherms from Lewatit MP 62 and Lewatit M 610 reisns [146], and YK and KK brown coals [64] were found to obey the Langmuir adsorption isotherm. Analysis of adsorption results obtained at various concentrations showed that the adsorption pattern on the resin followed Langmuir isotherms as seen Table 2. The maximum Cr(VI) adsorption at an optimum pH of 5.0 is 0.40 and 0.41 mmol/g of weight Lewatit MP 62 and Lewatit M 610 anion-exchangers respectively (Table 2). The Langmuir constant *b* can serve as an indicator of isotherm rise in the region of lower residual metal concentrations, which reflects the strength of the resin for the solute. It was found to decrease with increase in pH from 2.0 to 5.0, which implies that the removal of Cr(VI) at lower pH between pH 3.0 and pH 4 could be more complete than that at a high pH (around pH 5).

Factors influencing the adsorption rate are mainly, among others, initial concentration, pH, resin amount, shaking speed and temperature. The adsorption of Cr(VI) by resins is very high, in the range of 95–99 % at low initial concentrations, and reaches equilibrium very quickly. This indicates the possibility of the formation of monolayer coverage of the metal ions at the outer interface of resins and suggests that these ion-exchangers can remove most of the Cr ion from aqueous solution if their concentrations are low, below 100 ppm. At a fixed resin quantity, the amount sorbed increased with increasing concentration of solution, but the percentage of sorption decreased. In other words, the residual concentration of Cr(VI) ions will be higher at a higher initial concentration of the Cr(VI) under study. In the case of lower concentrations, the ratio of initial number of Cr(VI) moles to the available sorption sites is low and subsequently the fractional adsorption becomes independent of initial concentration [173]. At higher concentrations, however, the available sites of sorption become fewer and subsequently the removal of metals depends on the initial concentrations of Cr(VI). At higher initial concentration, it is not likely that metal ions are only adsorbed in a monolayer at the outer interface of anion-exchangers. As a matter of fact, the diffusion of exchanging ions within resin particles may govern the sorption rate at higher initial concentrations.

The adsorption capacity, $Q_0$ for Cr(VI) calculated from the Langmuir adsorption isotherm was found to be 0.40 and 0.41 and adsorption intensities *b;* 25234 and 35090 for the resins Lewatit MP 62 and Lewatit M 610, respectively. It was found that more than 95 % removal was achieved using these anion exchangers.

The Cr(VI) sorption on coals was also found to increase with the initial metal concentration, then it reached a maximum very quickly and thereafter it reached equilibrium. A maximum removal of about 85 % hexavalent Cr was observed at the initial metal concentration of $1x10^{-3}$ mol/L. The sorption equilibrium is established when the concentration of adsorbate in the bulk solution is in dynamic balance with that of the interface. The maximum adsorption capacity from the Langmuir isotherm equation was found to be 0.98 mmol/g coal for KK and 0.92 mmol/g coal for YK coals. The adsorption intensities, *b* for Cr(VI) calculated from the Langmuir adsorption isotherm was found to be 15.58 and 15.01 for the brown coals; YK and KK, respectively. It was found that more than 85 % removal was achieved using these coals.

The amount of metal ions adsorbed per unit mass of coal (i.e. adsorption capacity) increased with the initial concentration of metal ions, as expected. Linear least-squares fitted plots were obtained for the low concentration region, where the adsorption isotherms obeyed the Langmuir equation; both $Q_0$ and $b$ were evaluated and are given in Table 2.

The higher-rank coal can be seen to provide similar removal profiles, with only a small enhancement of metals. The observed trend in adsorption efficiency, where adsorption increases as the rank of the coal substrate decreases, is related to the number of surface sites available for adsorption on the substrate surface. Lower-rank coals possess higher surface functionality, and hence a greater number of sites are available for interaction with aqueous heavy metal ions [127, 151]. The data obtained for the adsorption isotherms from montmorillonite [141] and Lewatit S 100 [150] were found to obey the Langmuir adsorption isotherm. The adsorption capacities of adsorbents were shown in Table 2.

### 3.6. Effect of Adsorbent Dose

The effects of the adsorbent dose on Cr(III) uptake of the all adsorbents was investigated by varying the adsorbent dose. The percent adsorption increased from 39.1 % to 85.6 %, as the KK coal dose was increased (0.02-0.14 g/30 mL) after equilibrium time (15 min.). The resin dosage was varied from 0.1 to 1.0 g for Lewatit S 100, Chelex-100, and Lewatit TP 207 resins. It is clear that for the maximum removal of 10 mg/L of Cr(III) ion in 100 mL requires a minimum resin dosage of 100 mg for Lewatit S 100 resin. It is apparent that the equilibrium concentration in solution phase decreases with increasing resin doses for a given initial Cr concentration. For the YK coal, adsorption increased from 30.3 % to 88.3 % with similar doses. Increase in the adsorption with the coal dose can be attributed to increased surface area and the availability of more adsorption sites [164]. For montmorillonite, metal removals also increase when the amount of adsorbent was increased. This is due to the greater availability of the exchangeable sites or surface area at higher concentration of the adsorbent. This is an expected result because as the adsorbent concentration increases, the number of adsorbent particles surrounding the metal ions or ratio of adsorbent particles to metal ions increases, therefore, these particles attach more ions to their surfaces [80, 174].

It may be concluded that by increasing the adsorbent dose the removal efficiency increases but adsorption density decreases. The decrease in adsorption density can be attributed to the fact that some of the adsorption sites remain unsaturated during the adsorption process; whereas the number of available adsorption sites increases by an increase in adsorbent and this results in an increase in removal efficiency [20, 62, 156, 147]. As expected, the equilibrium concentration decreases with increasing adsorbent doses for a given initial Cr concentration, because for a fixed initial solute concentration, increasing the adsorbent doses provides a greater surface area or adsorption sites [129, 145, 147, 153-156].

Effect of the amount of clay on the adsorption for Cr(III) was studied at fixed temperature (20±1 °C). It is readily understood that the number of available adsorption sites increases by increasing the resin amount and it, therefore, results in the increase of removal efficiency.

The influence of adsorbent concentration on the removal efficiency of hexavalent Cr was studied at an initial pH of 3.0, 30±1 °C and initial metal concentration of $1x10^{-3}$ mol/L. When the brown coal amount increased in bulk solution, the reduction rate correspondingly increased. As a result, protons and electrons from brown coals were consumed in the reduction of Cr(VI). We concluded that similar results were observed by Lewatit MP 62, Lewatit M 610, KK, and YK coals for Cr(VI).

**Table 3. Thermodynamic parameters of adsorbents**

| Adsorbents | ΔH° (J.mol$^{-1}$) | ΔS° (J.K$^{-1}$.mol$^{-1}$) | ΔG° (J.mol$^{-1}$) | | | |
|---|---|---|---|---|---|---|
| | | | $T_1$;298.15 K | $T_2$;308.15 K | $T_3$;323.15 K | $T_4$;338.15 K |
| CEP-spr | -7.93 | 32.71 | -9576.38 | -10067.05 | -10557.71 | -11048.38 |
| b-DAEG-spr | -28.22 | 118.72 | -1286.45 | -1412.87 | -2077.23 | -2184.42 |
| Chelex-100 | -37.36 | -111.39 | -33158.24 | -34857.67 | -36557.11 | -38256.54 |
| Lewatit TP 207 | -46.98 | -11.36 | -33154.22 | -34853.94 | -36553.67 | -38253.39 |
| Lewatit S 100 | -103.23 | 383.88 | -112373.92 | -118132.13 | -123890.35 | -129649.57 |
| Lewatit MP 62 | -2.16 | 24.49 | -7177.26 | -7544.62 | -7911.98 | -8279.34 |
| Lewatit M 610 | -2.51 | 35.49 | -10402.48 | -10934.89 | -11467.29 | -11999.70 |

| **Initial metal concentration in solution phase, mol/L** | ΔG° (J. mol$^{-1}$) | |
|---|---|---|
| | ***YK brown coal*** | ***KK brown coal*** |
| $1x10^{-4}$ | -5063.27 | -3674.59 |
| $2x10^{-4}$ | -4594.07 | -4860.43 |
| $4x10^{-4}$ | -5019.02 | -6118.58 |
| $6x10^{-4}$ | -2136.80 | -5125.83 |
| $8x10^{-4}$ | -15.75 | -4471.45 |

## 4. CONCLUSIONS

- When the sorption equilibrium is known precisely, it is possible to predict the optimum conditions for a pre-concentration or removal of Cr(III) and Cr(VI) ions from aqueous solutions. The results here is indicate that all adsorbents we used can efficiently remove traces of Cr(III)/Cr(VI) present in aqueous solutions.
- The adsorption of Cr(III)/Cr(VI) ions was dependent on the initial concentrations of adsorbent, sorbate amount, pH, temperature, and time of contact.
- Chelex-100 and Lewatit TP 207 resins have a large collective adsorption with Cr(III) ion, TP 207 showing stronger binding and a strong chelating ability. The isotherm plots showed a high sorption for pH 4.5 for both of the resins. This is a desirable

feature of sorption systems as it implies efficient uptake of Cr(III) ion by the resins in dilute solutions. These chelating resins are thus attractive sorbents for the treatment of wastewater containing Cr(III) ion at trace levels. For a given Cr(III)-bearing wastewater it is possible to design a batch treatment system to remove Cr(III) ion at any pH within the range 3–6. A batch contactor requires about 2 h contact time to achieve 80–95 % removal percentages from aqueous solution. The application of the both resins to leather tanning, electroplating, textile and dyeing wastewater for adsorption of toxic Cr metal is economic.

- The results show that the metal uptake on CEP-sporopollenin and b-DAEG-sporopollenin chelating resins increased with metal ion concentration. As we can see, the Freundlich type isotherm of CEP-sporopollenin is more favorable than that of b-DAEG-sporopollenin. Significantly better are the results for Langmuir-type equation thus confirming that this simplified expression is valid not only in the cases when cations in liquid and resin phase are singly charged, but also when they have unequal charges and compete with each other. The percentage of adsorbed is at a maximum in the pH range 3-6, for a solution containing $1\times10^{-3}$ mol/L of Cr. In pH range of 1-2, there is little or no adsorption. The percent adsorption increases in the pH range of 3-6, showing the maximum adsorption at pH 5-6. The uptake of Cr(III) increases with increasing temperature, indicating that higher temperature favor the adsorption for CEP-sporopollenin, whereas the uptake of Cr(III) decreases with increasing temperature for b-DAEG-sporopollenin.
- Metal sorption is pH-dependent and maximum sorption for Cr(III) ion on Lewatit S 100 resin was found to occur at pH about 3 and the equilibrium conditions are attained after nearly 4 h.
- New alternative adsorbents for Cr(III) removal have been explored by using brown coals. Our study showed that the KK brown coal had lower adsorption efficiency than the YK brown coal at higher Cr(III) concentrations. Initial pH of 4.5 was optimum for maximum Cr(III) removal by the KK and YK brown coals. At higher and lower initial pH values than 4.5, adsorption of Cr(III) was decreased.
- The sorption of metal ions on montmorillonite was rapid during the first 10 min and the equilibrium was attained within 2 h. The adsorption of metal ions on clay decreases with decreasing pH and increasing temperature. At low pH values, the hydrogen ion competes with the heavy metals towards the superficial sites and, moreover, the $Si\text{-}O^-$ and $Al\text{-}O^-$ groups are less deprotonated and they form complexes with bivalent and trivalent ions in solution with greater difficulty.
- Cr(III) ion adsorption on the both brown coals, Chelex-100, Lewatit TP 207, Lewatit S 100 resins, and montmorillonite was described by the Langmuir isotherm models.
- The results show that clay, which has a very low economical value, may be used effectively in removal of Cr(III) ions from aqueous systems for environmental cleaning purposes. Also, montmorillonite clay minerals are relatively plentiful and they don't need any treatment. Therefore, montmorillonite may be used to remove trace amounts of heavy metals from aqueous solution. The brown coals and montmorillonite are easily available in the Anatolia, so can be used by small scale industries having low concentrations of Cr(III) in wastewater using batch reactors.

- Lewatit MP 62 and Lewatit M 610 resins have a large collective adsorption with Cr(VI) ion, Lewatit M 610 showing stronger binding. A strong-base and weak base anion-exchange resin were used to reduce trace concentrations (about 1 ppm) of Cr(VI) in aqueous solution to low levels, which are in the range 5-30 ppb. The isotherm plots showed a high sorption for pH 3-5 for both of the resins. The sorption is relatively quick and the process is very efficient especially for water containing low concentrations of pollutants. These anion-exchangers are thus attractive sorbents for the treatment of wastewater containing Cr(VI) ion at trace levels and at any pH within the range 2–6. A batch contractor requires about 1 h contact time to achieve 90–95 % removal percentages from aqueous solution.
- The adsorption of Cr(VI) by coals is very high, in the range of 80–85 % at low initial concentrations, and reaches equilibrium very quickly. This indicates the possibility of the formation of monolayer coverage of the metal ions at the outer interface of coals and suggests that these YK and KK coals can remove most of the Cr ion from aqueous solution if their concentrations are low, below $1x10^{-3}$ mol/L. At a fixed coal quantity, the amount sorbed increased with increasing concentration of solution, but the percentage of sorption decreased.
- The brown coals are not suitable for the anions since the carboxyl functionalities using the ion exchange and chelation mechanism able to bound only cations. Positively charged surface can form at very low pH, which probably a possible way for Cr(VI) anion sorption. Considering the structure of brown coal, during the interaction of Cr(VI) with the coal, Cr(VI) suppose to be reduced to Cr(III) and this metal ion will be bound to the skeleton. Still, the nature of metal binding onto brown coal has received extensive investigation, but a common conclusion on the exact mechanism has not yet to be reached. The data show that the removal efficiency of Cr(VI) increased with lowering pH (at pH values $< 3$). This observation is consistent with the hypothesis that the low pH renders the coal surface with relatively more positive and less negative charges, thus resulting in greater adsorption of the anions of Cr. Increasing the pH resulted in a decrease in the removal efficiency of Cr because at these pH values the high concentrations of $OH^-$ ions compete more effectively with the chromate anions for the positive surface charges present in the coal surface.
- The maximum Cr(VI) adsorption at an optimum pH of 3.0 is 0.92 mmol/g of weight for YK, 0.98 mmol/g of weight for KK brown coals, respectively. The adsorption of Cr(VI) by coals is very high, in the range of 80–85 % at low initial concentrations, and reaches equilibrium very quickly. We can conclude that the KK and YK brown coals may be an efficient material in wastewater containing Cr metal for purification.
- Analysis of adsorption results obtained at various concentrations on Lewatit MP 62, Lewatit M 610 anion exchange resins, YK and KK brown coals showed that the adsorption pattern on the coals followed Langmuir isotherm. The Langmuir constant $b$ can serve as an indicator of isotherm rise in the region of lower residual metal concentrations, which reflects the strength of the coal for the solute.
- The kinetic data obtained from the adsorption studies would be useful for the fabrication and designing of wastewater treatment plants.

## REFERENCES


[1] Rengaraj, S.; Yeon, K.-H.; Kang, S.-Y.; Lee, J.-U.; Kim, K.-W.; Moon, S.-H. Studies on adsorptive removal of Co(II), Cr(III) and Ni(II) by IRN77 cation-exchange resins. *J. Hazard Materials*, B92, (2002), 185-198.

[2] Liang, P.; Shi, T.; Lu, H.; Jiang, Z.; Hu, B. Speciation of Cr(III) and Cr(VI) by nanometer titanium dioxide micro-column and inductively coupled plasma atomic emission spectrometry. *Spectrochimica Acta Part B*, 58, (2003), 1709-1714.

[3] Cimino, G.; Passerini, A.; Toscano, G. Removal of toxic cations and Cr(VI) from aqueous solution by hazelnut shell. *Wat. Res.*, 34(11), (2000), 2955-2962.

[4] Narin, I.; Soylak, M.; Elçi, L.; Doğan, M. Determination of trace metal ions by AAS in natural water samples after preconcentration of pyrocatechol violet complexes on an activated carbon column. *Talanta*, 52, (2000), 1041-1046.

[5] Inglezakis, V.J.; Loizidou, M.M.; Grigoropoulou, H.P. Ion exchange studies on natural and modified zeolites and the concept of exchange site accessibility. *Journal of Colloid and Interface Science,* 275, (2004), 570-576.

[6] Adebowale, K.O.; Unuabonah, I.E.; Olu-Owolabi, B.I. Adsorption of some heavy metal ions on sulfate- and phosphate-modified kaolin. *Applied Clay Science,* 29(2), (2005), 145-148.

[7] Namane, A.; Mekarzia, A.; Benrachedi, K.; Belhaneche-Bensemra, N.; Hellal, A. Determination of the adsorption capacity of activated carbon made from coffee grounds by chemical activation with $ZnCl_2$ and $H_3PO_4$. *Journal of Hazardous Materials,* B119, (2005), 189-194.

[8] Ulmanu, M.; Maranon, E.; Fernandez, Y.; Castrillon, L.; Anger, I.; Dumitrıu, D. Removal of copper and cadmium ions from diluted aqueous solutions by low cost and waste material adsorbents. *Water, Air, and Soil Pollution,* 142, (2003), 357-373.

[9] Ali, I.; Aboul-Enein, H.Y. Speciation of arsenic and Cr metal ions by reversed phase high performance liquid chromatography. *Chemosphere*, 48, (2002), 275-278.

[10] Erdem, E.; Karapinar, N.; Donat, R. The removal of heavy metal cations by natural zeolites. *Journal of Colloid and Interface Science,* 280, (2004), 309-314.

[11] Kang, S.-Y.; Lee, J.-U.; Moon, S.-H.; Kim, K.-W. Competitive adsorption characteristics of $Co^{2+}$, $Ni^{2+}$, and $Cr^{3+}$ by IRN-77 cation exchange resin in synthesized wastewater. *Chemosphere*, 56(2), (2004), 141-147.

[12] Bayat, B. Comparative study of adsorption properties of Turkish fly ashes II. The case of Cr (VI) and cadmium (II). *Journal of Hazardous Materials,* B95, (2002), 275-290.

[13] Brown, P.A.; Gill, S.A.; Allen, S.J. Metal removal from wastewater using peat. *Wat. Res.*, 34(16), (2000), 3907-3916.

[14] Kubrakova, I.; Kudinova, T.; Formanovsky, A.; Kuz'min, N. Determination of Cr (III) and Cr (VI) in river water by electro thermal atomic absorption spectrometry after sorption preconcentration in a microwave field. *Analyst,* 119, (1994), 2477-2480.

[15] Tuzen, M.; Parlar, K.; Soylak, M. Enrichment/separation of cadmium(II) and lead(II) in environmental samples by solid phase extraction. *Journal of Hazardous Materials,* 121(1-3), (2005), 79-87.

[16] Balasubramanian, S.; Pugalenthi, V. Determination of total Cr in tannery waste water by inductively coupled plasma-atomic emission spectrometry, flame atomic absorption spectrometry and UV-visible spectrometric methods. *Talanta*, 50, (1999), 457-467.

[17] Raji, C.; Anirudhan, T.S. Batch Cr(VI) removal by polyacrylamide-grafted sawdust:kinetics and thermodynamics. *Wat. Res.*, 32(12), (1998), 3772-3780.

[18] Low, K.S.; Lee, C.K.; Ng, A.Y. Column study on the sorption of Cr(VI) using quaternized rice hulls. *Bioresource Technology*, 68, (1999), 205-208.

[19] Ma, W.; Tobin, J.M. Determination and modeling of effects of pH on peat biosorption of Cr, copper and cadmium. *Biochemical Engineering Journal*, 18, (2004), 33-40.

[20] Lakatos, J.; Brown, S.D.; Snape, C.E. Coals as sorbents for the removal and reduction of hexavalent Cr from aqueous waste streams. *Fuel*, 81, (2002), 691-698.

[21] Saraçoğlu, S.; Elçi, L. Column solid-phase extraction with Chromosorb-102 resin and determination of trace elements in water and sediment samples by flame atomic absorption spectrometry. *Analytica Chimica Acta*, 452, (2002), 77-83.

[22] Inglezakis, V.J.; Zorpas, A.A; Loizidou, M.D.; Grigoropoulou, H.P. The effect of competitive cations and anions on ion exchange of heavy metals. *Separation and Purification Technology*, 46(3), (2005), 202-207.

[23] Tamura, H.; Furuichi, R. Adsorption affinity of divalent heavy metal ions for metal oxides evaluated by modeling with the Frumkin Isotherm. *Journal of Colloid and Interface Science*, 195(1), (1997), 241-249.

[24] Korngold, E.; Belayev, N.; Aronov, L. Removal of chromates from drinking water by anion exchangers. *Separation and Purification Technology*, 33, (2003), 179-187.

[25] Üçer, A.; Uyanik, A.; Aygün, Ş.F. Adsorption of Cu(II), Cd(II), Zn(II), Mn(II) and Fe(III) ions by tannic acid immobilized activated carbon. *Separation and Purification Technology*, 47(3), (2006), 113-118.

[26] Uzun, I.; Güzel, F. Adsorption of some heavy metal ions from aqueous solution by activated carbon and comparison of percent adsorption results of activated carbon with those of some other adsorbents. *Turhish Journal of Chemistry*, 24, (2000), 291-297.

[27] Leinonen, H.; Lehto, J. Purification of metal finishing waste waters with zeolites and activated carbons. *Waste Management and Research*, 19, (2001), 45-57.

[28] Youssef, A.M.; El-Nabarawy, T.; Sarma, S.E. Sorption properties of chemically-activated carbons 1. Sorption of cadmum(II) ions. Colloids and Surfaces A: Physicochem. *Eng. Aspects*, 235, (2004), 153-163.

[29] Gulnaz, O.; Saygideger, S.; Kusvuran, E. Study of Cu(II) biosorption by dried activated sludge: effect of physico-chemical environment and kinetics study. *Journal of Hazardous Materials*, *120*(1-3), (2005), 193-200.

[30] Hardalaç, D.; Kaya, M.; Alkan, C. Bazı ağır metallerin aktifleştirilmiş bentonit tarafından adsorpsiyonu. *Türk Kimya Dergisi*, 19, (1995), 36-40.

[31] Li, Y.-H.; Di, Z.; Ding, J.; Wu, D.; Luan, Z.; Zhu, Y. Adsorption thermodynamic, kinetic and desorption studies of $Pb^{2+}$ on carbon nanotubes. *Water Research*, 39(4), (2005), 605-609.

[32] Elmacı, A.; Yonar, T.; Özengin, N.; Türkoğlu, H. Zn(II), Cd(II), Co(II) ve Remazol Turkish Blue-G boyar maddesinin sulu çözeltilerinde kurutulmuş Chara sp., Cladophora sp. ve Chlorella sp. türleri ile biyosorpsiyonun araştırılması. *Ekoloji*, 14, (2005), 24-31.

[33] Vijayaraghavan, K.; Jegan, J.; Palanivelu, K.; Velan, M. Biosorption of cobalt(II) and nickel(II) by seaweeds: batch and column studies. *Separation and Purification Technology,* 44(1), (2005), 53-59.

[34] Romero-González, J.; Peralta-Videa, J.R.; Rodríguez, E.; Ramirez, S.L.; Gardea-Torresdey, J.L. Determination of thermodynamic parameters of Cr(VI) adsorption from aqueous solution onto *Agave lechuguilla* biomass. *The Journal of Chemical Thermodynamics,* 37(4), (2005), 343-347.

[35] Sharma, P.; Kumari, P.; Srivastava, M.M.; Srivastava, Shalini. Ternary biosorption studies of Cd(II), Cr(III) and Ni(II) on shelled *Moringa oleifera* seeds. Bioresource Technology, In Press, Corrected Proof, Available online, (2006).

[36] Sarin, V.; Pant, K.K. Removal of Cr from industrial waste by using eucalyptus bark. *Bioresource Technology,* 97(1), (2006), 15-20.

[37] Acar, F.N.; Malkoc, E. The Removal of Cr(VI) from aqueous solutions by Fagus orientalis L. *Bioresource Technology,* 94, (2004), 13-15.

[38] Melo, D.M.A.; de Souza, J.R.; Melo, M.A.F.; Martinelli, A.E.; Cachima, G.H.B.; Cunha, J.D. Evaluation of the zinox and zeolite materials as adsorbents to remove $H_2S$ from natural gas. *Colloids and Surfaces A: Physicochemical and Engineering Aspects,* 272(1-2), (2006), 32-36.

[39] Sebe, G.; Pardon, P.; Pichavant, F.; Grelier, S.; De Jeso, B. An investigation into the use of eelgras (Zostera noltii) for removal of cupric ions from dilute aqueous solutions. *Separation and Purification Technology,* 38, (2004), 121-127.

[40] Horsfall, M. Jr.; Abia, A.A.; Spiff, A.I. Kinetic Studies on the Adsorption of $Cd^{2+}$, $Cu^{2+}$ and $Zn^{2+}$ ions from aqueous solutions by cassava (Manihot sculenta Cranz) tuber bark waste. *Bioresource Technology,* 97(2), (2006), 283-291.

[41] Deng, S.; Bai, R. Removal of trivalent and hexavalent Cr with aminated polyacrylonitrile fibers: performance and mechanisms. *Water Research,* 38(9), (2004), 2424-2432.

[42] Kocaoba, S.; Akcin, G. Removal and recovery of Cr and Cr speciation with MINTEQA2. *Talanta*, 57, (2002), 23-30.

[43] Yalcin, S.; Apak, R. Cr(III, VI) speciation analysis with preconcentration on a maleic acid-functionalized XAD soebent. *Analytica Chimica Acta,* 505, (2004), 25-35.

[44] Kurama, H.; Catalsarik, T. Removal of zinc cyanide from a leach solution by an anionic ion-exchange resin. *Desalination*, 129(1), (2000), 1-6.

[45] Anezaki, K.; Chien, X.; Ogasawara, T.; Nukatsuka, I.; Ohzeki, K. Determination of cadmium and lead in tap water by Graphite-Furnace Atomic Absorption Spectrometry after preconcentration on a finely divided ion-exchange resin as the pyrrlidinedithiocarbamate complexes. *Analytical Sciences,* 14, (1998), 523-527.

[46] Kutun Sungur, S.; Akseli, A. Separation and determination of rare earth elements by Dowex 2-X8 resin using sodium trimetaphosphate as elution agent. *Journal of Chromatography A*, 874(2), (2000), 311-317.

[47] Van Keulen, H.; Hollander, J. G.; Smit, J. A. Measurement and theoretical description of ion sorption and diffusion of water and counterions in a cation exchange resin bearing sulfonic acid groups. *Journal of Colloid and Interface Science,* 185, (1997), 119-130.

[48] Fernandez, F.M.; Tudino, M.B.; Troccoli, O.E. Automatic on-line ultratrace determination of Cd species of environmental significanve in natural waters by FI-ETAAS. *J. Anal. At. Spectrom.,* 15, (2000), 687-695.

[49] Menendez-Alonso, E.; Hill, S. J.; Foulkes, M.E.; Crighton, J.S. Speciation and rreconcentration of $Cr^{III}$ and $Cr^{VI}$ in waters by retention on ion exchange media and determination by EDXRF. *J. Anal. At. Spectrom.,* 14, (1999), 187-192.

[50] Appelblad, P.K.; Baxter, D.C.; Thunberg, J.O. Determination of metal-humic complexes, free metal ions and total concentrations in natural waters. *J. Environ. Monit.,* 1, (1999), 211-217.

[51] Chiarle, S.; Ratto, M; Rovatti, M. Mercury removal from water by ion exchange resins adsorption. *Water Research,* 34(11), (2000), 2971-2978.

[52] Yebra-Biurrun, M.C.; Bermejo-Barrera, A.; Bermejo- Barrera, M.P.; Barciela-Alonso M.C. Determination of trace metals in natural waters by Flame Atomic Absorption Spectrometry following on-line ion-exchange preconcentration. *Analytica Chimica Acta,* 303, (1995), 341-345.

[53] Ersoz, M.; Pehlivan, E.; Duncan, H.J.; Yildiz, S.; Pehlivan, M. Ion exchange equilibria of heavy metals in aqueous solution on new chelating resins of sporopollenin. *Reactive Polymers,* 24(3), (1995), 195-202.

[54] Ayar, A.; Yıldız, S. Sorption of purin and pyrimidine bases onto chelating resin. *Colloids and Surfaces A: Physicochemical and Engineering Aspects,* 201, (2002), 291-296.

[55] Ayar, A.; Gürten, A.A. Determination of the rate control step of purin and pyrimidine bases adsorption on cobalt(II)-CDAE-sporopollenin. *Colloids and Surfaces A: Physicochemical and Engineering Aspects,* 229, (2003), 149-155.

[56] Loureiro, J.M.; Costa, C.A.; Rodrigues, A.E. Recovery of copper, zinc and lead from liquid streams by chelating ion exchange resins. *Chemical Engineering Science,* 43(5), (1988), 1115-1123.

[57] Pesavento, M.; Biesuz, R.; Palet, C. Study of aluminium speciation in freshwaters by sorption on a chelating resin. *Analyst,* 123, (1998), 1295-1301.

[58] Baran, A.; Bıçak, E.; Hamarat Baysal, Ş.; Önal, Ş.Comparative studies on the adsorption of Cr(VI) ions on to various sorbents. *Bioresource Technology,* In Press, Corrected Proof, Available online, (2006).

[59] Kesenci, K.; Say, R.; Denizli, A. Removal of heavy metal ions from water by using poly(ethyleneglycol dimethacrylate-co-acrylamide) beads. *European Polymer Journal,* 38, (2002), 1443-1448.

[60] Grag, V.K.; Gupta, R.; Kumar, R.; Gupta, R.K. Adsorption of Cr from aqueous solution on treated sawdust. *Bioresource Technology,* 92, (2004), 79-81.

[61] Sciban, M.; Radetic, B.; Kevresan, Z.; Klanja, M. Adsorption of heavy metals from electroplating wastewater by wood sawdust. *Bioresource Technology,* In Press, Corrected Proof, Available online, (2006).

[62] Yabe, M.J.S.; Oliveira, E. Heavy metals removal in industrial effluents by sequential adsorbent treatment. *Advances in Environmental Research,* 7, (2003), 263-272.

[63] Gode, F.; Pehlivan, E. Adsorption of Cr(III) ions by Turkish brown coals. *Fuel Processing Technology,* 86(8), (2005), 875-884.

[64] Gode, F.; Pehlivan, E. Cr(VI) adsorption by brown coals. *Energy Sources,* Part A 28, (2006), 447-457.

[65] Kislenko, V.N.; Oliynyk, L.P. Treatment of humic acids with ferric, aluminum, and Cr ions in water. *Journal of Colloid and Interface Science*, 269, (2004), 388-393.

[66] Li, N.; Lee, H.K. Solid-phase extraction of polycyclic aromatic hydrocarbons in surface water Negative effect of humic acid. *Journal of Chromatography A,* 921, (2001), 255-263.

[67] Gezici, O.; Kara, H.; Ersöz, M.; Abali, Y. The sorption behaviour of a nickel-insolubilized humic acid system in a column arrangement. *Journal of Colloid and Interface Science,* 298, (2005), 381-391.

[68] Benhammou, A.; Yaacoubi, A.; Nibou, L.; Tanouti, B. Adsorption of metal ions onto Moroccan stevensite: kinetic and isotherm studies. *Journal of Colloid and Interface Science,* 282, (2005), 320-326.

[69] Wilson, K.; Yang, H.; Seo, C.W.; Marshall, W.E. Select metal adsorption by activated carbon made from peanut shells. *Bioresource Technology,* In Press, Corrected Proof, Available online, (2005).

[70] da Fonseca, M.G.; de Oliveira, M.M.; Arakaki, L.N.H. Removal of cadmium, zinc, manganese and Cr cations from aqueous solution by a clay mineral. *Journal of Hazardous Materials,* In Press, Corrected Proof, Available online, (2006).

[71] Altın, O.; Özbelge, H.Ö.; Doğu,T. Use of general purpose adsorption isotherms for heavy metal-clay mineral interactions. *Journal of Colloid and Interface Science,* 198, (1998). 130-140.

[72] Atanassova, I.; Okazaki, M. Adsorption-desorption characteristics of high levels of copper in soil clay fractions. *Water, Air, and Soil Pollution,* 98, (1997), 213-228.

[73] Inglezakis, V.I.; Loizidou, M.D.; Grigoropoulou, H.P. Equilibrium and kinetic ion exchange studies of $Pb^{2+}$, $Cr^{3+}$, $Fe^{3+}$ and $Cu^{2+}$ on natural clinoptilolite. *Wat. Res.,* 36, (2002), 2784-2792.

[74] Inglezakis, I.V.; Zorpas, A.A.; Loizidou, M.D.; Grigoropoulou, H.P. Simultaneous removal of metals $Cu^{2+}$, $Fe^{3+}$ and $Cr^{3+}$ with anions $SO_4^{2-}$ and $HPO_4^{2-}$ using clinoptilolite. *Microporous and Mesoporous Materials,* 61, (2003), 167-171.

[75] Petrus, R.; Warchol, J. Ion exchange between clinoptilolite and aqueous solutions of $Na^+/Cu^{2+}$, $Na^+/Cd^{2+}$ and $Na^+/Pb^{2+}$. *Microporous and Mesoporous Materials,* 61, (2003), 137-146.

[76] Lin, S.-H.; Juang, R.-S.; Heavy metal removal from water by sorption using surfactant-modified montmorillonite. *J. Hazard Mat.,* B92, (2002), 315-326.

[77] Jiang, J.-Q.; Cooper, C.; Ouki, S. Comparison of modified montmorillonite adsorbents part I: preparation, characterization and phenol adsorption. *Chemosphere*, 47, (2002), 711-716.

[78] Abollino, O.; Aceto, M.; Malandrino, M.; Sarzanini, C.; Mentosti, E. Adsorption of heavy metals on Na-montmorillonite. Effect of pH and organic substances. *Water Research,* 37, (2003), 1619-1627.

[79] Krishna, B.S.; Murty, D.S.R.; Jai Prakash, B.S. Thermodynamics of Cr(VI) anionic species sorption onto surfactant-modified montmorillonite clay, *J. Colloid Inter. Sci.,* 229(1), (2000), 230-236.

[80] Yavuz, Ö.; Altunkaynak, Y.; Güzel, F. Removal of copper, nickel, cobalt and manganese from aqueous solution by kaolinite. *Water Research,* 37, (2003), 948-952.

[81] Suraj, G.; Iyer, C.S.P.; Lalithambika, M. Adsorption of cadmium and copper by modified kaolinites. *Applied Clay Science,* 13, (1998), 293-306.

[82] Rauf, N.; Tahir, S.S. Thermodynamics of Fe(II) and Mn(II) adsorption onto bentonite from aqueous solutions. *J. Chem. Thermodynamics,* 32, (2000), 651-658.

[83] Kara, M.; Yuzer, H.; Sabah, E.; Celik, M.S.; Adsorption of cobalt from aqueous solutions onto sepiolite. *Water Research,* 37, (2003), 224-232.

[84] Chiron, N.; Guilet, R.; Deydier, E. Adsorption of Cu(II) and Pb(II) onto a grafted silica: isotherms and kinetic models. *Water Research,* 37, (2003), 3079-3086.

[85] Erdem, M.; Özverdi, A. Lead adsorption from aqueous solution onto siderite. *Separation and Purification Technology,* 42(3), (2005), 259-264.

[86] Conrad, K.; Hansen, H.C.B. Sorption of zinc and lead on coir. *Bioresource Technology*, In Press, Corrected Proof, Available online, (2006).

[87] Shukla, S.R.; Pai, R.S.; Shendarkar, A.D. Adsorption of Ni(II), Zn(II) and Fe(II) on modified coir fibres. *Separation and Purification Technology,* 47(3), (2006), 141-147.

[88] Ho, Y.-S. Effect of pH on lead removal from water using tree fern as the sorbent. *Bioresource Technology,* 96, (2005), 1292-1296.

[89] Farajzadeh, M.A.; Monji, A.B. Adsorption characteristics of wheat bran towards heavy metal cations. *Separation and Purification Technology,* 38, (2004), 197-207.

[90] Min, S.H.; Han, J.S.; Shin, E.W.; Park, J.K. Improvement of cadmium ion removal by base treatment of juniper fiber. *Water Research,* 38, (2004), 1289-1295.

[91] Montanher, S.F.; Oliveira, E.A.; Rollemberg, M.C. Removal of metal ions from aqueous solutions by sorption onto rice bran. *Journal of Hazardous Materials,* B117, (2005), 207-211.

[92] Jeon, C.; Höll, W.H. Application of the surface complexation model to heavy metal sorption equilibria onto aminated chitosan. *Hydrometallurgy,* 71, (2004), 421-428.

[93] Rojas, G.; Silva, J.; Flores, J.A.; Rodriguez, A.; Ly, M.; Maldonado, H. Adsorption of Cr onto cross-linked chitosan. *Separation and Purification Technology,* 44(1), (2005), 31-36.

[94] Wartelle, L.H.; Marshall, W.E. Quaternized agricultural by-products as anion exchange resins. *Journal of Environmental Management,* 78, (2006), 157-162.

[95] Bajpai, J.; Shrivastava, R.; Bajpai, A.K. Dynamic and equilibrium studies on adsorption of Cr(VI) ions onto binary bio-polymeric beads of cross linked alginate and gelatin. *Colloids and Surfaces A: Physicochem. Eng. Aspects,* 236, (2004), 81-90.

[96] Lalvani, S. B.; Hubner, A.; Wiltowski, T.S. Cr adsorption by lignin. *Environmental Technology,* 22, (2000), 45-56.

[97] Biskup, B.; Subotic, B. Kinetic analysis of the exchange processes between sodium ions from zeolite A and cadmium, copper and nickel ions from solutions. *Separation and Purification Technology,* 37(1), (2004), 17-31.

[98] Zamzow, M.J.; Eichbaum, B.R. Removal of heavy metals and other cations from wastewater using zeolites. *Separation Science and Technology,* 25(13-15), (1990), 1555-1569.

[99] Alvarez-Ayuso, E.; Garcia-Sanchez, A.; Querol, X. Purification of metal electroplating waste waters using zeolites. *Water Research,* 37, (2003), 4855-4862.

[100] Chwastowska, J.; Skwara, W.; Sterlinska, E.; Pszonicki, L. Speciation of Cr in mineral waters and salinas by solid-phase extraction and graphite furnace atomic absorption spectrometry. *Talanta*, 66, (2005), 1345-1349.

[101] WHO, 1996, Cr in drinking-water, Background document for development of WHO guidelines for drinking-water quality, 2nd ed. Vol.2. Health criteria and other supporting information. World Health Organization, Geneva.

[102] Hirata, S.; Honda, K.; Shikino, O.; Maekawa, N.; Aihara, M. Determination of Cr (III) and total Cr in seawater by on-line column preconcentration inductively coupled plasma mass spectrometry. *Spectrochimica Acta Part B,* 55, (2000), 1089-1099.

[103] Adria-Cerezo, D.M.; Llobat-Estelles, M.; Mauri-Aucejo, A.R. Preconcentration and speciation of Cr in waters using solid-phase extraction and atomic absorption spectrometry. *Talanta*, 51, (2000), 531-536.

[104] Mondal, B.C.; Das, D.; Das, A.K. Synthesis and characterization of a new resin functionalized with 2-naphthol-3,6-disulfonic acid and its application for the speciation of Cr in natural water. *Talanta*, 56, (2002), 145-152.

[105] Sankararamakrishnan, N.; Dixit, A.; Iyengar, L.; Sanghi, R. Removal of hexavalent Cr using a novel cross linked xanthated chitosan. *Bioresource Technology*, in press, available online, (2005).

[106] Yang, W.-P.; Zhang, Z.-J.; Deng, W. Simultaneous, sensitive and selective on-line chemiluminescence determination of Cr(III) and Cr(VI) by capillary electrophoresis. *Analytica Chimica Acta,* 485, (2003), 169-177.

[107] Zhu, X.; Hu, B.; Jiang, Z.; Wu, Y.; Xiong, S. Speciation of Cr(III) and Cr(VI) by in situ separation and sequential determination with electrothermal vaporization inductively coupled plasma atomic emission spectrometry. *Analytica Chimica Acta,* 471, (2002), 121-126.

[108] Dabrowski, A.; Hubicki, Z.; Podkoscielny, P.; Robens, E. Selective removal of the heavy metal ions from waters and industrial wastewaters by ion-exchange method. *Chemosphere*, 56, (2004), 91-106.

[109] Kotas, J.; Stasicka, Z. Cr occurrence in the environment and methods of its speciation. *Environmental Pollution*, 107(3), (2000), 263-283.

[110] Vitale, R.J.; Mussoline, CPC, G.R.; Rinehmer, K.A. Environmental monitoring of Cr in air, soil, and water. *Regulatory Toxicology and Pharmacology,* 26, (1997), S80-S85.

[111] Motomizu, S.; Jitmanee, K.; Oshima, M. On-line collection/concentration of trace metals for spectroscopic detection via use of small-sized thin solid-phase (STSP) column resin reactors. Application to speciation of Cr(III) and Cr(VI). *Analytica Chimica Acta,* 499, (2003), 149-155.

[112] Gopi Krishna, P.; Mary Gladis, J.; Rambabu, U.; Prasada Rao, T.; Naidu, G.R.K. Preconcentrative separation of Cr(VI) species from Cr(III) by coprecipitation of its ethyl xanthate complex onto naphthalene. *Talanta*, 63(3), (2004), 541-546.

[113] Mount, D.R.; Hockett, R. Use of toxicity identification evaluation methods to characterize, identify, and confirm hexavalent Cr toxicity in an industrial effluent. *Wat. Res.* 34(4), (2000), 1379-1385.

[114] Stasinakis, A.S.; Thomaidis, N.S.; Mamais, D.; Karivali, M.; Lekkas, T.D. Cr species behaviour in the activated sludge process. *Chemosphere*, 52, (2003), 1059-1067.

[115] Rengaraj, S.; Joo, C.K.; Kim, Y.; Yi, J. Kinetics of removal of Cr from water and electronic process wastewater by ion exchange resins:1200H, 1500H and IRN97H. *J. Hazard. Materials*, 102(2-3), (2003), 257-275.

[116] Elçi, L.; Arslan, Z.; Tyson, J.F. Flow injection solid phase extraction with Chromosorb-102: determination of lead in soil and waters by flame atomic absorption spectrometry. *Spectrochimica Acta Part B,* 55, (2000), 1109-1116.

[117] Köleli, N.; Kantar, Ç. Fosfat kayası, fosforik asit ve fosforlu gübrelerdeki toksik ağır metal (Cd, Pb, Ni, As) konsantrasyonu. *Ekoloji*, 14(55), (2005), 1-5.

[118] Pohl, P.; Prusisz, B. Pre-concentration of Cd, Co, Cu, Ni and Zn using different off-line ion exchange procedures followed by the inductively coupled plasma atomic emission spectrometric detection. *Anal. Chim. Acta, 502*, (2004), 83-90.

[119] Tuzen, M.; Soylak, M.; Elci, L. Multi-element pre-concentration of heavy metal ions by solid phase extraction on Chromosorb 108. *Analytical Chimica Acta,* 548, (2005), 101-108.

[120] Poole, C.F.; Gunatilleka, A.D.; Sethuraman, R. Contributions of theory to method development in solid-phase extraction. *Journal of Chromatography A,* 885, (2000), 17-39.

[121] Sarzanini, C.; Bruzzoniti, M.C. Metal species determination by ion chromatography. *Trends in Analytical Chemistry,* 20(6-7), (2001), 304-310.

[122] Lin, S.H.; Lai, S.L.; Leu, H.G. Removal of heavy metals from aqueous solution by chelating resin in multistage adsorption process. *Journal of Hazardous Materials,* B76, (2000), 139-153.

[123] Ferreira, L.M.; Loureiro, J.M.; Rodrigues, A.E. Sorption of metals by an amidoxime chelating resin. Part I: Equilibrium. *Sep. Sci. Technol.*, 33(11), (1998), 1585-1604.

[124] Lewatit Catalog, Removing traces, Bayer, 1998.

[125] Vural, U.S.; Ersoz, M.; Pehlivan, E. Ligand sorption kinetics of aromatic amines on new ligand-exchanger sporopollenin in cobalt ion form. *J. Applied Polymer Science,* 58, (1995), 2423-2428.

[126] Ersoz, M.; Vural, U.S.; Yildiz, S. Sorption of aromatic-amines on a new ligand exchanger of sporopollenin-bound $Co^{2+}$ ion. *Separation Science and Technology,* 30(18), (1995), 3555-3566.

[127] Karabulut, S.; Karabakan, A.; Denizli, A.; Yürüm, Y. Batch removal of copper(II) and zinc(II) from aqueous solutions with low-rank Turkish Coals. *Separation and Purification Technology,* 18, (2000), 177-184.

[128] Fotea, C.; D'Silva, C. The use of silane reagents as primers to enhance the adhesion of Cr tanned heavy-duty leather (Saltz leather). *International Journal of Adhesion and Adhesives,* 24, (2004), 1-7.

[129] Rengaraj, S.; Yeon, K.-H.; Moon, S.-H. Removal of Cr from water and wastewater by ion exchange resins, *J. Hazard. Mater.,* B 87(1-3), (2001), 273-287.

[130] Bishnoi, N.R.; Bajaj, M.; Sharma, N.; Gupta, A. Adsorption of Cr(VI) on activated rice husk carbon and activated alumina. *Bioresource Technology,* 91, (2004), 305-307.

[131] Lalvani, S. B.; Wiltowski, T.S.; Murphy, D.; Lalvani, L.S. Metal removal from process water by lignin. *Environmental Technology,* 18, (1997), 1163-1168.

[132] Laffert, C.; Hobday, M. The use of low rank brown coals as an ion exchange materials: 1. Basic parameters and the ion exchange mechanism. *Fuel*, 69(1), (1990), 78-83.

[133] Stuart, A.D. Selective cation exchange using acidic groups in coals and oxidized coals. *Fuel*, 65(7), (1986), 1003-1005.

[134] Murakami, K.; Ozaki, J.; Nishiyama, Y. Fuel Process. Effects of surface treatment on cation exchange properties of Australian brown coals. *Fuel Processing Technology,* 43(1), (1995), 95-110.

[135] Murakami, K.; Yamada, T.; Fuda, K.; Matsunaga, T.; Nishiyama, Y. The cation exchange properties of the heat-treated Australian brown coal in the water-organic compounds mixed solutions. *Fuel*, 76(12), (1997), 1085-1090.

[136] Eligwe, C.A.; Okolue, N.B. Adsorption of iron(II) by a Nigerian brown coal. *Fuel*, 73(4), (1994), 569-572.

[137] Baruah, M.K.; Upreti, M.C. Preferential uptake of Fe(III) by humic acid extracted from lignite. *Fuel*, 73(2), (1994), 273-275.

[138] Randle, K.; Hartmann, E.H. Applications of the continuous-flow stirred-cell(CFSC) technique. 2. The adsorption behavior of Na, Cs, Sr, Cu, Ni, and Pb on humic acids. *European Journal of Soil Science,* 46(2), (1995), 303-315.

[139] Murakami, K.; Yamada, T.; Fuda, K.; Matsunaga, T. Selectivity in cation exchange property of heat-treated brown coals. *Fuel*, 80, (2001), 599-605.

[140] Ranganathan, K. Cr removal by activated carbons prepared from Casurina equisetifolia leaves. *Bioresource Technology, 73*, (2000), 99-103.

[141] Gode, F. Heavy Metal Removal from Aqueous Solutions by Adsorption using Montmorillonite. *Asian Journal of Chemistry,* accepted.

[142] Gode, F.; Pehlivan, E. Sorption of Cr(III) onto b-DAEG-sporopollenin and CEP-sporopollenin. *Bioresource Technology,* online available, 2006.

[143] Sawalha, M.F.; Peralta-Videa, J.R.; Romero-González, J.; Gardea-Torresdey, J.L. Biosorption of Cd(II), Cr(III), and Cr(VI) by saltbush (*Atriplex canescens*) biomass: Thermodynamic and isotherm studies. *Journal of Colloid and Interface Science,* In Press, Corrected Proof, Available online, (2006).

[144] Namasivayam, C.; Senthilkumar, S. Adsorption of copper(II) by waste Fe(III)/Cr(III) hydroxide from aqueous solution and radiator manufacturing industry wastewater. *Separation Science and Technology,* 34(2), (1999), 201-217.

[145] Lacour, S.; Bollinger, J.-C.; Serpaud, B.; Chantron, P.; Arcos, R. Removal of heavy metals in industrial wastewaters by ion-exchanger grafted textiles. *Analytica Chimica Acta,* 428, (2001), 121-132.

[146] Gode, F.; Pehlivan, E. Removal of Cr(VI) from aqueous solution by two Lewatit-anion exchange resins. *Journal of Hazardous Materials*, 119(1-3), (2005), 175-182.

[147] Yu, L.J.; Shukla, S.S.; Dorris, K.L.; Shukla, A.; Margrave, J.L. Adsorption of Cr from aqueous solutions by maple sawdust. *J. Hazard. Materials*, B100, (2003), 53-63.

[148] Hu, Z.; Lei, L.; Li, Y.; Ni, Y. Cr adsorption on high-performance activated carbons from aqueous solution. *Sep. Purification Technol.,* 31, (2003), 13-18.

[149] Juang, R.S.; Shiau, L.D. Ion exchange equilibria of metal chelates of ethylenediaminetetraacetic acid (EDTA) with amberlite IRA-68. Industrial and Engineering Chemistry Research, 37, (1998), 555-560.

[150] Gode, F.; Pehlivan, E. Removal of Cr(III) from aqueous solutions using Lewatit S 100: The effect of pH, time, metal concentration and temperature. *Journal of Hazardous Materials,* 136(2), (2006), 330-337.

[151] Burns, C.A.; Cass, P.J.; Harding, I.H.; Crawford, R.J. Adsorption of aqueous heavy metals onto carbonaceous substrates. *Colloids and Surfaces A:Physicochemical and Engineering Aspects,* 155, (1999), 63-68.

[152] Mohan, D.; Chander, S. Single component and multi-component adsorption of metal ions by activated carbons. *Colloids and Surfaces A: Physicochemical and Engineering Aspects,* 177, (2001), 183-196.

[153] Pradhan, J.; Das, S.N.; Thakur, R.S. Adsorption of hexavalent Cr(VI) from aqueous solution by using activated red mud. *Journal of Colloid and Interface Science,* 217, (1999), 137-141.

[154] Cortina, J.L.; Miralles, N.; Aguilar, M.; Saster, A.M. Distribution studies of Zn(II), Cu(II) and Cd(II) with Levextrel resins containing di(2,4,4-trimethylpentyl)phosphinic acid. *Hydrometallurgy*, 40, (1996), 195-206.

[155] Namasivayam, C.; Yamuna, R.T. Studies on Cr(III) removal from aqueous solution by adsorption onto biogas residual slurry and its application to tannery wastewater treatment. *Water, Air, and Soil Pollution,* 113, (1999), 371-384.

[156] Dakiky, M.; Khamis, M.; Manassra, A.; Mer'eb, M. Selective adsorption of Cr(VI) in industrial wastewater using low-cost abundantly available adsorbents. *Advances in Environmental Research,* 6(4), (2002), 533-540.

[157] Üçer, A.; Uyanık, A.; Çay, S.; Özkan, Y. Immobilisation of tannic acid onto activated carbon to improve Fe(III) adsorption. *Separation and Purification Technology,* 44(1), (2005), 11-17.

[158] Donais, M.K.; Henry, R.; Rettberg, T. Cr speciation using an automated liquid handling system with inductively coupled plasma — mass spectrometric detection. *Talanta,* 49(5), (1999), 1045-1050.

[159] Biesuz, R.; Pesavento, M.; Gonzalo, A.; Valiente, M. Sorption of proton and heavy metal ions on a macroporous chelating resin with an iminodiacetate active group as a function of temperature. *Talanta*, 47(1), (1998), 127-136.

[160] Jones, P.; Nesterenko, P.N. High-performance chelation ion chromatography: A new dimension in the separation and determination of trace metals. *Journal of Chromatography* A, 789(1-2), (1997), 413-435.

[161] Muraviev, D.; Gonzalo, A.; Valiente, M. Ion exchange on resins with temperature-responsive selectivity: III. Influence of complex formation stoichiometry on temperature dependence of resin selectivity. *Journal of Chromatography* A, 868(2), (2000), 143-152.

[162] Mohan, D.; Gupta, V.K.; Srivastava, S.K.; Chander, S. Kinetics of mercury adsorption from wastewater using activated carbon derived from fertilizer waste. *Colloids and Surfaces A-Physcochemical and Engineering Aspects,* 177(2-3), (2001), 169-181.

[163] Ferreira, S.L.C.; de Brito, C.F.; Dantas, A.F.; de Araujo, N.M.L.; Costa, A.C.S. Nickel determination in saline matrices by ICP-AES after sorption on Amberlite XAD-2 loaded with PAN. *Talanta*, 48, (1999), 1173-1177.

[164] Gode, F.; Pehlivan, E.; A comparative study of two chelating ion-exchange resins for the removal of Cr(III) from aqueous solution. *Journal of Hazardous Materials,* 100(1-3), (2003), 231-243.

[165] Selvaraj, K.; Manonmani, S.; Pattabhi, S. Removal of hexavalent Cr using distillery sludge. *Bioreseource Technology,* 89, (2003), 207-211.

[166] Krishna, B.S.; Murty, D.S.R.; Jai Prakash, B.S. Thermodynamics of Cr(VI) anionic species sorption onto surfactant-modified montmorillonite clay. *J. Colloid Inter. Sci.,* 229(1), (2000), 230-236.

[167] Chong, K.H.; Volesky, B. Description of two-metal biosorption equilibria by Langmuir type models. *Biotechnol Bioeng.,* 47, (1995), 451-460.

[168] Yu, Q.; Kaewsarn, P. Binary adsorption of copper(II) and cadmium(II) from aqueous solutions by biomass of marine alga durvillaea potatorum. *Sep Sci Technol.,* 34(8), (1999), 1595-1605.

[169] Cooney, D.O. *Adsorption Design for Wastewater Treatment,* Lewis Publishers, London, 1999, 87.

[170] Kadirvelu, K.; Thamaraiselvi, K.; Namasivayam, C. Adsorption of nickel(II) from aqueous solution onto activated carbon prepared from coirpith. *Separation and Purification Technollogy,* 24(3), (2001), 497-505.

[171] Chanda, M.; Rempel, G.L.; Cr(III) removal by epoxy-cross-linked poly(ethylenimine) used as gel-coat on silica 1. Sorption characteristics. *Industrial and Engineering Chemistry Research,* 36(6), (1997), 2184-2189.

[172] Bailey, S.E.; Olin, T.J.; Bricka, R.M.; Adrian, D.D. A review of potentially low-cost sorbents for heavy metals. *Water Research,* 33, (1999), 2469-2479.

[173] Gupta, V.K.; Shrivastava, A.K.; Jain, N. Biosorption of Cr(VI) from aqueous solutions by green algae Spirogyra species. *Water Research,* 35(17), (2001), 4079-4090.

[174] Gode, F.; Erdoğan, S.; Özmert, S.; Pehlivan, E.; Kır, E. Equilibrium and kinetics studies for adsorption of copper on clay. *Ovidius University Annals of Chemistry,* 16(1), (2005), 111-114.

In: Hazardous Materials and Wastewater
Editor: Allison A. Lewinsky, pp. 309-359
ISBN 1-60021-257-3


*Chapter 10*

# The Application of Biosorption and Bioaccumulation of Toxic Metals in Environmental Pollution Control

***Katarzyna Chojnacka****
Institute of Inorganic Technology and Mineral Fertilizers,
Wrocław University of Technology, Wrocław, Poland

## ABSTRACT

Natural and controlled processes of biosorption and bioaccumulation may be efficiently used in the assessment of environmental pollution as well as in pollution prevention and cleaning the polluted environment. The property that is used in such pollution control processes is the ability of all types of biomasses to bind toxic metals, in different extent though, depending on morphology and physiology of an organism. If the processes are performed at controlled conditions, the efficiency can be greatly improved. If we would like natural processes to work for us, it would be indispensable to understand their mechanism, rules that govern them, we should know how to model their kinetics and equilibrium. Basically, we should be able to predict the course of the processes at given process parameters in order to design a complete technique and to make a practical use of it.

Biosorption and bioaccumulation differ when considering the mechanism and for this reason the potential applications will also be diverse. While in the process of biosorption, pollutants are bound to the surface of cellular wall of the biomass, in bioaccumulation contaminants are also transferred into cellular interiors. The latter can affect the metabolic functions of an organism. Biosorption can be performed by non-living biomass (materials of biological origin), but the essential condition for bioaccumulation to occur is that an organism should be metabolically-active.

---

* e-mail: katarzyna.chojnacka@pwr.wroc.pl

The present work discusses two main environmental applications of biosorption and bioaccumulation processes. The first is environmental biomonitoring performed by living organisms. The property that is used here is that the level of a given contaminant in biological tissues is related to the concentration of this substance in the environment. This is expressed by the assessment of bioaccumulation factors. The paper reports examples of the application of human hair and consumable tissues of animals, and various parts of plants (green parts as well as wood) in environmental pollution monitoring.

Biosorption as well as bioaccumulation can find an application in removal of contaminants from aqueous solutions. Biosorption by different materials of plant (microalgal biomass, aquatic plants, plant leaves, straw, grass) and animal origin (eggshells, bones) can be applied in industrial wastewater treatment processes. Also, bioaccumulation by aquatic organisms: microalgae, macroalgae and aquatic plants in metal ions removal from effluents can be used to bind metal ions simultaneously with nutrients.

An attempt to make the processes predictable is discussed here, including screening for good biosorbents/bioaccumulators, determining biosorption and bioaccumulation capacity as well as bioavailability as the function of the concentration of metals in the environment in which an organism is present, together with growth requirements, mathematical modelling, carrying out processes at different process conditions. This should constitute the basis for the assessment of standardized procedures, in the design of biosorption or bioaccumulation treatment plants and for environmental biomonitoring.

## 1. Introduction

Biosorption and bioaccumulation are processes instantaneously performed by all biomasses either living or dead. It is a natural property and the processes are carried out either by purpose or by mistake (toxic metal ions are taken up instead of essential ions). Capacity of the biomass to bind and concentrate toxic metals from solutions may create the fundamentals for a cost-effective technology for detoxification of industrial effluents [1], mainly from mining and electroplating industry or to recover precious metals from processing solutions [2]. The potential uses and mechanisms of biosorption and bioaccumulation for toxic metals control are shown in Figure 1.

Biosorption and bioaccumulation are significant stages of toxic elements cycles in the environment. Thanks to these processes, metals can be transferred from e.g. aquatic environment and become concentrated in bottom sediments. These phenomena have been used for decades in conventional biological wastewater treatment plants in which scattered soluble impurities (nutrients, toxic metals and organic compounds) are transferred from dilute soluble form into concentrated, condensed several fold in the biomass of activated sludge with the simultaneous use of processes of biosorption, bioaccumulation and biodegradation. While organic pollutants can be processed by the biomass with the use of all three processes, metal ions can only be utilized with biosorption and bioaccumulation – biodegradation is not possible. The advantage of activated sludge method is that the impurities become concentrated and transferred into solid state that facilitates their utilization.

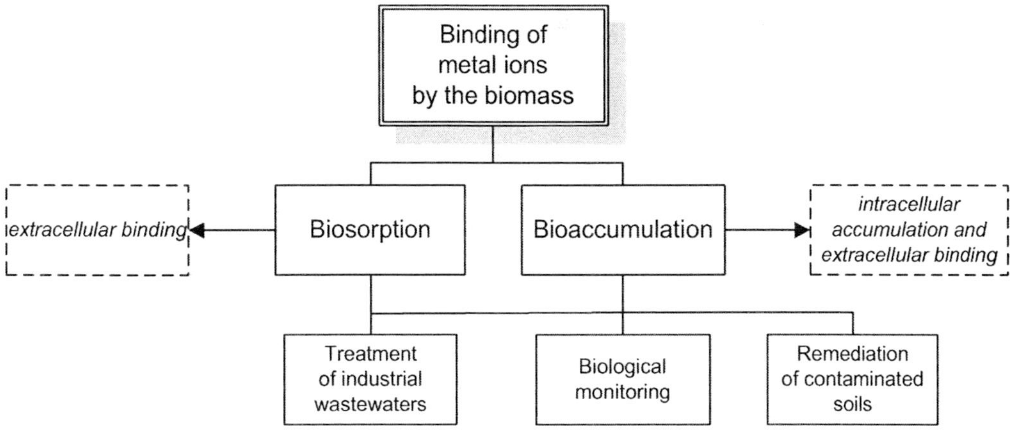


Figure 1. Binding of metal ions by the biomass.

The other practical aspect of the discussed processes is environmental pollution monitoring which bases on known bioaccumulation behavior of metals and organisms [3]. It has been observed that the concentration of toxic metals in biological tissues reflects their concentration in the environment over extended period of time. This theory created the fundamentals for biological monitoring – a technique that determines not only the degree of environmental pollution but also investigates bioavailability of toxic elements from the surrounding environment to living organisms and assessment of the risk posed by these contaminants.

# 2. Biosorption and Bioaccumulation of Toxic Metals : The Fundamentals of the Processes

## 2.1. Toxic Metals

A term "heavy metal" is commonly used to describe metals that are toxic. However, the term is rather used in causal language and has never been defined by any authority i.e. IUPAC [4]. There are numerous definitions of "heavy metals" classifying according to certain physical, chemical or biological properties, including density (specific gravity), atomic weight and number, other chemical properties, definitions without a clear basis other than toxicity or non-chemical definitions [4]. The majority of definitions have no relation with toxicity of these metals to living organisms since in fact no relation between density and atomic/molecular weight and toxicity has been observed [4]. In the present chapter, toxic metals are understood to be elements (not only metals) commonly used in industry and generally toxic to living organisms even at low concentrations [1], including As, Cd, Cr, Cu, Pb, Hg, Ni, Se, Zn [5]. Se and As are frequently named with a term "heavy metal" although these elements are not metallic [6]. However, when considering toxic properties, they are classified to the same group, similarly as the remaining toxic elements, and are named toxic metals [6].

It is always questionable whether a given element should be considered as toxic. There are metals that play only harmful role in living organisms, through enzyme inhibition or activation, damage of subcellular organelles, carcinogenicity, and effects on kidneys, nervous system, endocrine system, reproduction, and respiratory system [7]. They do not play any positive function in living organisms. This is so-called *toxic trio* that includes Hg, Pb and Cd – the mostly toxic elements used by the industry [8]. Another group includes metals that are needed for the proper functioning of an organism, in low quantities, but simultaneously at elevated levels are toxic [9]. In some cases the difference between essential and harmful concentration is very narrow (such as in the case of selenium). Sometimes, oxidation degree determines toxicity and essentiality of a given metal, e.g. Cr(III) is considered to be microelement, but Cr(VI) is toxic (carcinogenic) [6].

**Table 1. The levels of contaminants in raw effluents, maximum admissible concentration of contaminants in effluents and drinking water (American and European regulations) [mg/L]**

| Metal ion | Domestic wastewater[1] | Wastewater discharge limits | | Drinking water standards | |
|---|---|---|---|---|---|
| | | European[2] | American[3] | European[4] | American[5] |
| Ag | 0.003-0.01 | 0.1 | 1 | | |
| Al | 0.25-1 | | | 0.05 | |
| As | 0.001-0.005 | 0.1 | 2 | 0.05 | 0.05 |
| Ba | | 2 | | | 2 |
| Be | | | | | 0.004 |
| Cd | 0.001-0.004 | 0.1-0.2 | 1 | 0.005 | 0.005 |
| Co | 0.0005-0.002 | 1 | | | |
| Cr (total) | 0.01-0.004 | 0.5 | 2 | 0.05 | 0.1 |
| Cr(VI) | | 0.1 | | | |
| Cu | 0.03-0.1 | 0.5 | 5 | | 1.3 |
| Fe | 0.4-1.5 | | 100 | | |
| Hg | 0.001-0.003 | 0.05 | 0.05 | 0.001 | 0.002 |
| Mn | 0.04-0.15 | | | | |
| Ni | 0.01-0.04 | 0.5 | 5 | 0.05 | 0.1 |
| Pb | 0.025-0.08 | 0.5 | 2 | 0.05 | 0.015 |
| Sb | | | | 0.01 | 0.006 |
| Se | | 1 | | 0.01 | 0.05 |
| Sn | | 2 | | | |
| Th | | | | | 0.0005 |
| Zn | 0.08-0.3 | 2 | 5 | | |

[1] Typical content of metals in domestic wastewater [13]

[2] German requirements for the effluent concentrations in direct discharge into receiving water and indirect discharge into municipal sewers considered dangerous [14]

[3] Wastewater discharge limits (U.S., EPA); pretreatment standards [15]

[4] NPDW (national primary drinking water) regulations for inorganic chemicals MCLG (maximum contaminant level goal) [16]

[5] European Economic Community Standards (maximum admissible concentration) [17]

The presence of toxic metals in the scattered and bioavailable form in the environment is highly undesirable. Metals occur in nature in rocks, ores, soil, water, air in naturally low levels and in dispersed form [7]. We should bear in mind though that the total amount of toxic metals such as the majority of other elements present on the Earth has remained and will remain constant. The problem that arose with the industrialization was that metals were mobilized from their deposits (the forms that were not available to living organisms) by anthropogenic activity (making utensils and machinery, mining, smelting, welding, grinding, soldering and printing [7, 9]) and were transferred into bioavailable form and redistributed in such form in the environment [7]. Metals have been mainly mobilized by the branches of industry such as mining and metallurgy [10] and are generally commonly used by industry, agriculture and medicine [7]. The result was that toxic metals began to cycle in ecosystem at elevated concentrations between its biotic and abiotic components and also in the trophic chain becoming concentrated on the top of it in organisms of final consumers, frequently in human body [1]. Because it is impossible to degrade toxic metals, the only method of utilization is safe disposal – immobilization in a solid state and concentration into the form that will not undergo biological cycles and will be basically not available to living organisms [1].

Since the presence of toxic metals in the environment is not desired, their emission by the industry is regulated by law. Recently, environmental law has become more stringent (Table 1) [11]. It is difficult to employ conventional methods to remove metal ions below the level of "ppm's" since these methods if applied at such low concentrations cause that the methods become expensive and highly energy consuming [11]. These disadvantages of conventional technologies created the need to elaborate a new generation of efficient methods of environmental prevention (intervening at the impact source, in advance of pollutant event), protection (elimination of the effects of pollutant actions or minimization of these effects) and restoration (removing damages caused by previous actions) [12], as well as monitoring [9].

## 2.2. Biosorption

Biosorption is defined as the ability of materials of biological origin to bind e.g. toxic metals to the surface of cellular wall or membrane in the equilibrium process (Figure 2). More recently, it has been discovered that biosorption is the interaction between metal ions and functional groups present on the cell wall biopolymers of dead organisms [18]. For this reason, biosorption can be classified as a typical adsorption process. There are various classes of biosorbents: by-products or waste materials (with an intermediate sorption capacity), naturally grown and collected biomass [10, 19] (that is practically of no cost) and especially propagated biomass (with high sorption capacity and a possibility to become recovered [19] (e.g. *Spirulina* [20])). Biosorbents are materials of biological origin that can bind metals and organic compounds. Biosorbents can be of microbial, fungal, seaweed, plant or animal origin [21]. Practically, each type of biomass possesses metal-binding properties. This is determined by the chemical structure of biomolecules that all expose similar functional groups. Of course, there are differences in metal-binding capacities and binding mechanisms between sorbents.

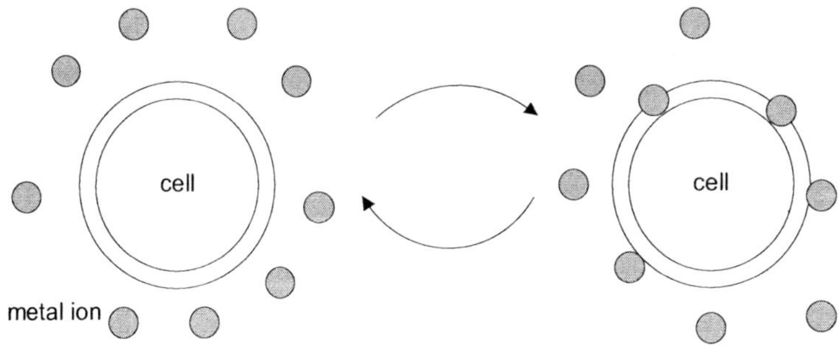


Figure 2. Biosorption of metal ions.

Fundamental for the complete design of the process is investigation of the mechanism. The reported mechanism of biosorption is very complex (a combination of ion-exchange, physical adsorption, surface complexation and surface microprecipitation [10, 21-23]), such as diverse is the structure of biological materials [10]. In the past, physical adsorption was thought to be the dominating mechanism. However, recent studies [2, 24-45] showed that the mechanism is similar to ion-exchange process with the functional groups (amino, carboxyl, phosphate, sulfate, hydroxyl groups) exposed by the cellular wall, which is composed mainly of polysaccharides, proteins and lipids [46], and for this reason biosorbents can be considered as weak acidic cation-exchangers [47-48]. Protons and metal cations are released from the biomass, and cations of other metal ions become bound by cellular surface [49]. Therefore, the biomass acts as organic polyelectrolyte. In the process, metal ions compete with protons for the binding sites [46] and for this reason pH is probably the mostly important process parameter. Since the biomass is not uniform and chemically defined material, there are some difficulties in modelling the process in terms of ion-exchange [18]. Recent literature reports that biomaterials can be characterized in terms of acid-base and metal based potentiometric titrations, used in studies of organo-metallic complexes [18].

Biosorption is a metabolically-passive process [2], which means that there is no need for the biosorbent to carry out metabolic functions during biosorption. For this reason, non-living biomasses are used. Process involving toxic metals and non-living biomass is much easier and cheaper since keeping biomass alive requires expenditures of nutrients and energy. Also, toxic effects that might be potentially posed by these metals can be avoided.

The mostly important process parameters include concentration of metal ions and the biomass, temperature, contact time, agitation and pH. The latter influences three significant parameters, such as speciation of metal ions and their solubility, as well as surface properties of the biomass [11]. The process may find a practical application only if its mechanism, process parameters, kinetics and equilibrium are known. Only then it will be reliable, predictable and possible to design and control [2]. In modeling of biosorption and bioaccumulation processes similar simplifications are used as in the case of microbial growth modelling (which is simplified to single enzymatic reaction).

Biosorption is a quick process. The equilibrium is usually reached within few minutes. Since the rate of the process is high, in some cases it is even difficult to select and fit the proper equation and to determine the order of the reaction due to errors related with quick sampling. The process is usually described with either first- [50], second- [51], pseudo-first- [52] or pseudo-second [52-53] order kinetic equation.

The equilibrium of biosorption can be described with either Langmuir [26, 54-56] or Freundlich [54-56] model (Figure 3). In studies on sorption behavior, both capacity and affinity of a given sorbent to sorbate are investigated [2]. If the dominating mechanism is ion-exchange, which means that a biosorbent possesses a finite number of cations binding sites, we can observe a plateau on isotherm graph. The equilibrium can be described with Langmuir equation, since it is more suitable for description of chemisorption process.

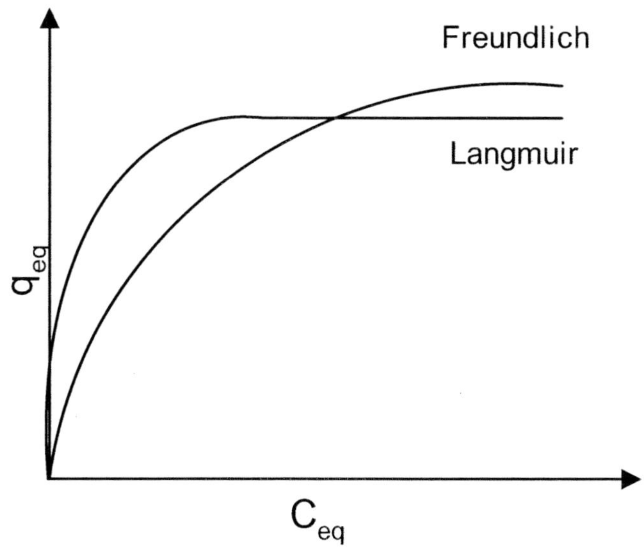


Figure 3. A comparison of Langmuir and Freundlich models.

It was observed that biosorption is usually a reversible process. It is possible to remove bound metal ions with weakly acidic solutions or with strong brine and thus to recover the biomass such as in classical cation-exchange process. Through the selection of the proper eluents, it is possible to restore biosorption properties of biosorbent and reuse it another biosorption cycle [57-58].

## 2.3. Bioaccumulation

Bioaccumulation is the process in which toxic metals or organic compounds become bound within the inner cellular structures. According to the definition [59], bioaccumulation is the process by which living organisms absorb and retain chemicals or elements from their surrounding environment. Other definitions say that bioaccumulation is the accumulation of a chemical either from a medium (usually water) directly or from the consumption of food containing the chemical [9]. Bioaccumulants are substances the concentration of which in living organisms increases as they are taken up from contaminated air, water or food [59].

The critical condition for bioaccumulation to occur is that the biomass should be alive. Bioaccumulation can be performed only by metabolically active cells [2] and therefore the process is metabolically controlled [59-61]. All living organisms throughout their lives accumulate essential, non-essential and toxic substances. All living beings developed pathways that protect them either from the excessive bioaccumulation (e.g. routes through which bioaccumulant enters an organism is blocked or bioaccumulant is excreted outside of an organism) or bioaccumulant is safely deposited within the cell to prevent incorporation into metabolic reactions (e.g. metal ions are bound by special chelating proteins that are rich in thiol groups, so-called phytochelatins in algae or metallothioneins [62]).

For simplification, bioaccumulation can also be described as absorption of metal ions by whole cells (not only by cellular surface as in biosorption). The route through which a bioaccumulant enters an organism is usually the same through which it takes up nutrients or food. In unicellular organisms, toxic metals enter through the transport channels, usually erroneously with essential elements (e.g. Ca(II) or Mg(II)). In plants, some toxic metals are transported from water and soil through the root system, some are acquainted foliarly (from atmospheric deposition) [63]. The route of transfer also depends on which environment is the mostly polluted with a given metal: water, soil or air. In animals, the routes through which toxic metals are bioaccumulated are mainly alimentary tract (food and water contaminated with toxic metals), but also from respiratory system (from polluted air) and through skin (contact exposure) [64]. In some cases the route of bioaccumulation determines toxic effect and protective response that is activated.

Bioaccumulation can find a dual application in pollution control. It can be used to monitor environmental pollution, since there is a correlation between bioaccumulation capacity (the concentration of metal in the biomass) and the concentration in the environment in pollution prevention (wastewater treatment) [65-70] and in cleaning the environment (soil remediation by bioaccumulating plants – so-called hyperaccumulators) [71]. If a method of environmental monitoring is elaborated, it is necessary to screen for good bioaccumulators – biomonitors [72-73]. The analysis of such biomass on the basis of known correlation between the level in the environment and biological material, enables to measure the concentration of a given pollutant in the environment (Figure 4). The mostly advantageous would be to find such biomonitors, the analysis of which would be non invasive nor painful to sample and would not cause any damage to an organism. An example of such biomonitor is human hair or fingernails.

The use of cosmopolitan organisms to monitor bioavailability of a given contaminant enables comparison of contaminants levels in different geographical regions. There are two types of bioaccumulating organisms distinguished: bioindicators that are used for identification and qualitative determination of environmental pollution and biomonitors that are used for quantitative analysis of contaminants. In the latter group, sensitive (optical type - morphological changes are observed) and accumulative biomonitors (ability to store contaminants is assessed) are distinguished. Also, the route of absorption may influence the subsequent distribution of the metal within the body [3, 12]. For instance in human, bioaccumulation occurs through inhalation, oral ingestion and dermal exposure [3].

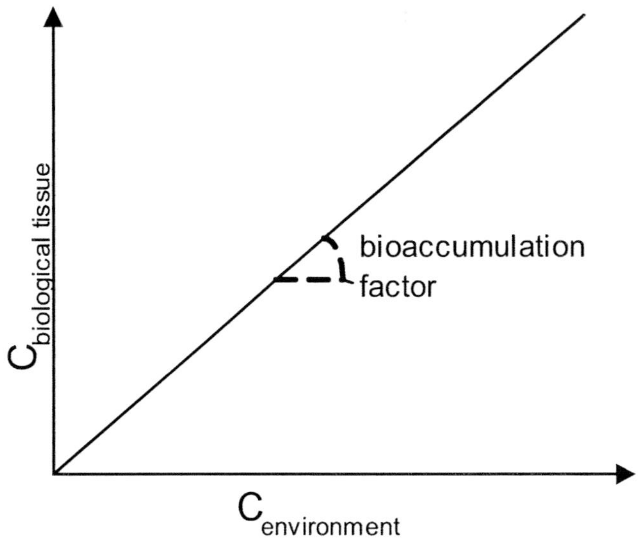


Figure 4. Correlation between the level of contaminant in environment and in biological tissues.

Bioaccumulation itself is the result of the equilibrium of biota intake and discharges from and into the surrounding environment [12]. The process depends on many factors: availability of elements, characteristics of an organism (species, age, state of health etc.) and parameters, such as temperature, available moisture, substrate characteristics, climatic factors, frequently being also parameters that influence metabolic functions of an organism [12]. Therefore, there are difficulties to elaborate standardized procedures and obtain reference values.

Quantitative biomarkers are used to identify toxic responses in populations exposed to very low concentrations in the environment [74]. Monitoring is the repetitive observation for defined purposes using comparable and standardized methods [75]. For this reason, it is very important to elaborate standardized procedures that would enable to monitor environmental pollution on the basis of analysis of contaminants levels in biological tissues.

Recently evolved, so-called bioaccumulation monitoring is the exposure assessment by measuring contaminant level in biota [75]. Since living organisms bioaccumulate toxic metals from the environment, their concentration in biological tissues is related with the concentration of a given pollutant in the environment in which they live and with the length of their life. The concentration of bioaccumulant in the biomass, when related to its concentration in the environment (bioaccumulation factor) (Figure 4) [76-77], is the measure of bioavailability of a given metal from the polluted environment in an organism of a given species. It was observed that the higher organization level an organism possesses, the more perfect protective system preventing bioaccumulation or excreting bioaccumulant it would have. E.g. in human body metals become concentrated in keratinous materials: hair or fingernails that are indicators of long-term exposure [3]. Basically, the lower organism is, the higher bioaccumulation capacity it possesses. Therefore, if we would like to employ bioaccumulation process as a method of toxic metals removal from the environment, it would be the mostly advantageous to employ the smallest and the mostly primitive organisms – microorganisms since they have the highest bioaccumulation capacity [19].

In bioaccumulation, two distinctive processes can be distinguished (Figure 5). In the first step, metal ions are bound to the surface of cells – the process is metabolically passive, it is identical with biosorption. In the second stage, metal ions are transported into the cellular interior. In order to perform this stage, cells must be metabolically active. In another stage, if nutrients are available, the concentration of the biomass increases. More metal binding sites and cellular interiors are available to metal ions. This enables to bind more metal ions when comparing with biosorption.

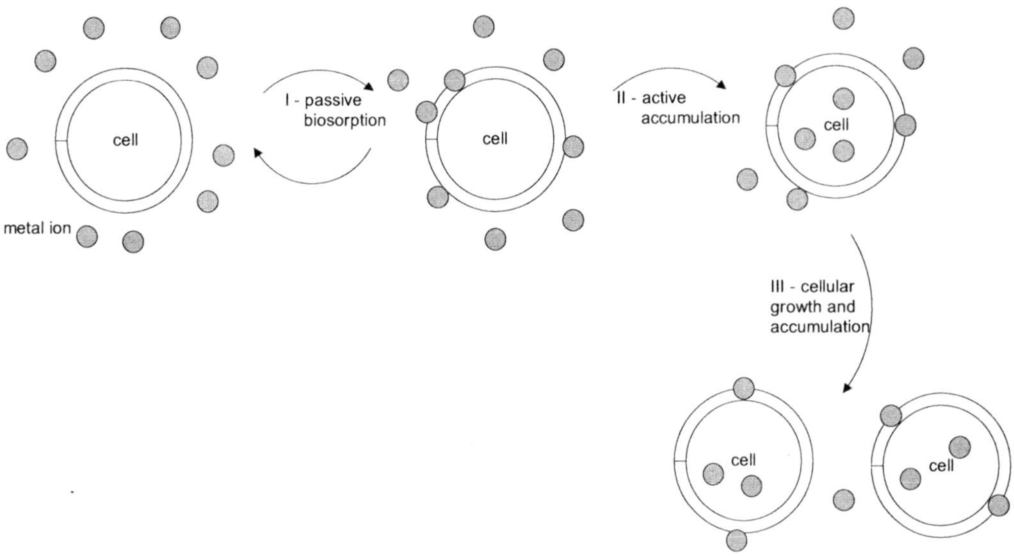


Figure 5. The stages of bioaccumulation of metal ions.

## 3. New Research in Biosorption and Bioaccumulation of Toxic Metals

The recent research on biosorption focuses mainly on removal of toxic metals commonly used in industrial processes, such as Cu [48, 65, 68, 78-82], Cd [18, 48, 80, 83-86], Pb [18, 48, 80, 83, 87-89], Cr [53-54, 61], Zn [52, 79], Ni [90] (Figure 6). There are few studies devoted Hg [84], Al [54] and Co, As, Th. In bioaccumulation literature concentrates mainly on Cd [77, 89, 91-92], Cu [60, 89], Hg [89], Pb [89], Cr [89], Zn [91], As and Ni (Figure 6).

Currently studied sorbents include: aerobic [8, 50] and anaerobic [90] granular biomass, bacteria (*Sphaerotilus natans* [12, 21, 82], *Pseudomonas putida* [79]), microalgae (*Chlorella vulgaris* [78], *Microcystis aeruginosa* [84]), macroalgae ((*Gelidium*) and algal waste [81, 83], *Sargassum muticum* [86]) fungi (*Cephalosporium aphidicola* [88], *Rhizopus arrhizus* [46, 55], *Neurospora crassa* [80]), biological products (extracellular polysaccharide (Pestan) produced by fungus *Pestalotiopsis* [87] and other fungal by-products (*Botrytis cinerea*) [85], cellulose/chitin bed [93], chaff [94]). Trends in current research include studies on kinetics [50, 52, 88] and equilibrium [12, 21, 51, 53, 82-83, 88, 84], influence of pH [8, 38, 81], ionic strength [12, 21, 81, 82] and temperature [81], including modelling [12, 21, 24, 51, 82-83].

Reports contain investigation of biosorption in single- and multi-metal systems [78]. A particular attention is paid to identification of the mechanism of the process [93], including i.e. metal-based potentiometric titration [19, 46]. Also, some modifications (pretreatments) of the biomass are tested [80], i.e. protonation [86]. The new process configurations are also studied extensively, including carrying out the process in membrane bioreactors [12].

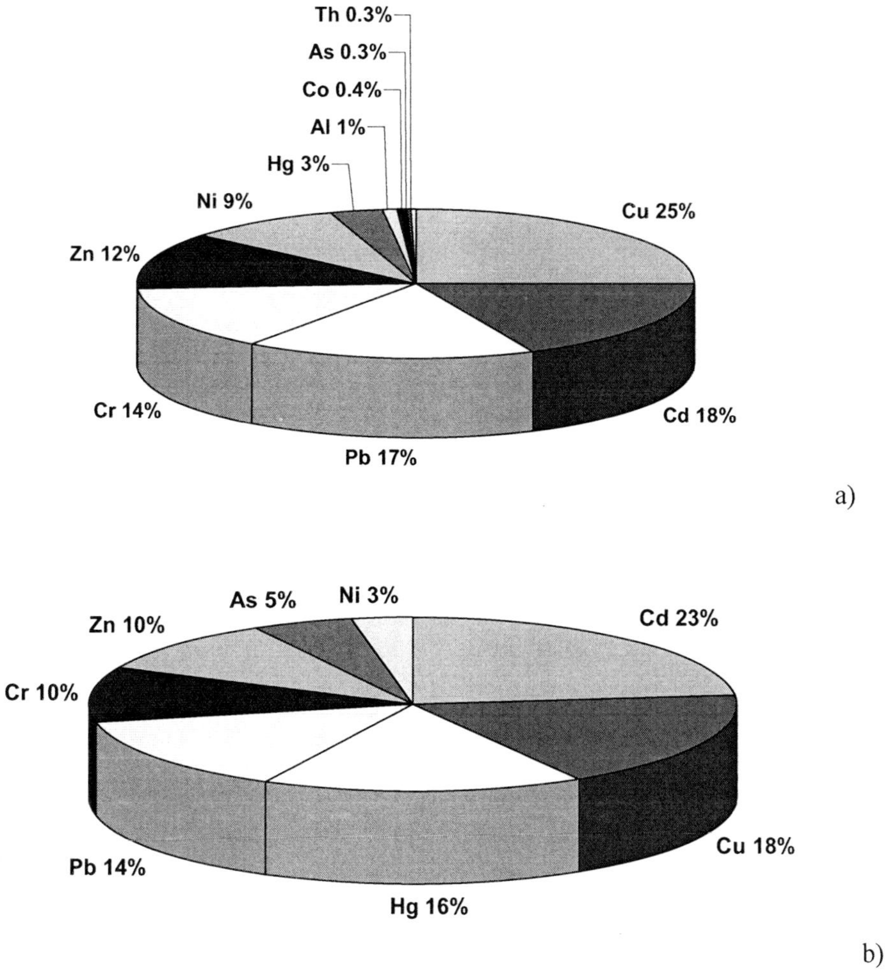


Figure 6. Current trends in research in biosorption (a) and bioaccumulation (b).

Bioaccumulators that are currently studied include: bacteria (genetically engineered *Escherichia coli* [66]), yeasts (*Kluyveromyces marxianus* [68]), microalgae (*Chlorella kessleri* [94]), plants (grass *Dactylis glomerata* [96]), aquatic plants (*Lemna minor*) [65], aquatic animals (Baikal seal (*Phoca sibirica)* [97], scleractinian coral (*Stylophora pistillata*) [98], tube worm *Riftia pachyptila* [89], soft tissues of mollusks *Patella aspera* [99] and zebra mussel (*Dreissena polymorpha*) [100], mangrove oysters (*Crassostrea rhizophorae*) [92], amphibian larvae [101], marine fish [69, 91], freshwater fish (European eel *Anguilla anguilla* [102]), multispecies monitoring of muscle tissue of fish [103]), terrestrial animals (earthworm *Lumbricus terrestris* [104-105], animals (rats [106-107]). The majority of studies concern environmental monitoring of aquatic environment with the use of bioaccumulating aquatic animals: invertebrates and vertebrates. Recently appeared also reports on monitoring by terrestrial plants and animals, including human (assessment of exposure by the analysis of hair and fingernails) [6]. Toxicity tests using metals bioaccumulation usually are performed on rats.

New research on bioaccumulation focuses on: monitoring of pollution of aquatic environment through the analysis of levels of toxic metals in tissues of aquatic organisms [89, 91, 101, 102-103], wastewater treatment [65-66, 68-69], bioaccumulation by plants grown on contaminated soils [96], bioaccumulation of metals from wastes by earthworms [105], inhalation toxicity tests by rats [106], effect of the presence of one metal on the accumulation of other toxic metals [91], the effect of solubility of toxic metals salts on their bioavailability [105], the assessment of relationship between bioaccumulation of toxic metals with environmental conditions and genetic variability [102], analytical studies on bioaccumulation of toxic metals in biological tissues [107].

## 4. Similarities and Differences between Biosorption and Bioaccumulation Processes

When comparing biosorption and bioaccumulation processes, bioaccumulation although performed in simpler installations, requires troublesome cultivation of the biomass in the presence of contaminants that may pose toxic effects to the biomass itself. The difference between the equilibrium of biosorption and bioaccumulation process is shown in Figure 7. Biosorption itself is the first stage of bioaccumulation. In the second stage, ions are furtherly transferred into cellular interior that results in the equilibrium shift towards lower concentration of metal ions. Table 2 discusses the differences between biosorption and bioaccumulation processes.

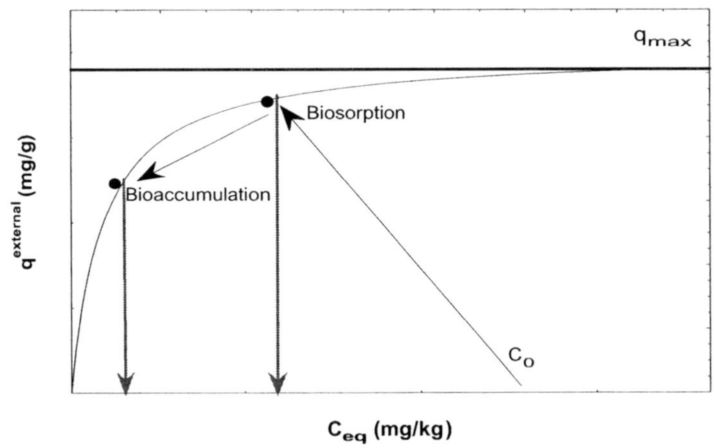


Figure 7. Equilibrium shift towards lower values of $C_{eq}$ in bioaccumulation process when comparing with biosorption.

Biosorption is an equilibrium process. The efficiency of biosorption is expressed in terms of biosorption capacity, which is the mass or molar equivalent of metal bound by the unit mass of a biosorbent [mg/g or meq/g]. The basic characterization of a biosorbent includes evaluation of maximum sorption capacity ($q_{max}$) and affinity ($b$ parameter in the case of Langmuir equation). The mostly advantageous in the case of both bioremoval and biobinding would be for a sorbent to have simultaneously high sorption capacity and affinity. In Figure 8 there are shown sorbents with different characteristics. Sorbent *1* possesses high affinity and high maximum binding capacity. This is the best sorbent. Sorbent *2* possesses high maximum sorption capacity but low affinity. Sorbent *3* – high sorption affinity but low maximum sorption capacity and sorbent *4* – low affinity and low maximum sorption capacity. The latter sorbent possesses the worse characteristics. In bioremoval processes of particular importance is affinity of a biosorbent. This is related with the potential applications of this process as polishing treatment in order to reduce the concentration of pollutants below the acceptable limits. For this reason the mostly desired characteristic would be high affinity of sorbent to sorbate, which is low equilibrium concentration of pollutant reached even at low equilibrium sorption capacities. For this reason, it is recommended to use sorbent possessing characteristics such as material *1* and *3* (Figure 8). If we are interested in other than bioremoval applications – biobinding (e.g. production of bioavailable feed supplements), of the primary importance would be the highest maximum sorption capacity since the goal is to bind possibly high quantities of metal ions by a unit of the biosorbent biomass. In this case it is recommended to use either sorbent *1* or *2*.

**Table 2. The comparison between biosorption and bioaccumulation process**

| Biosorption | Bioaccumulation |
|---|---|
| Passive process | Active process |
| Metals are bound with cellular surface | Metals are bound with cellular surface and interior |
| Adsorption | Absorption |
| Reversible process | Partially reversible process |
| Biomass is not alive | Biomass is alive |
| Nutrients are not required | Nutrients are required |
| Single-stage process | Double-stage process |
| The rate is quick | The rate is slow |
| Not controlled by metabolism | Controlled by metabolism |
| No danger of toxic effect | Danger of toxic effects caused by contaminants |
| No cellular growth | Cellular growth occurs |
| Intermediate equilibrium concentration of metal ions | Very low equilibrium concentration of metal ions |

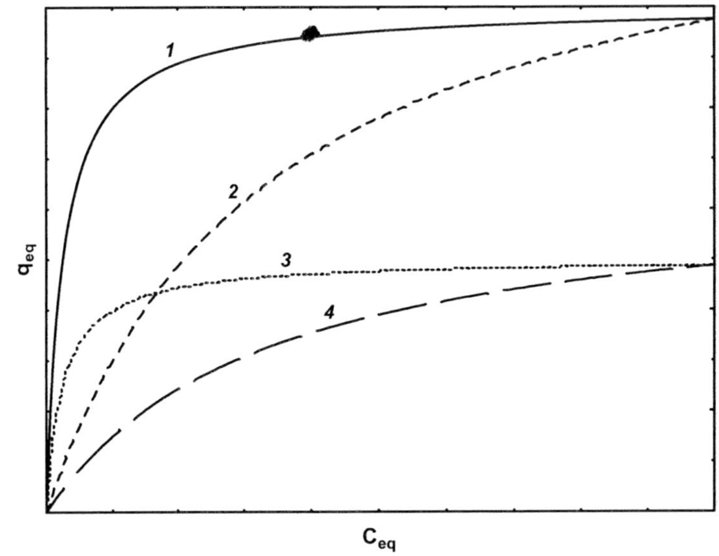


Figure 8. Isotherms for different types of sorbents.

# 5. Brief Information on Analytical Techniques in Determination of Fate of Toxic Metals in Biosorption and Bioaccumulation

Analytical techniques used in studies on biosorption and bioaccumulation processes should be very sensitive in determination of elements on the level of even ng/kg and should minimize interferences and matrix effects. These include modern instrumental techniques, such as atomic absorption spectrometry, atomic emission spectrometry, inductively coupled plasma emission spectrometry: ICP - Optical Emission Spectrometry (ICP-OES) and ICP – Mass Spectrometry (ICP-MS). It is very significant to use methods with very low detection limit, since in the majority of cases the concentration of elements is below 1 mg/kg. Prior the analysis, the studied materials are digested with mineral acids, preferably in microwave oven in closed vessels. This generates 50-100 times dilution. For this reason, it is advantageous to use methods with detection limit below μg/L, such as ICP-OES with ultrasonic nebulization or ICP with mass detector [108]. The analytical process should be controlled by the use of certified reference materials that have similar matrix as the analyzed material. E.g. in the analysis of human hair, Reference Material Human Hair was used [109].

It is recommended to perform semi-quantitative analysis of samples prior the quantitative analysis in order to evaluate the expected levels of analytes and to determine the main components. The knowledge of the levels of the main components is used in the preparation of calibration solutions. The mostly advantageous is to add the main components into the calibration solution. Such an approach minimizes matrix effects and also in some extent, interferences.

# 6. Using Bioaccumulation in Biomonitoring of Environmental Pollution

The processes of biosorption and bioaccumulation have found or will find in the nearest future an application in various areas related with environmental protection (Figure 9). Among these processes, two types of methods can be distinguished, which have common mechanisms, but the goals are different – bioremoval and biobinding. In the case of bioremoval, the goal is to achieve the highest possible removal of toxic metal ions via either biosorption or bioaccumulation processes. The objective is first of all to remove metal ions to the concentration below the level required by obligatory law, which is usually ca. 1 mg/L from large volume diluted solutions. This use of biosorption and bioaccumulation methods is recommended as the final, polishing step, before the final discharge of effluents.

Biobinding processes that can be applied in environmental protection include biomonitoring. In this case, the aim would be not to bind the highest amounts of metals but to assess the concentration and bioavailability of pollutant in the environment. Biobinding processes can also find another application – in the production of highly bioavailable mineral feed additives. In this case the goal would be to bind the highest quantities of metals via either biosorption or bioaccumulation.

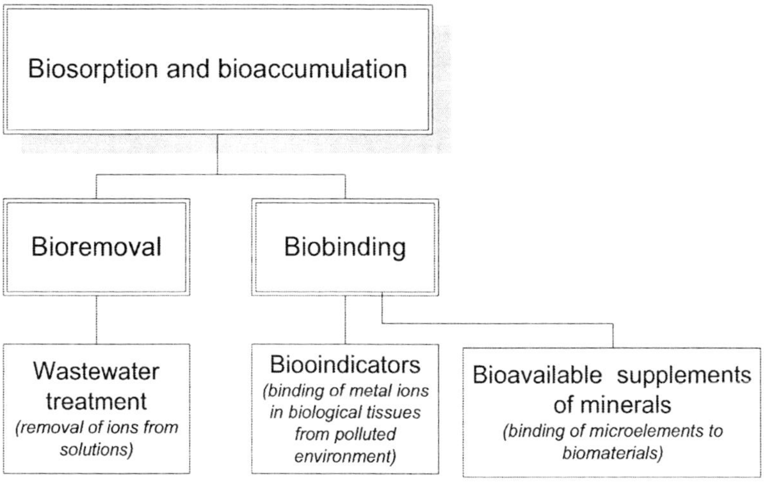


Figure 9. Biosorption and bioaccumulation understood as bioremoval and biobinding.

The content of toxic metals in tissues or in the biomass of each organism reflects the level of a given contaminant in the environment. But, on the other hand, not every organism has the potential to be a good biomonitor or biondicator. First of all, biomonitoring should be ethically correct. Sampling of biomonitor tissues should not be invasive, nor should not pose destruction or deterioration of an organism, nor pain. Another issue is that the concentration factor of a pollutant in the studied material should be high enough to be possible to determine after digestion and dilution of a sample. Criteria of good biomonitor fulfill particularly non-living tissues, such as human hair [109-111], animal fear or birds feather, as well as animal products, advantageously consumable, such as eggs and milk [112-123]. Also, plant tissues are good biomonitors of soil pollution [124-128]. The analysis of human hair provides the information on pollution of environment in which an organism lives, as well as contamination of food and drinks [109-111]. The level of soil pollution can be determined by the analysis of plant tissues [124]. It was also found that it is possible, through the use of specially elaborated extraction procedures to determine content of bioavailable forms of metals and on this basis (knowing so-called transfer factor) to predict the composition of plants that are to be cultivated on a given soil [124-128].

**Table 3. Descriptive statistics for element concentration in hair from an urban population group living in south-west Poland (Wrocław) (83 subjects) [110]**

| Element | Mean (mg/kg) | Std.Dev. |
|---|---|---|
| Ag | 0.395 | 0.78 |
| Al | 14.938 | 29.40 |
| As | 0.044 | 0.11 |
| Au | 0.049 | 0.14 |
| B | 2.041 | 2.09 |
| Ba | 2.025 | 2.53 |
| Be | 0.055 | 0.05 |
| Ca | 1087.8 | 970 |
| Cd | 0.114 | 0.14 |
| Co | 0.034 | 0.04 |
| Cr | 0.568 | 1.04 |
| Cu | 12.352 | 12.05 |
| Fe | 15.000 | 16.07 |
| Hg | 0.500 | 0.39 |
| Mg | 66.991 | 65.14 |
| Mn | 0.601 | 0.59 |
| Mo | 0.017 | 0.02 |
| Na | 217.330 | 268.78 |
| Ni | 0.838 | 1.13 |
| P | 132.031 | 271.84 |
| Pb | 1.046 | 1.39 |
| Pt | 0.0004 | 0.0003 |
| Sb | 0.455 | 0.32 |
| Se | 0.679 | 0.94 |
| Si | 57.51 | 145.9 |
| Sn | 1.205 | 0.92 |
| Sr | 2.882 | 3.69 |
| Ti | 1.491 | 2.28 |
| U | 0.208 | 0.41 |
| V | 0.092 | 0.12 |
| W | 0.002 | 0.00 |
| Zn | 156.48 | 74.5 |
| Zr | 0.568 | 1.77 |

## 6.1. Human Hair as Biomonitor of Environmental Pollution and Nutritional Status of Individuals

It is well known that the concentration of metals in biological tissues reflects their level in the environment. The property that is used in biomonitoring is that a mechanism that protects an organism from the excessive accumulation and toxic effects posed by a given contaminant is excretion outside an organism via different routes: urine and feces, sweat and also transfer to external non-living tissues, such as hair and nail or birds feather. Elements enter living organism via different routes: are transferred from air, water, foods, drugs, through skin, respiratory tract and gastro-intestinal tract. Metals are transported by blood and deposited in organs [110]. The elements that are present in hair can be divided into macroelements and trace elements. The latter group consists of toxic trace elements and essential trace elements (microelements) that comprises of major trace elements (Fe, Zn and Cu) and minor trace elements (Mn, Se, Cr, Co, Ni, Si, F, I) [109-111]. Table 3 presents the composition of hair from a population living in urban and industrialized area [110].

The elaboration of commonly used methods of environmental pollution monitoring through the analysis of biological tissues should preferably include studying materials that are sampled easily and painlessly and reflect an exposure history of a given organism. Human hair, for instance, fulfills these criteria.

Although human hair can be theoretically considered as reliable and convenient biological indicator of environmental pollution, there have been some problems encountered in the elaboration of reference values for the concentration of metals in human hair, in particular some problems with the standardization of the method occurred. It was found that the level of elements in hair is the function of age, sex, and hair color, as well as hair treatments, including hair coloring and undulation as well as other treatments that affect the structure of hair. Also sex, nutritional habits and smoking [111] influence the composition of hair. In studying the composition of hair it is also necessary to consider inter-element-interactions: antagonisms and synergisms [110]. It is also significant to distinguish between internal (originating from the body) and external deposition. There exist elaborated washing procedures (i.e. with water and acetone) [110] that enable to distinguish between the metal that entered an organism (was bioaccumulated) and was simply adsorbed by the external surface of hair.

Studies of inter-element interactions between elements in human hair were performed on the population living in an urban and industrialized region [110] in order to assess if a concentration of a given element is somehow related with the concentration of another. This approach enabled to point out on common occurrence of elements related with i.e. environmental exposure or dietary habits. Synergisms and antagonisms between elements were studied through determination of statistically significant linear correlation coefficients and also by linear multiple regression analysis [110]. On the basis of multielemental determination of 33 elements in human hair, the elements were grouped according to their concentration in hair (in decreasing order): Ca < Na < Zn < P < Mg < Si < Fe < Al < Cu < Sr < B < Ba < Ti < Sn < Pb < Ni < Se < Mn < Zr < Cr < Hg < Sb < Ag < U < Cd < V < Be < Au < As < Co < Mo < W < Pt. The concentration of Ca was above 1000 mg/kg. The concentration of Na, Zn and P was in the range 100-200 mg/kg. The level of Mg, Si, Fe, Al and Cu was above 10 mg/kg. The concentration of Sr, B, Ba, Ti, Sn and Pb was 1-2 mg/kg. The level of other metals was below 1 mg/kg [110]. The concentration of elements in hair of the studied population was compared with other studies. The differences in the level of Ag, Al, Ca and Na were encountered [110].

Studies on inter-element interactions showed the existence of the following statistically significant correlations, mainly of synergistic character: Mg-Ca, Mn-Ca, Sr-Ca,Sr-Mg, U-Na, Ni-Zn, Cd-Ni, Sb-Pt. It was found that there exist stronger interactions between industrially related elements when comparing with essential elements, i.e.: Cu-V, U-Ba, Fe-Mn and Al, Mn-Al and Sr, Mo-Au, Au-Pt, Be-Pt, Ba-Sb, Ag-Sb, Sr-Sb and W and W-Sb. It was possible to distinguish groups of elements that are mutually correlated, such as a group of (Sr, Sb, W) – the elements that are emitted by industry, including metallurgical works. Another group of (Ca, Mg, Mn, Sr) was negatively correlated with (Pt, W) [110]. Linear multiple regression analysis confirmed the following inter-element dependencies: Al=f(U, P, Mn), As=f(Zn, Fe) (Zn is an antagonist), Cu=f(V), Fe=f(Mn, As), Mg=f(Ca), Ca=f(Mg, Ba), Ni=f(Zn, Cd), Sb=f(Pt, Sn, W), Ti=f(Fe, Co) [110].

It was found that age of individuals has strong influence on elements concentration in human hair. The studied population was divided into four age groups: children (0-15 years old individuals), young people (15-25 years old), adults (25-45 years old), middle-aged people (45-65) (Table 4). The level of some elements did not depend on age of an individual: Ti, Se, Mn, Ni, Hg, Cd, As, Co. This group (except Co) was found to be related rather with environmental exposure or originated from food and drinking water. Hair of children (< 15 years) has a tendency to contain more Na, K, P, Al, B, Pb, Fe, Cr, Au, Pt, Sb, Be, W and less Ca, Cu, V, Sn, Be, Ag, Zr than hair of older people. Hair of individuals in the age 15-25 years (young people) has higher levels of Ca, Zn, Si, Mg, Sr and lower levels of Na and K than people from other age groups. Hair of adults in the age 25-45 years has higher level of Ca, Ag, Zr and lower level of Na and K. People from the oldest group (45-65 years) have higher level of Na, K, Cu, V, Sn and lower level of: Ca, Zn, Si, Mg, Sr, Ag, Zr in their hair [111].

**Table 4. The effect of age on the level of elements in hair [111]**

| Elements | Age group | | | |
|---|---|---|---|---|
| | 0-15 | 15-25 | 25-45 | 45-65 |
| Ca | ↓ | ↑ | ↑ | ↓ |
| Na, K | ↑ | ↓ | ↓ | ↑ |
| P, Al, B, Pb, Fe, Cr, Au, Pt, Sb, Be, W | ↑ | ↓ | ↓ | ↓ |
| Cu, V, Sn | ↓ | ↓ | ↓ | ↑ |
| Zn, Si, Mg, Sr | ↓ | ↑ | ↓ | ↓ |
| Ag, Zr | ↓ | ↓ | ↑ | ↓ |
| Ba | ↓ | ↓ | ↑ | ↑ |
| Ti, Se, Mn, Ni, Hg, Cd, As, Co | the level does not depend on age | | | |

↑ - higher concentration when compared with other age groups
↓ - lower concentration when compared with other age groups

Also, the concentration of elements of females' hair was compared with the level in males hair. It was found that hair of females contained significantly more Ag, W, Sr, Ca and more Mg, Si, Be, Cr, P and Mn. Hair of males contained higher concentrations of Sn, V, Au,

Sb, Pb, Zr, K, B, As, Na, Mo, Hg. It was found that the level of Al, Cd and Co did not depend on sex of an individual [111].

The composition of hair of tobacco-smokers was compared with hair of non-smokers. Surprisingly, it was found that hair of non-smokers did not contain higher levels of toxic metals, such as Cd. It was found that hair of non-smokers contained higher level of V, Se, P, Sr, Si, Ba, Na, Cr and hair of smokers contained more Zr, Mo and As. The level of the following elements was similar: Cd, K, Ag, Ca, Co and Be. Generally, hair of non-smokers contained more metals (both essential and non-essential). Literature reports that smokers, blood and urine contain elevated levels of toxic metals and this is the main route of detoxification of an organism [111].

The effect of hair treatments (e.g. coloring) and hair color on the elemental composition of hair was investigated. It was found that colored hair contained more Sr, Ba, Ca, Mg, W, Mo, Ag, Mn, but less V, Zr, Sb, Pb, As, Si, K and Hg. Colored hair contained almost twenty times higher level of metals than natural hair. Also, the level of elements in hair of different color (among naturally colored hair) was investigated. The highest concentration of Si was found in dark blond hair, Co – in blond hair, Au, Cd, Sb, Ni, Ag, Pb, Co, Cu – in dark hair, V, Pt – in auburn hair. Generally, the concentration of Ca was higher in dark hair than in fair hair. The level of Mn, Fe and Na was not the function of hair color [111].

The concentration of metals in hair of a given individual changes significantly along with the change of living habits. The elemental composition of hair of individuals that either lived together or were family related was investigated over extended period of time. The content of metals of the studied individuals was compared with the average composition of hair of people living in the studied urban area. The differences concerned mainly different living habits (Na, Si, Co, Fe, Mn, Zn) and local exposure or personal ability to accumulate a given metal (Pb, Cd, Ba). The concentration of the following elements did not depend on local exposure nor personal predispositions: Ag, Al, As, B, Cr, Pt, Se. Time profile from a given individual showed that the concentration of a given element changed several fold over the years with the change of living habits and environmental exposure. There were found similar tendencies in the accumulation of the majority of elements by people that lived together. It was found that the effect of living habits was stronger than the influence of either sex or family relationship. The concentration of the majority of elements was similar in people that lived together and thus consume similar food and drinks and live in the same environment. Of course, lifestyle of a particular subject is always individual. Also, inter-element interactions were observed between the elements that occur together: Ca-Mg, Fe-Mn, Na-K, Co-K, Au-Pt, Cd-Pb. Linear multiple regression analysis confirmed the existence of the following dependencies: Ca=f(Mn, Sr), Na=f(K, Mn), K=f(V, Ti). Groups of mutual statistically significant correlations were distinguished: (Fe, Mn, V), (As, Au, Pt, Zr), (Co, Pb, Sb) that resulted either from the common occurrence of these elements or synergistic effects between them [109].

It was found that hair of females contained more Ag, Sr, Ca, Mg, Cu, TI, P, Mn, but hair of males Mo, B, Co, Sn, Zr, Si, V, K, Pt, Au, As, Pb and Fe. No differences in the concentration of Ba, Cd, Zn, Ni, Hg, Cr, Se, Na, Al were observed. The level of the following elements was a function of living habits: Zr, Sn, Ag, Ti, As, Se, Co, Sb, Si, Sr, Ni, Ba, Mn, Cu, Mg, Cd, Pb, Ca, Na, Fe, P, V, B, Cr. No differences in the level of Mo, Zn, K, Hg, Al was denoted. Generally, it was found that hair of females contained more macroelements (Ca, Mg, P) and hair of males more microelements (Mo, B, Co). The level of toxic elements was similar and it was concluded that their level was rather related with local exposure and living habits. It was found that people that lived together have practically identical level of Hg, Al, Mo and Zn. When comparing with the whole population, the level of As and Al was similar. It was found that the whole population was exposed to As and Al and the level of these elements in hair was elevated when comparing with other populations. Generally, it was found that the strongest influence on elements level in hair was not posed by personal ability nor sex, but first of all by living habits and environmental exposure [109].

## 6.2. Bioavailability of Toxic Metals from Soils to Plants

Of particular concern is food contamination with toxic metals, such as As, Cd, Hg and Pb. Legal directives define the permissible content of elements in plants used for consumption purposes. Toxic metals are taken up by plants from air, water and soil. The presence of toxic metals, such as As, Cd, Hg and Pb in these environments results in their accumulation throughout the trophic chain. Also, some micronutrients, such as Cu, Cr, Ni, Zn are toxic at higher concentrations. Bioavailability of toxic elements to plants is affected by many factors, including soil and climatic conditions, plant species, agronomic management, active/passive transfer processes, sequestration and speciation, redox state, the type of plant root system and the response of plants to elements in relation to seasonal cycles [124].

Knowledge of bioavailability of toxic metals from soils to plants can be used in ecological risk assessment that relies on site-specific exposure models for the estimation of uptake of contaminants. There is the need to elaborate standardized procedures that would enable to predict before sowing, whether it is recommended to cultivate plants for consumption purposes on a given field. It is therefore necessary to assess techniques that would enable to estimate bioavailability of metals to plants. This would make it possible to predict the composition of plants before sowing, on the basis of the analysis of extractable fraction of metals in soil that is bioavailable to plants along with the assessment of soil-plant correlation coefficients. For many years, there have been attempts to obtain statistically significant soil-plant correlation coefficients, however without success, since the concentration of metals in plant was related with the total concentration in soil. This means that the presence of a given metal in soil at high concentrations would not necessarily mean that the level in plant would be exceeded. Extraction procedures were proposed as a method of simulation of transfer of metal ions from soil to plants. The proposed extraction tests were verified in field experiments on contaminated soils. The efficiency was confirmed by germination tests. In order to precisely determine the concentration of elements in soil, soil extracts and plants, it was necessary to use analytical techniques with a very low detection limits, such as inductively coupled plasma optical emission spectroscopy (ICP-OES) with ultrasonic nebulizer or inductively coupled plasma mass spectroscopy (ICP-MS). These experiments enabled to assess uptake and bioavailability of metals from soil and to investigate inter-element interactions [124, 129]. It was found that through the application of the proper extraction procedures it was possible to assess the portion of a given element that was soluble in the soil solution and thus was available to plants. It was shown that the highest soil-plant correlation coefficient was found in the case of extraction with 2 % w/v ammonium citrate and 0.1 M sodium nitrate solution. This means that extraction of soil with the use of such eluents and the analysis of the concentration of metals in the post-extraction solution would provide us with the information on bioavailability of metals from soil to plant [124].

The quantity of metal accumulated by plants from soil is defined as transfer factor (bioaccumulation coefficient) (*TF*):

$$TF = \frac{C_{Plant}}{C_{Soil}} \tag{1}$$

where $C_{Plant}$ is the concentration in plant and $C_{Soil}$ is the concentration of metal in soil. However, the equation can only be applied if there exists a statistically significant correlation between the concentration of metal in plant and soil (Figure 10). For this reason, $C_{Soil}$ would rather represent bioavailable concentration of metal in soil. In some cases, the dependence $C_{Plant}$=f($C_{Soil}$) does not intercepts at (0, 0). The equation should be expressed with the use of the following equation:

$$TF = a + \frac{C_{Plant}}{C_{Soil}} \tag{2}$$

If a≠0, a metal does not originate only from soil, but can be bioaccumulated by a plant from i.e. air deposition. If a=0, the metal was bioaccumulated by plant only from soil as illustrated by Figure 10 [124].

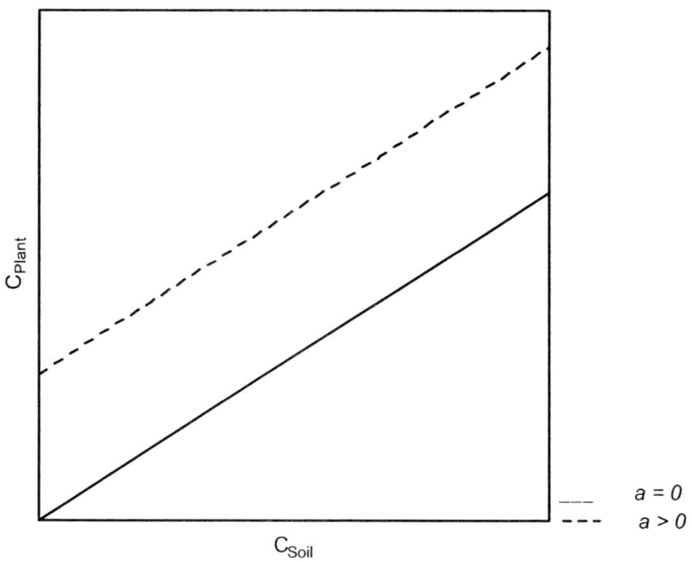


Figure 10. Soil-plant transfer of metal ions.

Soil-plant transfer factors were determined for various metals. Statistically significant dependencies between bioavailable soil fraction (extracted with sodium citrate) and the concentration in plant was detected in the case of As, Cd, Cr, Cu, Hg, Mn, Pb, and Zn. Only in the case of Ni no correlation was observed. Statistically significant correlation between total concentration in soil and plant was observed only for Cu. Also, transfer factors were calculated for elements that showed statistically significant correlations. The knowledge of *TF* enables to predict the composition of plants before sowing, on the basis of the analysis of concentration of bioavailable form of elements in soil [124]. Solubility (and thus concentration of bioavailable forms) of metals is not necessarily directly proportional to its total concentration in soil. It depends first of all on factors that affect solubility, such as pH and the concentration of $P_2O_5$ as well as on the presence of other elements (synergistic and antagonistic effects). Statistically significant correlations between the total content in soil and in plants were encountered only for potassium, the metal that forms soluble salts with almost all anions. Also, the experimental results showed that elevated pH and high concentration of $P_2O_5$ resulted in decreased bioavailability.

Transfer Factor is the measure of the extent to which a plant concentrates a given element in its biomass, as related to bioavailable fraction in soil. The highest *TF* was found for Zn (4000), afterwards Cu (120), Mn (44), Pb (33), Cd (31), As (17), Hg (4.5). Also, it was checked whether a given element was accumulated by plants only from soil (model parameter $a$=0). The following elements were taken up only from soil: As, Cd, Cu. In the case of Hg and Ni $a$>0, which meant that these elements were bioaccumulated also by air deposition. This is the explanation of low (although statistically significant) correlation coefficient for Hg and no statistical significance for Ni.

## 6.3. Wood Ash as Biomonitor of Environmental Pollution and Material that Decreases Bioavailability of Pollutants in Contaminated soils

Wood ash constitutes ca. 5 % of burnt wood. For this reason, wood ash can be considered as the concentrate of metals that were bioaccumulated by wood that became concentrated 20 times. In the combustion process, organic matter, nitrogen, sulfur, mercury and cadmium volatilize. The remaining macroelements (i.e. Ca), microelements and toxic metals become concentrated. For this reason, since the material is the concentrate, it is easier to determine the presence of trace elements in wood biomass.

It was found that wood ash is strongly alkaline material that can find an application as the substitute of lime, as soil deacidifying agent. Wood ash contains 35 % CaO, 9 % $K_2O$, 2 % $P_2O_5$ and 1.6 % MgO. These can be considered as macroelements of fertilizer significance. Wood ash contains also high levels of micronutrients that were bioaccumulated by wood and are of fertilizer value, such as Fe (0.5 %), Mn (0.1 %), B (350 mg/kg), Zn (230 mg/kg), Cu (170 mg/kg), Co (8 mg/kg), Mo (3 mg/kg). Also, wood ash contains toxic metals on the level related with the environmental pollution, such as Pb (30 mg/kg), Cd (8 mg/kg), As (20 mg/kg) [125].

The elemental composition of wood ash from different tree species was determined (Table 5) [130] and compared with soil. If used as soil amendment, wood ash should contain similar or lower level of toxic metals when compared with soil.

**Table 5. The elemental composition of ashes from various tree species and soil [130]**

| Element | Unit | Oak | Birch | Apple tree | Ash | Coniferous tree | Soil |
|---|---|---|---|---|---|---|---|
| Al | % | 0.849 | 0.233 | 0.147 | 1.76 | 0.717 | 0.635 |
| As | $mgkg^{-1}$ | 21.94 | 19.12 | 19.80 | 23.61 | 23.15 | 63.88 |
| B | $mgkg^{-1}$ | 344.87 | 510.65 | 257.36 | 520.88 | 274.27 | 24.27 |
| Be | $\mu gkg^{-1}$ | 535.08 | 382.72 | 254.72 | 473.03 | 601.18 | 468.63 |
| Bi | $mgkg^{-1}$ | 1.37 | 60.26 | ≤0.0005 | ≤0.0005 | 177.40 | 4.213 |
| Co | $mgkg^{-1}$ | 7.29 | 13.34 | 4.79 | 7.230 | 15.23 | 4.897 |
| Ca | % | 31.14 | 29.36 | 35.11 | 24.30 | 19.82 | 2.32 |
| Cd | $mgkg^{-1}$ | 7.82 | 23.27 | 1.674 | 3.23 | 16.03 | 0.266 |
| Ce | $mgkg^{-1}$ | 1.78 | 2.397 | 0.909 | 2.917 | 4.874 | 6.491 |
| Cr | $mgkg^{-1}$ | 75.90 | 16.30 | 5.79 | 5.69 | 82.82 | 25.33 |
| Cu | $mgkg^{-1}$ | 111.37 | 143.13 | 96.81 | 175.52 | 123.09 | 24.23 |
| Fe | % | 0.58 | 0.53 | 0.56 | 0.48 | 2.36 | 0.970 |
| Ga | $mgkg^{-1}$ | 7.75 | 84.77 | 66.26 | 87.89 | 84.62 | 77.37 |
| Ge | $mgkg^{-1}$ | 1.60 | 24.08 | 24.52 | 20.73 | 43.41 | 30.18 |
| Hg | $\mu gkg^{-1}$ | 6.30 | 0.01 | ≤0.0001 | ≤0.0001 | ≤0.0001 | 95.29 |
| K | % | 4.85 | 6.25 | 5.16 | 12.26 | 3.43 | 0.454 |
| La | $mgkg^{-1}$ | 1.11 | 2.31 | 0.666 | 1.99 | 10.39 | 3.1956 |
| Mo | $mgkg^{-1}$ | 2.807 | 0.960 | 1.280 | 0.419 | 1.95 | 1.741 |
| Mg | % | 0.96 | 2.51 | 5.16 | 2.77 | 1.35 | 0.251 |
| Mn | % | 0.999 | 1.691 | 0.191 | 0.236 | 0.775 | 0.0471.34 |
| Na | % | 0.21 | 0.16 | 0.19 | 0.24 | 0.86 | 0.122 |
| Nb | $\mu gkg^{-1}$ | 521.49 | 419.38 | 362.32 | 315.65 | 8107.00 | 1396.65 |
| Ni | $mgkg^{-1}$ | 87.66 | 37.80 | 46.98 | 102.47 | 53.11 | 10.84 |
| P | % | 0.85 | 1.97 | 1.15 | 2.19 | 2.11 | 0.169 |
| Pb | $mgkg^{-1}$ | 32.09 | 103.07 | 25.190 | 42.62 | 690.14 | 15.17 |
| Pd | $\mu gkg^{-1}$ | 653.23 | 448.54 | 580.49 | 718.64 | 538.22 | 566.67 |
| Rb | $mgkg^{-1}$ | 478.20 | 141.30 | 37.17 | 157.60 | 177.46 | 22.34 |
| Rh | $\mu gkg^{-1}$ | 20.44 | ≤0.0006 | 29.19 | 12.05 | 11.49 | ≤0.0006 |
| Sb | $mgkg^{-1}$ | 2.63 | 0.516 | 1.02 | 0.49 | 14.61 | 4.626 |
| Sn | $mgkg^{-1}$ | 1.76 | 0.351 | 0.346 | 0.230 | 2.321 | 1.37 |
| Th | $\mu gkg^{-1}$ | ≤0.0006 | ≤0.0006 | ≤0.0006 | ≤0.0006 | ≤0.0006 | 0.0015 |
| Ti | $mgkg^{-1}$ | 258.92 | 293.85 | 236.93 | 255.98 | 2941.45 | 675.02 |
| Tl | $\mu gkg^{-1}$ | 75.60 | 0.05 | 0.00 | 0.06 | 118.89 | 15.17 |
| U | $\mu gkg^{-1}$ | 28.80 | 15.51 | 6.45 | 17.97 | 39.98 | 47.48 |
| V | $mgkg^{-1}$ | 0.00 | ≤0.03 | ≤0.03 | ≤0.03 | 20.26 | 5.124 |
| Zn | $mgkg^{-1}$ | 227.70 | 4903 | 756.30 | 577.60 | 5020 | 117.22 |
| Zr | $mgkg^{-1}$ | 1.08 | 1.208 | 0.770 | 0.769 | 7.54 | 6.532 |

The composition of wood ash from various trees (oak, birch, acacia, ash tree and apple tree, coniferous tree) was analyzed in order to investigate which wood species bioaccumulated the highest quantity of a given metal. The concentration of various elements (macroelements, microelements and trace elements vary depending on tree species. The lowest concentration of both microelements and toxic metals (with the exception of Zn and B) was found in ash from fruit tree. Probably these elements are partially excreted to fruits. Wood ash from deciduous trees contained high level of Pb, As and Cu. It was found that wood ash from coniferous trees contained significantly higher levels of Pb (430 mg/kg) when compared with deciduous trees: birch (63 mg/kg), acacia (57 mg/kg), oak (39 mg/kg). The lowest level was found to be present in ash tree and apple tree (25 mg/kg). The level of As was similar in each species of tree and was on the level 25 mg/kg. Significant differences were observed in the concentration of Cd. The highest content of this metal was observed in the case of oak (29 mg/kg), afterwards in birch (13 mg/kg), coniferous tree (9 mg/kg). Significantly lower level was determined in acacia, ash tree and apple tree (2 mg/kg). The concentration of Hg was low in each wood ash – this element volatilized during combustion process. Also, the level of micronutrients was compared. The highest level of Fe contained coniferous trees (1.5 %), lower level was present in deciduous tree (0.4-0.6 %). The highest level of Mn was found in birch and acacia (1.6 %) as well as in oak (1.1 %). Lower level (0.2-0.4 %) was found in ash and apple tree as well as in coniferous tree. Ash and apple tree contained significantly higher levels of Zn (4800 and 5600 mg/kg, respectively). Coniferous tree and birch contained 3000 mg/kg. Oak and acacia 500 mg/kg. Very high level of Cu was found in coniferous tree (500 mg/kg). Deciduous trees, including oak, birch, acacia and ash tree contained 170 mg/kg. Significantly lower level was present in apple tree. The highest level of B was found in apple tree (800 mg/kg). The level of this element in wood ash from other trees was similar – on the level of 500 mg/kg. Birch ash was the richest in cobalt (24 m/kg), afterwards ash from coniferous tree (13 mg/kg), oak, acacia and ash tree (7 mg/kg). The concentration in apple tree was below 5 mg/kg. The highest concentrations of Mo was present in wood ash from oak, acacia and coniferous tree (above 3 mg/kg), lower levels were found in birch, ash tree and apple tree (below 2 mg/kg) [126].

Also, among wood ashes originating from different geographical regions differences in the elemental composition were encountered. For instance, the concentration of Pb in oak varied from 14 mg/kg (Silesia, Poland) to 87 mg/kg (Lower Silesia). The level of Cd from 1.4 mg/kg (Lesser Poland) to 52 mg/kg (Silesia). The concentration of As from 21 mg/kg in Lower Silesia and Central Poland to 40 mg/kg in Silesia. The level of Cr varied from 3 mg/kg (Greater Poland) to 76 mg/kg (Central Poland), Sn from 0.3 (Lower Silesia) to 75 mg/kg (Lesser Poland). It could be stated out that wood ash is a good indicator of environmental pollution. It is possible to elaborate a map of environmental pollution on the basis of analysis of wood ash from the same tree species from geographically different regions [127].

Agricultural studies showed that wood ash can be also used as soil amendment and deacidifying agent. It was found that can be used as the source of potassium and also neutralizing agent in the production of fertilizers. The fertilizer components present in wood ash have relatively low solubility due to alkaline pH and it could be stated out that it can act as slow release fertilizer. All the components are released slowly: macronutrients, micronutrients and also toxic metals. Toxic metals present in wood ash possess low solubility and thus low bioavailability. Moreover, wood ash can also be used to decrease bioavailability of toxic metals present in soil due to alkalization properties and thus will reduce the effects of environmental pollution [124, 128-132].

**Table 6. The comparison of the composition of wood ash and soil (a), extraction efficiency (b) and bioavailability (c) of elements from ash, soil and soil supplemented with ash [130]**

**(a)**

| Macronutrients and micronutrients the level of which was higher in wood ash when comparing with soil | Elements the level of which was similar in ash and in soil | Toxic elements the level of which was lower in ash when comparing with soil |
|---|---|---|
| Macronutrients:<br>K (15 times higher content in ash than in soil), Ca (13 times), Mg (11 times), P (10 times), Na (2.5 times)<br>Micronutrients: Cd, Pb, Bi, Ni, Tl. | Cr, Nb, La, Ti, Al, Pd, Ga Fe, Be, V, Sb, Ge, Sn, Mo | Hg (91 times), Zr (3 times), As (3 times), Ag (3 times), Ce (2.5 times), U (2 times) |

**(b)**

| The extraction efficiency of the following elements increased after addition of wood ash to soil | The extraction efficiency of the following elements reached maximum at intermediate level of ash supplementation (5-10 %) | The extraction of the following elements decreased with increase of ash supplementation | The extraction efficiency of the following elements did not change after addition of wood ash to soil |
|---|---|---|---|
| K, Sb, Mo, Rb, Ge, B, V, Ga, Na, Cr, U | Al, Cd, Zr, Ni, Nb, Co, Zn, Mn, Pb, La, Ce, Be, Rh | Ca, Mg, P, Tl | As, Cu, Fe, Ti, Bi, Sn |

**(c)**

| Availability of the following, elements increased with the increase of the level of ash supplementation | Availability of the following, elements decreased with the increase of the level of ash supplementation | Availability of the following, elements did not changed with the increase of the level of ash supplementation | The following elements were available from soil (extraction efficiency was higher than 5 %) | The following elements were available from ash (the level of supplementation 5 %) |
|---|---|---|---|---|
| Co, Zr, U, Sb, V, Cu, Ce, La, Ge, Ti, Nb, Fe, Sn, Ga, As, Pb, Cd, Pd, Ni, Mn, Zn | Rb, Ti, Cr, Al, Ca, P, Mg, Bi | Mo, K, B | Mo, K, Sb, V, Ti, Rb, Cr | Mo, K, Rb |

Bioavailability of nutrients and toxic elements was assessed. It was found that bioavailability of all metals, except potassium was very low i.e. Cd <2 %, Pb<1 %. The supplementation of ash into soil raises pH and this greatly reduces bioavailability. For this reason, ash can be considered as slow release fertilizer. The other advantage is that supplementation of wood ash reduces solubility of macronutrients, micronutrients and toxic metals and this reduces elution of these elements to ground waters (Table 6). The applicability of wood ash as soil supplement was confirmed by germination tests. It was found that the mostly advantageous is to supplement 5 % wood ash to soil [130]. The effect of supplementation with ash is increase of pH and this causes retention (immobilization) of macronutrients, micronutrients as well as toxic metals in soil and this reduces the intensity of their cycling in the environment and thus causes less damage [130].

## 6.4. Identification of Metal Cycles in Trophic Chain

The aim of studies that included balances of metals cycles in animal breeding was the assessment of bioavailability of elements supplemented in livestock feeds in order to assess distribution of microelements as well as to point out sites of potential accumulation of toxic metals in animals' organs and products. The level of macroelements, microelements and ultratrace elements was determined in feeds, physiological fluids, droppings, muscles, liver, kidneys, lungs and skin, as well as in livestock products, such as milk and eggs. Also, the composition of animal hair and feather was analyzed [113].

Bioindicative value posses cow milk, poultry eggs, some tissues and organs of livestock. Ducks, geese and hens are housed in many countries in extensive (free range) system. These animals are omnivorous and for this reason can be considered as a good biomonitor of environmental pollution [113].

The content of microelements and trace elements has recently been more widely studied, in particular in industrialized and polluted regions, milk is considered as a good biomonitor of pollution of agricultural environment. The concentration of the major mineral components in milk (Ca, P, K, Na, Mg,Cl, S) does not fluctuate. The total concentration of these elements in milk is 0.6 %. Also, the concentration of some trace elements is relatively constant, including Ni (0.03 mg/L), Fe (0.5 mg/L), Si (1.4 mg/L). The level of As, Cd, Hg and Pb strongly depends on environmental pollution and the level of these toxic metals should be monitored in milk [117] (Table 7).

The concentration of 38 microelements and trace elements in milk from cows kept in farms located in industrialized region was compared with milk from an ecological farm. The differences in the concentration of the following microelements and trace elements were encountered: I, Al, Rb, Ti, Mn, Ga, Se, Ge, Co, Cs, W, Th [117].

**Table 7. Permissible levels of toxic metals in some animal products [mg/kg or mg/L] [117]**

| Animal product | Pb | Cd | Hg |
|---|---|---|---|
| Milk and milk products | 0.02 | 0.01 | 0.01 |
| Meat of mammals and poultry | 0.10 | 0.05 | 0.02 |
| Animal fat | 0.10 | 0.02 | 0.01 |
| Liver | 0.50 | 0.5 | 0.05 |
| Kidney | 0.50 | 1.00 | 0.05 |
| Hen eggs | 0.30 | 0.05 | 0.02 |

Animal products from industrialized region influenced by the activity of copper smelter and refinery were analyzed for the content of toxic metals and their concentration was compared with the permissible limits. It was found that the concentration of Pb in milk exceeded the allowed standards. Average reported value was 0.03 mg/kg and the maximum value was 0.08 mg/kg. The concentration of Pb in blood was 0.04 mg/kg (maximally 0.2

mg/kg). Positive statistically significant correlation between the concentration of Pb in milk and blood was denoted. Also the content of Hg in milk was significantly and positively correlated with Hg level in blood. The highest concentration of Hg was denoted in liver (0.005 mg/kg), lungs (0.04 mg/kg) and muscles (0.003 mg/kg). The concentration of Hg in eggs from laying hens housed in a farm was 8 times lower when compared with the free range system [120].

The concentration of 37 trace elements in 25 samples of raw milk and full blood of cow milk was investigated with the use of ICP-MS technique (Table 8). Correlation coefficients between 18 elements in milk and blood were determined. Statistically significant correlation between the concentration of a given element in blood and milk were observed for the following metals: Rb, Ti, Mn, Ga, Ni, Ge, Mo, Sb. No statistically significant correlation between Cr, As and Se level in blood and milk was denoted. It was found that the concentration of the following elements was significantly higher in blood when compared to milk: Cu, Ga, Ge, Ti. Lower concentration of Ag, Th and Be was found in blood. The concentration of Cr, Se and Sb was similar in blood as it was in milk. Among the studied microelements and trace elements, milk contained the highest level of Zn (3.3 mg/L), afterwards I (0.6 mg/L), Rb (0.6 mg/L), Al (0.3 mg/L) and Ba (0.2 mg/L). Full blood contained Zn (2.6 mg/L), Cu (0.6 mg/L), I (0.4 mg/L) and Rb (0.16 mg/L). In milk, high variability of Cr, Cu, Ge, Mo, Rb and Se was observed. Similarly in blood: Al, Be, Cr, Mn, Mo, Ni and Rb. Microelements and trace elements that were found in milk and blood originated from soil, feed, water and air. Of course, the concentration of a given element in animal tissues depends on its bioavailability and on inter-element interactions, including synergisms and antagonisms. However, the mechanism of transport of minerals to milk has not been well understood yet. It is thought to be complex, consisting of simple diffusion, active or facilitated transport [115].

Eggs, similarly as milk, muscles and organs of animals, are indicator of environmental pollution. In particular, in the case of eggs from waterfowl birds (Table 8), it can be an indicator of water pollution. Not only the content of a given metal in the whole egg is considered, but also eggshell itself, as well as egg albumen and yolk possess bioindicatory values. Eggshells can be used first of all in biomonitoring of As and Cu, egg yolk to assess environmental pollution with As and Cd and egg albumen to bioindicate Cd, Hg, Pb and Cu. This theory was confirmed in industrialized regions and in eggs from so-called ecological farms. Eggs from industrialized area contained significantly more As, Hg and Cd and more Pb and Cu when compared with non-industrialized regions [113].

Physical and chemical eggs quality parameters from industrialized and non-industrialized regions from young (< 1 year old) ducks and geese from several farms were investigated. It was found that morphological characteristics were similar, including average mass, share of shell, albumen and yolk that were typical. Toxicological studies however showed diversified concentration of metals in various morphological parts as well as in whole eggs. Statistically significant differences were observed between polluted and ecological region. It was found that the highest concentration of As was present in geese eggshell (2.4 mg/kg) and the lowest in ducks egg albumen (0.003 mg/kg). The concentration of As in duck and geese eggs (whole eggs and in its elements) was up to 8 times higher in industrialized region. Geese eggs contained 25 % more As than ducks. Arsenic is easily absorbed by geese but is also easily and quickly excreted in droppings. The acceptable concentration of As in egg content (egg yolk and albumen) is 0.20 mg/kg. In industrialized regions was evaluated as 0.13 mg/kg and eggs from an ecological farm contained 0.07 mg/kg [113].

**Table 8. The average content of trace elements in cows milk and blood (μg/l) [126] and in eggs of waterfowl birds (μg/kg) [112]**

| Element | Cows | | Waterfowl birds | | | |
|---|---|---|---|---|---|---|
| | | | Geese | | Ducks | |
| | Raw milk | Full blood | Industrialized region | Agricultural region | Industrialized region | Agricultural region |
| Al | 276.5 | 66.8 | | | | |
| As | 32.8 | 24.6 | 14.6 | 8.2 | 75.1 | 32 |
| Ba | 217.2 | 69.8 | 1413 | 737 | 602 | 804 |
| Cd | 4.5 | 4.9 | 26.7 | 4.4 | 10.5 | 2.6 |
| Co | 6.5 | 4.6 | | | | |
| Cr | 84.9 | 77.6 | 23 | 4 | 5 | 8 |
| Cu | 79.6 | 620.8 | 596 | 510 | 797 | 653 |
| Ga | 64.5 | 104.3 | | | | |
| Ge | 28.2 | 70.8 | | | | |
| Hg | 1.56 | 2.03 | 55.1 | 10.5 | 137 | 39.7 |
| I | 559.6 | 387.9 | | | | |
| Mn | 74.3 | 27.7 | 240 | 122 | 308 | 223 |
| Mo | 11.5 | 5.4 | | | | |
| Ni | 61.0 | 13.0 | | | | |
| Pb | 17.4 | 31.2 | 113 | 20.6 | 75.1 | 42.9 |
| Rb | 579.0 | 163.7 | | | | |
| Sb | 5.5 | 5.3 | | | | |
| Se | 60.5 | 51.8 | 91 | 57 | 47 | 52 |
| Ti | 78.8 | 156.2 | | | | |
| V | 84.5 | 70.6 | | | | |
| Zn | 3307 | 2575 | 8280 | 9230 | 9777 | 8993 |

Waterfowl eggs are also good biomonitors of cadmium pollution. Similarly as in the case of As, geese eggshell contained the highest levels of this element (0.045 mg/kg) and the lowest level was present in albumen of ducks eggs. It was found that the concentration of Cd in eggs from industrialized region was up to 13 times higher when compared with ecological. Geese eggs contained averagely 5 times higher levels of Cd when compared with duck eggs. Similar dependencies were found in the case of Pb, Cu and Zn. It was found that eggshells are better biomonitor of environmental pollution than egg content, since the level of metals is several times higher [113]. Moreover, it was found that the level of toxic metals was higher in animals and their products kept in free range system when compared with battery system.

One issue is to measure the level of toxic metals in biological tissues as in biomonitors and the other is to elaborate methods that will reduce accumulation of these pollutants. A method that would enable to reduce bioaccumulation of toxic metals (including Pb, Cd and Hg) in body tissues and animal products was elaborated. The preventive detoxification preparation can be applied in cows and poultry breeding. Feeding animals with the prepared supplements inhibited transfer of toxic metals to animal products, including meat, eggs and milk. The supplement consisted of humic materials and aluminosilicates that have complexing, sorptive and ion-exchange properties and are capable to bind cations, such as Cd, Hg and Pb. The supplement consisted mainly of humic materials (humodetrynite – brown coal) and aluminosilicates (bentonite, baidelite). The preparation was also supplemented with selenium yeasts, plant oil, mineral additives (including chalk, phosphates and dolomites). The preparation is intended to be used in poultry and cows feeding. It was found that feeding livestock with humic-aluminosilicate preparation resulted in 23 % decrease Cd concentration and 33 % decrease of Hg concentration in milk and 13 % and 10 %, respectively, in whole blood of cows. In poultry, 29 % reduction of Cd concentration in muscles was observed, 50 % in liver and Pb decreased 34 % and 70 %, respectively. Also, the concentration of Pb and Cd in eggs was significantly lower [120].

**Table 9. The elemental composition of hens bones and feather [131]**

a) macroelements, %

| Macroelement | Bones | Feather |
|---|---|---|
| N | 4,07 | 10,4 |
| $P_2O_5$ | 18,5 | 0,457 |
| $K_2O$ | 0,911 | 0,331 |

b) microelements and toxic elements, mg/kg

| Element | Bones | Feather |
|---|---|---|
| Al | 28,4 | 119 |
| As | 1,06 | 0,68 |
| B | 0,720 | ≤ 0, 15 |
| Be | 0,333 | ≤ 0,002 |
| Cd | 0,100 | ≤ 0,009 |
| Co | 0,005 | 0,0009 |
| Cr | 3,31 | 0,78 |
| Cu | ≤ 0, 37 | 6,65 |
| Fe | 62,4 | 227 |
| Hg | 0,0064 | 0,0204 |
| Mg | 1048 | 182 |
| Mn | 1,73 | 10,2 |
| Mo | 0,26 | ≤ 0, 01 |
| Ni | ≤ 0,004 | ≤ 0, 04 |
| Pb | 0,286 | 4,08 |
| Ti | 2,41 | 3,38 |
| Tl | ≤ 0,0009 | ≤ 0,0009 |
| V | ≤ 0,02 | 0,48 |
| Zn | 53,5 | 105 |

Methods of animals detoxification should be based on either metal-binding properties of feed supplements or on antagonistic interactions between elements. Such an example of an antagonism is selenium and mercury. Mercury is present in milk products in the following concentrations: in milk 1.7 μg/L, muscles 1.5 μg/kg and in eggs 2.0 μg/kg. Selenium is present in the concentration 34 μg/L, 43 g/kg and 22 μg/kg, respectively. In an organism of animals and human selenides of toxic metals are formed and this is one of the method of their rendering harmless. Mercury selenide is insoluble and therefore if this compound is formed, mercury does not become incorporated into metabolic processes [116].

Also, hens feather can be considered as a good bioindicator of environmental pollution, similarly as human hair. Toxic metals are also deposited in bones. The composition of bones is also related with the level of environmental pollution, but the method of determination is invasive. Table 9 presents the elemental composition of hens heather and bones [131].

## 7. Using Biosorption and Bioaccumulation Processes in Wastewater Treatment

There are different classes of biosorbents (Figure 11). Low-cost sorbents include wastes, by-products and useless materials as well as naturally collected biomass. Sorption capacity of such sorbents is not high, but since the materials are practically of no cost, it is possible to work at high concentration of sorbent. Low-cost sorbents include use-less materials of plant and animal origin. The class of biosorbents of plant origin includes the biomass of terrestrial plants (leaves [132], grass and straw [133]) that are easy to harvest and are abundant in the environment and the biomass of aquatic plants (duckweed [134-135] and crystalworth [135-136]) and macroalgae (including *Cladophora* [137-138]). Biosorbents of animal origin are i.e. grinded waste eggshells [139] or animal bones [140].

High cost sorbents are materials specially propagated for sorption purposes. Since the costs are high, there will be also special requirements towards these materials. High-cost sorbents should simultaneously possess high sorption affinity and capacity (sorbent *1* on Figure 8) , so very low concentrations of these materials will bind high quantities of metals yielding simultaneously very low final concentration of metal ions (below 1 mg/L). For this reason, high-cost sorbent should simultaneously possess very high sorption capacity and affinity. The subsequent desired characteristic is regenerability of the sorbent, so the material can be used in several biosorption cycles. These criteria fulfill, e.g. microalgae *Spirulina sp.* that can incorporate up to 20 % mass metal ions at very low residual concentration of metal ions in solution [141].

Biosorption can be performed in an ordinary continuous stirred tank reactor (Figure 12). Since the rate of the process is quick, the residence time can be short (0.5-1 h). This enables to treat large volumes of wastewater in a reactor of small dimensions. It is advantageous to simultaneously recover and recycle the biomass. Biosorption can also be carried out similarly as typical ion-exchange, in columns. Biosorbent can be either directly packed or immobilized in a matrix. Working on at least two columns enables to switch between sorption and desorption (sorbent regeneration) process without the necessity of stopping the process. The possibility of sorbent regeneration is an economic solution. Very significant is to select the proper filtration device that enables to work at high fluxes over long period of time.

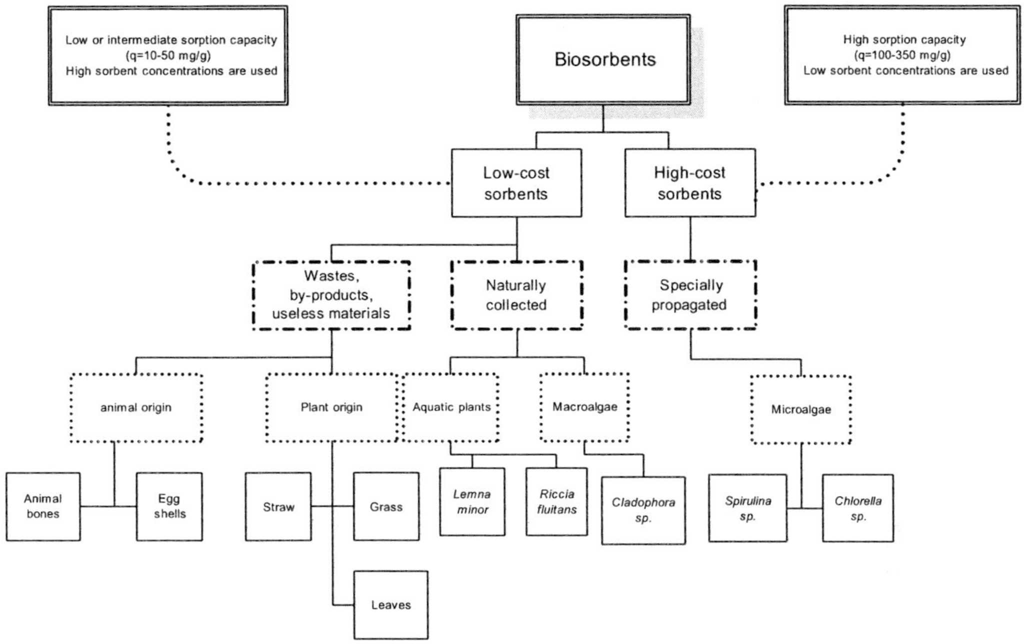


Figure 11. Low-cost and high-cost biosorbents.

The efficiency of the overall biosorption process can be examined through the evaluation of concentration factor between the metal concentrated in the biomass related to effluent or by the level of metal in eluent stream related to the concentration in raw wastewater. If we consider the first definition of concentration factor, the values reached are 10-1000, if we relate to the second definition the values would be 1-100. This means that starting i.e. with the initial concentration of metal ions 21 mg/L in raw wastewater, by using the quantity of sorbent 1 g/L, and with the concentration of metal ions in treated effluent 1 mg/L the metal would be concentrated almost 1000 times in the biomass. If we would like to furtherly regenerate the biomass, with e.g. acidic solution using 10 ml of eluent to wash 1 g biomass (assuming 100 % elution efficiency), we obtain solution containing 2000 mg/L metal ions, which means that the metal was concentrated almost 100 times when comparing it with raw wastewater. The solution of eluent containing metal ions can be furtherly treated with a conventional method, such as precipitation, ion exchange, electrodialysis, membrane separations.

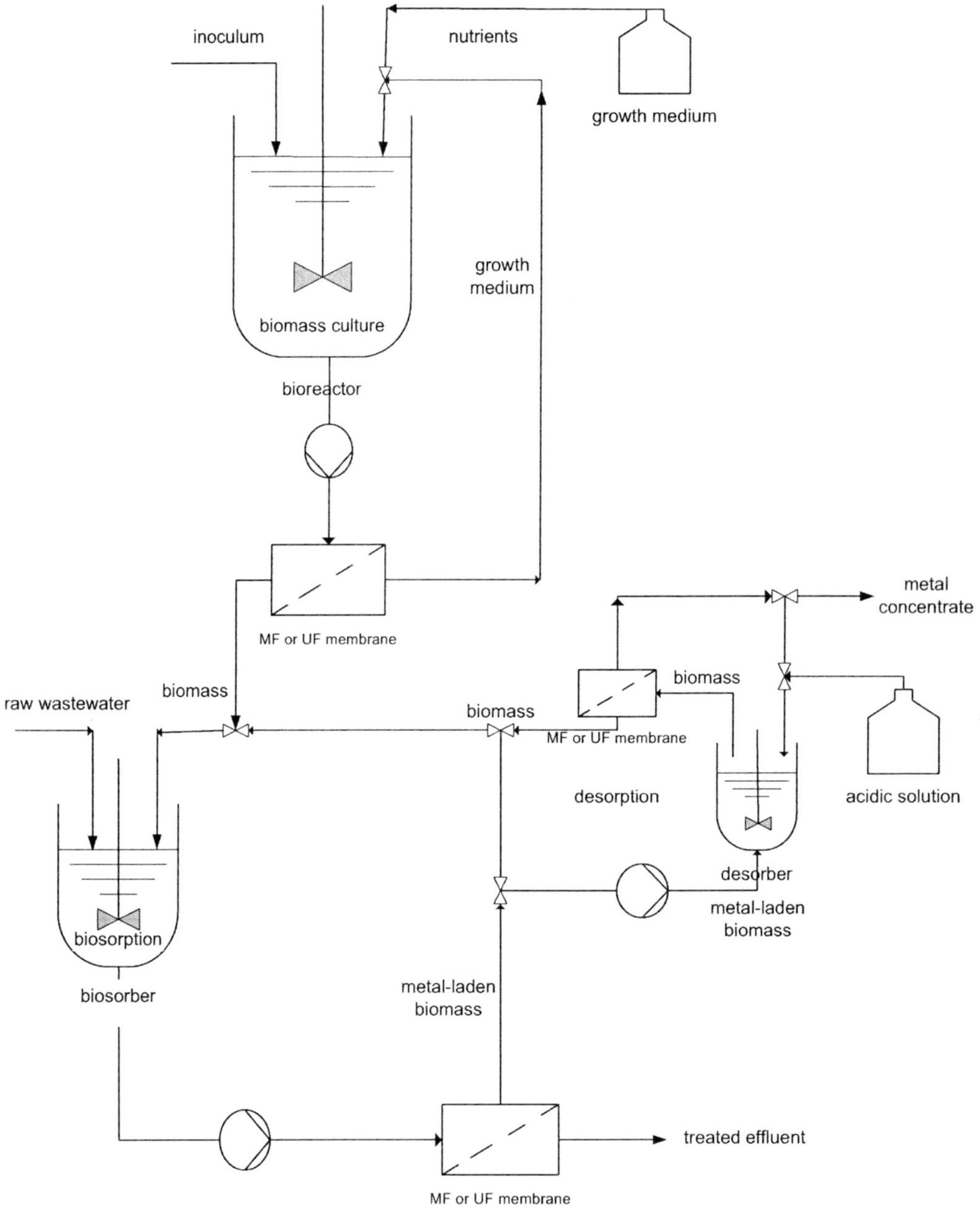


Figure 12. Schematic diagram of biosorption process.

If bioaccumulation is used as a method of wastewater treatment, it requires simpler installation when compared with biosorption, since both growth of cells and binding of metal ions occur simultaneously: in the same place and time, in a single bioreactor (Figure 13). In bioaccumulation it is possible to partially recover the biomass – metal ions bound passively with the cellular surface can be desorbed from the biomass, similarly as in biosorption. The disadvantage of the process is that the residence time in bioaccumulation is longer than in biosorption and that nutrients are required to sustain metabolic processes and cellular growth. A problem that might occur is that the presence of pollutants can inhibit cellular growth and thus biomass productivity.

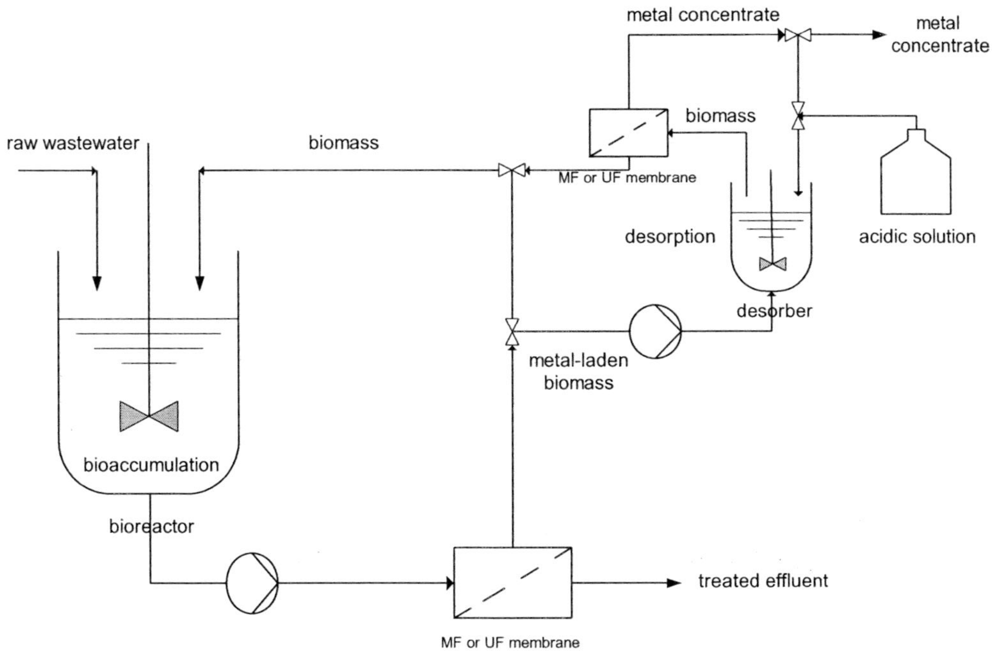


Figure 13. Schematic diagram of bioaccumulation process.

## 7.1. Wastewater Treatment by Low-cost Biosorbents

### *Eggshells*

Recent findings showed that eggshells can be considered as low cost biological sorbent. Waste eggshells are significant environmental problem and it would be advantageous to find an application for this material. Only in USA, 120 000 Mg of waste eggshells is annually generated and disposed to landfills. Eggshells mainly consist of calcium carbonate (85-95 %), magnesium carbonate (1.4 %), phosphorus (0.8 %) and organic matter. This material contains also traces of Na, K, Zn, Mn, Fe and Cu. It has been postulated that, similarly as calcite and calcareous soil, which are of similar chemical nature, eggshells should be good biosorbents with the dominating mechanism of biosorption as ion-exchange. It was found that in fact biosorption capacity of eggshells was high: 20-160 mg/g depending on conditions. The effect of process parameters: temperature (20-60 $^{o}$C), pH (3-5) and sorbate concentration (100-300 mg/L) on process kinetics was investigated. It was found that the equilibrium of the process was reached after 1 hour. Increase of pH resulted in increased sorption capacity and sorption rate. The highest differences were observed between pH 3 and 4. At temperature 20-40 $^{o}$C sorption capacity was not influenced by temperature, at higher temperature the efficiency decreased. The effect of temperature on sorption rate was adverse. On the basis of kinetic study, the proposed parameters for sorption were pH 5, temperature 20 °C, sorbate concentration 300 mg/L. This yields biosorbent capacity 56 mg/g at sorbent concentration 5 g/L. Also, the effect of pH, temperature and sorbent concentration on biosorption equilibrium was investigated. Similar dependencies as in kinetic experiments when considering process efficiency were determined. Increased sorbent concentration favored decreased sorption capacity. It was found that the best process parameters for biosorption by eggshells are pH 5 and 20 °C [139].

The experimental results were described with pseudo-second order kinetic equation, the model that is the most frequently used in modeling of biosorption kinetics. It was found that the process equilibrium strongly depends on sorbent concentration. A new mathematical model being a modification of Langmuir equation was proposed:

$$q_{eq}(C_{eq}, C_S) = q_m \left( \frac{bC_{eq}}{1 + bC_{eq}} \right) \left( 1 - \frac{C_S}{d + C_S} \right) \tag{3}$$

The first part of the equation is typical Langmuir equation, the second is an empirical dependence. If $C_S \to 0$, $\left( 1 - \frac{C_S}{d + C_S} \right) \to 1$ and $q_{eq} \to q_m \left( \frac{bC_{eq}}{1 + bC_{eq}} \right)$. At high sorbent concentration, the value of $\left( 1 - \frac{C_S}{d + C_S} \right)$ will be low, as well as the value of the overall sorption capacity. At a given sorbent concentration, the model would be alike Langmuir equation. Model parameters at various process conditions were determined. The model fitted the experimental data reasonably well [139].

On the basis of performed experiments it was concluded that it is possible to remove high quantities of metal ions with eggshells. It was possible to decrease the concentration of Cr(III) ions below the acceptable level (0.5 mg/L) at pH 5, 40 °C, initial concentration of Cr(III) ions 100 mg/L and sorbent concentration 15 g/L [139].

Since ion-exchange was hypothesized to be the mechanism of biosorption by eggshells, cation exchange capacity was determined by potentiometric titration and evaluated as 4.4 meq/g. It was found that cations are bound to carbonate groups. The value of sorption capacity was two times higher than of activated carbon, similar as of animal bones [140] and two times lower than for especially propagated biosorbent (*Spirulina sp.*) [141]. The problem however that should be resolved is sterilization of eggshells. But, since the mechanism is ion-exchange, it should be possible to recover the sorbent and use it several times [139].

### *Animal Bones*

Animal bones are another waste product that requires utilization. The material consists of 65-70 % inorganic substances including mainly hydroxyapatite $Ca_{10}(PO_4)_6(OH)_2$ and organic matter, mainly fibrous protein collagen. The postulated mechanism of sorption by hydroxyapatite was ion-exchange with $Ca^{2+}$ ions. The effect of sorption parameters on biosorption rate and efficiency was investigated. It was found that at initial concentration of

Cr(III) ions 200 mg/L and sorbent concentration 1-15 g/L, it was possible to remove 40-99 % of Cr(III) ions from the solution. It was found that increasing pH and temperature resulted in increase of sorption capacity. pH was found to be a very important process parameter since influenced protonation of metal binding sites, affected calcium sulphate solubility and metal speciation in the solution. Sorbent concentration had the influence on equilibrium pH and phosphates solubility. But this effect is not so strong as in the case of eggshells. Sorption capacity strongly increases with pH increase. Also, increasing temperature results in higher efficiency of the process up to 50 °C. Above this temperature, sorbent denaturation occurs. The rate of biosorption by animal bones is slightly slower than by eggshells. The equilibrium is reached after 2.5 h. The highest efficiency of the process was reached at pH and 50 °C (over 190 mg/g). At 20 °C sorption capacity was 60 mg/g [140].

***Biomass of Terrestrial Plants – Leaves, Grass and Wheat Straw***

Sorption of Cr(III) ions by dry leaves of *Forsythia intermediata* [132] grass and wheat straw [133] was investigated. The maximum experimentally determined biosorption capacity of all the sorbents was similar, within the range 5.5-25 mg/g depending on the process parameters (pH and temperature). The values of sorption capacity were not high, but since these materials are commonly abundant and practically of no cost, it is possible to improve wastewater treatment efficiency by increasing the concentration of the sorbent. The equilibrium was reached after 40-60 min. It was found that similarly as in the case of other sorbents, sorption capacity depended on temperature and pH. The best performance of this sorbent was reached at 50 °C, pH 5. Pseudo-second order model was used to describe kinetics of the process carried out by all the sorbents of plant origin. Langmuir equation was used in equilibrium modelling of leaves and grass and Freundlich equation in the case of wheat straw. The dominating mechanism of all the sorbents was mainly physical adsorption with maximum 20 % contribution of ion-exchange. Since the sorbents possess high heating value (20 kJ/g) and low ash content (7-9 %), after the combustion of metal-laden biomass it is possible to obtain both energy and ash being a concentration of metal [132-133].

## 7.2. Wastewater Treatment with High-cost Biosorbents

***Algae Spirulina sp.***

Microalgae *Spirulina sp.* were found to have very high sorption capacity – up to 200 mg/g. The kinetics of the process was described with reversible first-order kinetic equation and the process equilibrium with Langmuir model. It was found that cells cultivated at various process conditions had different sorption capacity [142]. The highest metal ions binding capacity possessed cells grown photoautotrophically (7 meq/g), afterwards heterotrophically and the lowest cells grown mixotrophically. Biosorption of three toxic metals to the biomass of this algae was carried out. It was found that cells bound the highest quantities of Cr(III) ions, afterwards Cu(II) and Cd(II). Adsorptive surface area of cells was low 0.3-1 $m^2/g$, however cation-exchange capacity of cells was very high (9 meq/g). For this reason, the mechanism of the process was identified as ion-exchange. It was found that cation-exchange capacity determined by potentiometric titration and biosorption capacity was directly proportional. Three groups capable of cation-exchange were identified on the cellular surface: carboxyl, phosphoryl and amino or hydroxyl group. The contribution of each group in biosorption was confirmed by chemical modification. It was found that cells with blocked groups exhibited lower biosorption capacity corresponding with the cellular concentration of the blocked group. Also, the possibility of desorption of bound ions from the biomass and subsequent reuse in another sorption cycle was investigated. The efficient sorbent that was identified was 0.1 M nitric acid. The application of 0.1 M acid solution assured almost complete desorption of bound cations from the biomass with the simultaneous restoration of sorption properties. Also, the effect of pH on biosorption performance was studied. This furtherly confirmed the participation of cation-exchange with functional groups in the biosorption process [141].

### 7.3. The Application of Bioaccumulation in Wastewater Treatment

Bioaccumulation is the property that can also be used in the removal of toxic metals from solutions. The method can find an application in treatment of industrial effluents. It is particularly advantageous to employ this treatment as the final, polishing step, since frequently the application of traditional methods does not make it possible to lower concentration of toxic metals below limits permissible by law. Since industrial effluents generally do not contain organic carbon source, it would be advantageous to employ photosynthetic organisms that can incorporate carbon to the biomass from carbon dioxide or carbonates. Bioaccumulation understood as wastewater treatment technique is the method of cultivation of cells in the presence of substance that is to be bioaccumulated. For this reason, only aquatic organisms can be employed in bioaccumulation-based wastewater treatment processes. The potential application in such methods can find microalgae – photosynthetic and aquatic organisms that posses high metal binding capacities, such as *Spirulina sp.* [143]. The alga was used in the treatment of effluents from copper smelter and refinery that contained toxic metal ions in trace concentrations (however in concentrations slightly exceeding the acceptable levels), contained ammonia nitrogen and phosphorus and carbon only in the form of carbonates. The studies showed that the treatment of this effluent via bioaccumulation by *Spirulina* was efficient and yielded effluent with the desired concentration of toxic elements (including Hg from 760 to 216 μg/L, cadmium from 3810 μg/L to 917 μg/L at initial concentration of cells 0.4 g/L. Also, the concentration of ammonia nitrogen (from 66 mg/L to 0.6 mg/L) and phosphorus was lowered [143]. The same effluent was also treated with natural zeolites [144]. Similar performance was achieved towards Hg ions, but the concentration of ammonia nitrogen has not decreased significantly in the process [144]. Concluding, bioaccumulation assures simultaneous removal of various pollutants, including toxic metal ions and nutrients (compounds of nitrogen and phosphorus).

## 8. Using Biosorption and Bioaccumulation in Integrated Processes

It is advantageous to combine biosorption and bioaccumulation into integrated production processes. It is possible to perform wastewater treatment with the simultaneous production of biological feed additives, if wastewaters do not contain high levels of toxic metals, such as Hg, Pb and Cd (Figure 14). Such an approach enables to manage waste sludges, which is metals-laden biomass, from wastewater treatment. Recently it was found that livestock feeds should be supplemented with deficient microelements. Therefore there is the need to elaborate a method of production of cheap and highly bioavailable mineral feed additives. Such a possibility offers the application of biological sludges from treatment of wastewater by biosorption and bioaccumulation techniques. Such an approach can solve several problems: wastewater treatment, utilization of waste biomass, production of commercially valuable product (mineral feed supplement) (Figure 14 a). If the treated wastewater contains toxic ions that are not microelements (Hg, Pb, Cd), the metals removed can be concentrated either in liquid-state form with the regeneration of the biomass (Figure 14 b) by elution or in solid-state form as the result of biomass ashing. The obtained biomass concentrates (either liquid or solid) can be used in the production processes of a given metal if it is possible (metal recovery) or can be utilized with conventional methods.

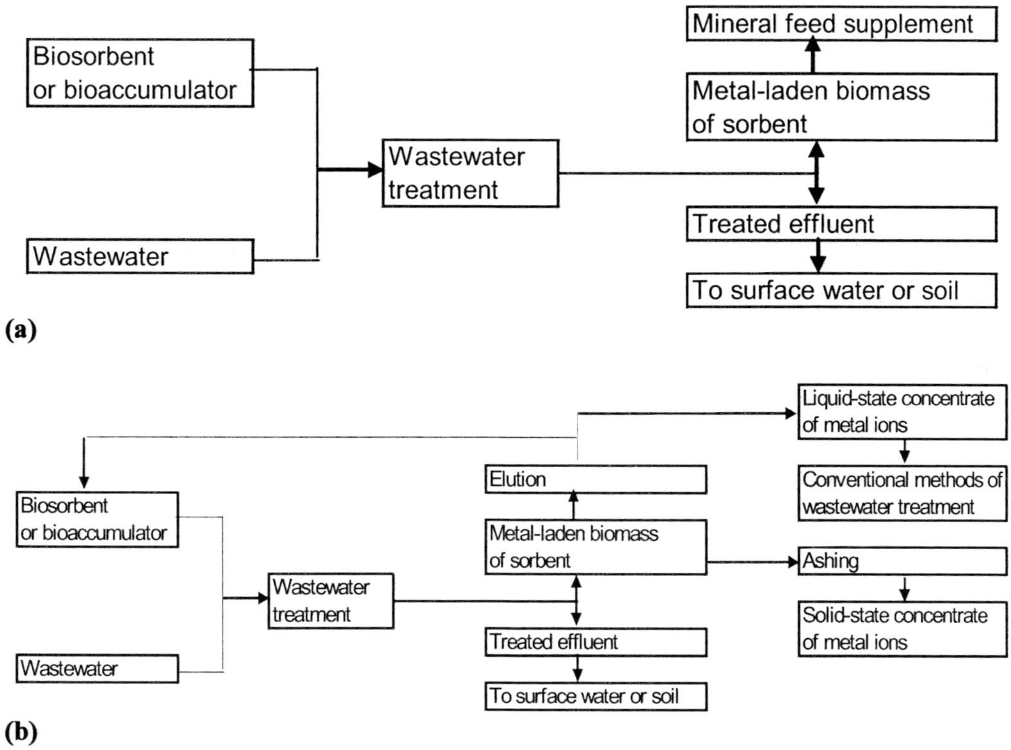


Figure 14. Integrated processes using biosorption or bioaccumulation processes; a) low level of Cd, Hg and Pb in treated wastewater, b) high levels of Cd, Hg and Pb in treated wastewater.

## 8.1 Environmental Biotechnology of Microalgae

Environmental biotechnology of microalgae is a new discipline that proposes integrated processes with the use of microalgae, including methods associated with environmental protection. In integrated processes, the biomass is cultivated in effluents containing biogenic compounds (C, N, P) with the use of $CO_2$ (e.g. from flue gases) and the energy of sun (Figure 15). The biomass produced can be used as animal feed, biosorbent, biofertilizer and in the production of chemicals (i.e. colorants). This enables to combine the processes of wastewater and flue gases treatment with the simultaneous production of low-cost product of commercial value [145].

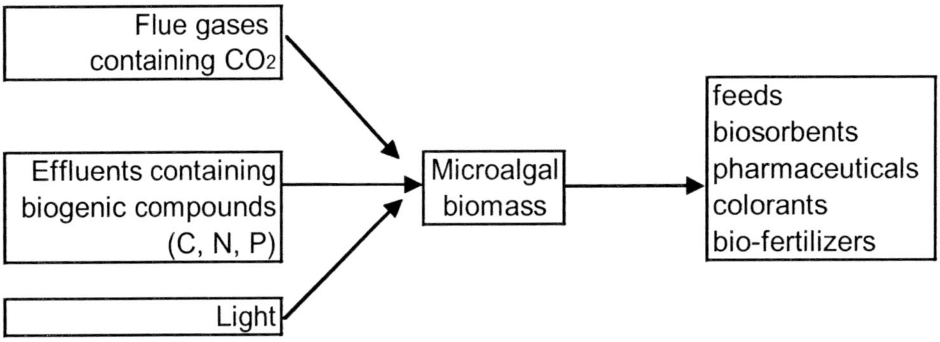


Figure 15. The concept of Environmental Biotechnology of Microalgae – integrated process of biomass culture and wastewater and flue gases treatment [145].

When designing such an integrated process, it is necessary to investigate the growth of microalgae and the effect of pollutants on growth kinetics and statics. Microalgae are organisms that are capable of using various sources of carbon (either inorganic or organic) and energy (either sunlight energy or energy stored in organic compounds). If algae use carbon dioxide and the energy of sun they grow photoautotrophically, if utilize organic compounds as the source of both carbon and energy – grow heterotrophically, if the observed growth is the sum of both processes – grow mixotrophically. A growth model in each metabolic mode was proposed [146] that consisted of the part that described the effect of organic carbon source (i.e. glucose) with typical Monod equation and the effect of light intensity with linear discontinuous model. The contribution of each constituting metabolism on the overall observed growth was investigated [147]. The possibility of microalgae to grow with the use of various metabolisms and different growth substrates makes their growth very flexible and enables them to adjust growth requirements to growth conditions. Also, the method of cells cultivation is significant since, as it was found, affects biosorption properties of cells [142].

The effect of growth parameters (light intensity, initial concentration of glucose) on metal ions binding capacity was investigated. It was found that cells cultivated at high light intensity and at low concentration of organic carbon source showed the highest metal binding capacity. Biomass cultivated with a single metabolism showed higher sorption capacity when compared with mixed metabolism, e.g. autotrophically grown cells possess two times higher capacity than mixotrophic biomass. It was found that with the increase of glucose in the growth medium, sorption capacity decreased in heterotrophic and mixotrophic metabolism. Increased light intensity favors synthesis of cells with higher biosorption capacity, however does not affect biosorption capacity of mixotrophically grown biomass. However, each type of *Spirulina* biomass possessed very high metal binding capacity (200 mg/g of Cr(III) ions for photoautotrophic cells). Since biosorption properties depend on growth conditions, when designing biosorption unit, significant would be not only high biomass productivity, but also high biosorptive quality of the produced biomass. It is necessary to take into consideration that at high glucose concentration, productivity of the biomass is higher, but sorption capacity would be lower and also biomass yield from substrate would be low [142].

There are some problems associated with photosynthetic cultivation of the biomass for biosorption purposes. While many classical physiological studies focus on the response of an organism to a single parameter, physical, chemical and biological parameters interact and vary independently in nature, thereby producing non-linear effects. Optimal conditions seldom occur in nature and thus physical factors, such as temperature, light or availability of nutrients and water regulate performance of species. Adaptation resulted in that populations tend to evolve characteristics that make them well suited to the particular conditions experienced over extended period of time [148].

In natural lakes, if the concentration of cells is high (exceeds ca. 1 g/L) the energy of sun would be available only to cells present in the top layer of the lake surface. The remaining biomass would experience deficiency of sun energy. Thus, cells tend to achieve low, but constant cells concentration, so the species can survive generations. In this case, self-control of cells concentration would be an adaptation strategy. While this characteristic is highly advantageous in the natural environment, it would cause severe problems in large scale biomass cultivation on commercial scale [148].

## 8.2. Using Biosorption and Bioaccumulation in the Production of Feed Supplements with Microelements for Livestock

Farm and industrial feeds do not meet livestock demands for minerals, in particular for microelements. For this reason, recently livestock feed has been supplemented with dietary additives containing these constituents. Impoverishments of cultivated land in mineral components as a result of intense agricultural production, caused that plants became deficient in micronutrients. This yielded mineral deficiencies in livestock that were manifested by decreased growth rate and health problems. Currently, minerals are supplied to livestock diet in the cheapest form – as inorganic salts. This can cause health problems to animals. Also, supplying minerals to livestock as inorganic salts is of environmental concern, since this form is characterized with low bioavailability and such minerals are rather of transit character. Thus, we could expect uncontrolled, elevated discharge of microelements (such as Zn, Cr, Cu) that are also considered as water pollutants, into manure. For this reason, there is a need to search for a new generation of mineral feed supplements that would be highly bioavailable and that would show no toxic effects and that would be economically advantageous [149].

Since all biological materials possess metal binding properties (characteristics that has been used so far in wastewater treatment and bomonitoring), theoretically, all biomaterials can be used as microelements carrying materials in supplying minerals to livestock diet. This would greatly improve bioavailability since biologically bound minerals would be similar to the natural form in which they are present in feeds. Also, biological materials to which microelements are bound possess additional nutritional value (vitamins, polyunsaturated fatty acids, high protein contents). It was found that supplementing i.e. algae in livestock diet would improve egg yolk color [149].

Also, aquatic plants enriched with microelements by biosorption and bioaccumulation can be used as mineral feed supplement. Microelements important in livestock feeding (Mn,

Zn, Cu, Cr) were bound to the biomass of commonly worldwide available freshwater floating aquatic plant *Lemna minor* (duckweed) [134] and *Riccia fluitans* (liverworth) [135]. The plants are commonly used as valuable feed and possess capability to concentrate metal ions from solutions several fold (1000-100000 times). These plants are also used in hydrobotanical wastewater treatment plants. It is also possible to combine treatment of wastewaters containing i.e. Cu or Zn ions with either algae or aquatic plants with the simultaneous production of mineral feed supplements. Such an approach enables to simultaneously solve environmental problem (wastewaters) and to produce commercially valuable product. Since photosynthetic organisms are used, there is no need for effluents to contain organic carbon source [134-135].

Biosorption and bioaccumulation in single- and multi-metal system was compared. It was found that aquatic plants can bind up to 8 % microelements in their biomass. The equilibrium was described with Langmuir equation. The mechanism was found to be ion-exchange mainly with K, Ca, Mg. It was found that the highest performance was reached in single-ion system [134-135]. Table 10 lists model parameters for equilibrium of biosorption and bioaccumulation in single- and multi-metal system [134].

**Table 10. Langmuir Parameters for biosorption and bioaccumulation by both non-growing and growing cells of L. minor [134]**

| | Single-ion system | Multi-ion system | | | |
|---|---|---|---|---|---|
| | Cr | Cr | Cu | Mn | Zn |
| *Biosorption* | | | | | |
| $q_{max}$ (meqg$^{-1}$) | 3.40 | 1.40 | 0.738 | 0.364 | 0.514 |
| b (lmeq$^{-1}$) | 0. 298 | 4.98 | 1.33 | 3.2 | 2.66 |
| $R^2$ | 0.983 | 0.977 | 0.964 | 0.991 | 0.992 |
| *Bioaccumulation by non-growing cells* | | | | | |
| $q_{max}$ (meqg$^{-1}$) | 2.46 | 0.773 | 0.0869 | 0.287 | 0.301 |
| b (Lmeq$^{-1}$) | 0.267 | 5.11 | 37.6 | 2.53 | 6.55 |
| $R^2$ | 0.932 | 0.977 | 0.948 | 0.964 | 0.972 |
| *Bioaccumulation by growing cells* | | | | | |
| $q_{max}$ (meqg$^{-1}$) | 1.72 | 1.26 | 0.116 | 0.321 | 0.326 |
| b (lmeq$^{-1}$) | 0.287 | 1.84 | 26.9 | 0.605 | 2.22 |
| $R^2$ | 0.937 | 0.952 | 0.958 | 0.967 | 0.982 |

## 9. CONCLUSION

Living organisms are known to interact with metals by a number of mechanisms, including binding of metal ions to cell wall and intracellular accumulation. In environmental pollution control, biosorption (by non living organisms) as well as bioaccumulation processes (by living organisms) can be used. Biosorption and bioaccumulation - based techniques and technologies can provide an alternative to conventional methods for removal of metal ions. The environmental applications of these processes include pollution prevention (wastewater treatment), pollution control (biomonitoring), decrease of damages caused by pollution through the purposeful reduction of availability of pollutants to living organisms (e.g. remediation of contaminated soils, deacidification of soils by the supplementation of wood ash, detoxification preparation for animals) with the use of knowledge on ionic synergisms and antagonisms.

It is important to understand the mechanism and rules governing biosorption and bioaccumulation in order to propose and implement useful techniques and technologies. In bioaccumulation of particular significance are studies on synergistic and antagonistic effects between elements. This should enable to elaborate techniques that would reduce bioavailability of toxic metals and would result in their excretion from an organism.

The analysis of the content of toxic elements in biomonitors, including animal products, such as eggs and milk, from so-called ecological agriculture and from industrial and urbanized regions showed that eggs and milk are good monitors of environmental pollution with toxic metals. It was found that these products originating from ecological agriculture contain significantly lower levels of pollutants. The analysis of animal products was shown to possess not only biomonitoring value but first of all enables to assess consumption usability of these products.

Both processes, biosorption and bioaccumulation, can be used in wastewater treatment [150-151]. In the case of biosorption, the process is analogical to conventional sorption or ion-exchange process. In bioaccumulation, also metabolic processes occur that involve transport of metal ions into cellular interior. The first step of bioaccumulation is identical with biosorption, but here also nutrients are removed (biogenic compounds, including compounds of carbon, nitrogen and phosphorus). Also, bioaccumulation enables to reduce the level of contaminants to lower concentrations than in biosorption.

Biosorption enables to solve significant environmental problem. The use of materials that require disposal or utilization (such as powdered eggshells or animal bones) in wastewater treatment, makes it doubly advantageous. If the level of toxic elements classified as *toxic trio* (Cd, Pb, Hg) in treated wastewater is low, metal-laden biomass containing toxic metals that are also microelements, such as Cu, Zn, Cr etc. can be used as highly bioavailable mineral feed additive with microelements.

Depending on the application, different characteristics of biosorbents and bioaccumulating organisms are important. In wastewater treatment processes the aim is to remove ions from solutions down to possibly low level, below the concentration required by law. The sorbents can be classified as high-cost (especially propagated biomass) and low-cost sorbents (waste or commonly available biomass, such as plant and animal materials). If applied in wastewater treatment processes, if sorbent material is cheap or possesses no value, it is possible to lower the concentration of pollutant by the introduction of high concentration of the sorbent. In this case the mostly important would be sorbent affinity to sorbent, not necessarily very high sorbent capacity. Different objectives are desired if a material is to be used as mineral feed supplement. In this case the mostly important will be high biosorption capacity. Of course, in both cases, it would be advantageous if a biosorbent possesses simultaneously high biosorption affinity and capacity.

Biosorption and bioaccumulation can be used in integrated production processes in which it would be possible to simultaneously perform wastewater treatment, utilization of a waste product that requires disposal (waste biomass, i.e. eggshells or animal bones) and use it as sorbent of pollutants from wastewater and furtherly to use metal-laden biomass as cheap and highly bioavailable mineral feed supplement for livestock with microelements. There can be proposed various variants of the process depending on the composition of effluents.

The methods proposed are consistent with the sustainable development policy [152]. The problem is of interdisciplinary character on the boundary of environmental biotechnology, chemistry, chemical technology, earth sciences, bioprocess engineering and agriculture, addressing problems such as bioremoval of contaminants from waters and for this reason, the elaboration of complete technologies requires cooperation of specialists from various disciplines.

## ACKNOWLEDGEMENTS

This work was financially supported by Polish Ministry of Science and Higher Education (grant No. 3 T09B 064 27).

## REFERENCES

[1] Volesky, B. and Naja, G. 2005. Biosorption: Application Strategies. International Biohydrometallurgy Symposium, Cape Town, South Africa (September 2005).

[2] Davis T. A., Volesky B., Mucci A., A review of the biochemistry of heavy metal biosorption by brown algae. *Water Research* 37(18), 2003, 4311-4330.

[3] Williams P.L., James R.C., Roberts S.M. (Eds.), *Principles of Toxicology: Environmental and Industrial Applications,* 2nd Edition, John Wiley & Sons, Ltd., 2000, New York.

[4] Duffus J.H., "Heavy metals" – a meaningless term? *Pure and Applied Chemistry,* 74(5), 793–807, 2002.

[5] Scott J. S., Smith P. G., *Dictionary of Waste and Water Treatment,* Butterworths, 1981, London.

[6] Kabata-Pendias A., *Trace Elements in Soils and Plants,* Third Edition, 2000, CRC Press, New York.

[7] Hodgson E., *A textbook of modern toxicology,* Third Edition, John Wiley & Sons, 2004, New York.

[8] Hui X., Yu L., Joo-Hwa T., Effect of pH on nickel biosorption by aerobic granular sludge. *Bioresource Technology* 97, 3, 2006, 359-363.

[9] Conti M.E., Cecchetti G. Biological monitoring: lichens as bioindicators of air pollution assessment – a review, *Environmental Pollution* 114, 2001, 471-492.

[10] Veglio F., Beolchini F., Removal of metals by biosorption: a review, *Hydrometallurgy,* 44, 1997, 301-316

[11] Diniz V., Volesky, B., Effect of counterions on lanthanum biosorption by *Sargassum polycystum*. *Water Research* 39, 2005, 2229-2236.

[12] Beolchini F., Pagnanelli F., Toro L., Vegliò F. Ionic strength effect on copper biosorption *by Sphaerotilus natans*: equilibrium study and dynamic modelling in membrane reactor. *Water Research* 40, 1, 2006, 144-152.

[13] Henze M., Trends in advanced wastewater treatment. *Water Science and Technology* 35, 10, 1997, 1-4.

[14] Bode H., Control of heavy metal emission from metal plating industry in a German river basin., *Water Science and Technology,* 38(4-5), 1998, 121-129.

[15] www.epa.gov.

[16] Cheremisinoff N.P., Handbook of Water and Wastewater Treatment Technologies, Marcel Dekker Inc., 2001, Newark.

[17] Liu D.H.F., Environmental Engineers' Handbook, CRC Press, 1999, New York.

[18] Naja G., Mustin C., Volesky B., Berthelin J., Stabilization of the initial electrochemical potential for a metal-based potentiometric titration study of a biosorption process. *Chemosphere* 62, 1, 2006, 163-170.

[19] Naja G., Mustin C., Volesky B., Berthelin J., A high-resolution titrator: a new approach to studying binding sites of microbial biosorbents, *Water Research,* 39, 2005, 579-588.

[20] Chojnacka K., Heavy metal ions removal by microagae *Spirulina sp.* in biosorption and bioaccumulation processes, Ph.D. Dissertation, Wrocław, 2003.

[21] Beolchini F., Pagnanelli F., Toro L., Vegliò F., Ionic strength effect on copper biosorption by *Sphaerotilus natans*: equilibrium study and dynamic modelling in membrane reactor. *Water Research* 40, 1, 2006, 144-152.

[22] Volesky B., Weber J., Park J.M., Continuous-flow metal biosorption in a regenerable *Sargassum* column. *Water Research* 37(2), 2003, 297-306.

[23] Schiewer, S., Volesky, B. 1999. Advances in Biosorption of Heavy Metals. In: Encyclopedia of Bioprocess Engineering, Flickinger, MC & Drew, SW, eds. Wiley, NY, 433-453.

[24] Schiewer S., Volesky B., Modelling of the proton-metal ion exchange in biosorption *Oceanographic Literature Review* 43, 6, 1996, 558.

[25] Ahuja P., Gupta R., Saxena R.K., $Zn^{2+}$ biosorption by *Oscillatoria anguistissima. Process Biochemistry* 34(1), 1999, 77-85.

[26] Cho D.H., Kim E. Y., Characterization of Pb2+ biosorption from aqueous solution *by Rhodotorula glutinis. Bioprocess and Biosystems Engineering* 25(5), 2003, 271 – 277.

[27] da Costa A.C.A., de Mesquita L.M.S., Tornovsky J., Batch and continuous heavy metals biosorption by a brown seaweed from a zinc-producing plant. *Minerals Engineering* 9(8), 1996, 811-824.

[28] Diniz V., Volesky B., Biosorption of La, Eu and Yb using *Sargassum* biomass. *Water Research* 39(1), 2005, 239-247.

[29] Figueira, M.M., Volesky, B., Ciminelli, V.S.T., Roddick, Felicity A., Biosorption of metals in brown seaweed biomass. *Water Research* 34(1), 2000, 196-204.

[30] Hak S. L., Volesky B. Interaction of light metals and protons with seaweed biosorbent. *Water Research* 31(12), 1997, 3082-3088.

[31] Juhasz A L., Smith E., Smith J., Naidu R., Biosorption of organochlorine pesticides using fungal biomass. *Journal of Industrial Microbiology & Biotechnology* 29(4), 2002, 163 – 169.

[32] Kapoor A., Viraraghavan T., Fungal biosorption — an alternative treatment option for heavy metal bearing wastewaters: a review. *Bioresource Technology* 53(3) 1995, 195-206.

[33] Kapoor A, Viraraghavan T., Heavy metal biosorption sites in *Aspergillus niger. Bioresource Technology* 61(3), 1997, 221-227.

[34] Kratochvil D., Volesky B., Biosorption of Cu from ferruginous wastewater by algal biomass. *Water Research* 32(9), 1998, 2760-2768.

[35] Kratochvil D., Volesky B., Multicomponent biosorption in fixed beds. *Water Research* 34(12), 2000, 3186-3196.

[36] Kratochvil D., Volesky B., Demopoulos G., Optimizing Cu removal/recovery in a biosorption column. *Water Research* 31(9), 1997, 2327-2339.

[37] Lo W., Chua H., Lam K.H., Bi S.P., A Comparative Investigation on the Biosorption of Lead by Filamentous Fungal Biomass. *Chemosphere* 39(15), 1999, 2723-2736.

[38] López A., Lázaro N., Priego J.M., Marqués A.M., Effect of pH on the biosorption of nickel and other heavy metals by *Pseudomonas fluorescens* 4F39. *Journal of Industrial Microbiology & Biotechnology* 24(2), 2000, 146 – 151.

[39] Matheickal J.T., Yu Q. Biosorption of lead(II) from aqueous solutions by *Phellinus badius*. *Minerals Engineering* 10, 1997, 947-957

[40] Matheickal J. T., Yu Q., Biosorption of lead(II) and copper(II) from aqueous solutions by pre-treated biomass of Australian marine algae. *Bioresource Technology* 69(3), 1999, 223-229.

[41] Palmieri, M.C., Volesky B., Garcia O., Biosorption of lanthanum using *Sargassum fluitans* in batch system. *Hydrometallurgy* 67(1-3), 2002, 31-36.

[42] Schneider I.A.H., Smith R.W., Rubio J., Effect of mining chemicals on biosorption of Cu(II) by the non-living biomass of the macrophyte *Potamogeton lucens*. *Minerals Engineering* 12(3), 1999, 255-260.

[43] Schneider, I.A.H., Rubio J., Smith R.W., Biosorption of metals onto plant biomass: exchange adsorption or surface precipitation? *International Journal of Mineral Processing* 62(1-4), 2001, 111-120

[44] Volesky B. Biosorption process simulation tools. *Hydrometallurgy* 71(1-2), 2003, 179-190.

[45] Yang J., Volesky B., Biosorption of uranium on *Sargassum* biomass. *Water Research* 33, 1999, 3357-3363.

[46] Naja G., Mustin C., Berthelin J., Volesky B., Lead biosorption study with *Rhizopus arrhizus* using a metal-based titration technique. *Journal of Colloid And Interface Science* 292, 2, 15, 2005, 537-543.

[47] Kratochvil D., Volesky B., Advances in the biosorption of heavy metals. *Trends in Biotechnology* 16(7), 1998, 291-300.

[48] Lodeiro, P., Cordero, B., Barriada, J.L., Herrero, R., Sastre de Vicente, M.E. Biosorption of cadmium by biomass of brown marine macroalgae. *Bioresource Technology* 96, 16, 2005, 1796-1803.

[49] Pankratz T.M., Environmental Engineers Dictionary, Lewis Publishers CRC Press, London, 2001.

[50] Liu Y., Yang S.F., Xu H., Woon K.H., Lin Y.M, Tay J.H., Biosorption kinetics of cadmium(II) on aerobic granular sludge. *Process Biochemistry* 38(7), 2003, 997 – 1001.

[51] Vilar V.J.P., Sebesta F., Botelho C.M.S., Boaventura R.A.R., Equilibrium and kinetic modelling of $Pb^{2+}$ biosorption by granulated agar extraction algal waste. *Process Biochemistry* 40(10), 2005, 3276-3284.

[52] Kargi F., Cikla S., Biosorption of zinc(II) ions onto powdered waste sludge (PWS): Kinetics and isotherms. *Enzyme and Microbial Technology* 38(5), 2006, 705-710.

[53] Loukidou M. X., Zouboulis A.I., Karapantsios T.D., Matis K.A., Equilibrium and kinetic modeling of chromium(VI) biosorption by *Aeromonas caviae*. *Colloids and Surfaces A: Physicochemical and Engineering Aspects* 242(1-3), 2004, 93-104.

[54] Ozdemir G., Baysal S. H., Chromium and aluminum biosorption on *Chryseomonas luteola* TEM05. *Applied Microbiology and Biotechnology* 64 (4), 2004,599 – 603.

[55] Zhou J. L., Zn biosorption by *Rhizopus arrhizus* and other fungi. *Applied Microbiology and Biotechnology 51*, 5, 31, 1999,686 – 693.

[56] Valdman E., Leite S.G.F., Biosorption of Cd, Zn and Cu by *Sargassum sp.* waste biomass. *Bioprocess Engineering* 22(2), 2000,171 – 173.

[57] Liu H.L., Chen B.Y., Lan Y.W., Cheng Y.C., Biosorption of Zn(II) and Cu(II) by the indigenous *Thiobacillus thiooxidans*. *Chemical Engineering Journal* 97(2-3), 2004,195-201.

[58] Vegliò F., Di Biase A., Beolchini F., Pagnanelli F., Heavy metal biosorption in binary systems: simulation in single- and two-stage UF/MF membrane reactors, *Hydrometallurgy* 66(1-3), 2002, 107-115.

[59] Pernfuβ, B., Pümpel, T., Schinner, F. Biosorption and Bioaccumulation of Silver by *Arthrobacter sp.* BP 7/26. *International Biodeterioration and Biodegradation* 37(3-4), 1996, 242-246.

[60] Kaduková J., Virčíková E. Comparison of differences between copper bioaccumulation and biosorption. *Environment International* 31(2), 2005, 227-232.

[61] Srinath T., Verma T., Ramteke P.W., Garg S.K., Chromium (VI) biosorption and bioaccumulation by chromate resistant bacteria. *Chemosphere* 48(4), 2002, 427-435.

[62] Kamer I., Douek J., Tom M., Rinkevich B., Metallothionein Induction in the RTH-149 Cell Line as an Indicator for Heavy Metal Bioavailability in a Brackish Environment: Assessment by RT-Competitive PCR. *Archives of Environmental Contamination and Toxicology* 45(1), 2003, 86 – 91.

[63] Raskin I., Smith R.D., Salt D.E., Phytoremediation of metals: using plants to remove pollutants from the environment. *Current Opinion in Biotechnology* 8(2), 1997, 221-226.

[64] Chapman P.M. Is Bioaccumulation Useful for Predicting Impacts. *Marine Pollution Bulletin* 34(5), 1997, 282-283.

[65] Kara Y. Bioaccumulation of Copper from Contaminated Wastewater by Using *Lemna minor*. *Bulletin of Environmental Contamination and Toxicology* 72(3), 2004, 467 – 471.

[66] Deng X., Wilson D.B., Bioaccumulation of mercury from wastewater by genetically engineered Escherichia coli. *Applied Microbiology and Biotechnology* 56(1-2), 2001, 276 – 279.

[67] Zhou H.Y., Cheung R.Y.H., Wong M.H., Bioaccumulation of organochlorines in freshwater fish with different feeding modes cultured in treated wastewater. *Water Research* 33(12), 1999, 2747-2756.

[68] Aksu Z., Dönmez G., The use of molasses in copper(II) containing wastewaters: effects on growth and copper(II) bioaccumulation properties of *Kluyveromyces marxianus*. *Process Biochemistry* 36(5), 2000, 451 – 458.

[69] Liang Y., Cheung R.Y.H., Wong M.H., Reclamation of wastewater for polyculture of freshwater fish: bioaccumulation of trace metals in fish. *Water Research* 33(11), 1999, 2690-2700.

[70] Macaskie L.E. Bioaccumulation of heavy metals, and application to the remediation of acid mine drainage water containing uranium. *Research in Microbiology* 148(6), 1997, 528-530.

[71] Chou, F.I., Chung H.P., Teng S.P., Sheu S.T., Screening plant species native to Taiwan for remediation of 137Cs-contaminated soil and the effects of K addition and soil amendment on the transfer of 137Cs from soil to plants. *Journal of Environmental Radioactivity 80*(2), 2005, 175-181.

[72] Dube B.K., Sinha P., Chatterjee C.Crop, Plants as Biological Tools for Assessing and Monitoring Agricultural Lands Inundated with Sewage and Sludge. *Bulletin of Environmental Contamination and Toxicology* 72(2), 2004, 429 – 436.

[73] Schuhmacher M., Domingo J.L., Agramunt M.C., Bocio A., Müller L., Biological monitoring of metals and organic substances in hazardous-waste incineration workers. *International Archives of Occupational and Environmental Health* 75(7), 2002, 500 – 506.

[74] Mutti A., Biological monitoring in occupational and environmental toxicology, *Toxicology Letters,* 108, 1999, 77-89.

[75] van der Oost, R., Beyer, J., Vermeulen, N.P.E., Fish bioaccumulation and biomarkers in environmental risk assessment: a review, *Environmental Toxicology and Pharmacology,* 13(2), 57-149, 2003.

[76] Lai K.M., Scrimshaw M.D., Lester J.N., Prediction of the bioaccumulation factors and body burden of natural and synthetic estrogens in aquatic organisms in the river systems. *The Science of the Total Environment* 289(1-3), 2002, 159-168.

[77] Mason R. P., Laporte J.-M., Andres S., Factors Controlling the Bioaccumulation of Mercury, Methylmercury, Arsenic, Selenium, and Cadmium by Freshwater Invertebrates and Fish. *Archives of Environmental Contamination and Toxicology* 38(3), 2000, 283 - 297.

[78] Al-Rub, F.A. Abu, El-Naas, M.H., Ashour, I., Al-Marzouqi, M. Biosorption of copper on *Chlorella vulgaris* from single, binary and ternary metal aqueous solutions Process Biochemistry 41, 2, 2006, 457-464

[79] Chen X.C., Wang Y.P., Lin Q.S., Ji Y.W., Wei X.C., Ying X., Biosorption of copper(II) and zinc(II) from aqueous solution by *Pseudomonas putida* CZ1. *Colloids and Surfaces B: Biointerfaces* 46, 2, 10, 2005, 101-107

[80] Kiran I., Akar T., Tunali S., Biosorption of Pb(II) and Cu(II) from aqueous solutions by pretreated biomass of *Neurospora crassa. Process Biochemistry* 40, 11, 2005, 3550-3558.

[81] Vilar V.J.P., Botelho C.M.S., Boaventura R.A.R., Influence of pH, ionic strength and temperature on lead biosorption by Gelidium and agar extraction algal waste. *Process Biochemistry* 40, 10, 2005, 3267-3275.

[82] Beolchini F., Pagnanelli F., Toro L., Vegliò F. Ionic strength effect on copper biosorption by *Sphaerotilus natans*: equilibrium study and dynamic modelling in membrane reactor. *Water Research* 40, 1, 2006, 144-152.

[83] Vilar V.J.P., Botelho C.M.S., Boaventura R.A.R., Equilibrium and kinetic modelling of Cd(II) biosorption by algae Gelidium and agar extraction algal waste. *Water Research* 40, 2, 2006, 291-302

[84] Chen J.Z., Tao X.C., Xu J., Zhang T., Liu Z.L., Biosorption of lead, cadmium and mercury by immobilized *Microcystis aeruginosa* in a column. *Process Biochemistry* 40, 12, 2005, 3675-3679.

[85] Akar T., Tunali S., Biosorption performance of *Botrytis cinerea* fungal by-products for removal of Cd(II) and Cu(II) ions from aqueous solutions. *Minerals Engineering* 18, 11, 2005, 1099-1109.

[86] Lodeiro, P., Rey-Castro, C., Barriada, J.L., Sastre de Vicente, M.E., Herrero, R. Biosorption of cadmium by the protonated macroalga *Sargassum muticum*: Binding analysis with a nonideal, competitive, and thermodynamically consistent adsorption (NICCA) model. *Journal of Colloid and Interface Science* 289, 2, 15, 2005, 352-358.

[87] Moon S.H., Park C.S., Kim Y.J., Park Y.I. Biosorption isotherms of Pb (II) and Zn (II) on Pestan, an extracellular polysaccharide, of *Pestalotiopsis sp.* KCTC 8637P. *Process Biochemistry* 41, 2, 2006, 312-316.

[88] Tunali S., Akar T., Özcan A.S., Kiran I., Özcan A., Equilibrium and kinetics of biosorption of lead(II) from aqueous solutions by *Cephalosporium aphidicola*. *Separation and Purification Technology* 47, 3, 2006, 105-112

[89] Ruelas-Inzunza, J., Páez-Osuna, F., Soto, Luis A. Bioaccumulation of Cd, Co, Cr, Cu, Fe, Hg, Mn, Ni, Pb and Zn in trophosome and vestimentum of the tube worm *Riftia pachyptila* from Guaymas basin, Gulf of California Deep-Sea Research Part I 52, 7, 2005, 1319-1323.

[90] Hawari, Alaa H., Mulligan, Catherine N., Biosorption of lead(II), cadmium(II), copper(II) and nickel(II) by anaerobic granular biomass. *Bioresource Technology* 97, 4, 2006, 692-700.

[91] Zhang L., Wang W.X., Effects of Zn pre-exposure on Cd and Zn bioaccumulation and metallothionein levels in two species of marine fish. *Aquatic Toxicology* 73, 4, 30, 2005, 353-369.

[92] Amaral C.R., Freitas Rebelo M.M., Torres P.M.J., Bioaccumulation and depuration of Zn and Cd in mangrove oysters (*Crassostrea rhizophorae*, Guilding, 1828) transplanted to and from a contaminated tropical coastal lagoon. *Marine Environmental Research* 59, 4, 2005, 277-285.

[93] Zhou D., Zhang L., Guo S., Mechanisms of lead biosorption on cellulose/chitin beads. *Water Research* 39, 16, 2005, 3755-3762.

[94] Han R., Zhang J., Zou W., Shi J., Liu H., Equilibrium biosorption isotherm for lead ion on chaff. *Journal of Hazardous Materials* 125, 1-3, 17, 2005, 266-271.

[95] Hassler C.S., Behra R., Wilkinson K.J., Impact of zinc acclimation on bioaccumulation and homeostasis in *Chlorella kesslerii*. *Aquatic Toxicology* 74, 2, 30, 2005, 139-149.

[96] Ortiz, O., Alcañiz, J.M., Bioaccumulation of heavy metals in *Dactylis glomerata* L. growing in a calcareous soil amended with sewage sludge. *Bioresource Technology* 97, 4, 2006, 545-552.

[97] Ciesielski T., Pastukhov M. V., Fodor P., Bertenyi, Z., Namiesnik J., Relationships and bioaccumulation of chemical elements in the Baikal seal (*Phoca sibirica*). *Environmental Pollution* 139, 2, 2006, 372-384.

[98] Ferrier-Pagès, C., Houlbrèque, F., Wyse, E., Richard, C., Allemand, D., Boisson, F. Bioaccumulation of zinc in the scleractinian coral *Stylophora pistillata* Coral Reefs 24, 4, 2005, 636 – 645.

[99] Cravo, A., Bebianno, M.J., Bioaccumulation of metals in the soft tissue of *Patella aspera*: Application of metal/shell weight indices. *Estuarine, Coastal and Shelf Science* 65, 3, 2005, 571-586.

[100] Zimmerman S., Messerschmidt J., von Bohlen A., Sures B., Uptake and bioaccumulation of platinum group metals (Pd, Pt, Rh) from automobile catalytic converter materials by the zebra mussel (*Dreissena polymorpha*). *Environmental Research* 98, 2, 2005, 203-209

[101] Unrine, J.M., Jagoe, C.H., Brinton, A.C., Brant, H.A., Garvin, N.T., Dietary mercury exposure and bioaccumulation in amphibian larvae inhabiting Carolina bay wetlands. *Environmental Pollution* 135, 2, 2005, 245-253.

[102] Maes, G.E., Raeymaekers, J.A.M., Pampoulie, C., Seynaeve, A., Goemans, G., Belpaire, C., Volckaert, F.A.M., The catadromous European eel *Anguilla anguilla* (L.) as a model for freshwater evolutionary ecotoxicology: Relationship between heavy metal bioaccumulation, condition and genetic variability. *Aquatic Toxicology* 73, 1, 1, 2005, 99-114

[103] Dusek, L., Svobodová, Z., Janousková, D., Vykusová, B., Jarkovský, J., Bioaccumulation of mercury in muscle tissue of fish in the Elbe River (Czech Republic): multispecies monitoring study 1991–1996. *Ecotoxicology and Environmental Safety* 61, 2, 2005, 256-267.

[104] Kozolkaya R. The role of different organic wastes on zinc bioaccumulation by earthworm *Lumbricus terrestris* L. (Oligochaeta) in successive Zn added soil. *Ecological Engineering* 25, 4, 2005, 322-331.

[105] T.R. Darling, Vernon T.G., Lead bioaccumulation in earthworms, *Lumbricus terrestris*, from exposure to lead compounds of differing solubility. *Science of the Total Environment,* The 346, 1-3, 15, 2005, 70-80.

[106] Monleau M., Bussy C., Lestaevel P., Houpert P., Paquet F., Bioaccumulation and behavioural effects of depleted uranium in rats exposed to repeated inhalations. *Neuroscience Letters* 390, 1, 16, 2005, 31-36.

[107] Huang R.J., Zhuang Z.X., Wang Y.R., Huang Z.Y., Wang X.R., Lee C., Frank S., An analytical study of bioaccumulation and the binding forms of mercury in rat body using thermolysis coupled with atomic absorption spectrometry. *Analytica Chimica Acta* 538, 1-2, 4, 2005, 313-321.

[108] Górecka H., Chojnacka K., Górecki H., The application of multielemental analysis emission and mass plasma spectrometry in the elaboration of new products and technologies and in pollution control, *Chemik*, 2006, 1, 60-63.

[109] Chojnacka K., Górecka H., Górecki H., The influence of living habits and family relationships on elements concentration in human hair, *Science of the Total Environment*, 2006, 366(2-3), 612-620.

[110] Chojnacka K., Górecka H., Chojnacki A., Górecki H., Inter-element interactions in human hair. *Environmental Toxicology and Pharmacology,* 2005, 20, 368-374.

[111] Chojnacka K., Górecka H., Górecki H., The effect of age, sex, smoking habit, hair color on the composition of hair, *Environmental Toxicology and Pharmacology,* 2006, 2006, 2, 52-57.

[112] Dobrzański Z., Chojnacka K., Górecka H., Chojnacki A., Wiśniewski J.: Chemical content of waterfowl eggs from industrial and agricultural regions. In: *Chemicals in*

*sustainable agriculture*. Ed. by H. Górecki, Z. Dobrzański, Paweł Kafarski. Prague: Czech-Pol Trade, 2003 (Chemistry for Agriculture 524-529).

[113] Dobrzański Z., Chojnacka K., Górecka H., Chojnacki A., Wiśniewski J., Waterfowl eggs as an indicator of pollution of rural environment. *Acta Agrophysica*. 2003, 1, 88, 395-401.

[114] Kołacz R., Dobrzański Z., Górecka H., Chojnacka K., Rudnicka A.: The content of lead and cadmium in milk and blood of cows kept in industrial and typically agricultural region. In: *New agrochemicals and their safe use for health and environment*. Ed. by H. Górecki, Z. Dobrzański, Paweł Kafarski. Prague Brussels: Czech-Pol Trade, 2004 (Chemistry for Agriculture, 312-316).

[115] Dobrzański Z., Górecka H., Opaliński Sebastian, Chojnacka K., Kołacz R. Trace and ultra-trace elements in cow's milk and blood. *Medycyna Weterynaryjna*. 2005, 61, 3, 301-304.

[116] Dobrzański Z., Górecka H., Kwaśnicki R., Barej R., Chojnacka K., Content of mercury and selenium in raw cow milk from Silesia Province. *Bromatologia i Chemia Toksykologiczna* 2004 37, 4, 347-351.

[117] Dobrzański Z., Kołacz R., Górecka H., Chojnacka K., Bartkowiak A., The content of microelements and trace elements in raw milk from cows kept in Silesian region. *Polish Journal of Environmental Studies* 2005, 14, 5, 145-149.

[118] Dobrzański Z., Górecka H., Chojnacka K., Górecki H., Kołacz R., Bartkowiak A., Opaliński S., The application of aluminosilicate-humin preparation for the reduction of cadmium accumulation in blood and milk of cows. In: *Development in production and use of new agrochemicals*. Ed. by H. Górecki, Z. Dobrzański, P. Kafarski. Prague Brussels: Czech-Pol Trade, 2005 (Chemistry for Agriculture, 288-292).

[119] Dobrzański Z., Górecka H., Chojnacka K., Górecki H., Kołacz R., Bartkowiak A.: The application of dietary aluminosilicate-humin mixture for the reduction of lead accumulation in blood and milk of cows. In: *Development in production and use of new agrochemicals*. Ed. by H. Górecki, Z. Dobrzański, P. Kafarski. Prague Brussels: Czech-Pol Trade, 2005 (Chemistry for Agriculture, 1025-1029).

[120] Dobrzański Z, Górecki H., Kołacz R, Górecka H., Opaliński S, Chojnacka K., Bartkowiak A., Elaboration of the technology and the recipe for the preventive-detoxication preparations, made on the basis of the humic and aluminosilicate materials, for farm animals housing in regions of ecological risk. *Polish Journal of Environmental Studies,* 2005, 14, 3, 95-99.

[121] Bartkowiak A., Opaliński S., Dobrzański Z., Chojnacka K., Górecki H., The effect of dietary humic-aluminosilicate preparation on the content of cadmium and lead in muscle and liver of hens housed in free range system. In: *Animals and environment. XIIth International Congress ISAH* [International Society for Animal Hygiene] 2005. Proceedings. Ed. A. Kryński, R. Wrzesień. Warsaw, [September 4-8th 2005]., 2. Warsaw: Bel Studio 2005, 28-31.

[122] Dobrzański Z, Kołacz R, Opaliński S, Górecka H., Chojnacka K., Trziszka T: The content of cadmium and lead in muscle and liver of laying hens housed in a copper industry region. In: *Proceedings of the XI European Symposium on the Quality of Eggs and Egg Products Organising Committee Symposia 2005*, Dutch Branch of the World's Poultry Science Association and the Working Groups 4 and 5 of the European Federation of the WPSA. Doorwerth, the Netherlands, 2005, 98-101 CD-ROM

[123] Górecki H., Chojnacka K., Dobrzański Z., Kołacz R., Górecka H., Trziszka T., The effect of phosphogypsum as the mineral feed additive on fluorine content in eggs and tissues of laying hens. *Animal Feed Science and Technology,* 2006, 128(1-2), 84-95.

[124] Chojnacka K., Chojnacki A., Górecka H., Górecki H., Bioavailability of heavy metals from polluted soils to plants. *Sci. Total Environ.* 2005, 337, 175-182.

[125] Górecka H., Chojnacka K., Dobrzański Z., Górecki H., Bioavailability of macronutrients, micronutrients and toxic elements from wood ashes used as fertilizer and soil deacidifying agent. *Polish Journal of Environmental Studies,* 2005, 14, 3, 14-18.

[126] Górecka H., Chojnacka K., Górecki H., Dobrzański Z., Wood fireplace ashes as the raw material for the fertilizers used in the fertilization of the homestead gardens. *Polish Journal of Chemical Technology* 2004, 6, 4, 15-19.

[127] Górecka H., Dobrzański Z. H., Górecki K., Chojnacka: Fireplace ashes as deacidification agent and fertilizer. In: New agrochemicals and their safe use for health and environment. Ed. by H. Górecki, Z. Dobrzański, Paweł Kafarski. *Prague Brussels: Czech-Pol Trade,* 2004 (Chemistry for Agriculture, 1-8).

[128] Górecka H., Chojnacka K., Dobrzański Z., Górecki H., Bioavailability of nutrients and heavy metals to plants from soil deacidified by wood ash. In: *Development in production and use of new agrochemicals.* Ed. by H. Górecki, Z. Dobrzański, P. Kafarski. Prague Brussels: Czech-Pol Trade, 2005 (Chemistry for Agriculture, 281-287.

[129] Górecka H., Chojnacka K., Hoffmann J., Górecki H., Micronutrients in wood ashes used for deacidification and fertilization of soils, *Chemik,* 2006, 1, 9-12.

[130] Górecka H., Chojnacka K., Górecki H., The application of ICP-MS and ICP-OES in determination of micronutrients in wood ashes used as soil conditioners, *Talanta,*2006, doi:10.1016/j.talanta.2006.05.061.

[131] Chojnacka K., Barańska M., Górecka H., Górecki H., Utilization of bones, feather and wood ash in the production of mineral fertilizers, *Przemysł Chemiczny,* 2006, 85(8-9), 30-35.

[132] Chojnacka K., Characteristics of Cr(III) sorption by dried leaves of *Forsythia intermediata* , *Chemistry for Agriculture* vol. 7, 2006, 256-303.

[133] Chojnacka K., Biosorption of Cr(III) ions by wheat straw and grass: a systematic characterization of new biosorbents. *Polish Journal of Environmental Studies* (in press).

[134] Chojnacka K., The application of multielemental analysis in the elaboration of technology of mineral feed additives based on *Lemna minor* biomass, *Talanta,* 2006, doi:10.1016/j.talanta.2006.05.063.

[135] Chojnacka K., The application of aquatic plants in the production of mineral feed additives, *Przemysł Chemiczny,* 2006, 85(8-9), 300-307.

[136] Walkowicz M., Chojnacka K., Biosorption and bioaccumulation of minerals by liverwort *Riccia fluitans. Pr. Nauk. Wydz. Chem. PWroc.* 2005, 3, 99-102.

[137] Michalak I., Chojnacka K., The application of macroalgae in animal feeding, *Polish Journal of Chemical Technology* (in press).

[138] Michalak I., Chojnacka K., Perspectives of the application of macroalgae as biological mineral supplements for livestock, *Chemistry for Agriculture* vol. 7, 2006, 456-461.

[139] Chojnacka K., Biosorption of Cr(III) ions by eggshells. *Journal of Hazardous Materials* B, 2005, 121, 167-173.

[140] Chojnacka K., Equilibrium and kinetic modelling of chromium(III) sorption by animal bones. *Chemosphere* 2005, 59, 315-320.

[141] Chojnacka K., Chojnacki A., Górecka H., The influence of growth parameters of microalgae *Spirulina sp.* on biosorption capacity of cells. *Inżynieria Chemiczna i Procesowa* 2004, 25, 3/1, 789-794.

[142] Chojnacka K., Chojnacki A., Górecka H., Biosorption of Cr3+, Cd2+ and Cu2+ ions by blue-green algae *Spirulina sp.*, kinetics, equilibrium and the mechanism of the process. *Chemosphere* 2005, 59, 75-84.

[143] Chojnacka K., Chojnacki A., Górecka H., Trace element removal by *Spirulina sp.* from copper smelter and refinery effluents. *Hydrometallurgy* 2004, 73, 1/2, 147-153.

[144] Chojnacki A., Chojnacka K., Hoffmann J., Górecki H., The application of natural zeolites for mercury removal: from laboratory tests to industrial scale. *Minerals Engineering* 2004, 17, 7/8, 933-937.

[145] Chojnacka K., Environmental biotechnology of microalgae. In: *New agrochemicals and their safe use for health and environment.* Ed. by H. Górecki, Z. Dobrzański, Paweł Kafarski. Prague Brussels: Czech-Pol Trade, 2004 (Chemistry for Agriculture, 719-722).

[146] Chojnacka K., Noworyta A., Evaluation of *Spirulina sp.* growth in photoautotrophic, heterotrophic and mixotrophic cultures. *Enzyme and Microbial Technology,* 2004, 34, 5, 461-465.

[147] Chojnacka K., Marquez-Rocha F.J. Kinetic and stoichiometric relationships of the energy and carbon metabolism in the culture of microalgae. *Biotechnology* 2004, 3, 1, 21-34.

[148] Chojnacka K., Noworyta A., Mechanisms of self-control of cells concentration in microalgal culture used in microbiological wastewater treatment. In: *Chemicals in sustainable agriculture.* Ed. by H. Górecki, Z. Dobrzański, Paweł Kafarski. Prague: Czech-Pol Trade, 2003 (Chemistry for Agriculture, 530-535).

[149] Chojnacka K., Górecka H., Praktyczne aspekty wykorzystania zdolności biosorpcyjnych i bioakumulacyjnych metali przez algi. In: *Mikroelementy w rolnictwie. Cz. 2. Warszawa*: [Wydaw. Nauk. PWN] 2004 (Zeszyty Problemowe Postępów Nauk Rolniczych, ISSN 0084-5477, 723-728).

[150] Chojnacka K., Bioaccumulation - the fundamentals of the process. In: *Development in production and use of new agrochemicals.* Ed. by H. Górecki, Z. Dobrzański, P. Kafarski. Prague Brussels: Czech-Pol Trade, 2005 (Chemistry for Agriculture, 329-335).

[151] Chojnacka K., Biosorption - the fundamentals of the process. In: *Development in production and use of new agrochemicals.* Ed. by H. Górecki, Z. Dobrzański, P. Kafarski. Prague Brussels: Czech-Pol Trade, 2005 (Chemistry for Agriculture, 336-341).

[152] Chojnacka K., Górecka H., Górecki H., Possibilities of technological application of renewable sources of phosphorus, *Przemysł Chemiczny*, 2006, 85(8-9), 420-426.

# INDEX

## A

## B

## C

## D

## E

## H



## I

## J



## K



## L



## M

## N

## O



## P

## Q



## R

## S

## V



## W



## X



## Y



## Z